Ecological Economics:
The Science and Management of Sustainability

ECOLOGICAL ECONOMICS:
The Science and Management of Sustainability

Edited by

Robert Costanza

Technical Editor: Lisa Wainger

Columbia University Press, New York

Columbia University Press

New York   Chichester, West Sussex

Copyright © 1991 Columbia University Press
All rights reserved

Library of Congress Cataloging-in-Publication Data

Ecological economics: the science and management of sustainability /
    edited by Robert Costanza
        p.   cm.
    Includes bibliographical references and index
    ISBN 0-231-07562-6
    ISBN 0-231-07563-4 (pbk.)
    1. Economic development -- Environmental aspects. I. Costanza,
Robert.
HD75.6.E29   1991
333.7 -- dc20   91 - 14453
                                                      CIP

Printed in the United States of America

c 10 9 8 7 6 5 4 3
p 10 9 8 7 6 5 4

ABOUT ISEE

The International Society for Ecological Economics (ISEE) is concerned with extending and integrating the study and management of "nature's household" (ecology) and "humankind's household" (economics). Ecological Economics studies the ecology of humans and the economy of nature, the web of interconnections uniting the economic subsystem to the global ecosystem of which it is a part. It is this larger system that must be the object of study if we are to adequately address the critical issues that now face humanity.

ISEE has active members in over 40 countries. It publishes the bimonthly journal *Ecological Economics* in conjunction with Elsevier Science Publishers, as well as a semiannual newsletter. ISEE holds major international conferences every other year in the even-numbered years. The first was in May 1990 in Washington, DC, the second will be in August 1992 in Stockholm, Sweden. In the odd-numbered years, smaller national and regional meetings are held. For further information about ISEE's activities, becoming a member of ISEE, or subscribing as an individual to *Ecological Economics* contact: Dr. Robert Costanza, Coastal and Environmental Policy Program, Center for Environmental and Estuarine Studies, University of Maryland, Box 38, Solomons, MD 20688-0038, USA. Telephone: 301-326-4281 FAX: 301-326-6342. Inquiries about institutional subscriptions to *Ecological Economics* and requests for sample issues should be addressed directly to Elsevier at: Ron Hayward, Elsevier Science Publishers, P. O. Box 330, 1000 AH Amsterdam, The Netherlands. Telephone: 20-5862911.

CONTENTS

PART II. ACCOUNTING, MODELING AND ANALYSIS

PART III. INSTITUTIONAL CHANGES AND CASE STUDIES

Incentives and Instruments:

Education:

Case Studies:

PREFACE

This book is intended to serve a number of purposes and audiences, including:

1. as a research agenda and as policy recommendations for developing ecologically sustainable economies, to be read by the environmentally and economically informed public, people in international development agencies, government environmental and economic management agencies, non-governmental organizations and other policy makers.
2. as a textbook or "sourcebook" of cogent readings in *ecological economics* that can serve as a basis for graduate courses.
3. as an academic text aimed at researchers in the field. Since the field is inherently multidisciplinary, the potential academic audience is broad, including: economists, ecologists, conservation biologists, public policy professionals, anthropologists, sociologists and others.

We have tried to limit jargon and define terms as they appear so that the text will be as accessible as possible to a broad audience. But, since this is a multiauthored work, the style varies considerably from chapter to chapter. We have not tried to force the different author's styles into a particular mold. We think the many styles and perspectives represented in this volume are more representative of the pluralistic nature of *ecological economics*. The introductory chapter synthesizes and summarizes these many perspectives and serves as a guide to the rest of the book.

The book is the product of a workshop held May 24 through 26, 1990, at the Aspen Institute's conference facilities on Maryland's Eastern Shore. The workshop immediately followed the first biannual conference of the International Society for Ecological Economics, held May 21 through 23, 1990, at the World Bank in Washington, DC. The theme of the conference was "The Ecological Economics of Sustainability: Making Local and Short-Term Goals Consistent with Global and Long-Term Goals." Over 370 participants attended the three-day conference in Washington, and thirty-eight invited participants attended the subsequent three-day workshop at Aspen.

The goals of the workshop were to consolidate a core of economists, ecologists and others into a nucleus of thinkers about *ecological economics*, and to produce a consensus on the state and goals of the emerging field of *ecological economics*, particularly as regards issues of sustainability. In particular, we wanted: (a) a working agenda for research, education, and policy for the coming decade to ensure sustainability; and (b) a set of policy guidelines and recommendations for decision makers. This consensus is presented in the book's introductory chapter. The remaining chapters of the book provide detailed elaboration on these themes from a number of different perspectives. Because of the workshop format, the papers in this book have benefited from more interaction between the authors than is usually found in a volume of contributed chapters. In addition to the general discussion of common themes at the workshop, each paper was formally reviewed by other workshop participants and not all of the papers were included in the book.

The process of facilitated policy dialogue was used to run the workshop. The policy dialogue process aims to create a meeting environment that is conducive to the free expression of the participants' ideas, and to the creative formulation of new approaches to problems. Responsibility for process support was provided by the following individuals:

Facilitators:
    Joy A. Bartholomew, Director, Washington Area Office, Center for Policy
        Negotiation, Port Republic, MD
    Paul DeLong, RESOLVE, The Conservation Foundation and World Wildlife Fund,
        Washington, DC
    Allison Gilbert, Free University, Amsterdam, The Netherlands
Synthesizers for the working group sessions:
    Robert Costanza, University of Maryland, Solomons, MD
    Herman Daly, The World Bank, Washington, DC
    AnnMari Jansson, University of Stockholm, Stockholm, Sweden
    Leon Braat, Free University, Amsterdam, The Netherlands
Support team:
    Ben Haskell, University of Maryland, Solomons, MD
    Laura Cornwell, University of Maryland, Solomons, MD

Acknowledgments

Major funding for the workshop and the conference, and for other expenses related to the preparation of this book was provided by the Pew Charitable Trusts. Additional funding was provided by the Jessie Smith Noyes Foundation, the World Bank and US AID. Additional sponsors of the conference included the Aspen Institute, the Coastal and Environmental Policy Program of the University of Maryland, the Center for Policy Negotiation, the Coastal Society, the International Society for Ecological Modeling, the World Wildlife Fund/Conservation Foundation, the Nature Conservancy, the Global Tomorrow Coalition and the World Resources Institute. We would also like to thank the following for their varied and valuable contributions: Laura Cornwell, Ben Haskell, Dana Flanders, Janet Barnes, Steve Viederman, Ed DeBellevue, Steve Tennenbaum, Paul Jivoff, Allison Gilbert, Paul DeLong, Betsey Henry, Patrick Hagan and Twig Johnson.

Robert Costanza
Solomons, Maryland

CONTRIBUTORS

E. O. A. Asibey	Environment Department, The World Bank
Joy A. Bartholomew	Washington Area Office, Center for Policy Negotiation
Kenneth E. Boulding	Institute of Behavioral Science, University of Colorado
Leon C. Braat	Institute for Environmental Studies, Free University, Amsterdam
Clóvis Cavalcanti	Institute for Social Research, Joaquim Nabuco Foundation, Brazil
Paul Christensen	Department of Economics, Hofstra University
Mary E. Clark	Center for Conflict Analysis, George Mason University
Colin Clark	Institute of Applied Mathematics, The Univ. of British Columbia
Cutler J. Cleveland	Dept. of Geography & CEES, Boston University
Robert Costanza	Center for Envir. and Estuarine Studies, University of Maryland
John H. Cumberland	Center for Envir. and Estuarine Studies, Univ. of Maryland
Herman E. Daly	Environment Department, The World Bank
Ralph C. d'Arge	Department of Economics, University of Wyoming
M. B. Dyson	Environment Department, The World Bank
Salah El Serafy	Economic Advisory Staff, The World Bank
Malte Faber	Alfred Weber Institute, University of Heidelberg, Germany
Stephen Farber	Economics Dept., Louisiana State University
Silvio O. Funtowicz	Joint Research Centre, Ispra, Italy
Robert Goodland	Environment Department, The World Bank
Bruce Hannon	Dept. of Geography, University of Illinois
Garrett Hardin	Dept. of Biological Sciences, Univ. of California, Santa Barbara
Richard B. Howarth	Energy and Resources Group, Univ. of California at Berkeley
Roefie Hueting	Central Bureau for Statistics, The Netherlands
AnnMari Jansson	Department of Systems Ecology, Univ. of Stockholm, Sweden
N. Marchettini	Department of Chemistry, University of Siena, Italy
Juan Martinez-Alier	Dept. of Economics, Univ. Autonoma de Barcelona, Spain
William J. Mitsch	School of Natural Resources, The Ohio State University
Richard B. Norgaard	Energy and Resources Program, Univ. of California at Berkeley
Bryan G. Norton	School of Social Sciences, Georgia Institute of Technology
Talbot Page	Environmental Studies, Brown University
Charles Perrings	Department of Economics, Univ. of California, Riverside
Henry M. Peskin	Edgevale Associates, Inc., Silver Spring, Maryland
J. C. Post	Environment Department, The World Bank
John L.R. Proops	Dept. of Economics and Management Science, Univ. of Keele, U.K.
Jerome R. Ravetz	Research Methods Consultancy, London, England
Clive Spash	Department of Economics, University of Wyoming
Ineke Steetskamp	Inst. for Envir. Studies, Free University, Amsterdam, The Netherlands
Enzo Tiezzi	Department of Chemistry, University of Siena, Italy
Robert Ulanowicz	Chesapeake Biological Laboratory, University of Maryland
Sergio Ulgiati	Department of Chemistry, University of Siena, Italy
James Zucchetto	National Academy of Sciences, Washington, DC
Tomasz Zylicz	Ministry of the Environment, and Warsaw University, Poland

Toward the end of the workshop, Kenneth Boulding read the following limerick, which he wrote to summarize our endeavor:

We need to make no apology
For thinking about world ecology,
For mere economics
Is stuff for the comics
Unless we can live with biology.

How can we achieve the facility
To encourage some sustainability
When all that it means
When it comes to our genes
Is to over expand our virility.

To stop the extinction of species
We must do something with our faeces
And we have to relieve
The air that we breath
Of the hot CO_2 that increases.

We need to do something in haste
About the production of waste
For if we do not
Then what have we got
But a world that is not to our taste.

Kenneth E. Boulding, May 26, 1990

GOALS, AGENDA, AND POLICY RECOMMENDATIONS FOR ECOLOGICAL ECONOMICS

Robert Costanza
Coastal and Environmental Policy Program
Center for Environmental and Estuarine Studies
University of Maryland
Box 38, Solomons, MD 20688-0038 USA

Herman E. Daly
Environment Department
The World Bank
1818 H. Street, NW
Washington, D.C. 20433 USA

and

Joy A. Bartholomew
Director, Washington Area Office
Center for Policy Negotiation
490 Chippingwood, Suite 1
Port Republic, MD 20676 USA

ABSTRACT

This introductory chapter: 1) Summarizes the state and goals of the emerging transdisciplinary field of *ecological economics*, particularly as regards issues of sustainability; 2) provides a working agenda for research, education and policy for the coming decade to ensure sustainability; 3) provides some policy guidelines and recommendations for achieving these goals.

   This chapter represents, to the extent possible, the "sense of the meeting" or consensus of the workshop which produced it. This does not mean that all the workshop participants agree with all that is said here; we can only offer one perspective. The following chapters by individual workshop participants elaborate the themes we describe and give more detailed and varied perspectives.

OVERVIEW OF THE BOOK

The book is divided into three major parts following this introductory chapter. Part I focuses on defining the basic world view of ecological economics, along with how (and why) it differs from conventional approaches. The ten papers in the section cover a broad range of perspectives. Boulding, Daly, and Hardin set the stage with incisive discussions of the root causes of the problems facing humanity and definitions of some basic ecological economic principles to build on. Page, Christensen, Norgaard and Howarth, and Norton offer perceptive insights into the problems of sustainability, discounting, and valuation. Martinez-Alier outlines some of the historical precedents for ecological economics. Funtowicz and Ravetz, and Perrings round out the section with their unique contributions on the role of uncertainty in an *ecological economic* world view and develop appropriate ways to deal with this uncertainty.

Part II of the book focuses on accounting, modeling, and analysis of ecological economic systems. It begins with El Serafy's discussion of the environment as capital. Peskin, Hueting, and Faber and Proops offer different perspectives on and methods for incorporating natural capital and services into national income accounting. Hannon and Ulanowicz extend and generalize these concepts to deal with ecosystems and combined ecological economic systems. Braat and Steetskamp offer a more elaborate modeling system for regional analysis and Cleveland rounds out the section with an analysis of resource scarcity from an ecological economics perspective.

Part III of the book deals with institutional changes necessary to achieve sustainability, and includes case studies. The first five papers in the section deal with incentives and instruments. Colin Clark offers an analysis of the perverse incentives that work against sustainability, while Costanza and Farber deal with methods to alter incentives to assure sustainability. Cumberland, and d'Arge and Spash apply these concepts to intergenerational transfers, while Zylicz attacks international transfers. Following the papers on transfers, two papers, by Mary Clark and Zucchetto, discuss the role of education in furthering the goals of ecological economics and sustainability. The section ends with five papers that offer case studies of ecological economic problems and approaches. Mitsch defines the field of ecological engineering and compares the experiences of the United States and China. Jansson takes an ecological economic look at the Baltic Sea region, Tiezzi et al. look at integrated agro-industrial ecosystems, and Cavalcanti looks at the Brazilian situation. Finally Goodland et al. offer a detailed analysis and policy recommendations for the management of moist tropical forests.

While the chapters overlap to some degree in their coverage of certain basic themes, the multiple perspectives enrich the reader's understanding of the pluralistic nature of *ecological economics*.

AN ECOLOGICAL ECONOMIC WORLD VIEW

Increasing awareness that our global ecological life support system is endangered is forcing us to realize that decisions made on the basis of local, narrow, short-term criteria can produce disastrous results globally and in the long run. We are also beginning to realize that traditional economic and ecological models and concepts fall short in their ability to deal with global ecological problems.

Ecological economics is a new *transdisciplinary* field of study that addresses the relationships between ecosystems and economic systems in the broadest sense. These relationships are central to many of humanity's current problems and to building a sustainable future but are not well covered by any existing scientific discipline.

By *transdisciplinary* we mean that *ecological economics* goes beyond our normal conceptions of scientific disciplines and tries to integrate and synthesize many different disciplinary perspectives. One way it does this is by focusing more directly on the problems, rather than the particular intellectual tools and models used to solve them, and by ignoring arbitrary intellectual turf boundaries. No discipline has intellectual precedence in an endeavor as important as achieving sustainability. While the intellectual tools we use in this quest are important, they are secondary to the goal of solving the critical problems of managing our use of the planet. We must transcend the focus on tools and techniques so that we avoid being "a person with a hammer to whom everything looks like a nail." Rather we should consider the task, evaluate existing tools' abilities to handle the job, and design new ones if the existing tools are ineffective. *Ecological economics* will use the tools of conventional economics and ecology as appropriate. The need for new intellectual tools and models may emerge where the coupling of economics and ecology is not possible with the existing tools.

How Is Ecological Economics Different from Conventional Approaches?

Ecological economics (EE) differs from both conventional economics and conventional ecology in terms of the breadth of its perception of the problem, and the importance it attaches to environment-economy interactions. It takes this wider and longer view in terms of space, time and the parts of the system to be studied.

Figure 1.1 illustrates one aspect of the relationship: the domains of the different subdisciplines. The upper left box represents the domain of "conventional" economics, the interactions of economic sectors (like mining, manufacturing, or households) with each other. The domain of "conventional" ecology is the lower right box, the interactions of ecosystems and their components with each other. The lower left box represents the inputs from ecological sectors to economic sectors. This is the usual domain of *resource* economics and environmental impact analysis: the use of renewable and nonrenewable natural resources by the economy. The upper right box represents the "use" by ecological sectors of economic "products." The products of interest in this box are usually unwanted by-products of production and the ultimate wastes from consumption. This is the usual domain of *environmental* economics and environmental impact analysis: pollution and its mitigation, prevention and mediation. *Ecological economics* encompasses and transcends these disciplinary boundaries. *Ecological economics* sees the human economy as part of a larger whole. Its domain is the entire web of interactions between economic and ecological sectors.

Table 1.1 presents some of the other major differences between *ecological economics* (EE) and conventional economics (CEcon) and conventional ecology (CEcol). These issues are covered in more detail and from a number of different perspectives in Part I of this book. The basic world view of CEcon is one in which individual human consumers are the central figures. Their tastes and preferences are taken as given and are the dominant determining force. The resource base is viewed as essentially limitless due to technical progress and infinite substitutability. Ecological economics takes a more holistic view with humans as one component (albeit a very important one) in the overall system. Human preferences, understanding, technology and cultural organization all co-evolve to reflect broad ecological opportunities and constraints. Humans have a special place in the system because they are responsible for understanding their own role in the larger system and managing it for sustainability. This basic world view is similar to that of CEcol, in which the resource base is limited and humans are just another (albeit seldom studied) species. But EE differs from CEcol in the importance it gives to humans as a species, and its emphasis on the mutual importance of cultural and biological evolution.

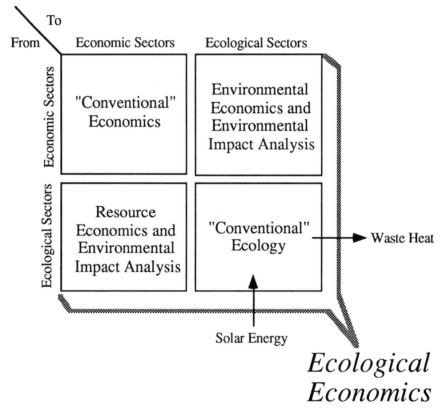

FIGURE 1.1 Relationship of domains of *Ecological Economics* and conventional economics and ecology, resource and environmental economics, and environmental impact analysis.

TABLE 1.1 Comparison of "Conventional" Economics and Ecology with *Ecological Economics*

	"Conventional" Economics	"Conventional" Ecology	Ecological Economics
Basic World View	**Mechanistic, Static, Atomistic** Individual tastes and preferences taken as given and the dominant force. The resource base viewed as essentially limitless due to technical progress and infinite substitutability	**Evolutionary, Atomistic** Evolution acting at the genetic level viewed as the dominant force. The resource base is limited. Humans are just another species but are rarely studied.	**Dynamic, Systems, Evolutionary** Human preferences, understanding, technology and organization co-evolve to reflect broad ecological opportunities and constraints. Humans are responsible for understanding their role in the larger system and managing it sustainably
Time Frame	**Short** 50 yrs max, 1-4 yrs. usual	**Multiscale** Days to eons, but time scales often define non-communicating sub-disciplines	**Multi-Scale** Days to eons, multiscale synthesis
Space Frame	**Local to International** Framework invarient at increasing spatial scale, basic units change from individuals to firms to countries	**Local to Regional** Most research has focused on smaller research sites in one ecosystems, but larger scales have become more important	**Local to Global** Hierarchy of scales
Species Frame	**Humans Only** Plants and animals only rarely included for contributary value	**Non-Humans Only** Attempts to find "pristine" ecosystems untouched by humans	**Whole Ecosystem Including Humans** Acknowledges interconnections between humans and rest of nature
***Primary* Macro Goal**	**Growth of National Economy**	**Survival of Species**	**Ecological Economic System Sustainability**
***Primary* Micro Goal**	**Max Profits** (firms) **Max Utility** (indivs) All agents following micro goals leads to macro goal being fulfilled. External costs and benefits given lip service but usually ignored	**Max Reproductive Success** All agents following micro goals leads to macro goal being fulfilled.	**Must Be Adjusted to Reflect System Goals** Social organization and cultural institutions at higher levels of the space/time hierarchy ameliorate conflicts produced by myopic pursuit of micro goals at lower levels
Assumptions About Technical Progress	**Very Optimistic**	**Pessimistic or No Opinion**	**Prudently Skeptical**
Academic Stance	**Disciplinary** Monistic, focus on mathematical tools	**Disciplinary** More pluralistic than economics, but still focused on tools and techniques. Few rewards for integrative work.	**Transdisciplinary** Pluralistic, focus on problems

The concept of *evolution* is a guiding notion for both ecology and ecological economics (see Boulding, this volume). Evolution is the process of change in complex systems through selection of transmittable traits. Whether these traits are the shapes and programmed behavioral characteristics of organisms transmitted genetically or the institutions and behaviors of cultures which are transmitted through cultural artifacts, books and tales around the campfire, they are both evolutionary processes. Evolution implies a dynamic and adapting nonequilibrium system, rather than the static equilibrium system often assumed in conventional economics. Evolution does *not* imply change in a particular direction (i.e., progress).

Ecological economics uses an expanded definition of the term "evolution" to encompass both biological and cultural change. Biological evolution is slow relative to cultural evolution. The price human cultures pay for their ability to adapt rapidly is the danger that they have become too dependent on short-run payoffs and thereby usually ignore long-term payoffs and issues of sustainability. Biological evolution imposes a built-in long-run constraint that cultural evolution does not have. To ensure sustainability, we may have to reimpose long-run constraints by developing institutions (or using the ones we have more effectively) to bring the global, long-term, multispecies, multiscale, whole systems perspective to bear on short-term cultural evolution.

The issue of humans' role in shaping the combined biological and cultural evolution of the planet is of critical importance. Humans are conscious of the processes of biological and cultural evolution and cannot avoid being anthropocentric. But in the long run, if humans are to manage the whole planet effectively, we must develop the capacity to take a broader *biocentric* perspective and to treat our fellow species with respect and fairness. We must also recognize that most natural systems are self-regulating and that the best "managerial strategy" is often to leave them alone.

The time frame, space frame, and species frame of EE all tend to be broader than CEcon and are more similar to the "frames" of CEcol. But there is an explicit recognition of the need for integrated, multiscale analysis. This view is also beginning to take hold in CEcol but it is all but absent from CEcon. In practice, CEcol all but ignores humans, CEcon ignores everything but humans, and EE tries to manage the whole system and acknowledges the interconnections between humans and the rest of nature. We must acknowledge that the human system is a subsystem within the larger ecological system. This implies not only a relationship of interdependence, but ultimately a relation of dependence of the subsystem on the larger parent system. The first questions to ask about a subsystem are: How big is it relative to the total system, how big can it be, and how big should it be? These questions of scale are only now beginning to be asked (see Daly, this volume).

The presumed goals of the systems under study are also quite distinct, especially at the macro (whole system) level. The macro goal of EE is sustainability of the combined ecological economic system. CEcol's macro goal of species survival is similar to sustainability, but is generally confined to single species and not the whole system. CEcon emphasizes growth rather than sustainability at the macro level. At the micro level, EE is unique in acknowledging the two-way interdependencies between the micro and macro levels. The conventional sciences tend to view all macro behavior as the simple aggregation of micro behavior. In EE, social organization and cultural institutions at higher levels of the space/time hierarchy ameliorate conflicts produced by myopic pursuit of micro goals at lower levels, and vice versa.

Perhaps the key distinctions between EE and the conventional sciences lie in their academic stances, and their assumptions about technical progress. As already noted, EE is transdisciplinary, pluralistic, integrative, and more focused on problems than on tools.

CEcon is very optimistic about the ability of technology to ultimately remove all resource constraints to continued economic growth. CEcol really has very little to say directly about technology, since it tends to ignore humans altogether. But to the extent that it has an opinion, it would be pessimistic about technology's ability to remove resource constraints because all other existing natural ecosystems that don't include humans are observed to be resource limited. EE is prudently skeptical in this regard. Given our high level of uncertainty about this issue, it is irrational to *bank on* technology's ability to remove resource constraints. If we guess wrong then the result is disastrous—irreversible destruction of our resource base and civilization itself. We should, at least for the time being, assume that technology will *not* be able to remove resource constraints. If it does, we can be pleasantly surprised. If it does not, we are still left with a sustainable system. EE assumes this prudently skeptical stance on technical progress.

A RESEARCH AGENDA FOR ECOLOGICAL ECONOMICS

To achieve sustainability, several steps are necessary including innovative research. This research should not be divorced from the policy and management process, but rather integrated with it. The research agenda for ecological economics that we suggest below is a snapshot, a first guess, intended to begin the process of defining topics for future ecological economic research rather than be the final word. The list of topics can be divided into five major parts: 1) sustainability: maintaining our life support system; 2) valuation of natural resources and natural capital; 3) ecological economic system accounting; 4) ecological economic modeling at local, regional, and global scales; and 5) innovative instruments for environmental management. Some background on each of these topics is given below, followed by a nonprioritized list of the major research questions.

Sustainability: Maintaining Our Life-Support System

Background
"Sustainability" does not imply a static, much less a stagnant, economy, but we must be careful to distinguish between "growth" and "development." Economic growth, which is an increase in quantity, cannot be sustainable indefinitely on a finite planet. Economic development, which is an improvement in the quality of life without necessarily causing an increase in quantity of resources consumed, may be sustainable. Sustainable growth is

an impossibility. Sustainable development must become our primary long-term policy goal (see Boulding, this volume, and Daly, this volume, for more on these ideas).

The most obvious danger of ignoring the role of nature in economics is that nature is the economy's life support system, and by ignoring it we may inadvertently damage it beyond it's ability to repair itself. Indeed, there is much evidence that we have already done so. Several authors have stressed the fact that current economic systems do not *inherently* incorporate any concern about the sustainability of our natural life support system and the economies which depend on it (e.g., Costanza and Daly 1987; Hardin, this volume, C. Clark, this volume). Pearce (1987) discusses the reasons for the inability of existing forms of economic organization (free market, mixed, planned) to guarantee sustainability. In an important sense, sustainability is merely justice with respect to future generations. This includes future generations of other species, even though our main interest may be in our own species.

Sustainability has been variously construed (cf. Pezzey 1989; World Commission on Environment and Development 1987) but a useful definition is the amount of consumption that can be continued indefinitely without degrading capital stocks—including "natural capital" stocks (see El Serafy, this volume). In a business, capital stock includes long-term assets such as buildings and machinery that serve as the means of production. Natural capital is the soil and atmospheric structure, plant and animal biomass, etc., that, taken together, forms the basis of all ecosystems. This natural capital stock uses primary inputs (sunlight) to produce the range of ecosystem services and physical natural resource flows. Examples of natural capital include forests, fish populations and petroleum deposits. The natural resource flows yielded by these natural capital stocks are, respectively, cut timber, caught fish, and pumped crude oil. We have now entered a new era in which the limiting factor in development is no longer manmade capital but remaining natural capital. Timber is limited by remaining forests, not sawmill capacity; fish catch is limited by fish populations, not by fishing boats; crude oil is limited by the accessibility of remaining petroleum deposits, not by pumping and drilling capacity. Most economists view natural and manmade capital as substitutes rather than complements. Consequently, neither factor can be limiting. Only if factors are complementary can one be limiting. Ecological economists see manmade and natural capital as fundamentally complementary and therefore emphasize the importance of limiting factors and changes in the pattern of scarcity. This is a fundamental difference that needs to be reconciled through debate and research.

Definitions of sustainability are also obviously dependent on the time and space scale we are using. Rather than trying to determine the *correct* time and space scale for sustainability we need to concentrate on how the different scales interact and how we might construct *multiscale* operational definitions of sustainability.

While acknowledging that the sustainability concept requires much additional research, we devised the following working definition of sustainability: *Sustainability* is a relationship between dynamic human economic systems and larger dynamic, but normally slower-changing ecological systems, in which 1) human life can continue indefinitely, 2) human individuals can flourish, and 3) human cultures can develop; but in which effects of human activities remain within bounds, so as not to destroy the diversity, complexity, and function of the ecological life support system.

Major Research Questions
- What do we mean by (and how do we quantify) "health" and "sustainability" in ecological and economic systems?
- What is the hierarchy (in time and space) of goals for these systems and how is sustainability defined at different levels in the hierarchy? What conflicts arise between setting overall system sustainability goals and providing subgroup, or cultural, autonomy?
- What are the sustainable levels of population and per capita resource use, and what are the paths to achieve these?
- What kinds of actions can benefit the future without harming the present?
- How can sustainability criteria be incorporated in quantitative indices of national income, wealth, and welfare? (See also "Ecological Economic Modeling at Local, Regional and Global Scales" below.)
- What is the degree of substitutability between natural and manmade capital, and ecological and economic services, and how does this influence sustainability?
- Do the basic assumptions underlying current economic and ecological paradigms need to be revised to incorporate sustainability criteria and what are the implications of alternative assumptions?
- How can basic ecological models and principles be incorporated into operational definitions of sustainability?
- What can we learn from the study of historical human societies and natural systems that have proven to be sustainable about the general characteristics of sustainable systems?
- How can we design better institutions and instruments to assure sustainability?
- What are the conditions by which international trade may be made both economically equitable and environmentally sustainable for all parties?

Valuation of Ecosystem Services and Natural Capital

Background
To achieve sustainability, we must incorporate ecosystem goods and services into our economic accounting. The first step is to determine values for them comparable to those of economic goods and services. In determining values, we must also consider how much of our ecological life support systems we can afford to lose. To what extent can we substitute manufactured for natural capital, and how much of our natural capital is irreplaceable (El Serafy, this volume)? For example, could we replace the radiation screening services of the ozone layer which are currently being destroyed?

Some argue that we cannot place economic value on such "intangibles" as human life, environmental aesthetics, or long-term ecological benefits. But, in fact, we do so every day. When we set construction standards for highways, bridges and the like, we value human life—acknowledged or not—because spending more money on construction would save lives. To preserve our natural capital, we must confront these often difficult choices and valuations directly rather than denying their existence.

Because of the inherent difficulties and uncertainties in determining values, ecological economics acknowledges several different independent approaches. There is no consensus on which approach is right or wrong—they all tell us something—but there is agreement that better valuation of ecosystem services is an important goal for ecological economics.

The conventional economic view defines value as the expression of individualistic human preferences, with the preferences taken as given and with no attempt to analyze their origins or patterns of long-term change. For goods and services with few long-term impacts (like tomatoes or bread) that are traded in well-functioning markets with adequate information, market ("revealed preference") valuations work well.

But ecological goods and services (like wetland sewage treatment or global climate control) are long-term by nature, are generally not traded in markets (no one owns the air or water), and information about their contribution to individual's well-being is poor. To determine their value, economists try to get people to reveal what they would be willing to pay for ecological goods and services in hypothetical markets. For example, we can ask people the maximum they would pay to use national parks, even if they don't have to actually pay it. The quality of results in this method depends on how well informed people are; it does not adequately incorporate long-term goals since it excludes future generations from bidding in the markets. Also, it is difficult to induce individuals to reveal their true willingness to pay for natural resources when the question is put directly. Contingent referenda (willingness to be taxed as a citizen along with other citizens, as opposed to willingness to pay as an individual) is superior to ordinary willingness to pay studies in this regard.

In practice, valuation or shadow pricing of environmental functions may require some collectively set quantitative standard. Then shadow prices can be calculated subject to the constraint represented by that standard (see Hueting, this volume).

An alternative method for estimating ecological values assumes a biophysical basis for value (see Costanza 1980; Cleveland et al. 1984; Costanza et al. 1989; Costanza, this volume; Cleveland, this volume). This theory suggests that in the long run humans come to value things according to how costly they are to produce, and that this cost is ultimately a function of how organized they are relative to their environment. To organize a complex structure takes energy, both directly in the form of fuel and indirectly in the form of other organized structures like factories. For example, a car is a much more organized structure than a lump of iron ore; therefore, it takes a lot of energy (directly and indirectly) to organize iron ore into a car. The amount of solar energy required to grow forests can therefore serve as a measure of their energy cost, their organization, and hence, according to this theory, their value.

The point that must be stressed is that the economic value of ecosystems is connected to their physical, chemical, and biological role in the long-term, global system—whether the present generation of humans fully recognizes that role or not. If it is accepted that each species, no matter how seemingly uninteresting or lacking in immediate utility, has a role in natural ecosystems (which *do* provide many direct benefits to humans), it is possible to shift the focus away from our imperfect short-term perceptions and derive more accurate values for long-term ecosystem services. Using this perspective we may be able to better estimate the values contributed by, say, maintenance of water and atmospheric quality to long-term human well-being. Obviously, these services are vital and of infinite value at some level. The valuation question relates to marginal changes, incremental tradeoffs between, say, forested land and agricultural land on a scale of hundreds of acres rather than hundreds of square miles. The notion of safe minimum standards championed by a few economists seems relevant to the protection of critical levels of natural capital against excess myopic marginal conversion, or large-scale conversion, into manmade capital. Of course, in a perfect system, marginal valuations would become prohibitive if the safe minimum standard were transgressed. But systems are far from perfect and redundancy in the interest of prudence is not extravagance.

Major Research Questions
- How do we measure the value of ecosystem services and natural capital? Under what conditions can values be translated to single scales e.g., money, utility or energy?
- Do measures based on subjective preferences (contingent valuation, contingent referenda, willingness to pay) have any relationship to values based on ecosystem functioning and energy flows?
- What is the appropriate discount rate to apply to ecosystem services?
- What (or where) are the thresholds of irreversible degradation for natural resources?

Ecological Economic System Accounting

Background
Gross National Product, as well as other related measures of national economic performance have come to be extremely important as policy objectives, political issues and benchmarks of the general welfare. Yet GNP as presently defined ignores the contribution of nature to production, often leading to peculiar results.

For example, a standing forest provides real economic services for people: by conserving soil, cleaning air and water, providing habitat for wildlife, and supporting recreational activities. But as GNP is currently figured, only the value of harvested timber is calculated in the total. On the other hand, the billions of dollars that Exxon spent on the Valdez cleanup—and the billions spent by Exxon and others on the more than 100 other oil spills in the last 16 months—all actually *improved* our apparent economic performance. Why? Because cleaning up oil spills creates jobs and consumes resources, all of which add to GNP. Of course, these expenses would not have been necessary if the oil had not been spilled, so they shouldn't be considered "benefits." But GNP adds up all production without differentiating between costs and benefits, and is therefore not a very good measure of economic health.

In fact, when resource depletion and degradation are factored into economic trends, what emerges is a radically different picture from that depicted by conventional methods.

For example, Herman Daly and John Cobb (Daly and Cobb 1989) have attempted to adjust GNP to account mainly for depletions of natural capital, pollution effects, and income distribution effects by producing an "index of sustainable economic welfare" (ISEW). They conclude that while GNP in the United States rose over the 1956-1986 interval, ISEW remained relatively unchanged since about 1970. When factors such as loss of farms and wetlands, costs of mitigating acid rain effects, and health costs caused by increased pollution, are accounted for, the US economy has not improved at all. If we continue to ignore natural ecosystems, we may drive the economy down while we think we are building it up. By consuming our natural capital, we endanger our ability to sustain income. Daly and Cobb acknowledge that many arbitrary judgments go into their ISEW, but claim nevertheless that it is less arbitrary than GNP as a measure of welfare. John Cobb and his group at Claremont have continued work on the index and their procedure is worth mentioning as a model for scholarly debate. Cobb sent the ISEW to a number of standard economists for criticism, offering an honorarium and contracting to publish their criticisms along with a revised version of the ISEW that would take account of their criticism, or else explain why that could not or should not be done. The result has been a fruitful interchange and better mutual understanding.

There are a number of additional promising approaches to accounting for ecosystem services and natural capital being developed (see El Serafy, Hannon, Hueting, Peskin, Faber and Proops, and Ulanowicz, this volume) and this area promises to be a major focus of research in ecological economics. The approaches are based on differing assumptions, but share the goal of attempting to quantify ecological economic interdependencies and arriving at overall system measures of health and performance. The economist Wassily Leontief (1941) was the first to attempt detailed quantitative descriptions of complex systems to allow a complete accounting of system interdependencies. Leontief's input-output (I-O) analysis has become a standard conceptual and applied tool in economic accounting. Isard (1972) was the first to attempt combined ecological economic system I-O analysis. Combined ecological economic system I-O models have been proposed by several other authors as well (Daly 1968; Victor 1972; Cumberland 1987). Ecologists have also applied I-O analysis to the accounting of material transfers in ecosystems (Hannon 1973, 1976, 1979, this volume; Costanza and Neill 1984; Costanza and Hannon 1989). We refer to the total of all variations of the analysis of ecological and/or economic networks as *network analysis*.

Network analysis holds the promise of allowing an integrated quantitative treatment of combined ecological economic systems and the "pricing" of commodities in ecological and/or economic systems (Costanza, this volume; Hannon, this volume; Costanza and Hannon 1989; Ulanowicz 1980, 1986, this volume; Wulff et al. 1989). This kind of analysis may provide the basis for a quantitative and general index of system health applicable to both ecological and economic systems.

Major Research Questions
• How can we create better systems of national, regional and global accounting to include natural resource depletion and ecological impacts?

- How can we develop systems for accounting for and managing transnational environmental impacts?
- How can we develop network based measures of system health that are applicable to both ecological and economic systems?
- How can we use network based measures of system interdependence (such as energy intensities) to evaluate components in both ecological and economic systems?

Ecological Economic Modeling at Local, Regional, and Global Scales

Background

Since ecosystems are being threatened by a host of human activities, protecting and preserving them requires the ability to understand the direct and indirect effects of human activities over long periods of time and over large areas. Computer simulations are now becoming important tools to investigate these interactions and in all other areas of science as well. Without the sophisticated global atmospheric simulations now being done, our understanding of the potential impacts of increasing CO_2 concentrations in the atmosphere due to fossil fuel burning would be much more primitive. Computer simulations can now be used to understand not only human impacts on ecosystems, but also our economic dependence on natural ecosystem services and capital, and the interdependence between ecological and economic components of the system (see, for example, Braat, this volume; Costanza et al. 1990).

Several recent developments make such computer simulation modeling feasible, including the accessibility of extensive spatial and temporal data bases and advances in computer power and convenience. Computer simulation models are potentially one of our best tools to help understand the complex functions of integrated ecological economic systems.

But even with the best conceivable modeling capabilities, we will always be confronted with large amounts of uncertainty about the response of the environment to human actions (see Funtowicz and Ravetz, this volume). Learning how to effectively manage the environment in the face of this uncertainty is critical (see Perrings, this volume).

The research program of ecological economics will pursue an integrated, multiscale, transdisciplinary, and pluralistic, approach to quantitative ecological economic modeling, while acknowledging the large remaining uncertainty inherent in modeling these systems and developing new ways to effectively deal with this uncertainty.

Major Research Questions

- What are appropriate model structures for a range of urban, agricultural, and natural subsystems, at several hierarchical scales?
- How can these models best be tested, scaled and integrated?
- How can existing data sources (i.e., remote sensing images, national accounting data) best be utilized in building, calibrating and testing ecological economic models at multiple scales?

- What role does biological diversity play in the health and sustainability of ecological economic systems?
- How can simulation modeling results best be used in system accounting and natural ecosystem valuation?
- What are the most appropriate roles of simulation, analytical, and optimization models? What should be their relationship to accounting frameworks?
- How are changes in the quality and cost of natural resources, i.e., rain forests, tropical seas, or grasslands to be measured? How do such changes affect economic welfare?
- Are there general system principles which govern the economy-ecology relationship?
- What viewpoints, modeling mechanisms system variables and other tools or techniques from economic models can be usefully applied to ecosystem models, and vice versa.
- How can intergenerational distribution be addressed analytically as well as ethically?
- What is the appropriate role of chaotic modeling in analyzing ecological economic problems with large degrees of uncertainty?
- How can we develop a philosophy of modeling which is open to the emergence of novelty and consistent with the evolutionary, dynamic, whole systems, multiscale paradigm?
- How do we model the interactions among local, regional, and global levels of ecological economic systems?

Innovative Instruments for Environmental Management

Background
Current systems of regulation are not very efficient at managing environmental resources for sustainability, particularly in the face of uncertainty about long-term values and impacts. They are inherently reactive rather than proactive. They induce legal confrontation, obfuscation, and government intrusion into business. Rather than encouraging long-range technical and social innovation, they tend to suppress it. They do not mesh well with the market signals that firms and individuals use to make decisions and do not effectively translate long-term global goals into short-term local incentives.

We need to explore promising alternatives to our current command and control environmental management systems, and to modify existing government agencies and other institutions accordingly. The enormous uncertainty about local and transnational environmental impacts needs to be incorporated into decision-making. We also need to better understand the sociological, cultural, and political criteria for acceptance or rejection of policy instruments.

One example of an innovative policy instrument currently being studied is a flexible environmental assurance bonding system designed to incorporate environmental criteria and uncertainty into the market system, and to induce positive environmental technological innovation (Perrings 1989; Costanza and Perrings 1990; Perrings, this volume).

In addition to direct charges for known environmental damages, a company would be required to post an assurance bond equal to the current best estimate of the largest potential future environmental damages; the money would be kept in interest-bearing escrow accounts. The bond (plus a portion of the interest) would be returned if the firm could show that the suspected damages had not occurred or would not occur. If they did, the bond would be used to rehabilitate or repair the environment and to compensate injured parties. Thus, the burden of proof would be shifted from the public to the resource-user and a strong economic incentive would be provided to research the true costs of environmentally damaging activities and to develop cost-effective pollution control technologies. This is an extension of the "polluter pays" principle to "the polluter pays for uncertainty as well." Other innovative policy instruments include tradeable pollution and depletion quotas at both national and international levels. Also worthy of mention is the newly emerging Global Environmental Facility of the World Bank that will provide concessionary funds for investments that reduce global externalities.

Major Research Questions
- What regulatory or incentive-based instruments are most appropriate for assuring sustainability?
- How can government and other institutions be modified to better account and respond to environmental impacts?
- What is the appropriate role for economic incentives and disincentives in managing ecological economic systems?
- What sociological, political, ethical, or other factors have limited acceptance of economic incentive-based instruments, and can these factors be addressed?
- How can we develop experimental economics in order to predict behavioral responses to new management instruments? What role might computer modeling play in this development?
- What is the impact of social security systems for limiting population growth?
- How do we equitably limit world population without oppressive programs?
- How do we develop mechanisms to lengthen the time horizons of institutions at all levels?
- What institutions are most effective at preserving the pool of genetic information; preserving the ecological knowledge of indigenous peoples; and facilitating cultural adaptations to environmental and/or technological change?
- What international institutions are available or necessary to assure local and global sustainability?
- Why are excise taxes on materials and energy (which are relatively effective and simple to conceptualize and design) so hard to implement politically, and can the obstacles to implementing these mechanisms be removed?

POLICY RECOMMENDATIONS

The following represents a limited set of policy recommendations on which the workshop participants reached general consensus. It is not prioritized, nor is it comprehensive, nor does it imply that all the participants were in complete agreement.

But it does represent the spectrum of policy recommendations that the workshop participants felt comfortable with as a starting point for further discussion.

Sustainability as the Goal

We should institute a consistent goal of sustainability in all institutions at all levels from local to global. We should strive to address prevailing values and decision-making processes by increasing the awareness of institutions and persons about ecological sustainability. We should promote long-term thinking, the use of a systems approach in decision-making, and use of "ecological auditors" (i.e., trained environmental professionals) by public and private institutions whose activities affect the environment.

For example, the World Bank is an important global institution that directly affects economic policy, and those policies severely affect the environment, especially in developing nations. We recommend that the bank and similar institutions require that all projects meet the following criteria: For renewable resources, the rate of harvest should not exceed the rate of regeneration (sustainable yield) and the rates of waste generation from projects should not exceed the assimilative capacity of the environment (sustainable waste disposal). For nonrenewable resources, the rates of waste generation from projects shall not exceed the assimilative capacity of the environment and the depletion of the nonrenewable resources should require comparable development of renewable substitutes for that resource. These are safe, minimum sustainability standards; and once met, the bank should then select projects for funding that have the highest rates of return based on other, more traditional economic criteria.

We recognize that this policy will be difficult at first, and that the policies will likely shift as more information is developed about managing for sustainability. However, there is a need for major institutions not only to affirm, but to operationalize the goal of sustainability, because of the global scope of their programs and because of the impact their example will provide for smaller institutions worldwide. We recognize that goal setting is an ethical issue, and that it is absurd to ignore the normative preconditions of policy, however necessary it may be to avoid mixing normative and positive statements in analysis. Both economists and ecologists, if they want to talk about policy, must offer much more explicit ethical support for their goals, whether sustainability or growth.

Maintaining Natural Capital to Assure Sustainability

A minimum necessary condition for sustainability is the maintenance of the total natural capital stock at or above the current level. While a lower stock of natural capital may be sustainable, given our uncertainty and the dire consequences of guessing wrong, it is best to at least provisionally assume that the we are at or below the range of sustainable stock levels and allow no further decline in natural capital. This "constancy of total natural capital" rule can thus be seen as a prudent minimum condition for assuring sustainability, to be abandoned only when solid evidence to the contrary can be offered. There is disagreement between technological optimists (who see technical progress eliminating all resource constraints to growth and development) and technological skeptics (who do not see as much scope for this approach and fear irreversible use of resources and damage to natural capital). By maintaining total system natural capital at current levels (preferably by using higher severance and consumption taxes), we can satisfy both the skeptics (since resources will be conserved for future generations) and the optimists (since this will raise the price of natural capital resources and more rapidly induce the technical change they predict). By limiting physical growth, only development is allowed and this may proceed without endangering sustainability.

Improving Our Use of Policy Instruments

We need to use a wide variety of policy instruments including regulation, property rights, permits, marketable permits, fees, subsidies and bonds to assure sustainability. Criteria for use of policy instruments are: equity, efficiency, scientific validity, consensus, frugality and environmental effectiveness. We should institute regulatory reforms to promote appropriate use of financial, legal and social incentives. We may use market incentives where appropriate in allocation decisions. In decisions of scale, individual freedom of choice must yield to democratic collective decision making by the relevant community.

Economic Incentives: Linking Revenues and Uses

We should implement fees on the destructive use of natural capital to promote more efficient use, and ease up on income taxes, especially on low incomes in the interest of equity. Fees, taxes and subsidies should be used to change the prices of activities that interfere with sustainability versus those that are compatible with it. This can be accomplished by using the funds generated to support an alternative to undesirable activities that are being taxed. For example, a tax on all greenhouse gases, with the size of the tax linked to the impact of each gas could be linked to development of alternatives to fossil fuel. Gasoline tax revenues could be used to support mass transit and bike lanes. Current policies that subsidize environmentally harmful activities should be stopped. For example, subsidies on virgin material extraction should be stopped. This will also allow recycling options to effectively compete. Crop subsidies that dramatically increase pesticide and fertilizer use should be eliminated, and forms of positive incentives should also be used. For example, debt for nature swaps should be supported and should receive much more funding. We should also offer prestigious prizes for work that increases awareness of, or contributes to, sustainability issues, such as changes in behavior that develop a culture of maintenance (i.e., cars that last for 50 years) or promotes capital and resource saving improvements (i.e., affordable, efficient housing and water supplies).

Ecological Economic Research

While economics has developed many useful tools of analysis, it has not directed these tools toward the thorny questions that arise when considering the concept and implementation of sustainability. In particular, we need to better understand preference formation, and especially time preference formation. We also needs to understand how individual time preferences and group time preferences may differ, and how the preferences of institutions that will be critical to the success or failure of sustainability are established. We have heretofore paid too little attention to ecological feedbacks. An understanding of these will be critical to the implementation of sustainability goals, whatever they may be. We need to concentrate on the valuation of important non-market goods and services provided by ecosystems. We need to better understand the effects of various regulatory instruments that can be utilized to attain sustainability. This may require experimental testing of behavior in a laboratory context. Most importantly, we need to study how positive sustainability incentives can be employed to induce reluctant participants to lengthen their time horizons and think globally about their resource policies.

We also need to develop an ecological history of the planet (to complement the existing human economic history) that would contain trends of resource use, development and exhaustion, changes in science and technology, etc. We should promote the use (as one of a bundle of decision-making tools) of broad benefit/cost analyses that includes the consideration of all market and non-market costs and benefits.

Ecological Economics Education

Our education system is currently characterized by overspecialization and disciplinary isolation. We need to develop transdisciplinary curricula and job and academic support systems for both specialists and generalists. This needs to be combined with an emphasis on the value of general education and personal development, versus the more narrow training of professional technical specialists.

We need to develop an ecological economics core curriculum and degree granting programs that embody the skills of both economics and ecology. This implies a curriculum with some blending of physical, chemical and biological sciences and economics. Within this curriculum quantitative methods are essential, but they should be problem directed rather than just mathematical tools for their own sake.

There is a need to develop a capacity for experimentation that provides ecological economics with a solid empirical base built upon creative and comprehensive theory. We need to develop extension programs that can effectively transfer information among both disciplines and nations.

We should promote at all levels education that weaves together fundamental understanding of the environment with human economic activities and social institutions, and promotes research that facilitates this interweaving process. Particularly, awareness by the media of the common benefits of sustainability should be promoted to insure accuracy in reporting, and the media should be encouraged to use opportunities to educate others through mechanisms such as special reports and public service announcements. We should promote education of broadly-trained environmental scientists, whose jobs will be to provide on-going environmental assessment as an addition to the decision-making processes of various institutions, and as an addition to the assessments now being provided by economic analysts. The ISEE and other international institutions can (and should) provide a vehicle to help students and others focus on "big picture" questions and problems.

Institutional Changes

Institutions with the flexibility necessary to deal with ecologically sustainable development are lacking. Indeed, many financial institutions are built on the assumption of continuous exponential growth and will face major restructuring in a sustainable economy. Many existing institutions have fragmented mandates and policies, and often have not optimally used market and non-market forces to resolve environmental problems. They also have conducted inadequate benefit/cost analyses by not incorporating ecological costs; used short-term planning horizons; inappropriately assigned property rights (public and private) to resources; and made inappropriate use of incentives.

There is a lack of awareness and education about sustainability, the environment, and causes of environmental degradation. In addition, much environmental knowledge held by indigenous peoples is being lost, as is knowledge of species, particularly in the tropics. Institutions have been slow to respond to new information and shifts in values, for example, concerns about threats to biodiversity or the effects of rapid changes in communications technologies. Finally, many institutions do not freely share or disseminate information, do not provide public access to decision making, and do not devote serious attention to determining and representing the wishes of their constituencies.

Many of these problems are a result of the inflexible bureaucratic structure of many modern institutions. Experience (i.e., Japanese industry) has shown that less bureaucratic, more flexible, more peer-to-peer institutional structures can be much more efficient and effective. We need to de-bureaucratize institutions so that they can effectively respond to the coming challenges of achieving sustainability.

REFERENCES

Cleveland, C. J., R. Costanza, C. A. S. Hall, and R. Kaufmann. 1984. Energy and the United States Economy: A Biophysical Perspective. *Science* 225:890-897.

Costanza, R. 1980. Embodied Energy and Economic Valuation. *Science* 210:1219-1224.

Costanza, R. and H. E. Daly. 1987. Toward an Ecological Economics. *Ecological Modelling* 38:1-7

Costanza, R. and B. M. Hannon. 1989. Dealing with the "Mixed Units" Problem in Ecosystem Network Analysis. In F. Wulff, J. G. Field, and K. H. Mann, eds., *Network Analysis of Marine Ecosystems: Methods and Applications*, pp. 90-115. Coastal and Estuarine Studies Series. Heidelberg: Springer-Verlag.

Costanza, R. and C. Neill. 1984. Energy Intensities, Interdependence, and Value in Ecological Systems: A Linear Programming Approach. *Journal of Theoretical Biology* 106:41-57

Costanza, R. and C. H. Perrings. 1990. A Flexible Assurance Bonding System for Improved Environmental Management. *Ecological Economics* 2:57-76.

Costanza, R., S. C. Farber, and J. Maxwell. 1989. The Valuation and Management of Wetland Ecosystems. *Ecological Economics* 1:335-361.

Costanza, R., F. H. Sklar, and M. L. White. 1990. Modeling Coastal Landscape Dynamics. *BioScience* 40:91-107

Cumberland, J. H. 1987. Need Economic Development be Hazardous to the Health of the Chesapeake Bay? *Marine Resource Economics* 4:81-93.

Daly, H. 1968. On Economics as a Life Science. *Journal of Political Economy* 76:392-406.

Daly, H. E. and J. B. Cobb, Jr. 1989. *For The Common Good: Redirecting the Economy Toward Community, the Environment, and a Sustainable Future.* Boston: Beacon.

Hannon, B. 1973. The Structure of Ecosystems. *J. Theo. Biology* 41:535-46.

Hannon, B. 1976. Marginal Product Pricing in the Ecosystem. *J. Theo. Biology* 56:256-267.

Hannon, B., 1979, Total Energy Costs in Ecosystems. *J. Theo. Biology,* 80:271-293.

Isard, W. 1972. *Ecologic-Economic Analysis for Regional Development.* New York: The Free Press.

Leontief, W. 1941. *The Structure of American Economy, 1919–1939.* New York: Oxford University Press.

Pearce, D. 1987. Foundations of an Ecological Economics. *Ecological Modelling.* 38:9-18.

Perrings, C. 1989. Environmental Bonds and the Incentive to Research in Activities Involving Uncertain Future Effects. *Ecological Economics* 1:95-110.

Pezzey, J. 1989. Economic Analysis of Sustainable Growth and Sustainable Development. Environment Department Working Paper No. 15. Washington, D.C.: The World Bank.

Ulanowicz, R. E.: 1980. An Hypothesis on the Development of Natural Communities. *J. theor. Biol.* 85: 223-245.

Ulanowicz, R. E. 1986. *Growth and Development: Ecosystems Phenomenology.* New York: Springer-Verlag.

Victor, P. A. 1972. *Pollution, Economy and Environment.* Toronto: University of Toronto Press

World Commission on Environment and Development. 1987. *Our Common Future.* Oxford: Oxford University Press.

Wulff, F., J. G. Field and K. H. Mann. 1989. *Network Analysis of Marine Ecosystems: Methods and Applications.* Coastal and Estuarine Studies Series. Heidelberg: Springer-Verlag.

Part I

Developing an Ecological Economic
World View

2

WHAT DO WE WANT TO SUSTAIN? ENVIRONMENTALISM AND HUMAN EVALUATIONS

Kenneth E. Boulding
Institute of Behavioral Science
University of Colorado
Campus Box 483
Boulder, Colorado 80309 USA

ABSTRACT

The sustainability of biological evolution on this planet, punctuated as it seems to have been by catastrophes from which it recovered, is a fascinating story of which we have very imperfect knowledge. There seems little doubt, however, that the human race is an ecological catastrophe, simply because its intelligence has enabled it to spread over the whole planet and to produce very large numbers of artifacts which have an impact on biological populations. Most evolutionary catastrophes seem to have been followed by an increase in the complexity of organisms, perhaps because of the empty niches created by the catastrophe itself, and one wonders whether this is a possible guide to the future.

The unique characteristic of the human race is its capacity for forming complex images not only of its immediate environment but of the whole planet, and its ability to put evaluations on these larger images. Environmentalism as an ideology involves applying human evaluations to the possible future states of the total planet. If these evaluations are adverse, suggesting a worsening of the state of the planet, then the question arises, what changes in human behavior and institutions may result? We already see some changes in individual behavior, as in concerns for recycling and biodegradability and the cutting back on fluorocarbons because of the ozone effect. These changes affect only a small part of the human race and a critical question is whether the learning can be much further extended to change the behavior of those whose actions are constrained only by their perception of their immediate benefit. It is worth asking what new institutions might be necessary to achieve these objectives, even though the answers may not be easy to find.

The role of economic institutions in this process is obviously of great importance, but also hard to identify. An important question here is the extent to which the institutions of society feed back on the perceived welfare of decision makers and the overall effect of their decisions on the total system. Market institutions have the advantage that decisions which are not favorably regarded by potential purchasers, have fairly rapid feedback. Under central planning, feedback is much slower and disastrous decisions can be made without the consequences falling on the decision maker. This perhaps may account for the present disillusionment with central planning. It may be that we are looking for some optimum combination of political or threat power, economic power, and what I have called "integrative power," to lead us either away from catastrophe or towards a situation in which catastrophe will actually sustain the evolutionary process.

INTRODUCTION

The concept of sustainability has a good many possible meanings. Perhaps the most extreme meaning is that of "static equilibrium." A diamond at any temperature below its dissolution temperature is a good example of this. Static equilibrium is simply the absence of change, though, of course, there may be change in location over time. Another meaning of sustainability would be dynamic equilibrium, which can have two meanings: One is that of a system in motion in which the parameters of the motion are constant, such as the solar system, though there seems to be some possibility of chaos even in this Newtonian paradise. Another example would be a system with some sort of throughput, in what might be called "cybernetic equilibrium," for instance, an ecosystem in a stable environment, such as a pond or a forest. Here there is birth and death, cooperative and competitive relationships of populations, but there is a state of the system that can at least be called a quasi-equilibrium, in which a change in any population would lead to the restoration of its original value.

None of these concepts might correspond to sustainable development or evolution, which would not exhibit any equilibrium, but in which a process of ongoing change has some stable patterns or parameters. Thus, there is fairly general agreement at the moment that the universe is expanding, though whether it has been doing so from some initial "big bang" may be somewhat in doubt. In this case, sustainability comes close to meaning predictability, at least at the human level.

When it comes to the larger concept of evolution, however, the problem of what we mean by sustainability is by no means clear. We can certainly detect a "time's arrow" in the evolution of the universe in terms of complexity, both chemical and biological, and also in terms of knowledge and control. Sustainability here presumably means continuing to follow the "time's arrow" of the universe, and especially, of course, of the earth. Certainly, if a planetoid hit the earth and destroyed all life on it, the previous process would not have been sustainable.

The "time's arrow," however, especially in biological evolution on earth, is by no means a simple, steady process. There is some evidence that the catastrophes which have occurred in the past, which tend to separate one geological age from the next, have had a positive role to play in the evolutionary process itself. It is an interesting, though unsolved, problem as to whether a mutational equilibrium is possible, that is, an ecosystem in which all mutations are adverse, so there cannot be any production of new species. We have never actually observed this, but the theory of "punctuated equilibrium" (Somit and

Peterson, 1989) suggests that, over quite long periods, evolutionary change slows down, moving perhaps towards a mutational equilibrium. Then some improbable event, whether an external catastrophe or some improbable mutations, creates new niches, new species, and perhaps widespread extinction of old species, after which things settle down again as evolution slows down.

This process is observable also in societal evolution. Over long periods, nothing much seems to happen, as in the paleolithic, and then some improbable events start evolution up at a faster pace, after which it again sometimes slows down. Islam, for instance, is a very interesting example of a culture begun by a very improbable event, namely the arrival of Mohammed, which, in its first few centuries, led to the formation of a new civilization, many social innovations, and an acceleration of human learning. But after the 15th century—one could almost put a date on it of 1453, when the Turks conquered Constantinople—Islam stagnated and Europe, with its expansion into the Americas and the south temperate zone, almost literally exploded into science, world settlements, and eventually outer space.

The development both of DNA and of *Homo sapiens* may also have been very improbable events that happened. It is a fundamental principle that no matter how improbable an event may seem, if we wait long enough, it will happen. The probability that something that might be called a "billion-year flood" will happen at sometime within a billion years is about 63% and in four billion years is about 98%.

HUMAN INFLUENCE ON THE RATE OF ECOLOGICAL CHANGE

However it happened, there is little doubt that the development of *Homo* (and *Mulier*) *sapiens* shifted evolution on this planet into a new gear and proved ecologically catastrophic for many older species. *Homo sapiens* has an extraordinary capacity for increasing knowledge, for the transmission of knowledge from one mind into another through language and symbols, and for the transformation of knowledge into "know-how," which is a genetic factor. The DNA in my fertilized egg "knew how" to make me and humans now "know how" to make airplanes, which we did not a hundred years ago. These capacities have led to profound changes in the evolutionary process. *Homo sapiens* is the first species to invade virtually all the known ecosystems of the world, including now the deepest levels of the oceans, and even the moon. We have been where other forms of life have rarely penetrated—the poles, the highest mountains, the most lifeless deserts.

Furthermore, we have produced a very large number of species of human artifacts—tools, weapons, buildings, cities, roads, automobiles, airplanes—each of which occupies a niche in what might be called the larger ecosystem, which includes both biological and human artifacts. And by occupying these niches, human artifacts certainly reduce the populations of previous biological artifacts, with many going to zero and becoming extinct. We cut down forests, we plant crops, we domesticate biological species and change them, we transport biological species all over the world—eucalyptus trees from Australia; potatoes and corn from the Americas; horses, pigs, and cows from Europe; and so on. Before the human race, the world consisted of a large number of virtually isolated ecosystems, such as the Oceanic Islands, the Australian continent, the high mountains. The human race has made the world virtually a single ecosystem, which has led to an unprecedented rate of extinction of biological species. Something like ecological catastrophes, of course, have happened from time to time, even before the human race. What is unique about the present era is the accelerating rate at which the human race has created ecological change, and now climatic change.

Humans have also transformed the food and energy chains. Before the human race, solar energy was virtually the only primary form of energy used by the biosphere, with some few exceptions in volcanic vents. Plants were able to capture a small amount of the solar energy that falls on them by photosynthesis. (Bacteria also play a role in this process). Plants in turn opened up niches for herbivores, and herbivores for carnivores. Energy throughputs and materials cycles, and sometimes material throughputs like soil formation and erosion, have been essential for the development of life. Now the human race comes along and taps a new form of energy in fossil fuels, which feed automobiles and airplanes, and greatly increase the transportation facilities for living creatures. The use of fossil fuels, of course, is relatively recent, dating back not much more than 300 or 400 years. Their use continues to expand, almost exponentially.

In earlier periods, the human race utilized solar energy in the form of wind and waterfalls, as well as the burning of wood and other combustibles like dung produced by living creatures. Under these circumstances, the growth of both population and artifacts was fairly slow. With the use of fossil fuels, however, there has been a spectacular explosion in the number of humans, which has now more than doubled in one lifetime. We have also expanded, in even greater proportion than the population, the production of human artifacts, including many that are toxic or deleterious to life. The real question is not, can the biosphere adapt to change (which it has done constantly), but rather, how can it adapt to an enormous increase in the rapidity of change? That question is really worrisome. Another critical question is: Can the human race adapt to the changes that it is producing so rapidly? We cannot really get away from the fact that we are biological creatures and part of the biosphere, and that catastrophes in the biosphere inevitably involve us.

HUMAN EVALUATION OF THE WORLD

Another unique characteristic of the human race is its greatly expanded capacity for evaluation. Evaluation is very old, almost contemporaneous with life. Even the amoeba values "food" more highly than "not food," which it rejects. Before the human race, however, all evaluations were related simply to the immediate environment of the valuer. The capacity of the human race for images of the whole world over which humans can put evaluations represents a new evaluation process in the evolutionary pattern. The behavior of humans is not determined solely by their immediate environment, although this remains important, but relates to their image of a whole world that is seen as getting "better" or "worse." Modern environmentalism consists of the application of human evaluations to the total state of the planet, even to the solar system; (we don't go much beyond that). This is a kind of generalized accounting problem. We have an image of the state of the world as an enormous position statement. We put rough values on each item and we come out with some kind of rough "bottom line," even though this may be qualitative rather than quantitative. We are interested in human activity, over which we have some kind of control, and are concerned that such activity should make the world "better" in terms of "mature" human evaluations.

In the economy, because of the phenomenon of exchange, exchange ratios, and price structures, the evaluation process takes on a quantitative form in terms, say, of monetary units. This is what accounting is all about. This is applicable only to a limited part of the total world system, but the general principles which it involves can be applied to the whole. We are always making these qualitative, as well as quantitative, evaluations. It is, for instance, by an extraordinarily complex qualitative accounting process, involving ranking rather than numbers, but also involving the appraisal of proportions and colors, that we decide that one face is more beautiful than another.

There is a problem here of cost accounting. The production of one thing involves the consumption and destruction of others. Wheat is destroyed when it is ground into flour, flour is destroyed when it is baked into bread. All production is how we get from the genotype to the phenotype, from the egg to the chicken, from the idea in the architect's mind to the house. All production involves costs and circulation of materials and the utilization of different forms of energy. I have argued (Boulding 1981) that these non-genetic factors are *limiting* factors, the absence of any one of which prevents the realization of a genetic potential. These limiting factors may be very diverse. In the biosphere, for instance, in the tundra, the limiting factor is probably energy, in the Sahara it is water, in the tropical forest it may be just space, or in some cases, time. We have paid far too little attention to what the limiting factors are. It is the *most* limiting factor which is important. It is the first fence that we come to that stops us.

If production is to be continued, there must, of course, be reproduction, especially of the genetic factors. This takes two forms: biogenetic reproduction (biogenetics), by which DNA molecules, genes, and biological species reproduce themselves; and what might be called "noogenetic" reproduction, which is the learning process by which a structure in one mind is transmitted to another one, in language and symbols, or even just by observation. This is not unknown in prehuman species, but is of overwhelming significance in the human race, where the biogenetic structure seems to have changed very little in the past 50,000 years. Learning, indeed, is the one thing that offers hope for the human future. In prehuman species the learning process is a fairly minor element in the survival pattern of species, though it has some importance in primates. In the human species it is overwhelmingly significant, and whether we survive or not depends on whether we can learn how to do it. And there is no non-existence theorem which says that we cannot learn how to survive.

LEARNING FROM OUR MISTAKES

What the environmental problem is all about then is: Can the human species, by a learning process, change the evolutionary process over time so that it produces at any one time the best possible world that is in human power to make? The best possible world at any particular time may, of course, be worse than the last one. We may not be able to prevent deterioration, but by the learning process we can slow down deterioration and accelerate improvement.

The human learning process in regard to "facts" as well as values, purposes, and meanings, is quite closely related to the structure of human and social institutions, although the exact way in which these institutions affect the learning process is by no means easy to detect. Human learning begins almost without exception in the family, where the child grows up learning the language, the culture, and the values of his or her parents and other relations and friends. Schools, colleges, and universities are important in widening knowledge of the world outside the family—to other places, other languages, other cultures, other times. Religious and community institutions, professional societies, and so on, again, widen the sphere of knowledge, although they may sometimes put limits on it when knowledge outside the existing culture is discouraged. All the various institutions of learning help to create the images of preferences and values which are so important in decision making.

Economic and political institutions are also a very important part of the learning process and may profoundly affect the degree to which error in images is detected and corrected. The important question here is: To what extent are the results of decisions fed back to decision makers in a reasonably accurate way so that they can learn from their mistakes?

We learn more from failure of expectations than we do from success. Success tends to reinforce our existing images of the world on which we based our previous decisions. Failure, however, does not always produce the elimination of error. It may change our image of the world in directions that do not remove error but even increase it. Failure may lead to panic or irrational decisions or even to the destruction of self respect. There are problems here also of "fashion," that is, the infectiousness of certain beliefs and images of the world. The wild fluctuations in both the financial and commodity markets and the collapse of the savings and loans institutions suggest that fashion within a subculture can be very destructive where it leads to images of the world that are in error. "War moods" are another example of infectious images that can lead to disaster. There is also the problem of the self-justification of images. If everybody thinks the prices of stocks are going to rise, they will rise. Such a rise, however, cannot go on forever because the perception will gain ground that prices are "high," which will lead to the expectation of a "fall," which will be followed by a fall, and when prices are perceived as "low" they will rise, at which point the cycle starts again. Again, belief in an enemy creates one, which reinforces the belief.

Under many circumstances, however, market institutions do provide feedback to the decision maker which leads to a correction of error. When the Ford Motor Company produced the famous Edsel, it learned very soon that it had made a big mistake and it turned to making more popular models of cars. Firms that produce commodities that are perceived widely as having adverse environmental consequences may be impelled to change their products in the face of the withdrawal of demand for the old ones. The problem here is whether what might be called "social accounting" can, through the choices of demanders, make an impact on the financial accounting which is so important in business decisions. Some pressure may come from businesses themselves, where the decision makers may be influenced by their image of the impact of their own products on the environmental state of the world. Some sense of pride or shame may induce them to sacrifice profits up to a point in order to benefit the environment. The impact of the demanders, however, is likely to have more impact than that of the suppliers.

It is one of the virtues of democratic political institutions that they operate, as it were, as a market in political ideas, and those which are perceived as not working may be rejected when it comes to an election. In dictatorial political institutions, the feedback from error is much less. An error indeed can perpetuate itself and even accumulate until there is some crisis or revolution, as we have seen recently in Eastern Europe, especially in Romania. Dictatorships generally emerge out of the military institution, and military institutions, again, are very insensitive to the correction of error in their images of the world. This is simply because they specialize in threat, which tends to destroy communication between the threatener and the threatened. The experience of many African countries in this regard, especially, at the moment, Liberia, is a horrifying testimony to the painful slowness of the political learning process.

The events of the last few months of 1989 revealed that central planning can have disastrous environmental consequences. Decisions at the local level are governed almost entirely by the demand of the planner to make local enterprises conform to the plan and produce the planned quantities of particular goods. The central planners, like corporate executives, whom they may somewhat resemble, usually live far from the scenes of environmental devastation. Mistakes in central planning, therefore, are much less likely to be corrected than are similar mistakes in the market—and even the political—economy, especially where there is an aroused public opinion. Some of the present disillusionments with central planning may well be related to the belated discovery of its disastrous environmental consequences.

POWER TO BRING ABOUT CHANGE

Those who are concerned with the future of the planet in its totality have to ask themselves: What are the sources of power in the human species directed towards changing the future in a desirable direction? It is not enough to preach "we must do this, we must do that." Such preaching is perhaps a necessary prerequisite to action, but it does not necessarily produce the action required. The Brundtland Report (World Commission on Environment and Development 1987) occasionally falls into this trap. We do need to study very carefully, therefore, the structure of power in the human race over the future, which involves not only particular decisions but also the overall images of the world in the minds of the decision makers, and the learning process by which these images are created. I distinguish three major categories of power, each of which is capable of many subdivisions (Boulding 1989). The first is threat power, which begins by the threatener saying, in effect, "You do something I want or I'll do something you don't want." The results depend partly on the credibility of the threatener and the threatener's capacity for injuring the threatened party, but also on the actions of the threatened party, which can vary greatly, resulting in quite complex and unpredictable systems. The threatened party may submit, defy, issue a counterthreat which may lead to deterrence, may run away, or may even develop what might be called "disarming behavior," which absorbs the threatener into a larger system with the threatened. Each of these possibilities produces very different kinds of systems. Threat power is usually associated with political structures—governments, national states, and so on. It is significant in the form of sanctions in legal systems, although legal decisions are supposed to be made virtually in the absence of threat. A judge, for instance, is supposed to be influenced only by the evidence presented in the court, not by threats or by bribes.

Both economists and psychologists are apt to think that expected rewards are much more likely to influence human behavior than expected punishments. This is perhaps partly because people tend to blame themselves if they fail to achieve expected rewards, and so try to learn from the failure, whereas they attribute a [failed] penalty to the penalizer, and hence do not change their own behavior. In the last few decades, we have seen an increase in laws, police, crime, and prisons, almost going hand in hand, which suggests that deterrence is not very effective in many cases. National defense expenditures are usually justified in terms of deterrence these days, but the evidence from history indicates that deterrence can be stable only in the short run and that it will always eventually break down. Deterrence indeed is only stable in the short run because it has a positive probability of breaking down, otherwise it would cease to deter. Anything that has a positive probability will eventually happen if we wait long enough. National defense, therefore, justified by deterrence, is just like living in the flood plain of a 25- or 50-year flood. This is not to say that there is no role for threat in social life—it would be hard to collect taxes without it—but it is most effective when it is highly specific, with threatened penalties spelled out and associated with particular acts, like breaking the speed limit or not paying income taxes. National defense threats are not entirely credible and are vague—"You do something unspecifiedly nasty to me, and I will do something unspecifiedly nasty to you." It is not surprising that such threats are both rather ineffective and very costly to all parties and break down rather frequently.

The second form of power is economic power, that which the rich have more of than the poor, but which we all possess in some degree. Economic power, again, is a very complex phenomenon created by a combination of inheritance and various learning processes, especially the learning of improved methods of production and trade. Enrichment frequently requires capital accumulation, which means producing more than we consume. It also includes complex relations of property involving legitimation on the one hand and occasionally something like theft, ransom or tribute on the other. Purchasing power—the power to exchange—is an important aspect of decision. It does depend on having something to exchange. Free exchange is fundamentally subject to the veto of either of the parties, and demand has a great influence on what is supplied. Widespread boycotts of environmentally offensive products may be quite effective. Economists tend to go for taxes on pollution rather than prohibitions of it, as these will encourage changes in production processes that will lessen pollution more cheaply than would be the result of mere prohibition by threat.

A third form of power is what I have called "integrative power." This is the power of legitimacy, loyalty, respect, affection, love, truth, and so on. In many ways this is the most fundamental form of power, as neither threat power nor economic power is very effective in the absence of legitimacy. Unless the threat which enforces a tax system, for instance, is accepted widely as legitimate, the tax system will collapse, as George III discovered in the American colonies. One can make a good case for the proposition that if there is any single system in society on which all the other systems depend, which there may not be, the best candidate is the structure of legitimacy. In the 20th century, for instance, we have seen both the collapse of the legitimacy of financial markets and private business in the socialist countries and its partial restoration with the collapse of the legitimacy of the centrally planned economies. We have also seen the collapse of the legitimacy of empire, which was almost unquestioned until the 1930s in the imperial powers. Then it somehow became clear that empires were both economic burdens and moral burdens and their legitimacy collapsed very rapidly; and it did not take very long for the empires themselves to be disbanded. We may see something like this happening now in the Soviet Union. It will be surprising if it does not happen in China. It is a paradox that the use of threat up to a certain point may create legitimacy; beyond a certain point, threat destroys legitimacy almost completely. The rise and decline of the legitimacy of religious organizations and structures is again extraordinarily puzzling. With the nuclear weapon and the long-range missile, we have now gotten to the point where the legitimacy of national defense is seriously threatened, and it is beginning to dawn on people that national defense no longer gives national security.

CONCLUSIONS

When it comes to problems like the sustainability of the present patterns of economic growth, which were hardly questioned 20 years ago, the integrative power of truth will emerge even though it may take a long time. The population explosion, the using up of fossil fuels, the impact of the excrements of human society on global warming, the ozone layer, and so on, are "truths" which are increasingly hard to deny. And in the solution of these problems—if there is one—we will undoubtedly have to make some use of threat power—taxation, regulations, and so on—and some use of economic power. But the major element unquestionably will be integrative power, based first on the widespread knowledge that we all live on the same fragile planet, which we have now seen from outer space and to which we owe a common loyalty and affection. Unless this view is very widespread, legitimacy will not be granted to those frequently painful processes which may be necessary to prevent catastrophe.

Integrative power, again, can be created through a learning process. This may also involve a teaching process. How we learn and how we teach about things like the world environment which are not obvious in ordinary daily experience, is an important problem that environmentalists must give a great deal of attention to. These three major forms of power interact with each other in very complex ways. The more aware we are of these interactions, the more likely we are to act effectively. One consideration that leads to at least a modest optimism is the reflection that the learning capacity of the human mind is a very long way from being exhausted and that erroneous images and beliefs are a little more likely to change than are truer ones. The great task is to speed up this process of the elimination of error. The model of the scientific community here is important. It has created a subculture in which the orderly elimination of error is encouraged. How we transfer this quality into political and economic life is an important question.

In changing the world, and especially in changing human evaluations and images, the skills and quality of leadership may be a crucial factor. Bad leaders (like Hitler and Stalin) whose own images and values are in deep error, can produce disastrous consequences when they have effective skills of leadership. It is very important that "good" leaders have effective skills. This is a matter that requires much further research and thought.

REFERENCES

Boulding, K. E. 1981. *Evolutionary Economics*. Beverly Hills, Cakif.: Sage Publications.
Boulding, K. E. 1989. *Three Faces of Power*. Newbury Park, Calif.: Sage Publications.
Somit, Al and Steven A. Peterson. In press. *The Punctuated Equilibrium Debate: Scientific Issues and Implications*. Ithaca: Cornell University Press. Originally published in the *Journal of Social and Biological Structures* (Special Issue), (April/July 1989), 12 (2/3).
The World Commission on Environment and Development. 1987. *Our Common Future*. New York: United Nations.

3

ELEMENTS OF ENVIRONMENTAL MACROECONOMICS

*Herman E. Daly**
Environment Department
The World Bank
1818 H. Street, NW
Washington, DC 20433 USA

ABSTRACT

Environmental economics is traditionally treated as a subdivision of microeconomics. The focus is on getting prices right for optimal allocation. It is argued that in addition there is a neglected macroeconomic question, that of optimal scale, that is independent of optimal allocation, and must be served by an independent policy instrument. The issue of optimal scale, reasons for its neglect, definition, and policy implication are discussed.

INTRODUCTION

Environmental economics, as it is taught in universities and practiced in government agencies and development banks, is overwhelmingly microeconomics. The theoretical focus is on prices, and the big issue is how to internalize external environmental costs to arrive at prices that reflect full social marginal opportunity costs. Once prices are right, the environmental problem is "solved"—there is no macroeconomic dimension. Cost/ benefit analysis in its various permutations is the major tool for estimating full-cost prices. So in practice as well as theory we remain within the domain of microeconomics. There are, of course, good reasons for environmental economics to be closely tied to microeconomics and it is not my intention to argue against that connection. Rather, I ask if there is not a neglected connection between the environment and macroeconomics.

---

* The views here presented are those of the author and should in no way be attributed to the World Bank.

A search through the indexes of three leading textbooks[1] in macroeconomics reveals no entries under any of the following subjects: *environment, natural resources, pollution, depletion.* One of the three does have an entry under "resources," but the discussion refers only to labor and capital, which, along with efficiency, are listed as the causes of growth in GNP. Natural resources are not mentioned. Evidently GNP growth is thought to be independent of natural resources. Is it really the case, as prominent textbook writers seem to think, that macroeconomics has nothing to do with the environment? What historically has impeded the development of an environmental macroeconomics? If there is no such thing as environmental macroeconomics, should there be? Do parts of it already exist? What needs to be added? What policy implications are visible?

The reason that environmental macroeconomics is an empty box lies in what Thomas Kuhn calls a "paradigm," and what Joseph Schumpeter more descriptively called a "preanalytic vision" (Schumpeter 1954). As Schumpeter emphasized, analysis has to start somewhere—there has to be something to analyze. That something is given by a preanalytic cognitive act that Schumpeter called "vision." One might say that vision is what the "right brain" supplies to the "left brain" for analysis. Whatever is omitted from the preanalytic vision cannot be recaptured by subsequent analysis. Schumpeter is worth quoting at length on this point:

In practice we all start our own research from the work of our predecessors, that is, we hardly ever start from scratch. But suppose we did start from scratch, what are the steps we should have to take? Obviously, in order to be able to posit to ourselves any problems at all, we should first have to visualize a distinct set of coherent phenomena as a worthwhile object of our analytic effort. In other words, analytic effort is of necessity preceded by a preanalytic cognitive act that supplies the raw material for the analytic effort. In this book, this preanalytic cognitive act will be called Vision. It is interesting to note that vision of this kind not only must precede historically the emergence of analytic effort in any field, but also may reenter the history of every established science each time somebody teaches us to *see* things in a light of which the source is not to be found in the facts, methods, and results of the pre-existing state of the science.

The vision of modern economics in general, and especially of macroeconomics, is the familiar circular flow diagram. The macroeconomy is seen as an isolated system (i.e., no exchanges of matter or energy with its environment) in which exchange value circulates between firms and households in a closed loop. What is "flowing in a circle" is variously referred to as production or consumption, but these have physical dimensions. The circular flow does not refer to materials recycling, which in any case could not be a completely closed loop, and of course would require energy which cannot be recycled at all. What is truly flowing in a circle can only be abstract exchange value—exchange value abstracted from the physical dimensions of the goods and factors that are exchanged. Since an isolated system of abstract exchange value flowing in a circle has no dependence

---

[1]See the following: R. Dornbusch and S. Fischer, *Macroeconomics*, 4th ed. (New York: McGraw-Hill, 1987); R. E. Hall and J. B. Taylor, *Macroeconomics*, 2nd ed. (New York: W. W. Norton, 1988); R. J. Barro, *Macroeconomics*, 2nd ed. (New York: Wiley and Sons, 1987).

on an environment, there can be no problem of natural resource depletion, nor environmental pollution, nor any dependence of the macroeconomy on natural services, or indeed on anything at all outside itself (Daly 1985).

Since analysis cannot supply what the preanalytic vision omits, it is only to be expected that macroeconomics texts would be silent on environment, natural resources, depletion and pollution. It is as if the preanalytic vision that biologists had of animals recognized only the circulatory system, and abstracted completely from the digestive tract. A biology textbook's index would then contain no entry under "assimilation" or "liver." The dependence of the animal on its environment would not be evident. It would appear as a perpetual motion machine.

Things are no better when we turn to the advanced chapters at the end of most macroeconomics texts, where the topic is growth theory. True to the preanalytic vision the aggregate production is written as $Y = f(K,L)$, i.e., output is a function of capital and labor stocks. Resource flows (R) do not even enter! Nor is any waste output flow noted. And if occasionally R is stuck in the function along with K and L it makes little difference since the production function is almost always a multiplicative form, such as Cobb-Douglas, in which R can approach zero with Y constant if only we increase K or L in a compensatory fashion. Resources are seen as "necessary" for production, but the amount required can be as little as one likes!

What is needed is not ever more refined analysis of a faulty vision, but a new vision. This does not mean that everything built on the old vision will necessarily have to be scrapped, but fundamental changes are likely when the preanalytic vision is altered. The necessary change in vision is to picture the macroeconomy as an open subsystem of the finite natural ecosystem (environment), and not as an isolated circular flow of abstract exchange value, unconstrained by mass balance, entropy and finitude. The circular flow of exchange value is a useful abstraction for some purposes. It highlights issues of aggregate demand, unemployment and inflation that were of interest to Keynes in his analysis of the Great Depression. But it casts an impenetrable shadow on all physical relationships between the macroeconomy and the environment. For Keynes, this shadow was not very important, but for us it is. Just as, for Keynes, Say's Law and the impossibility of a general glut cast an impenetrable shadow over the problem of the Great Depression, so now the very Keynesian categories that were revolutionary in their time are obstructing the analysis of the major problem of our time. Namely, what is the proper scale of the macroeconomy relative to the ecosystem?

Once the macroeconomy is seen as an open subsystem, rather than an isolated system, the issue of its relation to its parent system (the environment) cannot be avoided. The obvious question is, how big should the subsystem be relative to the overall system?

THE ENVIRONMENTAL-MACRO ECONOMICS OF OPTIMAL SCALE

Just as the micro unit of the economy (firm or household) operates as part of a larger system (the aggregate or macroeconomy), so the aggregate economy is likewise a part of a larger system, the natural ecosystem. The macroeconomy is an <u>open subsystem of the ecosystem</u> and is totally dependent upon it, both as a <u>source</u> for inputs of low-entropy matter-energy and as a <u>sink for outputs</u> of high-entropy matter-energy. *The physical exchanges crossing the boundary between the total ecological system and the economic subsystem constitute the subject matter of environmental macroeconomics.* These flows are considered in terms of their scale or total volume relative to the ecosystem, not in terms of the price of one component of the total flow relative to another. Just as standard macroeconomics focuses on the volume of transactions rather than the relative prices of different items traded, so <u>environmental macroeconomics focuses on the volume of exchanges that cross the boundary between system and subsystem</u>, rather than the pricing and allocation of each part of the total flow within the human economy or even within the nonhuman part of the ecosystem.

The term "scale" is shorthand for "<u>the physical scale or size of the human presence in the ecosystem</u>, as measured by population times per capita resource use." Optimal *allocation* of a given scale of resource flow within the economy is one thing (a microeconomic problem). Optimal *scale* of the whole economy relative to the ecosystem is an entirely different problem (a macro-macro problem). The micro allocation problem is analogous to allocating optimally a given amount of weight in a boat. But once the best relative location of weight has been determined, there is still the question of the absolute amount of weight the boat should carry. This absolute optimal scale of load is recognized in the maritime institution of the Plimsoll line. When the watermark hits the Plimsoll line the boat is full, it has reached its safe *carrying capacity.* Of course, if the weight is badly allocated, the water line will touch the Plimsoll mark sooner. But eventually as the absolute load is increased, the watermark will reach the Plimsoll line even for a boat whose load is optimally allocated. Optimally loaded boats will still sink under too much weight—even though they may sink optimally! It should be clear that optimal allocation and optimal scale are quite distinct problems. The major task of environmental macroeconomics is to design an economic institution analogous to the Plimsoll mark—to keep the weight, the absolute scale, of the economy from <u>sinking our biospheric ark.</u>[2]

The market, of course, functions only within the economic subsystem, where it does only one thing: it solves the allocation problem by providing the necessary information and incentive. It does that one thing very well. What it does not do is solve the problem of optimal scale and of optimal distribution. The market's inability to solve the problem

what is being distributed ? ① *"inputs" (resources) are being depleted*
② *"outputs" (waste) may exceed the capacity of the natural ecosystem to absorb it.*

---

[2]Any analogy has its limits, and the Plimsoll line is used here mainly to clarify the difference between optimal allocation and optimal scale. But the analogy might be pressed just a bit more regarding the obvious difficulty of determining just where to draw the analogous line for the economy. Drawing the line on a ships bow seems comparatively easy, and indeed it is. But carping academic relativists can point out that there would be different Plimsoll lines for fresh and salt water; that the line is not just a physical measurement but involves some social judgment of acceptable risk; that the technical design of the ship will influence the position of the line, etc. Yet in spite of all these difficulties we do manage to draw a reasonable line somewhere to the immense benefit of generations of seafarers. Likewise for the economy, it is more important that a limit be placed somewhere than that the limit be accurate.

prices ⇒ Allocation of resources

Optimal scale: Definition?

Sustainable scale?

of just distribution is widely recognized, but its similar inability to solve the problem of optimal or even sustainable scale is not as widely appreciated.[3]

An example of the confusion that can result from the nonrecognition of the independence of the scale issue from the question of allocation is provided by the following dilemma.[4] Which puts more pressure on the environment, a high or a low discount rate? The usual answer is that a high discount rate is worse for the environment because it speeds the rate of depletion of nonrenewable resources and shortens the turnover and fallow periods in the exploitation of renewables. It shifts the allocation of capital and labor towards projects that exploit natural resources more intensively but it restricts the total number of projects undertaken. A low discount rate will permit more projects to be undertaken even while encouraging less intensive resource use for each project. The allocation effect of a high discount rate is to increase throughput, but the scale effect is to lower throughput. Which effect is stronger is hard to say, although one suspects that over the long run the scale effect will dominate. The resolution to the dilemma is to recognize that two independent policy goals require two independent policy instruments. We cannot serve both optimal scale and optimal allocation with the single policy instrument of the discount rate (Tinbergen 1952). The discount rate should be allowed to solve the allocation problem, within the confines of a solution to the scale problem provided by a presently nonexistent policy instrument that we may for now call an "economic Plimsoll line" that limits the scale of the throughput.

Policy Goals

Allocation

Equity

Scale

Economists have recognized the independence of the goals of efficient allocation and just distribution and are in general agreement that it is better to let prices serve efficiency, and to serve equity with income redistribution policies. Proper scale is a third independent policy goal and requires a third policy instrument. This latter point has not yet been accepted by economists, but its logic is parallel to the logic underlying the separation of allocation and distribution. In pricing factors of production and distributing profits the market does, of course, influence the distribution of income. Providing incentive requires some ability to alter the distribution of income in the interests of efficiency. The point is that the market's criterion for distributing income is to provide an incentive for efficient allocation, not to attain justice. And in any case, historical conditions of property ownership are major determinants of income distribution and have little to do with either efficiency or justice. These two values can conflict, and the market does not automatically resolve this conflict. The point to be added is that there are not just two, but three, values in conflict: allocation (efficiency), distribution (justice), and scale (sustainability).

---

[3]This can be illustrated in terms of the familiar microeconomic tool of the Edgeworth box. Moving to the contract curve is an improvement in efficiency of *allocation*. Moving along the contract curve is a change in *distribution* which may be deemed just or unjust on ethical grounds. The *scale* is represented by the dimensions of the box, which are taken as given. Consequently, the issue of optimal scale of the box itself escapes the limits of the analytical tool. A microeconomic tool cannot be expected to answer a macroeconomic question. But, so far, macroeconomics has not answered the question either—indeed has not even asked it. The tacit answer to the implicit question seems to be that a bigger Edgeworth box is always better than a smaller one!

[4]See, for example, David Pearce et al. 1989, p. 135.

Microeconomics has not discovered in the price system any built-in tendency to grow only up to the scale of aggregate resource use that is optimal (or even merely sustainable) in its demands on the biosphere. *Optimal scale, like distributive justice, full employment or price level stability, is a macroeconomic goal.* And it is a goal that is likely to conflict with the other macroeconomic goals. The traditional solution to unemployment is growth in production, which means a larger scale. Frequently the solution to inflation is also thought to be growth in real output, and a larger scale. And most of all, the issue of distributive justice is "finessed" by the claim that aggregate growth will do more for the poor than redistributive measures. Conventional macroeconomic goals tend to conflict, and certainly optimal scale will conflict with any goal that requires further growth once the optimum has been reached.

HOW BIG IS THE ECONOMY?

In the past, it has not been customary to consider the macroeconomy as a subsystem of a larger ecosystem. As long as the human economy was infinitesimal relative to the natural world, then sources and sinks could be considered infinite, and therefore not scarce. And if they are not scarce then they are safely abstracted from economics. There was no need to consider the larger system since it imposed no scarcities. This was a reasonable view at one time, but no longer. As Kenneth Boulding says, when something grows it gets bigger! The economy has gotten bigger, the ecosystem has not. How big has the economy become relative to the ecosystem?

Probably the best index of the scale of the human economy as a part of the biosphere is the percentage of human appropriation of the the total world product of photosynthesis. Net primary production (NPP) is the amount of solar energy captured in photosynthesis by primary producers, less the energy used in their own growth and reproduction. NPP is thus the basic food resource for everything on earth not capable of photosynthesis. Vitousek, et al. (1986)[5] calculate that 25% of potential global (terrestrial and aquatic) NPP is now appropriated by human beings. If only terrestrial NPP is considered, the fraction rises to 40%. Taking the 25% figure for the entire world, it is apparent that two more doublings of the human scale will give 100%. Since this would mean zero energy left for all nonhuman and nondomesticated species, and since humans cannot survive without the services of ecosystems (which are made up of other species), it is clear that two more doublings of the human scale is an ecological impossibility, although arithmetically possible. Furthermore, the terrestrial figure of 40% is probably more relevant since we are unlikely to increase our take from the oceans very much. Total appropriation of the terrestrial NPP can occur in only a bit over one doubling time. Perhaps it is theoretically possible to increase the earth's total photosynthetic capacity

---

[5]The definition of human appropriation underlying the figures quoted includes direct use by human beings (food, fuel, fiber, timber), plus the reduction from the potential due to ecosystem degradation caused by humans. The latter reflects deforestation, desertification, paving over, and human conversion to less productive systems (such as agriculture).

somewhat but the actual trend of past economic growth is decidedly in the opposite direction.

Assuming a constant level of per capita resource consumption, the doubling time of the human scale would be equal to the doubling time of population, which is on the order of 40 years. Of course economic growth currently aims to increase the average per capita resource consumption and consequently to reduce the doubling time of the scale of the human presence below that implicit in the demographic rate of growth. The greenhouse effect, ozone layer depletion, and acid rain all constitute evidence that we have already gone beyond a prudent Plimsoll line for the scale of the macroeconomy.

[handwritten left margin: Policy alternatives / 1. Manage Allocation balance / 2. Manage scale / 3. Manage ecosystem to ↑ sustainable scale]

[handwritten: But balance - allocation - affects sustainable scale. So we respond by redistributing or by adjusting scale. Or by environmental mgt that affects sustainable scale.]

COWBOY, SPACEMAN, OR BULL IN THE CHINA SHOP?

If one starts from the vision of the economic process as an open subsystem of a closed finite total system, then the question of how big the subsystem should be relative to the total system is hard to avoid. How then have we managed to avoid it? In two ways: first, by viewing the economic subsystem as infinitesimally small relative to the total system, so that scale becomes irrelevant because it is negligible; second, by viewing the economy as coextensive with the total system. If the economy includes everything, then the issue of scale relative to a total system simply does not arise. These polar extremes correspond to Boulding's colorful distinction between the "cowboy economy" and the "spaceman economy." The cowboy of the infinite plains lives off of a linear throughput from source to sink, with no need to recycle anything. The spaceman in a small capsule lives off of tight material cycles and immediate feedbacks, all under total control subservient to his needs. For the cowboy, scale is negligible; for the spaceman, scale is total. There is no material environment relative to which scale must be determined; there is no ecosystem, only economy. In each of these polar cases, the only problem is allocation. Scale is irrelevant.

It is only in the middle ground between the cowboy and the spaceman that the issue of scale does not get conflated with allocation. But, as Boulding realized, the middle ground happens to be where we are. Between the cowboy and the spaceman economies is a whole range of larger and smaller "bull-in-the-china-shop economies" where scale is a major concern. We are not cowboys because the existing scale of the economy is far from negligible compared to the environment. But neither are we spacemen, because most of the matter-energy transformations of the ecosystem are not subject to human control either by prices or by central planning. In a finite system subject to the conservation of mass, the more that is brought under our economic control, the less remains under the spontaneous control of nature. As our exactions from and insertions back into the ecosystem increase in scale, the qualitative change induced in the ecosystem must also increase, for two reasons. The first is the first law of thermodynamics (conservation of matter-energy). The taking of matter and energy out of the ecosystem must disrupt the functioning of that system even if nothing is done to the matter and energy so removed. Its mere absence must have an effect. Likewise, the mere insertion of matter and energy into an ecosystem must disrupt the system into which it is newly added. This must be the case

[handwritten bottom: Paper idea: Agro-ecosystem and sustainability - policy alternatives for managing sustainability in the Salinas Valley. Policy alternatives, cost of policy, setting and measuring "sustainable scale."]

even without appealing to any qualitative degradation of the matter and energy thus relocated. The second reason is the second law of thermodynamics which guarantees that the matter-energy exacted is qualitatively different from the matter-energy inserted. Low-entropy raw materials are taken out, high-entropy wastes are returned. This qualitative degradation of the matter-energy throughput, along with the purely quantitative dislocation of the same, induces changes in the ecosystem which to us are surprising and novel because our information and control system (prices) assumes nonscarcity (nondisruptablity) of environmental source and sink functions. Economic calculation is about to be overwhelmed by novel, uncertain and surprising feedbacks from an ecosystem that is excessively stressed by having to support too large an economic subsystem (Perrings 1987).

How big should the subsystem be relative to the total ecosystem? Certainly this, the question of optimal scale, is the big question for environmental macroeconomics. But since it is such a difficult question, and since we cannot go back to the cowboy economy, we have acquired a tendency to want to jump all the way to the spaceman economy and take total control of the spaceship earth. The September 1989 special issue of *Scientific American* entitled "Managing Planet Earth" is representative of this thrust. But, as environmentalist David Orr points out, God, Gaia or Evolution was doing a nice job of managing the earth until the scale of the human population, economy and technology got out of control. Planetary management implies that it is the planet that is at fault, not human numbers, greed, arrogance, ignorance, stupidity, and evil. We need to manage ourselves more than the planet and our self-management should be, in Orr's words, "more akin to child-proofing a day-care center than to piloting spaceship earth." The way to child-proof a room is to build the optimal scale playpen within which the child is both free and protected from the excesses of its own freedom. It can enjoy the light and warmth provided by electrical circuits beyond its ken, without running the risk of shorting out those circuits, or itself, by experimenting with the "planetary management technique" of teething on a lamp cord.

Our manifest inability to centrally plan economies should inspire more humility among the planetary managers who would centrally plan the ecosystem. Humility should argue for the strategy of minimizing the need for planetary management by keeping the human scale sufficiently low so as not to disrupt the automatic functioning of our life support systems, thereby forcing them into the domain of human management. Those who want to take advantage of the "invisible hand" of self-managing ecosystems have to recognize that the invisible hand of the market, while wonderful for allocation, is unable to set limits to the scale of the macroeconomy. Our limited managerial capacities should be devoted to institutionalizing an economic Plimsoll line that limits the macroeconomy to a scale such that the invisible hand can function in both domains to the maximum extent. It is ironic that many free marketeers, by opposing any limit to the scale of the market economy (and therefore to the increase in externalities), are making more and more inevitable the very central planning that they oppose. Even worse is their celebration of the increase in GNP that results as formerly free goods become scarce and receive a price. For allocation it is necessary that newly scarce goods not continue to have a zero price—no one disputes that. The issue is that, for all we know, we might have been better off to remain at the smaller scale at which the newly scarce goods were free and their proper allocative price was still zero. The increase in measured national income and wealth resulting as formerly free goods are turned into scarce goods is more an index of cost than of benefit, as was recognized by the classical economist Lauderdale back in 1819 (Lauderdale 1819; Foy 1989).

A GLITTERING ANOMALY

Optimal scale of a single activity is not a strange concept to economists. Indeed, microeconomics is about little else. An activity is identified, be it producing shoes or consuming ice cream, and a cost function and a benefit function for the activity in question are defined. Good reasons are given for believing that marginal costs increase and marginal benefits decline as the scale of the activity grows. The message of microeconomics is to expand the scale of the activity in question up to the point where marginal costs equal marginal benefits, a condition which defines the optimal scale. All of microeconomics is an extended variation on this theme.

When we move to macroeconomics, however, we never again hear about optimal scale. There is apparently no optimal scale for the macro economy. There are no cost and benefit functions defined for growth in scale of the economy as a whole. It just doesn't matter how many people there are, or how much they each consume, as long as the proportions and relative prices are right. But if every micro activity has an optimal scale, then why does not the aggregate of all micro activities have an optimal scale? If I am told in reply that the reason is that the constraint on any one activity is the fixity of all the others and that when all economic activities increase proportionally the restraints cancel out, then I will invite the economist to increase the scale of the carbon cycle and the hydrologic cycle in proportion to the growth of industry and agriculture. I will admit that if the ecosystem can grow indefinitely then so can the aggregate economy. But, until the the surface of the earth begins to grow at a rate equal to the rate of interest, one should not take this answer too seriously.

The total absence in macroeconomics of the most basic concept of microeconomics is a glittering anomaly, and it is not resolved by appeals to the fallacy of composition. What is true of a part is not necessarily true for the whole, but it can be and usually is unless there is some aggregate identity or self-cancelling feedback at work. (As in the classic examples of all spectators standing on tiptoe to get a better view and each canceling out the better view of the other; or, in the observation that while any single country's exports can be greater than its imports, nevertheless the aggregate of all exports cannot be different than the aggregate of all imports). But what analogous feedback or identity is there that allows every economic activity to have an optimal scale while the aggregate economy remains indifferent to scale? The indifference to scale of the macroeconomy is due to the preanalytic vision of the economy as an isolated system—the inappropriateness of which has already been discussed.

As an economy grows, it increases in scale. Scale has a maximum limit defined either by the regenerative or absorptive capacity of the ecosystem, whichever is less. However, the maximum scale is not likely to be the optimal scale. Two concepts of optimal scale can be distinguished, both formalisms at this stage, but important for clarity.

1. **The anthropocentric optimum.** The rule is to expand scale, i.e., grow, to the point at which the marginal benefit to human beings of additional manmade physical capital is just equal to the marginal cost to human beings of sacrificed natural capital. All nonhuman species and their habitats are valued only instrumentally according to their capacity to satisfy human wants. Their intrinsic value (capacity to enjoy their own lives) is assumed to be zero.

2. **The biocentric optimum.** Other species and their habitats are preserved beyond the point necessary to avoid ecological collapse or cumulative decline, and beyond the point of maximum instrumental convenience, out of a recognition that other species have intrinsic value independent of their instrumental value to human beings. The biocentric optimal scale of the human niche would therefore be smaller than the anthropocentric optimum.

The definition of sustainable development does not specify which concept of optimum scale to use. It is consistent with any scale that is not above the maximum. Sustainability is probably the characteristic of optimal scale on which there is most consensus. It is a necessary, but not sufficient, condition for optimal scale.

STEPS ALREADY TAKEN IN ENVIRONMENTAL MACROECONOMICS

What has macroeconomics contributed to environmental economics so far? As we have seen, the textbooks make no claim to any contribution whatsoever, but that is too modest. National Income Accounting is a part of macroeconomics, and there has been an effort to correct our income accounts for consumption of natural capital. Current national accounting conventions also treat environmental clean up costs as final consumption rather than intermediate costs of production of the commodity whose production gave rise to those costs (Hueting 1980; Leipert 1986; Repetto 1987; Ahmad et al. 1989). Traditional national income accountants have not exactly been in the forefront of the effort to correct these two errors, and may even be said to be dragging their feet. However, the conservatively motivated and impeccably orthodox attempt to gain a closer approximation of true Hicksian income (maximum available for consumption without consuming capital stock) will surely make this effort an important foundation of environmental macroeconomics.

Inter-industry or input-output analysis is also a useful tool of environmental analysis, although it is hard to classify it as either micro or macro. But because of its close relation to national accounts, let us call it macro and credit it as an existing part of environmental macroeconomics. Certainly, it has been important in elucidating total (direct and indirect) requirements of materials and especially energy that must be extracted from the environment in order to increase any component of the economy's final bill of goods by some given amount.

CARRYING CAPACITY AS A TOOL OF ENVIRONMENTAL MACROECONOMICS

Many resist the application of the concept of carrying capacity to human beings. Certainly the concept is easier to apply to animals than to humans. For animals, carrying capacity can be considered almost entirely in terms of population. This is because per capita resource consumption for animals is both constant over time (animals do not experience economic development), and constant across individual members of the species (animals do not have rich and poor social classes). The latter is not to say that animals are egalitarian. Clearly there exist dominance hierarchies and territoriality. But these inequalities are mainly related to reproduction, not to large differences in per capita consumption. Also for animals, technology is for all practical purposes a genetic constant, while for humans it is a cultural variable. For human beings, we cannot speak of carrying capacity in terms of population alone, but must specify some average level of per capita consumption ("standard of living"), some degree of inequality in the distribution of individual consumption levels around that average, and some given level or range of technology. A great deal of human and nonhuman suffering could be avoided by employing the carrying capacity concept in environmental macroeconomics. The case of Paraguay provides an example.

Paraguay's greatest environmental advantage has been its small population (some 3 million in 1982, and close to 4 million today). At the current 2.5% annual rate of population growth (doubling time of 28 years or roughly one generation), this advantage is rapidly disappearing. Furthermore, this environmental advantage has historically been considered an economic disadvantage. Demographic pressures are exacerbated by the fact that all public lands available for colonization have been distributed. In the future, land cannot be made available to some citizens without taking it away from others. Also fractioning of landholdings into uneconomic small land holdings is driven by population growth and the practice of equal inheritance.

There is very little concern about population growth. Traditionally, the goal has been to increase the population by bringing in colonists to settle the land. After the disastrous war of the Triple Alliance, Paraguay was left in 1875 with only something like 220,000 people. It is quite understandable that pronatalist views should be overwhelmingly dominant. Some prominent leaders have conjectured that Paraguay could support 20 million people with no difficulty. Yet a 1979 FAO study (PNUD/FAO/SFN 1979) concluded that "the agricultural frontier has already exceeded the limits of desirable development in most of the Eastern Region," and that continued expansion would be profoundly destructive of the ecosystem.

At the time (1979) that the FAO study said limits had been reached, Paraguay's population was 3 million. Thus, their implicit estimate of carrying capacity was around 3 million, neglecting the Chaco. The prominent leaders' estimate of 20 million is larger by almost a factor of seven. It is quite important to narrow this range of difference as a precondition for any sensible economic planning and policy. A few back-of-the-envelope calculations can be very useful.

How many people could be supported by the ecosystem of the Chaco, if it were as densely populated as the Oriente? Multiplying the population density of the Oriente times the area of the Chaco (18.6 persons/Km^2 times 247,000 Km) gives 4,594,200 people. Viewed in this way, 5 million is an absurdly high overestimate, because it assumes that the carrying capacity of the Chaco equals that of the Oriente. There are good reasons why 98% of all Paraguayans live in the Oriente and only 2% in the Chaco.

A better estimate of carrying capacity for the Chaco can be derived by taking the most successful colony in the region, the Mennonites, calculating their population density, and then generalizing that to the whole region. There are (in 1987) 6,650 Mennonites living on 420,000 hectares, giving a density of 6650/420,000 = 0.0158 persons/ha. To get persons per km^2 we multiply by 100, the number of hectares in one km^2, giving 1.58 persons/km^2. That density times the total area of 247,000 km^2 gives 390,260 or roughly 400,000 persons, not even half a million. The two estimates differ by an order of magnitude and it is extremely important to plan on the basis of the more realistic number.

It is obvious that the second is more realistic. But it is very crude, and more information is needed. The Mennonites themselves have unused land and think that they could double their number on their existing land (which at their 2% population growth rate they will do in 35 years). Perhaps our estimate should be 800,000. On the other hand, our calculation assumes that the Mennonites have average Chaco land when in fact it is better than average by virtue of the fact that they got there first. The calculation also assumes that other settlers could do as well as the Mennonites which is doubtful for several reasons. First, the Mennonites brought with them the peasant traditions of Europe, which are absent among Paraguayan colonos (colonizers). They also had a strong community of mutual aid and support, as well as help from European and American Mennonites. That community cohesion cannot be assumed for new colonists. And we must remember that it took the Mennonites, 60 years of hard work and sacrifice to reach their present level. All things considered, it would be difficult to match their productivity, and consequently even 400,000 may well be an overestimate for Chaco carrying capacity, especially if ranching rather than agriculture is the best use of much of the non-Mennonite area, as seems likely. Water, rather than soil quality, is the limiting factor, so naturally one thinks of large irrigation projects. The Mennonites are extremely skeptical of irrigation in the Chaco and are convinced that it would ruin the soil by salinization (raising the level of existing salt closer to the surface). Drip irrigation and minimum tillage methods seem promising to them.

Even these very crude calculations are enough to allow us to dismiss the 20 million estimate as whimsy and the 5 million estimate as highly unrealistic within the time frame of one generation. On the basis of technologies and investment capacities likely to be available to Paraguay over the next 28 years, any estimate of Chaco carrying capacity over half a million faces a heavy burden of proof. Since population is projected to increase by about 4 million over this period, it is clear that there is a strong likelihood of overshooting the carrying capacity of the Chaco. Optimistic speculations about undreamed of technologies a century from now may prove true, but would not change the impasse posed by the next doubling in the next 28 years.

A country such as Paraguay that is unwilling to countenance population policy must plan for roughly one more doubling of the population (to 8 million) over the next generation. The very low population density of the Chaco makes it the "obvious" place to put the 4 million new people. Land conflicts in the Oriente are already becoming violent. The stage is set for a large, expensive settlement program of the type witnessed in the Brazilian Amazon. The likelihood of failure due to ecological reasons is very high. The ecosystems of the Chaco and the Amazon are very different, but the common feature is the political unwillingness to respect the ecological reasons for the historically low population density. Politically, the colonization of the Chaco will be seen as the way to: minimize already serious land conflict in the Oriente, postpone dealing with population control, maintain temporarily the mirage of progress and optimism, and offer a great national project to galvanize public support. Against such political advantages, realistic estimates of carrying capacity over the next generation may not be very persuasive. But such a study is a precondition for any realistic plan. It is an elementary but very important contribution of environmental macroeconomics. Nothing is more uneconomic than to waste resources in the pursuit of an impossible goal.

POLICY IMPLICATIONS OF ENVIRONMENTAL MACROECONOMICS

Optimal scale is not well defined at present, but one characteristic at least is known—the optimal scale must be sustainable. Our attention then naturally becomes focused on how to limit scale to a sustainable level, thereby giving the sustainable development discussion a bit more of a theoretical foundation than it has had to date. From there, we can begin to investigate operational principles of sustainability and of environmental macroeconomics, such as those summarized below.

1. The main principle is to limit the human scale (throughput) to a level which, if not optimal, is at least within carrying capacity and therefore sustainable. Once carrying capacity has been reached, the simultaneous choice of a population level and an average "standard of living" (level of per capita resource consumption) becomes necessary. Sustainable development must deal with sufficiency as well as efficiency, and cannot avoid limiting scale. An optimal scale (in the anthropocentric sense) would be one at which the long-run marginal costs of expansion are equal to the long-term marginal benefits of expansion. Until we develop operational measures of cost and benefit of scale expansion, the idea of an optimum scale remains a theoretical formalism, but a very important one. The following principles aim at translating this general macro level constraint to micro level rules.

2. Technological progress for sustainable development should be efficiency-increasing rather than throughput-increasing. Limiting the scale of resource throughput (raising resource prices) would induce this technological shift. A high tax on energy would go a

long way in this direction. Both technological optimists and pessimists should agree on a policy of high resource prices: the pessimists in order to limit the growth of throughput and the related environmental stress; the optimists in order to provide incentives for the very resource-efficient technologies in which they have so much faith.

3. Renewable resources, in both their source and sink functions, should be exploited on a profit-maximizing sustained yield basis and in general not driven to extinction (regardless of the dictates of present value maximization), since they will become ever more important as nonrenewables run out. Specifically this means that: a) harvesting rates should not exceed regeneration rates; and b) waste emissions should not exceed the renewable assimilative capacity of the environment.

4. Nonrenewable resources should be exploited, but at a rate equal to the creation of renewable substitutes. Receipts from the exploitation of a nonrenewable resource should be divided into an income component and a capital component. The division is made such that by the end of the life expectancy of the nonrenewable a new renewable asset will have been built up by the annual investment of the capital component. The annual sustainable yield from that renewable asset must be equal to the income component of the nonrenewable that was being consumed annually from the beginning. El Serafy (1989) has shown how this separation of capital from income can be calculated in the context of national income accounting. But the principles are quite general and are applicable to the project level as well. The capital component will be larger, the shorter the life expectancy of the nonrenewable and the lower the rate of growth of the renewable asset. Nonrenewable investments should be paired with renewable investments and their sustainable joint rate of return should be calculated on the basis of their income component only, since that is what is perpetually available for consumption in each future year. If a renewable resource is to be partially divested, then the same pairing rule should apply to it as to a nonrenewable resource. Thus, the mix of renewable resources would not be static, but there would be a compensating renewable investment for every divestment.

Perhaps there are other principles of sustainable development as well, and certainly those listed above need to be refined, clarified, and made more consistent between the micro and macro levels. But these four are both an operational starting point and a sufficient political challenge to the present order. Will the nations seeking sustainable development be able to operationalize a concept from which such "radical" principles follow so logically? Or will they, rather than face up to the scale limits (population control and/or per capita consumption limits) required in order to live on income, revert to the cornucopian myth of unlimited growth, rechristened as "sustainable growth"? It is easier to invent bad oxymorons than to develop the environmental macroeconomics of sustainability.

Will environmental macroeconomics be able to shift the primary attention of standard macroeconomics away from "full-employment without inflation via an ever-growing GNP and ever-tighter planetary management," towards defining the optimal scale of the macroeconomy; from the spaceman economy towards the playpen economy? Can we draw a "Plimsoll line" at which quantitative growth must cease and give way to qualitative development as the dynamic path of human betterment? Can macroeconomics serve sustainable development rather than unsustainable growth?

Acknowledgments

Helpful comments on this manuscript were received from M. Clark and R. Hueting.

REFERENCES

Ahmad, Y. J., S. El Serafy, and E. Lutz, eds. 1989. *Environmental Accounting for Sustainable Development*. Washington, D.C.: The World Bank and United Nations Environment Programme.

Daly, H. E. 1985. The Circular Flow of Exchange Value and the Linear Throughput of Matter-Energy: A Case of Misplaced Concreteness. *Review of Social Economy* 43(3): 279-297.

El Serafy, Salah. 1989. The Proper Calculation of Income from Depletable Natural Resources. In Yusuf J. Ahmad, Salah El Serafy, and Ernst Lutz, eds. *Environmental Accounting for Sustainable Development*. Washington, D.C.: World Bank, pp. 10-18.

Foy, George. 1989. Public Wealth and Private Riches: Past and Present. *The Journal of Interdisciplinary Economics* 3:3-10.

Hueting, Roefie. 1980. *New Scarcity and Economic Growth*. Amsterdam: North Holland.

Lauderdale, J. M. 1819. *An Inquiry Into the Nature and Origin of Public Wealth and Into the Means and Causes of its Increase*. 2d. Ed. Edinburgh: Archibald Constable.

Leipert, Christian 1986. Social Costs of Economic Growth. *Journal of Economic Issues* 20(1): 109-31.

Pearce, D. W., A. Markandya and E. B. Barbier. 1989. *Blueprint for a Green Economy*. London: Earthscan.

Perrings, C. A. 1987. *Economy and Environment*. New York: Cambridge University Press.

PNUD/FAO/SFN. 1979. Capacidad de Uso de los Suelos, Uso Actual y Tendencias, y el Desarrollo del Sector Forestal. Documento de Trabajo 30:13.

Repetto, Robert. 1987. *Natural Resource Accounting for Indonesia*. World Resources Institute: Washington D.C.

Schumpeter, Joseph 1954. *History of Economic Analysis*. New York: Oxford University Press, p. 41.

Tinbergen, Jan. 1952. *On the Theory of Economic Policy*. Amsterdam: North Holland.

Vitousek, Peter M., Paul R. Ehrlich, Anne H. Ehrlich, and Pamela A. Matson. 1986. Human Appropriation of the Products of Photosynthesis. *BioScience* 34 (6):368-373.

4

PARAMOUNT POSITIONS IN ECOLOGICAL ECONOMICS

Garrett Hardin
Department of Biological Sciences
University of California
Santa Barbara, CA 93106 USA

ABSTRACT

In data-rich ecology and economics, purely empirical research is inefficient, and even misleading. E. T. Whittaker has shown the necessity of such "impotence principles" as the Second Law of Thermodynamics These cannot be proved true; they place the burden of proof on those who deny them. They range from the Second Law of Thermodynamics to Parkinson's Law. Acceptance of them is the "default position" of a progressive science.

"No free lunch" belongs to a family of conservation laws that traces back to Epicurus, 3rd century B.C. The theoretical development that came to physics in the 19th century had to wait until the 20th in economics. The delay was caused by two factors: the apparent magic of technology, and the illusion that usury creates wealth. It remained for Soddy (who was championed by Daly) to show that only debt can increase exponentially.

Conventional measures of income and wealth are built on the myths of GNP and GDP. Many people are now working to replace these myths with material truths. Until this replacement is made, all economic policy is built on sand—hence the conflict between environmentalism and economics. This paper offers a concise roster of the default positions that should guide a truly ecological economics.

INTRODUCTION

We are told that the disciplines of ecology and economics are about to be married. We are not surprised. The two should be compatible: it has often been noted that they share a common etymology, a derivation from the Greek word for "house" or "household." Moreover, the disciplines have for some time been in a relationship of the sort defined by the U.S. Bureau of the Census as POSSLQs—"Persons of the Opposite Sex Sharing Living Quarters." It's about time that the union be regularized. As we celebrate this new relationship it would be well to take stock of the dangers that lie in wait for the new couple. Chief among these dangers is language, which can divide as easily as it can unite.

"The official function of language is to facilitate thought and communication. One of its unofficial functions, just as real, is to *prevent* thought and communication" (Hardin 1972, p. 66). When language is used to discourage thinking, this fact is never, of course, announced by the speaker or writer (who may be honestly unaware of what he is doing). Repression rules. One sign of repression, long recognized as such by psychiatrists, is *logorrhea*—verbal diarrhea, the pathological multiplication of words. Since science is both literate and numerate, psychological defenses in the exposition of science can also take a form that we may call *arithmorrhea*—the multiplication of numbers and statistics beyond useful bounds. Both pathologies result in an information overload, where the word "information" is understood in the limited sense in which it is used in "information theory." Proliferation of narrowly defined information can easily interfere with the acquisition of real information. Simply put, information can be the enemy of wisdom.

An example should make the point clear. Less than a decade ago, the World Fertility Survey was completed. In scores of nations, women were polled to learn how many children they wanted, how many they expected to have, etc. The project cost more than 50 million dollars. One of the few people to publicly question whether the money was well spent was the editor of *Population and Development Review*, Paul Demeny, who (in the following slightly abbreviated extract) asked: "What differential policy implications follow from the finding that in Nigeria among women with 6 living children 8.2% did not want additional children, whereas in Kenya the corresponding percentage was 25.5 and in Morocco 68.8? Arguably none" (Demeny 1988, p. 470). In the "hard sciences" of physics and chemistry, even a slight change in a constant may produce great changes in practical conclusions, because physical constants are part of a tightly-knit theoretical fabric. Demeny's statement makes it clear that this is not the case in demography. In passing, we should note that the common habit of expressing demographic statistics to the first decimal (e.g., 8.2% instead of plain 8%) falsely implies a significance such statistics seldom have.

The uncertainty afflicting attitude statistics extends far beyond the well-understood uncertainties caused by sampling error. Demeny pointed out that an appropriate change in the way the polling question was asked, or a different set of instructions to the survey takers, could have produced a figure that was drastically different from the recorded figure of 8.2%—say, 30% or 55%. This devastating criticism applies to the entire World Fertility Survey. The general public is justified in concluding that the 50 million dollars was wasted.

Why, then, was the survey carried out? Perhaps because demographers were envious of "Big Science," in which physicists secure grants of hundreds of millions of dollars (Weinberg 1967). Perhaps they saw the WFS as a way to train graduate students in research (even research into the useless?). Perhaps they sought to diminish unemployment in the profession. Or perhaps they were afraid to ask, "What *is* the population problem?"

They may be running for cover because of the publication of a book with the title *The War Against Population* (Kasun 1988).

Most social scientists stand in awe of the physical sciences, in which (it is assumed) no statement is accepted without the most rigorous proof. Weighing in against this assumption are the statements of several of the most successful practitioners of physics and chemistry, which clearly show that some universally held scientific commitments would not stand up to the battering of a competent adversary in a court of law. Not only that, but the attitudes of respectable scientists include a generous measure of intolerance toward those who question the foundations of science. Since the physical sciences have prospered from this attitude, we need to know its justification. It is just possible that ecological economics can also benefit from adopting the same attitude.

THE MANAGEMENT OF PRODUCTIVE INTOLERANCE

By the middle of the 19th century, physicists were sure that perpetual motion machines were impossible. The argument was enshrined in the second law of thermodynamics, which said that in any closed system, entropy (disorder) always tends to increase. Professional confidence in this generalization did not discourage amateurs from proposing an endless stream of purported perpetual motion machines. Responding to such claims physicists had to choose between wasting their time and being rude. After a few experiences, most scientists chose to be rude. Legislators, less confident of the power of reason and experiment, sometimes wasted their time listening to enthusiasts.

In 1917, an Armenian named Garabed Giragossian, at the time of America's entry into the First World War, petitioned Congress to examine his perpetual motion machine. The House voted 234 to 14 to appoint a committee of experts in physics to evaluate the claims. They soon uncovered a childish error in the so-called Garabed machine, and the matter was dropped. The following year the U.S. Patent Office announced that henceforth it would refuse to consider proposals for perpetual motion machines (Alder 1986).

No doubt many laymen thought the bureau's announcement bespoke an unscientific orientation. Surely scientists are not prejudiced? Should not the minds of real scientists be open to all new suggestions? The distinguished physicist Arthur Eddington presented the case for limiting tolerance:

> The law that entropy always increases—the second law of thermodynamics—holds, I think, the supreme position among the laws of Nature. If someone points out to you that your pet theory of the universe is in disagreement with Maxwell's equations—then so much the worse for Maxwell's equations. If it is found to be contradicted by observation—well, these experimentalists do bungle things sometimes. But if your theory is found to be against the second law of thermodynamics I can give you no hope; there is nothing for it but to collapse in deepest humiliation. (Eddington 1928, pp. 74-75)

The physicist went on to say that this exaltation of the second law is not unreasonable. In 1877, Ludwig Boltzmann had shown how the second law could be derived from the laws of probability; thus was its provenance moved from the world of empiricism (where doubt is always reasonable) to the mathematical world, where doubt is unreasonable. Said Eddington: "The chance against a breach of the second law . . . can be stated in figures which are overwhelming."

In 1942, the mathematical physicist E. T. Whittaker returned to this problem, making the entropy law a member of a special class of statements. He identified the following as supremely true statements:

"It is impossible to derive mechanical effect from any portion of matter by cooling it below the temperature of the coldest of the surrounding objects"; or the postulate of Relativity, "It is impossible to detect a uniform translatory motion, which is possessed by a system as a whole, by observations of phenomena taking place wholly within the system"; . . . or the postulate of Imperfect Definition in quantum mechanics, "It is impossible to measure precisely the momentum of a particle at the same time as a precise measurement of its position is made." Each of these statements, which I propose to call *Postulates of Impotence*, asserts the impossibility of achieving something, even though there may be an infinite number of ways of trying to achieve it. A postulate of impotence is not the direct result of an experiment, or of any finite number of experiments; it does not mention any measurement, or any numerical relation or analytical equation; it is the assertion of a conviction of the mind, that all attempts to do a certain thing, however made, are bound to fail. (Whittaker 1942, p. 168)

A popular view of science surely holds that the attitude embodied in the views of Eddington and Whittaker is what one would expect of metaphysicians rather than scientists; but careful observation shows that the behavior of the most productive scientists is consistent with the Eddington-Whittaker model (Holton 1986, p. 8). Some scientific principles are so basic that scientists refuse to waste time giving a "fair" hearing to contradictory proposals. The most basic principles are held, as Whittaker said, as *"a conviction of the mind"* rather than a mere distillate of a wealth of experiments. In the statements of Eddington and Whittaker, we hear an echo of what Martin Luther said on April 18, 1521, on the eve of the Diet of Worms: "Hier stehe ich! Ich kann nicht anders. Gott helfe mir." *Here I stand! I cannot do otherwise. God help me.*

The connotations of the language used (or not used) need to be noted. There is something arrogant, and perhaps egotistical, about claiming to utter "self-evident truths" or "fundamental principles." A "postulate of impotence" is much less arrogant, less egotistical. Luther's claim to do no more than *stand* at a certain position is even more humble. Lutheran humility is detectable in the statements of Eddington and Whittaker, who admit that their hard-won statements are no more than convictions of the (human) mind. Some will say that scientists should not behave in this way, but creative scientists do. Their humility (from one point of view) and their inflexibility (from another) pay off.

This "conviction of the mind" can be traced all the way back to Epicurus in the 3rd century B.C.: "Nothing is created out of that which does not exist: for if it were, everything would be created out of everything with no need of seeds. And again, if that which disappears were destroyed into that which did not exist, all things would have perished, since that into which they were dissolved would not exist" (Bailey 1926, p.21). From the

Epicurean conviction, came the laws of the conservation of matter and energy, without which there would be no true physical sciences. Economics also draws on this conviction, which is embedded in the popular saying that "There's no such thing as a free lunch." From modesty (or diffidence?) economists have not yet, I believe, stated this law in formal academic terms, but it nonetheless guides—and should guide—economic investigations.

Can it be *proved* that there are no free lunches? That matter and energy are conserved? That a perpetual motion machine is impossible? Probably not, if one uses the word "prove" as it might be used in a court of law. But only fools and charlatans waste their time (and ours) claiming to have invented free lunches or perpetual motion machines.

PARAMOUNT POSITIONS AND THE BURDEN OF PROOF

We need a name for propositions that furnish the necessary foundation stones of a rational science. Like it or not, the connotations of a term affect the ease with which people will accept it. Whittaker's "postulates of impotence" has not been widely adopted, perhaps because both nouns are burdened with distracting connotations. It has been noted (by a man, if a sexist side-remark is permitted) that the word "impotence" is likely to arouse anxiety in 50% of the population.

The word "postulate" can also be objected to. As used in mathematics (Whittaker's first field of training) the word "postulate" is a weak word in the sense that it implies that we are free to accept it or reject it. Much the same may be said of the possible substitute, "axiom." But, by Whittaker's own description, postulates of impotence are rooted in a strong "conviction of the mind." A word stronger than "postulate" is needed. We note that the author himself later introduced a variation when he spoke of *principles* of impotence (Whittaker 1958, p. 59). (Was it just absent-mindedness that caused the substitution? Or a linguistic change of heart?)

"Law" is another old word that might do. Cicero spoke of *lex naturae*, the law of nature. The generalized term was given more specific meaning in the 17th century, beginning with Newton's "Laws of Motion." But the legalistic odor of "law" that met with favor in the 19th century came to be disparaged in the twentieth. The word seems designed to put a permanent end to doubt: this is more than most scientists intend, particularly when it comes to the laws (postulates, principles, or whatever) of the behavioral sciences. On the other hand, we don't want to encourage merely adversarial nitpicking.

The essential point of the naming problem is to place the burden of proof on those who might otherwise lead serious investigators down paths that have in the past proven unproductive. We need defenses against people who suffer from logorrhea and arithmorrhea, for they will always be with us. For their own sake, as well as for the advancement of science, they should be discouraged. The attempt to pursue analysis to its ultimate roots leads to an infinite regress. We must resist this Hamlet-like impulse in order that the world's work can be carried forward.

For those few statements that seem, after long acquaintance and many experiences, to bespeak a "conviction of the mind," I propose the term *paramount position*. Identifying a paramount position does not free it forever from examination: it merely announces, in firm tones, that *the burden of proof* falls on anyone who asserts a contrary proposition. The scientific mind is not forever closed: it is merely intolerant of wasting time on the proposals of attention-seeking amateurs who are too lazy to master the simplest fundamentals of a science.

ESSENTIAL PARAMOUNT POSITIONS OF ECOLOGICAL ECONOMICS

What follows is tentative. I have tried to arrange what I regard as the essential paramount positions of ecological economics in a hierarchy of importance; but I do not claim much for the ordering. The result cries out for criticism; and it needs to be enlarged.

To begin with, ecological economics must accept all the well-established paramount positions of the physical sciences, e.g., the laws of conservation of matter and energy. From biology it takes the idea of natural selection as a pervasive and inescapable process. To discourage logorrhea and arithmorrhea I have, for the most part, couched the positions in "folksy" language.

1. *The world available to the human population is limited to Earth.* Wishfully thinking men and women have, over the centuries, tried to escape from mundane problems by dreaming (in succession) of Heaven, of unoccupied frontiers, and (lately) of a science-fictional "Space." In the distant future, a few individuals may be sent off to inoculate some distant solar systems with *Homo sapiens*, but billions of human beings will have to be left behind to make the best they can of the limited resources provided by this planet. Such a conclusion is no doubt a great disappointment to science-fiction enthusiasts. Let us hope that "Space" is the last of the trans-mundane Providences to divert human beings from facing human problems on Earth.

2. *There's no such thing as a free lunch.* This is implied by Epicurus's remarks.

3. *The First Law of Human Ecology: "We can never do merely one thing"* (Hardin 1963). Economists have referred to the same idea under the title of the "Law of Unintended Consequences." A painful and constant awareness of this principle, vague though its statement may be, can save would-be reformers from many mistakes.

4. *The Second Law of Human Ecology: "There's no away to throw to."* Until the invasion of public consciousness by the environmental movement, many economists implicitly presumed an "away" for the disposal of "side effects." This presumption blinded people to seeing the obvious. Tacit features of the definition of private property also had a shielding effect; as Dan McKinley has pointed out, the conventional definition of private property "includes the smokestack but not what comes out of it" (McKinley 1969). When it comes to the wastes generated by nuclear reactors, rhetorical tactics succeeded in suppressing corrective action for forty years. (The difficulty is still not solved.) Malthus's problem of finding sufficient "subsistence"—food, principally—has been

replaced by the much more formidable challenge of finding sufficient absorptive capacity for the exponentially growing inventory of human-generated wastes.

It is worth noting that the preceding principles follow from the ecologizing of economics. Conventional economics has for long worked implicitly with an analytical universe of resources, forces and restrictions that included little more than man and his artifacts. The input of solar energy was taken for granted (that is, it was ignored much of the time). Within this intellectually limited world, it was easy to speak of the "creation of wealth," a concept completely at variance with the thrust of the second law: when the total system is taken into account, wealth continually decays (Soddy 1933). The thrust of 20th-century ecology into cost-benefit analysis has forced economists to admit that mundane wealth is never created; it is merely rearranged, with varying consequences. The required attitudinal change is still not complete. Paul Ehrlich tells how, at a planning meeting that took place in Stockholm, economists kept proposing solutions to the population problem that, on analysis, proved to imply a violation of the second law. Biologists and physicists were not backward in pointing this out. "Finally, in frustration, one of the economists blurted out, 'Who knows what the second law of thermodynamics will be like in a hundred years?'" (Ehrlich 1981, p. 28).

I have never encountered a biologist who thought that the solution of our problems required the confuting of the second law. Perhaps not many economists anticipate such a refutation either; but the assumption of textbook economics that growth will go on forever (or that growth is *normal*, while a "steady state" is *abnormal*) may well encourage subconscious repudiations of the conservative thought of Epicurus. The language of commercial economics is, of course, even more disturbing.

5. *The Third Law of Human Ecology: "(Population) x (Per Capita Impact) = (Total human impact on the environment)."* This can be called the *Impact Equation*. One might suppose that this principle is so obvious as to not need saying, but not so. For instance, a professor of philosophy has claimed that "Pollution results not from our numbers . . . but from our lifestyles and our rate of consumption" (Sagoff 1980, p. 315). I think this remarkable assertion can be understood only in the light of two attitudes now widespread among the self-styled "liberals" of our time. The first is the *ethnofugal* attitude, the preference for distant, bizarre and miserable cultures over our own. By condemning the using of resources by rich nations like ours while holding poor nations guiltless in generating larger populations, many liberals justify international charity. The ethnofugal attitude (which Edwin Arlington Robinson depicted so well in his poem "Miniver Cheevy") is the opposite of ethnocentrism . . . . The word "ethnocentrism" was coined by William Graham Sumner for use in his *Folkways* in 1907. "Ethnofugalism" has yet to make its way into any English dictionary.

Another attitude that is all too common among otherwise well-educated people today is a strong distaste for numbers. To such people, 1 is no different from 1,000,000,000; only by recognizing this assumption can one understand the philosopher's statement that the size of a population has nothing to do with the amount of pollution it generates. Innumerate critics sometimes go to astonishing lengths to make sure that no one takes population problems seriously. Taking the lead from intellectuals of this stamp, a popular magazine recently published a 2600-word analysis of deforestation which concluded: "Ultimately, the only sure way to stop deforestation is to minimize the things that cause it—lack of fuel, fodder and farmland for the third of humanity that live on the edge of starvation" (Vanishing forests, 1980). There was not a word about population, the other factor in the Impact Equation.

6. *Scale effects, though sometimes compensable, are inescapable.* An example will make the point clear. For hundreds of years men tried to fly by their own muscle power, without success and some loss of life. The quantitative relations between muscle power, air resistance, gravity, etc., that make it possible for a bat to fly simply don't "scale up" to a mammal the size of a large man. The exponents of the variables are unequal. Ultimately, structural materials were created that were strong enough to permit a small man to fly (with difficulty). Compensation was found in the nature of the materials.

A given process may enjoy "economies of scale" over one range of variation and "diseconomies of scale" over another. For understandable reasons, economists have emphasized only the former. Political scientists are not yet sufficiently impressed with scale effects. When they become more so they may be able to draw up a calibrated schedule of successors to simple democracy, which clearly suffers from diseconomies of scale.

CULTURAL CARRYING CAPACITY, THE FRAMEWORK OF RATIONAL ANALYSIS

"Carrying capacity" is the fundamental basis for demographic accounting. Illiterate farmers and herdsmen the world over understand carrying capacity, but it seems to be beyond the comprehension of myriads of our own "animal lovers," highly "educated" though they may be (Hardin 1982). People who object to the rational solution of game management problems are not so much highly educated as highly *schooled*; and their schooling is narrowly urban and includes no daily experience with fields, pastures and non-human animals.

Many economists have totally rejected the concept of carrying capacity. Remarkably, a committee of economists, at the behest of the National Research Council, in 1986 issued a discussion of population growth and economic development in which neither the word *nor the concept* of "carrying capacity" played a role (National Research Council 1986). A careful (but uncritical) reading of the text would lead one to conclude, Herman Daly said, that "Apparently limited natural resources can be substituted by "artificial resources," which are expandable without limit," which leads to the final conclusion that "carrying capacity is infinitely expandable" (Daly 1986, p. 584). Epicurus must be turning over in his grave. The publication of the National Research Council committee is as astonishing as would be a textbook on accounting that stoutly maintained that there is no necessity for assets and liabilities to balance.

The error of the committee is understandable. The carrying capacity of a territory for a population of non-human animals can be fairly easily determined, because we demand only that the population survive indefinitely. But when we come to consider *Homo sapiens* our standards rise, we ask not only that human beings survive, but that they enjoy a reasonable repertoire of amenities. But what is "reasonable"? Should the repertoire include meat? Automobiles? Television? Wilderness? Dune buggies? Rock concerts? Round-the-world cruises? Obviously, the question of "capacity" has entered the realm of *values*.

By long tradition, economists have avoided value questions like the plague. The evasion is prudent: but it does not justify an implicitly anti-Epicurean assumption that resources can be created out of nothing. The issue of limits must be restated in terms of the *cultural carrying capacity* of a territory for human beings, living the sort of life they want to live (which is, of course, a matter of opinion). The fact that there are as many "capacities" as there are opinions does not justify the presumption that the concept of capacity has no meaning, or that any capacity is infinite. The variation is encompassed by the following generalization:

7. *Cultural carrying capacity and the standard of living are inversely related.* The higher the standard of living, the fewer the number of people that can enjoy it. Amenities like solitude and wilderness can be enjoyed by very few, and this fact creates serious problems of distribution and justice (Hardin 1986, 1988).

Though human beings take account of much more than energy as they try to define the "quality of life," energy alone can be used to illustrate the importance of quantity. Remaining alive while working moderately requires about 2,300 kilocalories per adult per day. Americans use about 230,000 k-cal per person per day—some 100 times the minimum. If all the world's people lived at the American cultural level of energy consumption, the energetic carrying capacity of Earth would be only one one-hundredth as great as it would be if everyone lived at the absolute minimum level. Recognition of this fact leads most people to accept the next paramount position of human ecology:

8. *The maximum is not the optimum.* Calculus offers such a simple way of determining maxima that it is easy to forget that the mathematical answer has little to do with the problem of the human optimum. It is the rare economics paper that does not implicitly assume that the maximum Gross National Product is an optimum of some sort, though Kenneth Boulding, Herman Daly, and others have eloquently pointed out the short-comings of this measure of felicity. For a statesman to try to maximize the GNP is about as sensible as for a composer of music to try to maximize the number of notes in a symphony. Maximizing the "quality of life" is a deep and subtle problem. This problem lies in the domain of ecological economics.

9. *"The greatest good of the greatest number" is nonsense.* The theory of partial differential equations tells us that we cannot maximize for more than one variable at a time. Since the time-honored utilitarian ideal is mathematical nonsense it must be practical nonsense also. We need to find other approaches to the problem of the optimum. We must, in a word, decide whether we want to maximize the number of human beings on Earth, or to maximize their *average*—not their total—well-being.

The emergence of nuclear energy as a possible and quasi-unlimited source of energy led many to think that the average well-being might be moved upward, almost without limit. With a different emphasis it has been assumed that unlimited energy would solve all population problems. But the realization of nuclear energy's potential for creating long-lived and lethal by-products has changed our perceptions. Everything hinges on the reliability of machine-human couplings. Half a century's experience has led to the following paramount position:

10. *Attempts to create perfectly reliable machine-human couplings are inescapably self-defeating.* At first, the spasmodic occurrence of nuclear accidents was explained away as part of the "learning curve" of the new technology. The role of human nature was largely ignored. But experience has shown that the more perfect a machine-human coupling is, the more boring the job becomes for people. The more boring the task, the lower the intelligence of those who will accept the job. The longer an operator continues in the same position, the less alert he is. The longer the record of accident-free operation, the more difficult it is to imbue new recruits with the necessary conviction that accidents *are* possible. Industrial education deteriorates. Thus it comes about that perfection in the machine selects for imperfection in human performance. *We get what we select for.* Therefore, sooner or later, the "unthinkable" accident happens. It can only be viewed as optimistic that, since Chernobyl, the "unthinkable" has come to be thinkable by an ever increasing number of people. (Another Chernobyl will probably clinch the change.)

What all this adds up to is a paramount position that is not descriptive but normative in nature, namely:

11. *The 11th Commandment of Human Ecology: "Thou shalt not transgress the carrying capacity."* Carrying capacity transgressed is carrying capacity reduced. Presumably the policy goal is to specify a sustainable carrying capacity, whether the subject be cattle in a pasture or human beings in a nation. In the animal example, too many cattle in a pasture trample the soil and selectively eat up the "sweet grass," resulting in a competitive advantage for weeds—"weeds" by bovine standards—the following year. Soil erosion progressively reduces plant growth in subsequent years. Unless the overpopulation is corrected the carrying capacity diminishes year by year. The same principles, *mutatis mutandi*, govern the human exploitation of cultural carrying capacity.

In trying to match environmental resources to the demands of populations, the language used makes progress either more or less likely. The final paramount position is purely a definitional one, but a definition chosen to bias perceptions greatly influences the probability of success in matching resources to demands.

12. *Every shortage of supply is equally a longage of demand.* According to the Oxford English Dictionary the word "shortage" was first used in 1868. The word "longage," still not in any dictionary, did not appear in print until 1975. (See Hardin 1978.) The difference is no mystery. Language is a weapon for controlling the thoughts of others; using "shortage" suggests that we should search for ways to make money by marketing a commodity. "Longage," by contrast, does not easily generate marketing opportunities. Those who offer help in reducing human demands are often regarded as marginal members of the community—ministers, psychiatrists, ecologists, and the like. Still more aberrant (and less welcome) are those who recommend an actual reduction in the size of the population that figures in the Impact Equation. Understandably, fears of draconian measures are aroused.

But a reduction in population need not necessarily involve the liquidation of people. The Impact product can be reduced by attrition, which takes place whenever a controlled birth rate falls below the normal death rate. Of course, even this possibility is viewed with horror by conventional moralists and "Growthmanship" economists. The taboo operating against thinking in terms of longages is great. This is unfortunate because—*no "shortage" reported in a growing population can be cured by increasing the supply.* The only constructive action is a decrease in demand. The word "longage" is more than a mere inversion of the word "shortage." Using this hitherto unfamiliar word can prepare the mind to think creatively about the problem of reducing human misery.

REFERENCES

Alder, K. 1986. The Perpetual Search for Perpetual Motion. *American Heritage of Invention & Technology.* 2(1):58-63.

Bailey, C. 1926. *Epicurus, The Extant Remains.* Oxford: Clarendon.

Daly, H. E. 1986. Comments on "Population Growth and Economic Development." *Population and Development Review* 12:582-585.

Demeny, P. 1988. Social Science and Population Policy. *Population and Development Review* 14:451-479.

Eddington, A. S. 1928. *The Nature of the Physical World.* New York: Macmillan.

Ehrlich, P. R. 1981. An Ecologist Standing Up Among Seated Social Scientists. *CoEvolution Quarterly* 31:24-35.

Hardin, G. 1963. The Cybernetics of Competition. *Perspectives in Biology and Medicine* 7:58-84.

Hardin, G. 1972. *Exploring New Ethics for Survival.* New York: Viking.

Hardin, G. 1978. *Stalking the Wild Taboo.* 2d. ed. Los Altos, Calif: William Kaufmann.

Hardin, G. 1982. Sentiment, Guilt and Reason in the Management of Wild Herds. *Cato Journal* 2:823-833.

Hardin, G. 1986. Cultural Carrying Capacity: A Biological Approach to Human Problems. *BioScience* 36:599-606.

Hardin, G. 1988. Wilderness, a Probe into "Cultural Carrying Capacity." *Population and Environment* 10:5-13.

Holton, G. 1986. *The Advancement of Science, and Its Burdens.* Cambridge: Cambridge University Press.

Kasun, J. 1988. *The War Against Population.* San Francisco: Ignatius.

McKinley, D. 1969. Personal communication.

National Research Council, Committee on Population: 1986. *Report of the Working Group on Population Growth and Economic Development.* Washington, D.C.: National Academy Press.

Sagoff, M. 1980. The Philosopher as Teacher? *Metaphilosophy* 11:307-325.

Soddy, W. 1933. *Wealth, Virtual Wealth and Debt.* 2d ed. New York: Dutton.

Vanishing Forests. 1980. *Newsweek* (Nov. 24).

Weinberg, A. M. 1967. *Reflections on Big Science.* Cambridge, Mass.: M.I.T. Press.

Whittaker, E. T. 1942. Some Disputed Questions in the Philosophy of the Physical Sciences. *Proceedings of the Royal Society of Edinburgh* 61:160-175.

Whittaker, E. T. 1958. *From Euclid to Eddington.* The Tarner Lectures, 1947. New York: Dover.

5

SUSTAINABILITY AND THE PROBLEM OF VALUATION

Talbot Page
Environmental Studies
Brown University
Providence, Rhode Island 02912 USA

ABSTRACT

Biologists have taught us in the last 150 years is that there are no sharp boundaries between us (humans) and other species. Further, the notion of the individual itself has blurred, both from below and from above, from studies of genetics and evolutionary biology and from studies of primate and other animals' social behavior. These two insights are usually offered as positive, descriptive, or explanatory statements about the way the world is. But they have important implications for our value theories and policy prescriptions. One approach is to extend the traditional moral concepts of rights, duties, utility, etc., from humans to other species. But a problem with this approach is that these traditional moral concepts are individualistic. An alternative approach is to run the extension the other way and ask what happens to the traditional moral concepts when the individual and species lines blur.

The purpose of this paper is to explore the alternative approach. I suggest that for the world biologists describe, less highly individuated value concepts become relatively more useful and important, for example in concepts of opportunity and sustainability. It also becomes clear why some policy questions, which are often viewed as problematic in current economic analysis, are indeed problematic in the traditional framework. These problems include optimal population, species preservation, and intergenerational equity.

INTRODUCTION

"Sustainability" means different things to different people. In *Conservation and Economic Efficiency* (1977), I defined a sustainable economy as one in which the resource base is kept intact over generational time. In this view, the various broad aggregates of resources (e.g., energy, metals, wood, soils, water) are managed to balance depletion with renewal (technological and otherwise) and balance waste generation with environmental capacity for assimilation. And since, roughly speaking, every time the population doubles the resource base halves, sustainability eventually requires stabilizing the population.[1] We do not normally use "congestion" to describe problems of overpopulation, but this problem fits the definition of crowding against a limited resource. Thus sustainability requires controlling each of the three main types of environmental problems: depletion, pollution and congestion. In *Conservation and Economic Efficiency,* I attempted to define more clearly this elusive concept, and offer a justification for it, while suggesting some policy instruments for achieving it.

In the earlier book, the justification was based on an intergenerational equity argument. In this paper, instead of equity considerations, I will focus on the basis of valuation. More specifically, I explore how some of the things we have learned from biologists[2] are influencing the way we think about the problem of valuation. The path leads toward justification of sustainability as a policy goal, but "justification" is too grand a word for what I do in this paper. More modestly, I will suggest that "lessons from biology," by influencing the way we think about valuation, strengthen the normative appeal of sustainability as a policy goal and help explain the increased interest in it.

The paper's genesis came with an idea of reinterpreting *extensionism* (extensionism has to do with enlarging our value system to include referents besides people). I began thinking about the motivation for extensionism in terms of the blurring of species boundaries. This led me to a second lesson from biology which seemed to undermine extensionism and challenge the more prominent value theories. I then began to consider the general question of how ideas from science influence value theories. I addressed this question with four rather simple "lessons from biology" and one (two-tier) value system.

The four lessons, although controversial in places, are all well known and by now commonplace. Being commonplace, their impact on value theory is easy to overlook. But not only are these lessons affecting the way we think about valuation, but it also seems sensible to rethink value theories in light of them.

THE LESSONS

Lesson One: The Fading of the Species Boundary

The first lesson from biology is that there is no sharp boundary between humans and other animals. This is an old lesson, but with accumulating evidence from behavioral and

---

[1] The idea is to strike a balance between defining the categories broadly, to allow substitution within a category, but still sufficiently narrowly, to characterize various "essential" categories. As preferences and technologies change, not every resource type needs to be kept intact. But even if preferences and technologies change, we will not be able to do without energy, and it is unlikely that we will be able to do without soils or metals either, for the foreseeable future (in other words for a sensible planning horizon). Land per capita, and hence population, is another essential resource category.

[2] I include ecologists and other life scientists with biologists here.

genetic studies it has been pressed home, especially in the last few decades. I think that the impact of this lesson has been more pronounced in moral philosophy than in economics. (Moral philosophy can be defined as the study of value theories.) In the last three centuries in the moral philosophy of the West there have been, and still are, two main approaches to value theories. Both are directly challenged by this first lesson from biology, the blurring of species boundaries.

In both strands of moral philosophy—one Kantian or deontological and the other utilitarian—a basic way of proceeding is to identify "morally relevant characteristics" and then accord special treatment to those beings with the identified characteristics. In the first strand, Kant found the special characteristic to be rationality, the capacity to understand and be committed to rational argument. More modern deontological philosophers have found the special characteristic as having "interests," "dignity," "autonomy," or a "complex psychological life." In the second strand, Bentham, an early utilitarian, found the special characteristic to be sentience, the ability to feel pain and pleasure.

Thus far in the story, the first lesson from biology is not directly at odds with either of the two stands. Both Kant and Bentham, among others, said that their special characteristic (rationality or sentience) was not necessarily limited to humans and thus other beings might require special treatment.

However, in both approaches to moral philosophy "morally relevant characteristics" have been interpreted as dividing lines between humans and other beings. Besides the "morally relevant characteristics" just mentioned, other candidate characteristics have been explicitly used as boundaries: for example: "only humans have true language;" "only humans use tools;" "only humans have consciousness;" "only humans have continuing life plans;" "only humans are self-aware."

Typically and with increasing success, biologists have delighted in showing that these defining characteristics are not sharply defined conceptually and not limited to humans empirically.

A response to the first lesson—the fading of the species line—is *extensionism*, a way of thinking fairly common now in environmental philosophy. The idea is roughly this: in extensionism we take the traditional moral concepts, such as duty, utility, rights, autonomy, and justice, and apply them to an extended field of reference. The extended field of reference is defined in the traditional way, by the special characteristics of moral relevance. But, and here is the new departure, the extensionists accept the larger and more diffuse boundaries of the morally relevant characteristics and apply the traditional moral concepts to non-humans.

In the hands of Singer (1979), extensionism takes the form of broadening the application utilitarianism from humans to other conscious and sentient animals. (His book, *Practical Ethics*, is particularly interesting to teach from, because it systematically develops the extensionist approach and jars students along the way.) In the hands of Tom Regan (1985), extensionism takes the form of broadening the application of rights from humans to other animals. Stone (1974), in his introduction to *Should Trees Have Standing?*, notes that Darwin viewed "man's moral development . . . [as] a continual extension in the objects of his 'social instincts and sympathies.'" The extension went from an individual person and his close circle, to other humans, to the lower animals. Stone himself then follows an extensionist path in his argument to provide legal standing for trees and other natural objects. Leopold (1949) begins "The Land Ethic" by recalling that Odysseus treated his slave-girls as property (he killed them), but that the history of ethics has been a gradual extension to broader categories of conduct. Bentham was a pioneer extensionist. After saying that "Can they suffer?" is the foundational question of moral consideration, he went on to suggest that other animals could suffer and thus their sufferings should be balanced on the same scale as our own.

In brief, as biologists have increasingly blurred the species line, environmental philosophers have developed extensionism. The widely used reader in environmental philosophy, *People, Penguins, and Plastic Trees* (VanDeVeer and Pierce 1986), which relies heavily on extensionism, is an example of this development. However, while the first lesson has encouraged extensionism, the second lesson undermines it.[3]

Lesson Two: The Fading of the Individual's Boundary

The second lesson is that just as species boundaries are fading, so too are the boundaries defining individuals. It used to be that the concept of an individual human being was unproblematic, but now the concept is challenged from below and from above.

From below, at the genetic level, we are told that the genes are the real individuals and the real units of competition and natural selection. The picture one gets from Dawkins' (1978) *The Selfish Gene* is that we humans are mere campsites where the genes congregate for the night (a "generation") and then reshuffle, depending on chance and fitness, and move on to other campsites. Moreover, a gene is not a single thing either. It is a pattern of itself and all its replicas. Thus bits and pieces of "me" are presently camping out in my sister, my cousins, people I have never met, chimpanzees, and even blades of grass. Worse yet, genes are not even individual and defined patterns either. Instead, they are lengths along chromosomes, arbitrarily "specified" for the convenience of geneticists and evolutionary biologists. It also appears that bits of human enzymes can become parts of viruses, making them potent human diseases; and parts of viruses become incorporated into human chromosomes.

I realize that not everyone agrees with Dawkins and that some of what I just said is controversial. Nonetheless, I think we can say that the concept of an individual human being becomes less "individuated" when approached from the perspective of evolutionary biology.

From above, at the social level, the boundaries of the individual are fading as well. De Waal (1982, 1989) describes in detail how chimpanzees and other primates are like us in social behavior in their coalition-forming, aggression, and peacemaking. One of the strongest observations that comes out of his two books is that society creates the individual, the individual does not create society. This impression is so strong, I think,

---

[3] Others, including Callicott, Rolston, Hargrove, Sagoff, and Norton, have criticized the approach of extensionism. I am indebted to Bryan Norton for comments on this.

because other primate societies are in some ways strikingly different from our own and in some ways strikingly similar and this gives us a distance and perspective that is hard to achieve when thinking about our own society. De Waal's primate societies also seem simpler than our own (part of this may be that we are still unaware of their complexities), and this makes the major themes stand out more sharply. In any case, in de Waal's studies, individuals are not fixed and predefined but individual behaviors and the individuals themselves emerge from social interaction.

The analogous point is that human individuals are also created and maintained by social interaction. The biologists have renewed this lesson and given it a fresh perspective.

A response to the second lesson is to search for a value theory which does not rely foundationally on the *predefined (exogenous) individual*. As we shall see in a moment, this lesson challenges extensionism, and traditional moral philosophy as well.

Extensionism says take traditional moral concepts and apply them to an extended field of reference. But traditional moral concepts are highly individuated. By this, I mean that they are closely tied to notions of the individual for their definition. For example, the concept of utility has a referent. We speak of utility to someone. When there is no well-defined, pre-existing referent, the definition of utility slips away.[4]

Similarly, with the concept of rights there is a referent, and again the referent is a predefined individual. And in the usual story, there are other predefined individuals in the background. There is an individual who is a potential rights violator and there are other individuals employed by the state who intervene to protect the rights holder.

Other traditional moral concepts, such as duty, obligation, autonomy, and (even) justice, tend to be strongly individuated as well. This reliance on the individual in moral philosophy is not surprising, for many of the normative concepts which we now take for granted were developed during and after the Age of the Enlightenment, when individualism was a primary focus.

Thus the extensionism practiced by Singer, Regan, and other environmental philosophers moves in a difficult direction. It takes the traditional moral concepts and applies them to contexts where individualism is even more problematic.

But the challenge is broader and applies to much of the current discussion in value theory. For example, the present shift in moral philosophy from utilitarianism to rights, as described by Hart (1979), is no solution either. The rights concepts are as deeply individuated as the utility ones. Mainstream welfare economics and cost-benefit analysis are also based on the concept of a predefined, exogenous individual. For example, ideas of Pareto optimality and potential Pareto improvement require predefined reference groups for application.

On a more practical level, we can see if we are on the right track by listing some of the environmental problems that we are having the most trouble conceptualizing. These problems include "optimal" population size, intergenerational equity and the "discount

---

[4] We do speak of social utility or the public interest, but normally in economic analysis these concepts are interpreted as aggregations of individual utilities. This is what we mean when we speak of methodological individualism, which is often taken to be a starting point for economic welfare analysis.

rate problem," species extinction, and the proper treatment of animals. A common thread is that for all these problems we do not have prior existing individuals with fixed, exogenous preferences (or values or rights). This common thread suggests a diagnosis: part of the difficulty we are having in thinking about these troublesome problems is that we are approaching non-individuated problems with highly individuated tools. It suggests that our heavily individuated value theories may be part of the problem. The mismatch between the nature of the problems and the tools used to address them may be adding to our difficulty in conceptualizing the problems.

Lesson Three: Bad News

Economics used to be the dismal science; environmental science is now taking its place. Over the last twenty years, biologists (along with ecologists, demographers, climatologists, and others) have amassed much evidence of species extinction, soil depletion, forest loss, minerals depletion, carbon dioxide and CFC buildup, pollution, and so on. In fact, a principal difference between this round of concern over sustainability and the last one twenty years ago is a much stronger documentation of the bad news.[5] Another difference is that we have gone through almost another doubling period since that time in both population density and materials and energy flows, so there is more bad news to report.

There is, however, much disagreement over how to interpret the empirical evidence of non-sustainability. Julian Simon and Harold Barnett, among others, argue that things are going well and the resource base is being kept intact, except for a few minor problems for which we can make substitutions. Resource and technological optimists may even go so far as to argue that future generations are going to be so much better off than we in the present, that we should take actions to make us better off at the expense of the future, to make things more equal.

Even without attempting to resolve this controversy, it is clear from the biologists' evidence that future sustainability is in doubt. The doubt suggests that we need a value theory which includes concepts of sustainability and provides a framework for defining and analyzing policy instruments to achieve it.

How strong or weak the policy instruments should be depends on the strength of the empirical evidence. If one interprets the evidence of non-sustainability as strong, it makes sense to recommend major policy action. If one takes the empirical evidence less seriously, then it suggests minor policy action (or none). Of the four "lessons from biology," this third one is the one most directly pointed toward policy action. (I suspect that part of the reason why this lesson is the most controversial is because some of the suggested policy actions are not only strong but also very costly (e.g., major control of carbon dioxide emissions).

For the purposes of this paper, I do not need to evaluate the controversy. The accumulation of empirical evidence and the existing doubt suggests the need for a value theory

---

5 *Limits to Growth* was criticized for its empty models, which were highly aggregative models with little empirical fitting. At the time, Carl Kaysan wrote a particularly caustic review entitled "Garbage In, Garbage Out."

including sustainability and a place for policy instruments to achieve it. I will attempt to sketch a value theory and its framework for policy instruments. For the purposes of the paper, I do not need to take the further step of recommending specific strengths of the policy instruments. To do that would require evaluating the weight of the empirical evidence, a major task beyond the scope of the paper. In sum, the existence of the doubt motivates the development of a value framework and a place for sustainability instruments; specific evaluation of the evidence is a prerequisite for specifying the strengths of the instruments. I attempt the first but not the second of the two tasks.

But whether the biologists' evidence is assessed heavy or light, the lesson reinforces what economists have learned in principle about idealized markets. Just as it is unwarranted to expect markets to automatically maintain some desirable distribution of income or some specific price level, there is no guarantee that market incentives will maintain the resource base intact.[6] Correspondingly, just as there need to be policy instruments, such as the income tax with a particular degree of progressivity built into it, to achieve a desired balance of income distribution, in principle (and in general) policy instruments are needed to achieve desired balances of depletion and renewal, waste generation and assimilation, and population stability. We might be lucky and find that sustainable balances are happening automatically, but in general we cannot count on luck.

The third lesson challenges the optimistic view that a sustainable balance between depletion and renewal would be struck automatically by market forces, without explicit policy attention. (This view is commonly held in resource economics and found explicitly in Barnett and Morse's *Scarcity and Growth*.)

The challenge from the third lesson is to construct a value theory that allows us to define a "good" balance between depletion and renewal, waste and assimilation, and population and land. More than that, the challenge is to develop a framework in which the possible policy actions (large or small) can be placed. And in conjunction with the second lesson, the challenge is to develop a system of valuation for the possible policy actions which does not foundationally rely on the predefined individual.

Lesson Four: A Long Time Horizon

Different systems move along different time scales. Geology works in the millions of years; economics in the tens of years; biology from a few minutes to a few centuries; evolutionary biology from a few years to millions of years. Appropriate time scales depend on how long it takes for things to happen in the subject area.

Market processes are notoriously myopic. With typical market discount rates and opportunity costs of capital, the future disappears (for decision purposes) after a few decades. Political processes are similarly myopic. In response to election pressures, political horizons can be from a few years to a few months. An important reason why environmental

---

[6] In principle we expect a particular resource price to rise on anticipation of future scarcity of the resource, and in consequence for there to be increased incentives for substitution and technological renewal. But these market incentives do not imply that the renewal will be forthcoming or that it will be sufficient to maintain the resource base intact.

problems are becoming larger and more difficult to manage is that more of these problems happen over longer time scales and thus they are increasingly misfitted to traditional economic and political institutions.

In contrast to the short horizon of economic and political institutions, worldwide population doubling periods are on the order of 30 or 40 years. Optimistic demographers tell us that the world population might be stabilized at about 15 billion in 75 years. (The traditional time horizon for demographers and other long-term forecasters seems to be about 75 years, the closest time for a stabilized population, and also the period when global warming might be strongly felt.) Given present stresses, this period 75 years from now is likely to be a transition period of hard times. For example, it is estimated that currently one species is becoming extinct every hour; 75 years from now the rate is likely to be much higher. However, if the instruments for population stabilization are in place during this period, they might remain in place for the next 200 years, gradually bringing the population back down to a more comfortable 2 billion (thus it might take 275 years to get back to where we were 50 years ago). A shift to sustainable management of energy and material flows, combined with increasing standards of living in the non-Western world, may also be a long-term prospect.

To economists used to thinking with ten-year horizons, the great uncertainty of the future may make it seem unsound to look far ahead. But an advantage of the longer view (besides possibly being less dismal) is that it suggests different strategies toward option preservation, including species preservation, that would be pointless under a permanent population of 15 to 20 billion.

The fourth lesson from biology is that the appropriate time scale for understanding resource and population stability is much longer than we are used to. The challenge to a value theory, from this lesson, is to find ways of valuing the distant future, perhaps over three centuries or more.

A TWO-TIERED VALUE THEORY

The task in this section is to sketch a value theory which is compatible with the four lessons from biology. But first, I will say an introductory word on the connection between "science" and "value." I have phrased the lessons from biology mostly, but not entirely, as positive claims.[7] There is much current attention on how normative ideas influence positive ideas (conceptions as to what is). In this paper, we turn our attention to consider the impact of positive (empirical) ideas on normative ideas (value theory).

I do not think that normative ideas come directly from positive ones, but that normative ideas come from other normative ideas, until we get to foundational ones. The appeal

---

[7] Positive claims are empirical statements that attempt to describe how the world is. In contrast, normative claims are statements about how the world (or a model or theory) should be. Normative claims are statements of evaluation. To use the two words does not mean that we must believe that there is a clean dichotomy between normative and positive. Even if the two levels of discourse are interwoven, some statements may be more focused on what *is* and some more focused on *evaluation*.

of the foundational normative ideas comes, if it comes at all, from the whole system of normative ideas and their connection to the world, including our (positive) ideas about how the world is. Thus the lessons from biology are not the direct grounding of our normative concepts, but they shape our world view and indirectly shape how we think about valuing things. As the view of what the world *is* changes, so too, indirectly, may our approach to valuing things. In this section, I describe one possible path of response.

We begin with the challenge from the second lesson, to construct a value theory that does not rely foundationally on predefined individuals. A few examples may help clarify the challenge and a possible response.

In the drafting of the U.S. Constitution, Southerners argued with Northerners about whether slaves should be counted in establishing the number of representatives for each state. Southerners, who did not want the slaves to vote, nonetheless wanted to count their number to increase the number of Southern representatives in Congress. Northerners argued that white Southerners should not get extra representation on the basis of non-voting slaves. In the crudest terms, Southerners wanted to increase their representation; Northerners wanted to limit it. The compromise was the infamous "three-fifths rule" which counted five slaves as equal to three whites in apportioning the number of representatives.

In contrast, there was a more general consensus on an amendment to protect requiring "due process." Part of the contrast can be explained in terms of the difference between particular, known preferences and generalized interests. In the "three-fifths" debate, the Southerners had particular and well-defined preferences. They knew what they had to gain directly and immediately from counting slaves for representation. Similarly, the Northerners had particular and well-defined preferences in opposition. Both Northerners and Southerners knew, or thought they knew, how the "three-fifths" rule would affect their children and grandchildren.

The framers knew less about the particulars of how they, their children and grandchildren would fare under the proposed amendment requiring due process. In place of specific gains and losses and well-defined preferences, the framers had generalized interests in making the justice system work fairly. Their grandchildren might be defendants or plaintiffs. The parts of the Constitution which have stood up to time better are (at least sometimes) where the framers looked to general interests and "abstracted" individuals; the parts that fared worse are where they looked to well-defined preferences of "predefined" individuals.

What is important for our purposes, in the amendment requiring due process, is that we *need* to know little about the particular preferences of ourselves or our grandchildren, except that they will have a general interest in the fairness of the criminal justice system. No matter who they are, they are likely to benefit from the amendment.[8]

For a second example, consider how the Edgeworth box is used in economics to illustrate the decision problem faced by two people. The first decision is to apportion the

---

[8] The idea here is similar to Rawls' "original position" and "veil of ignorance." Barry also stresses how different considerations vary with the level of abstraction in the decision process.

endowment of resources between the two. The second is for the two to negotiate, arriving at the contract curve if they are efficient.

Again, different considerations come into the two decisions. In the second decision, once the endowment is apportioned, it is usual to analyze the negotiation as narrowly self-seeking. Here, the two individuals are predefined, know their preferences, and act on them. But in the first decision, setting the endowments, the considerations may be more general and abstracted. It may not matter what particular preferences the two individuals have. The endowment may be set on the grounds of fairness; the decision may not be analyzed at all. It may be sufficiently abstracted to be outside the model altogether.

For a final example, consider the street version of basketball called buckets. In buckets, the team that scores gets to put the ball into play again. In contrast with regular basketball, where the team scored upon gets to put the ball back into play, the buckets rule is a decided advantage to the scoring team. Also in buckets, the team that wins a game gets to play again, against a challenging team. These two rules together are to the advantage of the good players, who get to play longer and more often.

It may not be a surprise to note that the good players are the ones who tend to define the rules in buckets. The good players know who they are and they know what rules favor them. In regular basketball, the rule makers are not the players but the league owners and commissioners. Thier interests are less oriented to the specific strengths and preferences of particular players (there is some of this) and more towards getting close and exciting games. (Thus the best draft choices go to the worst teams and not the best.) Again, the point is that at different levels of abstraction from the particulars of a decision, there are different concerns. At the more abstracted levels, it is less important to have predefined individuals with specific preference structures.

Turning our attention from these examples to the problem of sustainability, we need to ask "how much sustainability?" and "in what form?" We do not know the specific preferences of those in the next generation (and following generations). More fundamentally, their specific preferences, identities and even existence are formed by our actions. They are endogenous to our decisions. Future people are not individuals in the sense of having predefined preference structures. Like de Waal's chimpanzees, they will become defined as individuals by the environment we provide them. But whomever they may come to be, future people are likely to want and need water, soil, energy, and other "essential" resources.

The above examples suggest that a two-tier value theory is consistent with the second lesson.[9] (Such consistency is not a justification of the value theory, but it may increase

---

[9] The two-tier system I am developing in this paper is contractarian. An alternative response to the second lesson is communitarian. In the latter view, a value is placed on a central concept which is also not individuated. The central concept becomes a context within which individuals can emerge and at least some of the traditional normative concepts can have a role. The central concept is the community. The idea has been developed by political philosophers such as Walzer, Sandel, and Taylor. Interestingly, it appears that these forms of communitarianism were developed as critiques of liberalism (the two-tier view of this paper is more supportive). If we follow the communitarian approach, we could simply add on the

its normative appeal.) The first tier provides the circumstances under which the second tier—comprised of legislatures, courts, and markets—operates. For example, the U.S. Constitution is in the first tier. This instrument incorporates such normative ideas as "rule of law," "equality before the law," "Madisonian balance of separated powers." In its living form, it also incorporates institutional means, such as voting rules and provision for interpretation through courts.

The institutional environment for markets is also in the first tier. The first tier includes normative ideas of property rights, contracts, and non-fraudulent information. The first tier includes the macro-economic prerequisites for micro-economic markets to flourish. These prerequisites include (reasonably) stable prices, (reasonably) full employment, (reasonably) even distribution of income. The first tier also incorporates the macro-economic policy instruments to implement these and other prerequisites.

The second tier includes "ordinary" legislatures, courts, markets and individual behavior, all shaped by the circumstances of the first tier. Within the favorable circumstances provided by the first tier, the liberal state can develop and be sustained. The idea is that the liberal state is a good but somewhat fragile thing, and will not survive and flourish under just any circumstances. It needs some nurturing conditions. The constitutional rules of the game affect the players. By providing basic opportunities, such as education and health, by treating people as having continuing identities, and by holding people accountable to their actions and contracts, people become more autonomous. Thus, the liberal theory goes, the proper conditions of the first tier encourage the individual to develop.

Hume's view of justice is a two-tier view. To Hume, justice is a good thing but it is not to be taken for granted. Only under certain circumstances does the question of justice arise at all, and only in certain circumstances can justice survive. The sustaining circumstances are, for Hume, *moderate scarcity*. With too little scarcity, there is no issue of justice, and with too much, justice cannot be maintained.

The circumstances are not important in themselves (intrinsically). We do not place value directly on *moderate scarcity*. Yet the circumstances of justice are primary to justice (which we may value directly) because the circumstances create the conditions for justice to exist. The first tier (circumstances of justice) creates the conditions for the second (justice).[10]

In defining the first tier we do not need, nor do we have, the predefined autonomous individual. The causality goes the other way. Once we have the first tier (if all goes well), the individual emerges in the second tier. Foundationally, what gets the first tier going is the concept of generalized interests, not particular preferences of particular indentified individuals. Thus this two-tier value theory does not depend foundationally on the predefined individual and is consistent with the challenge from the second lesson. As the individual develops in the second tier, under the circumstances of the first, the traditional

---

sustainability principle and interpret it as another condition of a viable (intergenerational) community.

[10] Rawls, Barry and Scanlon have developed two-tier views. Leopold, Norton, and Daly have also developed two-tier views for environmental management.

normative concepts (utility, rights, etc.) become available. As with the liberal state, the traditional concepts of moral philosophy are not automatically appropriate, but become appropriate under favorable conditions.

This two-tier value theory provides a framework for thinking about sustainability. As a circumstance for a viable liberal state, we add the sustainability principle to the first tier. Analogous to the Humean system, where moderate scarcity is a prerequisite for justice, preserving the resource base is a prerequisite for the individual to emerge (and survive) in the second tier. We do not define sustainability on the basis of preferences of particular individuals living today, instead we define sustainability on the basis of generalized interests, in particular a generalized interest in intergenerational equity. The approach avoids circularity because the referent of intergenerational equity is not a particular, identified individual, but resources. The key question is what is a fair division of resources across generational time. The easiest and perhaps purest concept of intergenerational equity is an equal division of resource capacity across generational time. This division is achieved by keeping the resource base intact. If the resource base is kept intact, it provides for a world of equals across generational time. To apply this idea we do not need to know whether or not a particular person will be alive 100 years from now. It is true that if there will be a population of 50 billion people 100 years from now it is unlikely that the resource base will be kept intact, but this prediction does not depend on the identity of particular individuals.

As a value theory, the two-tier view is consistent with the other lessons from biology. The fourth lesson says that the value system should accommodate long-term time horizons. Traditionally, in economic analysis the value of sustainability is based on discounting procedures, leading to short-term time horizons. But in the two-tier view sketched here, we do not use discounting to define and evaluate sustainability. We use concepts of sustainability to set constraints on markets. Once we have done this, we discount at normal rates.[11] The situation is similar to how we view property rights. We do not define property rights on the basis of today's distribution of market power and specific individuals' preferences. We use concepts of property rights to define and create markets. Within these (and other) constraints, market evaluations guide resource allocations. The constraints of the first tier (e.g. property rights and sustainability) are oriented toward the long term and indefinite future, and are thus compatible with the fourth lesson.

The third lesson says that market, political, and legal processes cannot be expected to automatically provide sustainability, and in fact may be far from doing so. If sustainability is to happen, it needs to be provided for through explicit policy choice and action. The two-tier value system allows for this view and provides a framework for the policy goal and its implementation.

And finally, the first lesson says that the value system should not depend critically on sharp species boundaries. If we think that other species have general interests (e.g., survival), then preservation of the resource base, in particular stabilization and gradual reduc-

---

[11] One response to the mismatch of time scales is recommend that lower discount rates be used in the evaluation of long term environmental harms. The two-tier approach differs by saying: keep the discount rate as it is but adjust the circumstances in which it operates.

tion of human population, works toward these general interests. If we think that other species do not have general interests, or we do not know how to value them—or if we consider only our own general interests—it still makes sense to preserve the resource base intact with stabilized population (less than 15 billion). In this way, sustainability, as a part of a two-tier value system, does not depend critically on sharp species boundaries.

IMPLEMENTATION

So far I have said that differing considerations (e.g., broad equity versus specific preference satisfaction) vary with the level of abstraction in the decision process, and I have used this idea to define and separate the two tiers. As a practical matter, the separation is not complete. Actual people wrote the Constitution and specific preferences were a part of the "three-fifths" rule. But four other ideas, besides the level of abstraction, help separate the two tiers, conceptually. If applied in the design of policy instruments, they help distinguish the two tiers on a practical level as well.

The first of the ideas of separation is a principle of *non-interference*. The idea is that instruments implementing the conditions of the first tier should be broad, avoiding minute control in the day-to-day workings of markets and the ordinary second tier institutions. For example, the progressive income tax, in its role as an instrument of income redistribution, can be viewed as a first tier instrument. The tax in its ideal simple form takes its bite from broad income classes and does not affect micro-decision making by individuals (except in its inescapable general disincentive to earn income). In contrast the actual income tax is "corrupted" by its complicated code with special privileges affecting day-to-day individual decision making in almost countless specific ways.

According to the non-interference principle the broad-based general tax is likely to be better than the present system of narrowly specified, special tax clauses. The practical rationale for having broader, more generalized instruments is to avoid the distortions of individual incentives associated with narrower clauses. But there is a further consequence. When the non-interference principle is followed, so that broader more generalized instruments are used in the first tier, the first tier becomes more separated from the second.

Another form of separation arises when *different decision procedures* are established for each tier. The Constitution is subject to amendment by actual people (not abstracted ones), but with an amendment process requiring more time and consensus than ordinary legislation, people are encouraged to take a longer view, setting aside narrow interests in favor of broader concerns of justice and the prerequisites of the liberal state. With a similar motivation, judges are sometimes given life terms of office. Different decision procedures further delineate the two tiers.

The two tiers are separated in another way as well. We can view the first tier as an *open system* that encloses the second tier. Viewed this way, it seems that, as a matter of principle, it is not possible to specify the instruments of the first tier exactly and in closed form. The inexactness of open systems can be illustrated by the problem of designing research and development (R&D) programs.

For a given budget of a firm, say 10% of the capital budget, resources to various research projects can be allocated on the basis of expected return. Someone may then ask if the 10% figure is the right one. Some research budget can be allocated to optimizing the research budget. But now, the question is: How do we decide the right amount to spend optimizing the research budget? Some resources can be allocated to settle this question. The problem is that we need information to prioritize research, we need additional types of information to set priorities for the priority process, and so on back.

We are in an open system. As far as I know, there is no exact answer to the problem of setting research budgets. Nonetheless, research budgets and allocations get made. They are made on rules of thumb. If the total research budget is 10% of the capital budget, the amount spent on prioritizing the research budget might be less than 1% of the total capital budget. A firm may use such a simple rule, see how things go, and readjust after ten years or so. The process is inexact in principle, but even inexact it is better to have some research budget than none.

Similarly, in the two-tier value system, the first tier is open and we should not expect to define instruments for sustainability and other first tier goals, in closed form, exactly.[12] In general terms we can talk about keeping the resource base intact but we may not be able to specify exactly the best category boundaries of "essential" resources, the degrees of substitution among them, the "optimal" level of population, and so on. These questions are analogous to finding the "best" research budget. As a matter of principle, we may have to make do with rules of thumb.[13]

Finally, it is important to observe that in designing instruments for the first tier, we are asking actual people in the present, with specific interests, to take actions benefiting general interests of the future. To some extent the specific interests of those in the present coincide with the general interests of the future (or people in the present may be farsighted) and there is no sacrifice from those who have power, the actual people in the present. But many in the present will see policy steps toward sustainability as sacrifices. To the extent that people in the present have a limited willingness to sacrifice their spe-

---

12 Daly compares the sustainability principle with the Plimsoll line on ships. This line, at the bow, marks the maximum safe loading and the number of feet between that line and the waterline. The line sets the decision environment for the loading of cargo, which can be done more exactly within this environment, once the environment is set.

13 Some recent work on the modeling may help in thinking about open and closed systems and compatibility of differing time scales. At the Santa Fe Institute's Economics Research Program models are being developed with positive feedbacks and hierarchies. In the two-tier system sketched in this paper, the second tier has potentially positive feedbacks (e.g., population). This tier under some circumstances may settle down, or it may go haywire. Even though the second tier is not fully predictable, there may be relative areas of stability and instability. The task of the first tier is to guide the second tier, increasing its probability of staying within safe bounds. To do this requires matching up the differing time scales. Considerations having to do with the biological and physical time scales help in the design of the first tier. Once within the environment of the first tier, behavior resulting from myopic economic and political institutions becomes consistent with the long term goals. The first tier is specified with the long-run processes in mind, the second tier takes care of itself. Thus the purpose of the sustainability instruments is to make the world safe for myopia.

cific known interests for the general, abstract, unknown interests of the future, it is sensible to select policy instruments that have high *leverage*. By this, I mean policy instruments which cost little to the present but whose effects accumulate over time to have large impact on sustainability in the future.

The above observations help guide the selection of sustainability instruments. To achieve the sustainability principle, three flows must be controlled: materials, energy, and people (population numbers). There are simple, "non-interfering" instruments to control the first two, for example energy taxes, severance taxes and other virgin materials taxes. These taxes are special forms of excise taxes and are simple to collect, with little administrative interference to the economy. At the same time, they can have large systematic effects on the whole economic system, affecting source flows, conservation, process design, recycling, and technology. The taxes can become a way of matching depletion with technological renewal, and their effects accumulate over time.

The taxes could be large. For example, an excise tax on gasoline of $1 a gallon would bring the price of gasoline closer to the level of the price in Europe. Over a period of years, there would be an effect on the design of cars and fuel efficiency. The revenues of this tax might be on the order of $85 billion. The revenue could be used to offset payroll taxes and part of the income tax. By offsetting regressive taxes, the net package (the gasoline tax and its offset) could be made progressive. The package would reduce "deadweight social losses" by decreasing taxes on things we want (work being discouraged by payroll and income taxes).[14] Similarly other virgin material taxes could become an important source of revenue, replacing much of the present tax system.

It is more difficult to identify instruments to control population with minimal interference and high leverage. C. K. Varshney reports that the most successful population programs in India are those which involve the education of women. Educational programs lower the birth rate in three ways: they postpone the time of family formation, they encourage women to enter the "formal sector economy," and they increase the family decision power of women, who tend to want smaller families than the men. Given that education is generally considered good social policy for reasons other than population control, educational programs which coincidentally lower birth rates are likely to be more acceptable than other population programs, many of which are more directly coercive.

A second alternative is also non-coercive at the level of the individual parent and works by increasing freedom, at least to the parents. The technology of sex selection is made available to prospective parents and they are allowed to choose their children's sex. According to Lucile Newman, in some countries this possibility of choice can have profound effect on both the birth rate and the sex ratio. In countries with strong preferences toward male children (or preferences toward having families with at least one male child), there will be smaller families and in the smaller families there will be a smaller propor-

---

14 Ordinary taxes, which have the primary purpose of raising revenue, have the undesired effect of altering the price structure. This effect leads to the "deadweight social losses" analyzed by Harberger. Pollution and other externality taxes—along with virgin material taxes established for the sustainability principle—are designed to alter (or "correct") the price structure and thus avoid Harberger's social welfare triangles.

tion of girls. While this policy alternative may be effective in lowering birth rates now and a generation from now, policy people are understandably concerned that it might have disruptive social effects and they are cautious about it. A third alternative would make social security payments (old age support) inversely proportional to the number of children one has. The above three possibilities are nontraditional and little used (of the three, the first seems preferable). Other options are the traditional policies of providing sex education, subsidizing and promoting contraceptives, and making abortion available.

The point here is that instruments for the implementation of the sustainability principle exist—for energy, materials, and people—and some of these instruments satisfy the criteria of noninterference and leverage. The instruments are flexible in that they could be implemented in strong or weak forms. At present, sustainability instruments tend to be weak. For example, for virgin material taxes to be effective they need to be broadbased throughout the world. Yet we have the reverse situation in the United States where there are still large subsidies on the extraction of virgin material taxes (functioning towards an "insustainability principle"), and many countries have strongly pro-natalist subsidies. Thus if people see the need for action and can overcome collective action problems, options exist to greatly strengthen implementation of the sustainability principle.

SUMMARY AND CONCLUSION

The first lesson from biology is the fading of species boundaries. This lesson has led environmental philosophers toward extensionism, in which traditional normative concepts are applied more broadly to nonhumans. But in light of the second lesson—the fading of the individual's boundaries—this response seems headed in the wrong direction. The second lesson suggests that we run extensionism the other way. Instead of taking the traditional normative concepts and applying them to the enlarged field of reference, we should take the more connected view of the biologists and apply it to the traditional normative concepts. In doing so, these concepts, being strongly individuated, become more problematic. The difficulty is not just with extensionism, but also with emerging rights views in moral philosophy and traditional economic welfare analysis.

In further response to the second lesson, we searched for normative concepts which were not dependent on predefined individuals. Intergenerational equity is one such concept, and I developed a version of this idea as the sustainability principle. I placed this principle in a two-tier value system in which the first tier provides the circumstances of the second. The circumstances are chosen to support the liberal state and the emergence of the autonomous individual. In the second tier, traditional normative concepts, depending on the individual, become available.

In the old days, we did not need a sustainability principle, since we neither had the technology nor the human numbers to rapidly deplete the resource base. Now we do.[15]

---

[15] See Barry (1977) for a discussion of how traditional theories of political philosophy have been unable to cope with the growth of power of the present generation to harm future generations.

And now because of this shift in power we need an explicit sustainability principle to circumscribe its scope. The task of establishing sustainability instruments is similar to that of setting up an educational system or a criminal justice system or other requisites of a liberal state. I doubt that these things can be done in principle exactly, but we can still think about them in a systematic way.

Acknowledgments

I wish to thank Bryan Norton, Bruce Hannon, Malte Faber, Doug McLean and Alan Strudler for helpful comments.

REFERENCES

Barnett, H. J. and C. Morse. 1963. *Scarcity and Growth: The Economics of Natural Resource Availability*. Baltimore: Johns Hopkins University Press.

Barry, B. 1977. Justice Between Generations. In Weaker and Raz, eds., *Law, Morality, and Society: Essays in Honour of H. L. A. Hart*. Oxford: Claudias Press.

Hart, H. L. A. 1979. Between Utility and Rights. *Columbia Law Review* 79:828-46.

Dawkins, R. 1978. *The Selfish Gene*. New York: Oxford University Press.

De Waal, F. 1989. *Chimpanzee Politics*. Baltimore: Johns Hopkins University Press.

De Waal, F. 1989. *Peacemaking Among the Primates*. Cambridge, Mass.: Harvard University Press.

Leopold, A. 1966. The Land Ethic. In *The Sand County Almanac*. New York: Ballantine.

Norton, B. Forthcoming. Aldo Leopold and the Search for an Integrated Theory of Environmental Management. In *The Unity of Environmentalism*. New York, Oxford University Press, ch. 3.

Page, T. 1977. *Conservation and Economic Efficiency*. Baltimore: Johns Hopkins University Press.

Singer, P. 1979. *Practical Ethics*. Cambridge: Cambridge University Press.

Stone, C. 1974. *Should Trees Have Standing? Toward Legal Rights for Natural Objects*. Los Altos, Calif.: William Kaufmann.

VanDeVeer, D., and C. Pierce, eds., 1986. *People, Penguins, and Plastic Trees*. Belmont, Calif.: Wadsworth.

6

DRIVING FORCES, INCREASING RETURNS AND ECOLOGICAL SUSTAINABILITY

Paul Christensen
Department of Economics
Hofstra University
Hempstead, NY 11550 USA

ABSTRACT

The neoclassical economic theory which dominates resource and environmental analysis and policy is based on atomistic and mechanistic assumptions about individuals, firms, resources, and technologies which are inappropriate to the complex and pervasive physical connectivity of both natural and economic systems. For example, this theory entirely neglects 1) any physical specification of the materials, energy, and information inputs (including the "machines" which transform energy, materials and information), 2) the physical connectivity (complementarity) between these inputs within production techniques, and 3) the sequential nature of production activities (geophysical and biological production systems and the stages of extraction, processing, and fabrication in economic systems). It also neglects 4) the implications of energetic and information processes for the non-equililbrium, self-reinforcing behavior of economic systems. Partly for ideological reasons and partly for reasons of mathematical tractability, neoclassical theory has confined itself within a world characterized by diminishing returns and negative or self-limiting feedback. Since the industrial revolution, economic activity has combined technological and information replication and innovation with the large-scale exploitation of environmental and "stock" resources (including fossil fuels). As Adam Smith, Alfred Marshall, Allyn Young, Gunnar Myrdal, and Nicholas Kaldor well knew, internal or sectoral economic relations are often characterized by increasing returns, positive feedback or self-reinforcing processes of disequilibrium growth.

The older classical tradition of economic theory, which was set out from a nascent materials and energetic foundation, developed an asymmetric analysis of diminishing returns in extraction and agricultural sectors and increasing returns in manufacturing. The latter was based on the assumption that machines and skills could be replicated over time. A reinterpretation of this framework from a biophysical and information perspective can provide a "new" interpretation of the operation of production and market processes in relation to resource availability (and environmental sustainability). Since resource prices in this approach are driven by macro demand and technology and move with the business cycle, the view that market prices provide an index of scarcity is considerably weakened. Increasing returns in the economic core reinforces the view that resource, economy, and environmental interactions must be managed to preserve ecosystem integrity and function. A physical approach to resource and environmental management is given support against the excessive reliance of neoclassical theory on extending the sphere of market interactions and private property rights.

ECOLOGICAL VERSUS NEOCLASSICAL ECONOMICS

Why do we need an "ecological economics" and not just more attention to resource values and environmental viability within standard theory? Neoclassical economics lacks any representation of the materials, energy sources, physical structures, and time-dependent processes that are basic to an ecological approach. Worse, it is inconsistent with the physical connectivity and positive-feedback dynamics of energy and information systems.

In contrast to the materials-based approach of classical theory (which developed from physiological sources), neoclassical economists reconstructed economic theory using the methods, concepts, and mathematical tools of nineteenth-century analytical mechanics. This led to a view of the economy as a system of exchange where atomistic agents optimally deploy a finite bundle of substitutable resources to meet a set of predetermined ends. Scarcity, generalized diminishing returns, and negative feedback ensure a conception of an economy which is self-regulating.

This is also an economy which is self-sustaining. In the field theory metaphor of mechanics, each point of Euclidean space is permeated by a field of active forces which provides a source of motion for otherwise inert matter. But what is the source of motion when this spatial model is transferred to economics? In neoclassical production theory, each input is assumed to be incrementally productive. Materials, energy resources, physical connectivity, and the structures and organization of real world production are ignored. Diminishing marginal productivity ensures in turn the equilibrium structure of the theory. Such theory is inappropriate for the material, energy, and information systems of evolved life forms and their support systems.

A biophysical perspective emphasizes the materials, energy resources, technologies, and information processes underlying economic activity. It emphasizes the self-reinforcing feedback mechanisms resulting from the exploitation of large stocks of materials and fuels, advancing technology, increasing returns in production, and the competitive advantage gained by innovators. The positive-feedback dynamics of such systems raises new challenges for resource and environmental theory and policy.

WHERE ECONOMICS WENT WRONG

Classical economics traces to the physiological analogies which Thomas Hobbes (1651) had taken over from Aristotle and Harvey (biological generation, nutrition, circulation, and sensitivity). Like living bodies, an economy depends on the materials and foodstuffs extracted from the land and sea and taken through various stages of processing and fabrication. In the classical model subsequently developed by Petty, Cantillon, Quesnay, Smith, Malthus, Ricardo and others, only land could produce food and materials. As Malthus (1815) put it, the "machinery" of the land could do something that no industrial machine could do, i.e., produce food and raw materials.

Manufacturing, it was recognized, was a materials-processing activity that obeys the law of conservation of mass (Diderot 1751; Malthus 1803, p. 672). Manufactured products require material inputs, labor, tools or machines, and the energetic materials required by the laborers and other engines. Thus, an increase in output in manufacturing requires a proportional increase in raw materials (Senior 1836, pp. 82-83). Individual capital goods lack productivity independent of other inputs (Mill 1848, pp. 63-64).

This physical perspective provided the background for the classical conception of capital. Adam Smith divided capital into fixed and circulating components (the former included land, buildings, machines, etc., and the latter the raw materials, intermediate materials, and foodstuffs in the production pipelines). Labor and machines required materials and food. Fixed and circulating capital were *complementary* (Smith 1937, bk. 2, p. 267).

Like the physiocrats, the classicals failed to link agricultural productivity to the materials and energy resources which were essential to a crop. This link would have made the dependence of economic production on geological and biological production explicit. All production whether agricultural or industrial depends on materials, energy and "machines" (and gets those flows of materials and energy from other sectors). In agriculture, these come from the soil, sun and metabolic cycles of the biosphere driven by the sun.

In the neoclassical revolution (post-1870), this physical approach to production, to capital goods, and to manufacturing as a materials-processing activity disappeared. The neoclassicals developed a new vision of the economy based on analytical mechanics. They rejected the classical cost-of-production theory of value and formulated value theory in terms of scarcity and a diminishing marginal utility (corresponding to the scarcity and diminishing marginal productivity of the land). The difficulty, of course, was what to do with the materials, food, tools, and machines which were produced and not given at the outset as an original endowment (and the support this gave to the cost-of-production approach). The result was a static equilibrium theory of prices based on diminishing returns and negative feedback.

By adopting the mechanical model, economists implicitly adopted the principle of energy conservation (between potential and kinetic energy) inherent in the laws of motion. Mirowski (1989) suggests that it was the idea of energy, transposed to utility, which provided the inspiration behind the neoclassical revolution. Neoclassicals simply substituted utility for energy in the equations of analytical mechanics. Treating utility like energy provided economics with a powerful metaphor for individual action, a rigorous set of mathematical techniques (the calculus of variations), a theory of economizing (in the principle of least effort), and a theory of optimality.

Qualms as to whether the early neoclassicals began their revolution by using the energy metaphor are certainly justified. Walras, in particular, was more influenced by a

Newtonian-type conception of an identity between economics and celestial mechanics. But it is highly plausible that Jevons may have used the metaphor; he was preoccupied with an economy of self-interest and maximum effect, was well acquainted with nineteenth century energy physics, and had written a famous book on the importance of coal in the British economy (Jevons 1864) plus articles on the influence of the sun's fluctuations on agriculture. Jevons also understood the dual meaning of Liebig's definition of civilization as an economy of power and frequently quoted him.

In any case, once the initial conceptions and formalism of analytical mechanics were adopted in economics, the logic of the development of the theory was inexorable. It was only a short step to the treatment of utility in terms of a field of forces where it is assumed that an individual knows his or her potential desires at every possible state of the economy and can move to any point in consumption space without changing the strength of desire at other points. Utility is thus independent of the path by which the state of the economy is reached. History, culture, and experience have no effect on individual desires (i.e., tastes are given and not affected by learning). Maximizing utility from a given budget, an individual achieves a state of equilibrium which demonstrates least-effort (optimal) properties.

The energy metaphor is full blown in Edgeworth:

. . . the particular hypothesis adopted in these pages (is) that Pleasure is the concomitant of Energy. *Energy* may be regarded as the central idea of Mathematical Psychics (economics); *maximum energy* the object of the principle investigations in that science. (1887, p. 9).

The development of the neoclassical theory of production was a much more protracted process. It was not until the mid-1890s that the full marginal productivity theory of production was developed (by Wicksteed). Materials, energy, and path-dependent technologies had to be first cleared away. Despite his awareness of the importance of coal and the sun in production, Jevons reduces capital to "nothing more" than the subsistence of workers (1871, p. 226). Menger (1871) assumes a law of "variable (input) proportions" universally applicable to all sectors. Walras assumes that an individual capital good yields a flow of output just like "a field grows a crop year after year" (1874, p. 213), and ignores the flows of materials and energy that are converted into a crop and the materials and the energy needed in industrial processes. He eliminates raw materials by vertically aggregating production into an instantaneous one-stage transformation of original factors into final output (pp. 237-240). Once capital goods are assumed to be like land in terms of yielding positive marginal products, it is merely one more step to assuming that the economy does not need land or resources since capital can substitute for resources (see Solow 1978).

In the field theory of production, output is obtained from varying combinations of inputs which are combined together like objects in a picnic basket (Arrow and Hahn 1971). Production techniques are defined not in terms of physical logic but by input proportions. A firm knows all possible input combinations it might use (depending on relative prices) and can move anywhere in factor space without affecting future productivity. Production alters neither factors nor processes. Technologies are path-independent and parametrically defined. A change in technology is treated as a change in the strength of the field and affects all input combinations (Mirowski 1989, ch. 5).

That resources conserve their productive potential no matter what input proportions are employed, is a consequence of the conservation of energy assumption that follows from the laws of motion. Any combination of inputs will yield output. Any input (capital) can substitute, albeit with diminishing returns, for another (land or a scarce material) and maintain the level of output. Energetic resources are not separately defined because all inputs (and any combination of inputs) are assumed to be inherently productive. To specify materials and energy would challenge the internal logic and structure of neoclassical theory. The energetic (and nutritive) requirements, physical connectivity, technological and social learning, emotional bonding, and positive-feedback dynamics of real world production systems will not fit the theory.

TOWARD A BIOPHYSICAL-ORGANIZATIONAL ECONOMICS

Production Inputs and Interdependence

As in classical economics, the starting point of a biophysical approach is production: finding and extracting materials and energy, upgrading materials, and producing the machines and services which reproduce the economy and its components through time (Cleveland et al. 1985). The inputs of an ecological economics are not land, labor, and capital or *n* unspecified heterogeneous inputs (as in neo-Walrasian theory) but flows of materials, energy, and information and the engines, machines, and workers organized to process materials, energy, and information.

Economies, like organisms and biological systems, run on matter and energy. Energy provides the "driving force" of production transformations. But it has to be linked to to a technological and organizational capability. Organisms and economies evolve physical structures (cells, chloroplasts, machines, etc.) for these transformations. Production depends upon the genetic and cultural "know-how" encoded in material and energy forms and upon material and energy use:

All production . . . whether of the chicken from the egg or the automobile from designs and blueprints, originates from a genetic factor which might be called "know-how." The ability of the genetic factor to realize its potential depends on a number of limiting factors of available energy in different forms, materials of different kinds, space, and time. (Boulding 1986, p. 8)

These inputs—flows of materials, energy, and information and the informative agents which act on flows—play physically different roles in production and need to be separately specified.

Biological and economic processes obey physical principles: the first and second laws of thermodynamics (conservation of energy and materials and the entropy law) and the principles governing individual material and energetic transformations (Ayers 1978, ch. 3). These principles underlie the physical connectivity which characterizes biological and economic systems and make ecology a science of relationship: "The virtue of the thermodynamic approach to evolution is its ability to connect life *ecologically* to the rest of nature through shared matter and energy flows" (Wicken 1983, p. 442).

Ecological connections shape production choices, technology, cost structures, and the timing and dynamics of economic interaction, including the operation and adjustment of markets. Thus, an ecological economics emphasizes not just the dependence of an economy on its environment but also the internal physical connectivity and interdependence within an economy.

Technological Change, Resources, and Growth

A biophysical approach gives a much greater emphasis than neoclassical theory to technology and to the physical and organizational logic underlying technological change. Neoclassical theory treats technology as preanalytical data, technical choice as a response to changes in factor prices, and technological change as a shift of the production function.

The inadequacy of the neoclassical theory of technical change has been persuasively argued by Dosi (1984, 1988). Technological innovation, he writes, involves "the solution of problems, for example, transforming heat into movement, shaping materials in certain ways, producing compounds with certain properties." It involves an ongoing mutual dependence between the development of fundamental science, technological research, and innovation. It is shaped by historical experience and policies, moves along technologically defined avenues (it is not a simple response to relative prices), is cumulative, and involves irreversibilities, nonconvexities, and dynamic increasing returns (Dosi 1988; Landau and Rosenberg 1986; Arthur 1990).

These features of technical change are in fundamental conflict with the core hypotheses of neoclassical theory. Dosi traces these problems to the idea of production possibility sets and the idea that technology is a malleable black box. But these difficulties must be located at a deeper theoretical level. They are a product of the field metaphor which assumes as Mirowski tellingly puts it:

> that the technology field is path-independent, so that the firm can go anywhere in the field (within the stated constraints) from any other point, and actually end up producing the anticipated output. Further, this has to be true for each and every technology, past, present, or future. (1989, p. 315).

Real world processes are not reversible. They dissipate available energy and lead to learning and the emergence of new arrangements and qualities. Dissipation forces learning (for survival) and permits it, since available energy must be present to permit experimentation and the successful consolidation of new methods. Technological changes are thus closely linked with material and energy use. They are the product of thermodynamic forces and the kinetic mechanisms that are employed to exploit available energy and materials (Wicken 1987, ch. 4). They are nondeterministic. They have histories, they possess emergent properties and they create (and foreclose) new possibilities.

The close connection between resource exploitation and technological change is strongly supported by historical evidence. The industrial revolution in Britain was based on the exploitation of large stocks of coal to replace increasingly scarce supplies of charcoal. The large-scale utilization of mineral fuels in iron smelting and fabrication and other heat processes sparked a technological revolution which connected in "ever-widening concentric circles" innovations in steam power, metallurgy (primarily iron), machine tools, precision engineering, transportation, textiles and other manufacturing production (Rosenberg 1979; Musson 1981; Pollard 1981).

Similar clusterings of innovations made possible by the exploitation of low-cost and abundant sources of energy occurred in the development of mechanized production in the United States in the mid-nineteenth century, electrification and electromechanical production at the end of the century, the internal combustion engine in the early twentieth century, and oil-based synthetic chemicals more recently. New processes created new products, used new resources, and created new forms of pollution.

Technology has been at the heart of economic growth in the West. Historical studies (Smith 1981) and current research (Borrus et al. 1988; Landau 1990) indicate the importance of leadership and organization (including government support of research and development) in innovation. Technology creates and requires intrafirm, interfirm, and interindustry linkages. It is characterized by flows of information and personnel between industry, research labs, and universities and between applied and basic science. This research carries an important message about the treatment of technology and technological change in resource and environmental economics. Prices alone have not been the major instrument for stimulating and managing technologies and innovation; they have been managed by organizational not market coordination. In resource and environmental policy where the public goods characteristics of technology are pervasive, it certainly cannot be presumed that markets and prices are the only or the best instrument for managing technological change.

Production Returns, Costs, and Economy Dynamics
Diminishing Returns

The theory of production returns is the bridge between the "physiology" of production processes and a theory of how the economic system evolves through time. Adam Smith's (1776) vision of a progressive extension of the division of labor appeared to imply a world of increasing returns, certainly, at least, for industry. James Anderson (1801) extended this argument to agriculture, arguing for increasing—not diminishing—returns in intensive cultivation (Hollander 1903). This was clearly at odds with Malthus' population principle and he soon formulated a law of diminishing returns (Malthus 1803, 1815).

The law of diminishing returns, which the classicals applied in agriculture but not manufacturing, has been one of the most fundamental propositions of economic theory. The neoclassicals extended it along with an assumption of general scarcity to all sectors, resulting in upward-sloping marginal cost and supply curves in all markets. Thus, a shock to a market sets in motion processes of adjustment via price changes which restore market equilibrium (negative feedback). Increasing returns, as Arthur notes, would destroy this predictable world of equilibrium and "the notion that the market's choice is always best" (1990, p. 94). Scarce resources and diminishing returns are crucial to neoclassical theoretical and policy conclusions.

A biophysical approach offers a physical foundation for the theory of diminishing returns. In the traditional formulation of diminishing returns, adding labor to a plot of land, all other inputs constant, yields positive additions to output although the size of the additional output gets progressively smaller (returns diminish). Neoclassical theorists have made no attempt to explain why returns are initially positive. In the biophysical view, the farmer is also adding more corn plants (photosynthetic machines of a given capacity), utilizing more of an excess stream of sunlight, obtaining more groundwater, more nutrients from the soil, and CO_2 from the atmosphere. Positive returns to labor (and tools) on a fixed plot of land, assuming no change in basic technology, is possible because of the on-site availability of extra material and energy resources which can be channelled into a crop. Farming a plot more intensively draws more of these inputs into the production flow. Returns are positive in agriculture because of the presence of "hidden" resources whose use is increased as the flow of output is increased. Returns diminish because one or more of those resources, including space to intercept solar flow, have limited on-site availability. This is Liebig's law of the "limiting factor."

Increasing Returns

Following Babbage (1835) who had given particular attention to inanimate power and power-based machines in industry, the post-Ricardian classical economists such as Senior (1836) and Mill (1848) characterized the manufacturing sectors of the economy in terms of increasing returns. For example, doubling labor and capital would more than double output. Marshall (1920) also applied increasing returns to manufacturing although he recognized that changing the scale of production would change the proportions of machinery to skilled and unskilled labor. We would add that it would also involve changes in energy use and the qualities of materials required (see Gold 1982).

Static micro theory has, via Marshall, incorporated a limited version of increasing returns into its assumption of U-shaped cost curves for both the short and long run. These curves are first downward and then upward sloping. Empirical studies of industry give little support to this construction. It has been adopted to preserve the equilibrium assumptions of the theory. As J. R. Hicks wrote, "Unless we can suppose . . . that marginal costs generally increase with output . . . the basis on which economic laws can be constructed is shorn away (Hicks 1939).

The importance of increasing returns (and positive feedback) continued to be emphasized by a few economists, most notably Young (1928) and Kaldor (1972, 1979). It has recently resurfaced in international trade theory (Helpman and Krugman 1985), in industrial organization, and in economic history (Chandler 1977). It has also been taken up by mathematical economists who are exploring the parallels between nonlinear positive-feedback processes in economics and in physics (see Arthur 1989, 1990).

Kaldor (1972) recalled Young's (1928) emphasis on the cumulative and reciprocal interaction between economies of scale, decreasing unit costs, and the extension of the market. He later explored (Kaldor 1976) a model integrating increasing returns and cost-plus prices in industrial sectors with demand-determined commodity prices. Given different conditions of reproduction in these sectors, high rates of growth in industry will eventually lead to temporary supply constraints and sharp price increases in primary sectors. Cost-push inflation and the demand constraints of tight budgets produced the stagflation of the 1970s. The prevalence and importance of self-reinforcing mechanisms for both short-run macro events and long-term patterns of growth and evolution open a rich and largely unexplored field of inquiry. For evidence of the historical importance of technology and economies of scale, see Chandler (1977).

Arthur (1990) argues the importance of nonlinear linkages and self-reinforcing mechanisms in knowledge-based industries. Firms, regions, or nations that gain high volume and experience in high-technology manufacturing reap the advantages of lower costs and higher quality. The experience gained in one sector can be transferred to related industries and products. Japan used an initial investment in precision instruments to mass-produce consumer electronics and then the integrated circuits that went into these products. New generations of chips are revolutionizing machine and computer design and in turn are being used to manufacture the chips.

As Arthur points out, policies appropriate to a world of diminishing returns are inappropriate to a world of increasing returns and self-reinforcing feedback. Current debates about resource and environmental policy are conducted in terms of an underlying equilibrium model characterized by inappropriate adjustment mechanisms and viability assumptions.

The Maximum Power Economy

A crucial condition of an industrial economy operating through time is its ability to obtain flows of low entropy energy and materials. As Alfred Lotka (1925) noted, any organism that discovers how to take advantage of unused energy running over a dam gains a selective advantage over other organisms. There are, he observes, two tendencies in resource use. One is to use resources more efficiently. The second is to exploit unused resources. The way that modern economies have overcome renewable resource limits by exploiting large stocks of fossil fuels and high-quality materials is an example of the second.

Conditions of limited resources and primitive technology would be expected to foster cultural practices which establish and maintain equilibrium populations. These "primitive" techno-cultures appear to have evolved a considerable array of behaviors for minimizing environmental stress (Rappaport 1984; Perrings 1985). The shift from a renewable resource base to exploitation of large but exhaustible stocks of fossil fuel has permitted long-term economic growth. Industrial technologies use materials and energy to produce machines, know-how, and so on. Increasing returns accelerate this positive-feedback loop.

An economy with increasing returns which uses maximum power generation puts increasing stress on the resources and absorptive capacity of local ecosystems and on the biosphere as a whole. The basic question is how to control and manage such an economic system and reconcile it with the sustainability criteria of natural ecosystems that have evolved under conditions of resource limits and that are vulnerable to potentially irreversible degradation.

As Hall et al. point out, Lotka's principle implies "that people who wish to survive, and prosper in a competitive environment (such as our business world) must be interested in not only the efficiency but also the rate at which something is done" (1986, p. 65).

An economy that operates at high rates of return extends that pressure to low productivity sectors. If that pressure is experienced via higher demand, an increase in resource prices may lead to increased exploitation of resources. Resource sustainability may be jeopardized, resulting in environmental degradation and the reorganization of the system at a lower level of productivity. Also, the influence of temporarily high resource prices on conservation will not be sustained if high resource prices lead to a macroeconomic contraction.

ECONOMIC MANAGEMENT AND ECOSYSTEM SUSTAINABILITY

To the extent that an economy works according to the logic of increasing returns, it will need to be managed. If this is true for macroeconomics and industrial and trade policy, it is particularly compelling for environmental resources and systems. The demise of a firm or industry may be costly for a region and its population but the know-how is probably preserved elsewhere and is reproducible, if at some cost. The loss of ecosystems, particularly to the extent they include irreplaceable physical and genetic structures, is much less acceptable. Sustainability is, then, a basic goal of ecology, ethics, and economics.

But what does sustainability of renewable resources mean? I assume it means maintaining intact the productivity, diversity, and biological organization of a given system at something like current levels. But given that a system will be exploited for its sustainable yield, how will productivity be maintained if inputs are not recycled? Maintenance of diversity poses even more thorny problems. Ecological systems maintain their productivity by importing energy to recycle nutrients. Harvesting a sustainable yield from a system when nutrients are not returned is ecologically impossible. In sustainable agriculture, productivity levels are maintained (at considerable loss of diversity) by crop rotation, green manures, etc., but there is still some loss of elemental nutrients to crops and to soil erosion from mechanical cultivation. Since nutrients are being removed from the resource system, nutrients have to be put back. They have to to come from somewhere. This suggests that even sustainable resource systems will need a subsidy from nonrenewable stock resources. Does this mean moving natural systems towards agricultural practice? What implications will this have for maintaining diversity and organization?

Thedifficulties with the sustainable yield concept have emerged most clearly in fisheries ecology. According to Larkin maximum sustainable yield exploitation (MSY) results in a large number of expected and unexpected complications, such as: spawning populations becoming dominated by first time spawners, possible reduction of egg quality, increase in catastrophic risk to populations through a failure in egg or larval survival, and the removal of less productive components of natural populations. "It may be necessary," he writes, "to compromise MSY in order to preserve genetic variability" (1977, p. 4). Harvesting desirable species increases numbers of less desirable species, as is too evident in fisheries history. The fisheries case points to the complexity of species interaction in ecological systems. "It does not seem likely," Larkin adds, "that an MSY based on the analysis of the historic statistics of a fishery is really attainable on a sustained basis."

Similar degradation appears in natural forests and rangeland where crops are taken without nutrient restoration. Opening up a system that has been previously closed creates a net loss which will eventually affect productivity. Renewable resource systems cannot be maintained without replenishment. This implies that maintaining productivity with harvesting will require some reverse subsidies from exhaustible resources and that we must be even more cautious about maintaining productivity of "renewable" resources than is at first apparent in the term.

So sustainability is not theoretically sustainable...

CONCLUSION

According to neoclassical theory, a market economy is an atomistic isolated entity which is self-regulating and self-sustaining. The production side of this argument rests on a conception of resources and technology which presumes the productivity of individual productive factors. The inherent productivity which was presumed for land (Quesnay and Ricardo) was generalized to capital goods (Walras and Wicksteed). This view reflected the conceptual and mathematical framework of nineteenth century analytical mechanics which presumes the presence of source (a gravitational field of forces, etc.) which can be taken as given. Capital in this framework can literally substitute for natural resources since all inputs are assumed to be inherently productive. This takes no account of: thermodynamic sources and their depletion; the interdependence of materials, energy, and environmental support structures; the limits of environmental systems; or the contributions and limits of social systems. The conceptual and mathematical framework of analytical mechanics are inappropriate to the tasks of economic theory and policy.

A biophysical organizational approach to ecological economics starts from a recognition of the environmental, technological, individual and social sources and support systems of productivity. Economies are constructed historically in relation to environmental, technological, and social possibilities. In contrast to the diminishing returns assumptions of neoclassical theory, a biophysical approach recognizes the operation of increasing returns in materials-energy transformations and knowledge-based processes. An increasing returns economy puts enormous pressure on environmental resources and systems which are subject to limits, depletion, and destruction. Production and consumption choices are not path independent. We must make hard choices about the technologies and institutions we choose to employ. Obviously, market-based signals and policies will be vital in spurring those choices. But technological and social policies are also crucial and these cannot be meaningfully evaluated from an atomistic and mechanistic framework based on nineteenth century physics. An increasing returns, positive-feedback economy must be managed, regulated, and coordinated. Preservation of sustainability is a primary condition of economic and social development. Economic development must proceed in a way that preserves ecological viability. This requires the development of an ecological, economic, and social framework for analysis and policy, an ecological economics.

Acknowledgments

I would like to thank Carl Carlossi, Thomas von Foerster, and Bronislau Czarnocha for discussions about ecology and physics, and Jerome Ravetz and Leon Braat for helpful criticisms of the paper. They are not responsible for the uses to which I have put their help.

REFERENCES

Anderson, J. 1801. A Calm Investigation of the Circumstances That Have Led to the Present Scarcity of Grain in Britain. London: J. Cumming.
Arrow, K. and F. Hahn. 1971. *General Competitive Analysis*. San Francisco: Holden Day.
Arthur, W. B. 1989. Competing Technologies, Increasing Returns, and Lock-in by Historical Events. *Economic Journal* 99(394):116-131.
Arthur, W. B. 1990. Positive Feedbacks in the Economy. *Scientific American* 263:92-99.
Ayres, R. U. 1978. *Resources, Environment, and Economics: Applications of the Materials/Energy Balance Principle*. New York: Wiley Interscience.
Babbage, C. 1835. *On the Economy of Machinery and Manufactures*. London: C. Knight, 1932.
Borrus, M., L. Tyson, and J. Zysman. 1988. Creating Advantage: How Government Policies Shape International Trade in the Semiconductor Industry. In P. Krugman, ed., *Strategic Trade Policy and the New International Economics*, pp. 91-113. Cambridge, Mass.: Massachusetts Institute of Technology Press, 91-113.
Boulding, K. 1986. What Went Wrong With Economics? *American Economist* 30(1): 5-12.
Christensen, P. 1981. Land Abundance and Cheap Horsepower in the Mechanization of the Antebellum United States Economy. *Explorations in Economic History* 18:309-329.
Chandler, A. D. 1977. *The Visible Hand: The Managerial Revolution in American Business*. Cambridge, Mass.: Harvard University Press.
Cleveland, C., R. Costanza, C. A. S. Hall, R. Kaufman. 1985. Energy and the U.S. Economy: A Biophysical Perspective. *Science* 225:890-897.
Diderot, D., 1751. Art. In: D. Diderot (ed), Encyclopédie, 1 Paris, Briasson. Reprinted: H. Diekmann (ed.), *Oeuvres Complètes de Diderot*, Paris, Hermann, 1975, 5:495-509.
Dosi, G. 1984. *Technical Change and Industrial Transformation*. London: Macmillan.
Dosi, G. 1988. Sources, Procedures, and Microeconomic Effects of Innovation. *Journal of Economic Literature* 26:1120-1171.
Edgeworth, F. Y. 1887. *Mathematical Psychics*. London: Kegan Paul.
Gold, B. 1982. Changing Perspectives on Size, Scale, and Returns. *Journal of Economic Issues*. 20:5-33.
Hall, C. A. S., C. J. Cleveland and R. Kaufman. 1986. *Energy and Resource Quality: The Ecology of the Economic Process*. New York: Wiley-Interscience.
Helpman, E. and P. Krugman. 1985. *Market Structure and Foreign Trade*. Cambridge, Mass.: Massachusetts Institute of Technology Press.
Hicks, J. 1939. *Value and Capital*. Oxford: Clarendon Press.

Hobbes, T. 1651. *Leviathan*. C. B. Macpherson, ed., New York: Penguin, 1968.

Hollander, J. H. 1903. Introduction to T. R. Malthus, *The Nature and Progress of Rent*. Baltimore: Johns Hopkins Press.

Jevons, W. S. 1864. *The Coal Question; an Enquiry Concerning the Progress of the Nation, and the Probable Exhaustion of our Coal Mines*. London: Macmillan.

Jevons, W. S. 1871. *The Theory of Political Economy*. London: Macmillan. Reprinted by Penguin Books, 1970.

Kaldor, N. 1972. The Irrelevance of Equilibrium Economics. *Economic Journal* 82.

Kaldor, N. 1976. Inflation and Recession in the World Economy. *Economic Journal* 86.

Landau, R. and N. Rosenberg, eds., 1986. *The Positive Sum Strategy*. Washington, D.C.: National Academy Press.

Landau, R. 1990. Capital Investment: Key to Competitiveness and Growth. *Brookings Review* 8(3):52-56.

Larkin, P. A. 1977. An Epitaph for the Concept of Maximum Sustained Yield. *Transactions of the American Fisheries Society* 106:1-11.

Lotka, A. 1925. *Elements of Physical Biology*. Baltimore: Williams and Wilkins. Reprinted as *Elements of Mathematical Biology*. New York: Dover.

Malthus, T. R. 1803. *An Essay on the Principle of Population*. 2d. ed. London: J. Johnson.

Malthus, T. R. 1815. An Inquiry into the Nature and Progress of Rent. Reprinted in *The Pamphlets of Thomas Robert Malthus*. New York: Kelley, 1970.

Marshall, A. 1920. *Principles of Economics*. 8th Ed. London: Macmillan.

Menger, C. 1871. *Grundsatze du Volkswirtschafts-Lehre, Principles of Economics*. Glencoe, Ill.: Free Press. Trans. 1950.

Mill, J. S. 1848. *Principles of Political Economy*. 7th ed. London: Longmans, Green, 1871.

Mirowski, P. 1989. *More Heat than Light: Economics as Social Physics*. Cambridge: Cambridge University Press.

Musson, A. E. 1981. British Origins. In O. Mayr and R. C. Post, eds., *Yankee Enterprise: The Rise of the American System of Manufactures*. Washington, D.C.: Smithsonian Institution Press.

Perrings, C. 1985. The Natural Economy Revisited. *Economic Development and Cultural Change* 33: 829-850.

Pollard, S. 1981. *Peaceful Conquest: the Industrialization of Europe*. Oxford: Oxford University Press.

Rappaport, R. A. 1984. *Pigs for the Ancestors: Ritual in the Ecology of a New Guinea People*. 2d. ed. New Haven: Yale University Press.

Rosenberg, N. 1979. Technological Interdependence. *Technology and Culture* 50:25-50. Reprinted *Inside the Black Box: Technology and Economics*. New York: Cambridge University Press, 1982.

Senior, N. 1836. *An Outline of the Science of Political Economy*. Reprinted 1965 New York: Augustus M. Kelley.

Smith, A. 1776. *An Inquiry into the Nature and Causes of the Wealth of Nations*. New York: Modern Library. 1937.

Smith, M. R. 1981. Military Entrepreneurship. In O. Mayr and R.C. Post, eds., *Yankee Enterprise: The Rise of the American System of Manufactures*. Washington, D.C.: Smithsonian Institution Press.

Solow, R. 1978. Resources and Economic Growth. *American Economist* 2: 5-11.

Walras, L. 1874. *Elements of Pure Economics*. Trans.: W. Jaffe trans., 1954. London: George Allen and Unwin.

Wicken, J.S., 1987. *Evolution, Thermodynamics, and Information: Extending the Darwinian Program*. New York and Oxford, Oxford University Press.

Wicken, J. S. 1988. Thermodynamics, Evolution, and Emergence: Ingredients for a New Synthesis. In B. Weber, D. Depew, and J. D. Smith, eds., *New Perspectives in Physical and Biological Evolution*.

Wicksteed, P. H. 1894. *An Essay on the Co-ordination of the Laws of Distribution*. London: Macmillan.

Young, A. 1928. Increasing Returns and Economic Progress. *Economic Journal* 38:527-549.

7

SUSTAINABILITY AND DISCOUNTING THE FUTURE

Richard B. Norgaard and Richard B. Howarth
Energy and Resources Program
Room 100 Building T-4
University of California at Berkeley
Berkeley, California 94720 USA

ABSTRACT

Though high interest rates discourage the long term management of slow growing resources (forests) and the protection of long term environmental assets (biodiversity), high interest rates also discourage investment in projects which transform environments (dams) and in projects which are necessary to extract resources (oil wells). Thus, the relationship between interest rates and conservation (protecting the interests of future generations), and sustainable development is ambiguous.

We show that the conservationist's dilemma results from a misspecification of the problem. Economists heretofore have not distinguished between decisions concerning the efficient use of this generation's resources and decisions concerning the reassignment of resource rights to future generations. All decisions over time have been simply treated by economists as investment questions, as if all resources were always this generation's resources. We properly specify the economic questions involved, clearly distinguishing between equity and efficiency, and show why discounting is appropriate with respect to the efficient use of this generation's resources but is inappropriate when this generation is primarily concerned with redistributing resource rights to future generations. Further, we show that when rights are reassigned between generations, interest rates themselves change. We conclude that the assignment of rights to the future is the instrument of conservation and sustainability; interest rates are derivative.

INTRODUCTION

Conservationists have long been concerned that economic theory rationalizes discounting the future. A dollar next year is worth only 91 cents today if one can put 91 cents in the bank and have it earn 10% interest. From the perspective of a utility-maximizing individual or profit-maximizing firm, discounting the future makes sense. Discounting also seems highly descriptive of how individuals and firms behave. Whether or not discounting should take place, conservationists and now the broader academic and lay community concerned with sustainable development accept this conceptual argument for one of the same reasons economists do—because it has considerable explanatory power. However, since a positive discount rate means that effectively no weight is given to resource use or welfare beyond a generation hence, discounting appears to be inconsistent with sustainability. Thus, many people are questioning the prescriptive stance of the economics profession that not only individuals and firms should discount the future but also that discounting is appropriate for public decision making. The resolution of the apparent contradictions between sustainability and discounting will ameliorate the conflicts between people with economic and ecological world views and further the evolution of an ecological economics.

The discount rate is defined as the return on foregone present consumption that is sacrificed to secure future consumption. In an ideal economy with no market failures, the discount rate is equal to the rate of interest or return on capital investment. Economists have generally approached choices of use of resources and the environment over time as investment decisions and argued that net present value criteria should dominate policy decisions. Some have defined "optimal" resource allocation strictly in terms of Pareto efficiency,[1] side-stepping questions of distributional equity altogether (Fisher 1981).

Nevertheless, economists have not overlooked the implications of the rate of discount on future generations, resource use, and environmental management. When confronted with the possibility that the prevailing rate of interest may be "too high" to ensure the welfare of future generations, they have debated the merits of "violating" efficiency criteria by picking an alternative discount rate for the sake of intergenerational equity (Batie 1989). Tension has arisen in the economic literature on resource management with respect to wildlands (Krutilla and Fisher 1975), energy policy (Arrow and Lind 1970), and biological diversity (Fisher 1982). Economists in the World Bank are now struggling with the paradox of discounting as they try to formulate evaluation criteria to promote sustainable development (Markandya and Pearce 1988). In this paper, we argue that the controversy concerning the appropriate rate of discount for public decisions arises, in part, from economists having forgotten how distributional equity and allocational efficiency interrelate.

In an earlier paper (Howarth and Norgaard 1990a), we developed a simple model of resource allocation in a pure exchange economy that shows how the efficient intertemporal path of resource use depends on the distribution of resource rights between generations. In this paper, we extend that model to show how the rate of interest, rather than being an appropriate instrument to affect intergenerational equity, derives from the distribution of economic assets between generations. In the next part of this paper, we review the dis-

---

[1] Pareto efficiency is the economic state in which goods are allocated such that no individual can be made better off without making others worse off. The resources have been exchanged such that users are getting the most out of the resources.

count rate controversy. Then we document how the controversy is due to an over-emphasis on resource use as efficient investment rather than as intergenerational distribution. We show that the rate of interest itself is a function of distribution of wealth between generations. We discuss policy implications of our analysis and in the last section, address the broader significance of our analysis to the evolution of an ecological economics.

THE CONSERVATIONIST'S DILEMMA

Faced with the pervasive use of discounting, conservationists have typically preferred lower discount rates over higher rates because lower rates favor the management of slow growing trees, the protection of biodiversity, and the conservation of exhaustible resources. Yet conservationists have also argued for the use of high discount rates when higher rates make projects with deleterious environmental consequences appear uneconomic. Conservationists in the United States, for example, have joined with the economics profession to argue that water development projects should be evaluated using discount rates based on market rates of interest rather than the lower interest rate on tax-exempt government bonds (Reisner 1986). Conservationists want to be able to select the discount rate strategically, but both high and low rates can favor conservation interests. This is the conservationist's dilemma.

No doubt there exists an economist who has never experienced the slightest moral qualm over discounting the benefits to be received or the costs to be borne by future generations. The literature, however, reflects considerable frustration over discounting per se and the conservationist's dilemma in particular. To be sure, most of the literature concentrates on the determination of the "correct" rate of discount. This literature stresses how market rates might best be adjusted to account for the distortions of taxes on savings and investment (Krutilla and Eckstein 1958, ch. 4; Hirshleifer, De Haven and Milliman 1960, ch. 6; Baumol 1968; Stern 1977), public good qualities of transfers to future generations (Marglin 1963), differences between private and social rates of time preference (Marglin 1963), differences between private and public risk (Arrow and Lind 1970; Lind 1982), distortions in factor prices due to trade restrictions (Little and Mirrlees 1974), and the influence of capital mobility and lags in macroeconomic adjustments (Lind 1990). Though the bulk of the literature struggles primarily with the correctness of the discount rate rather than the correctness of discounting, most authors acknowledge the question of future generations and rationalize how the determination and use of the correct rate of discount may benefit not only ourselves but our descendants as well.

Environmental, land, and resource economists have perhaps had the greatest difficulty rationalizing discounting the future. These overlapping subdisciplines evolved out of the work of scholars who addressed the conservation worldviews of Gifford Pinchot and John Muir on their own terms (Gray 1915; Ely and Wehrwein 1940; Ciriacy-Wantrup 1952; Scott 1955). The relationship between conservation and the interest rate was obvious to Gray (1913, p. 515) who argued: "The primary problem of conservation expressed in economic language is the determination of the proper rate of discount on the future with respect to the utilization of our natural resources." These subdisciplines evolved around the examination of intrinsic failures of markets rather than their potential for success (Castle 1965; Kelso 1977). Historically, this emphasis on market failure was an exploration of the limits of market reasoning (Norgaard 1985), a stance that was more conducive to identifying why the social rate of discount should diverge from the private rate or for questioning discounting altogether than for rationalizing discounting the future. Perhaps more than others, these subdisciplines continue to evolve around a rich mixture of classical, institutionalist and positivist philosophies (Castle et al. 1981; Nelson 1987; Maxwell and Randall 1989). Thus, scholars from these subdisciplines still pursue classical questions with respect to the role of the state that stem from their understanding of land and resource markets (Schmid 1987; Bromley 1989) long after the other subdisciplines of economics were reduced to positivism, mathematics and weak empirical confirmations of market theory (McCloskey 1985).

From this rich evolutionary base, however, neoclassical environmental economists eventually advanced arguments that helped rationalize discounting (Krutilla and Fisher 1975). Early environmental, institutional and land economists explored what they saw as largely inherent contradictions in world views of economists and conservationists. The new environmental economists referenced the historic concerns and then pushed development of economic thinking in directions which made these concerns less significant. We argue that these advances in economic thought which once brought some reconciliation have become obstacles to understanding conservation and sustainable development.

First, the new environmental economists, responding to the concerns of the 1960s and early 1970s with the loss of environmental amenities, argued that conservationists should join economists in promoting the inclusion of environmental costs in project analysis. The irreversibility of most environmental transformations and the value of retaining future options provides an argument for putting an extra value on environmental factors. Since environmental impacts often extend indefinitely into the future, low discount rates will give them the greatest influence on the benefit-cost ratio. This means there is no reason why conservationists concerned with the negative consequences of development should favor high rates over low. While the internalization of externalities had certainly been discussed before, the new environmental economists began to make substantial progress in environmental valuation for the purposes of internalization (Lin 1976).

Second, neoclassical economists developed the idea that technological advance favors availability of new resources over time and production of material goods more than it favors provision of environmental services and amenities (Smith 1972). This means that resource and commodity prices will decline over time relative to environmental services and amenities. Findings by Barnett and Morse (1963) seemed to confirm that resource prices were falling. This argument is simply a sophisticated way of appealing to the conventional wisdom that progress makes future generations better off except with respect to environmental amenities. With commodity and environmental price trends explicitly incorporated in project analyses, environmental factors accrue more weight, especially at low discount rates. Though price trends since the 1960s have not supported technological optimism (Slade 1982; Hall and Hall 1984), the argument still attracted adherents (Porter 1982). The theoretical basis of this argument is under serious challenge (Norgaard 1990).

Lastly, neoclassical environmental economists acknowledge with the rest of the discipline that the present generation must value future generations, for people regularly leave estates for their children and donate to causes that protect the access of future generations to resources and the environment. Thus the interests of future generations are considered to some extent, perhaps to a considerable extent, in the decisions of present generations. This argument is usually coupled with the observation that since each generation has lived better than the one before, this mechanism must be adequate.

Each of these arguments is an extension of neoclassical economics or is consistent with findings developed from that theory, each favors conservation relative to prior conventions of project analysis, and, most importantly, none challenges the existing arguments for discounting or the methods for determining the discount rate. This combination of arguments gave considerable voice to conservation objectives, and the interesting conceptual and empirical elaboration entailed in environmental and option valuation kept environmental economists busy. Thus attention was diverted from the relationship between discounting and intergenerational equity for two decades.

Nevertheless, the paradox between discounting and intergenerational equity has crept onto the environmental economic agenda again in a more virulent form. The question of our obligation to future generations was picked up and nurtured in the 1980s by philosophers and theologians (Partridge 1981), environmental ethicists and scientists (Norton 1986), and legal scholars (Stone 1987). A new journal, *Environmental Ethics*, dealt with the question regularly. But more importantly, the broad acceptance of the sustainability of development as a priority in our thinking indicates the demise of two centuries of faith in the idea that progress will take care of posterity (WCED 1987; Norgaard 1988).

One of the earliest forays of the new Environment Department of the World Bank was to commission an analysis of "Environmental Considerations and the Choice of the Discount Rate in Developing Countries" (Markandya and Pearce 1988). This study concluded that the choice of the discount rate was extremely important, and even more so when environmental factors are considered. On the other hand, the study simply ended with a restatement of the conservationist's dilemma. In light of the difficulties of addressing the issues of sustainability through the choice of the discount rate, Markandya and Pearce conclude that separate criteria should be established for sustainability and that economic and sustainability criteria need to be used together in project analysis. This dual approach seems to be an emerging consensus both within academic economics (Tietenberg 1988; Batie 1989; Pearce and Turner 1990) and development agencies (Pezzey 1989; Daly 1990). We argue, however that this uneasy compromise between economic and environmental methodologies and criteria deserves deeper theoretical examination.

THE MISFRAMING OF THE FUTURE

While we are staunch advocates of methodological pluralism (Norgaard 1985; Norgaard and Dixon 1986; Norgaard 1989), we also contend that each analytical technique should be used as straight-forwardly and effectively as possible. In this respect, neoclassical environmental economic thought has led the profession astray. In our earlier article (Howarth and Norgaard 1990a), we documented that, of the many articles describing resource allocation over time under alternative assumptions with respect to market structure, resource availability, technology, and information using the Hotelling model, none systematically examined the role of the assignment of resource rights and other forms of wealth across generations on efficient allocation. While the profession occasionally notes that a "resource allocation can be intertemporally efficient and yet be perfectly ghastly" (Dasgupta and Heal 1979, p. 257), economists have typically failed to mention that less ghastly distributions of resources across generations can be efficiently allocated.

The situation is similar with respect to discounting. The economics profession has been concerned with both the welfare of future generations and the efficient allocation of resources. The possibility that public issues with respect to the use of resources over time are issues of the distribution of resource rights or other forms of wealth across generations has been contemplated but not pursued. Rather, economists on approaching the question of intergenerational distribution have typically reverted to arguments with respect to the choice of discount rate. Stiglitz (1974, p. 139), for example, argues:

> There is, of course, no presumption that the intertemporal distribution of income which emerges from the market solution will be "socially optimal" (although in the absence of market failure, the market allocation will be Pareto optimal), just as there is no presumption that the distribution of income among individuals at any moment is "socially optimal." But this is a problem which is not peculiar to the allocation of natural resources over time; indeed, if there were no other sources of "market failure" and if the government correctly controls the rate of interest (our underlining), then there would be no objection to the competitive determination of the rate of utilization of our natural resources.

Robert Solow (1974, p. 10) also raised the issue of intergenerational distribution and then retreated to an argument with respect to the discount rate: "The intergenerational distribution of income or welfare depends on the provision that each generation makes for its successors. The choice of social·discount rate is, in effect, a policy decision about that intergenerational distribution." Krutilla and Fisher (1975, p. 76) note the possibility that questions of future generations might be handled as distributional issues, but do not pursue this possibility: "the present generation might prefer for reasons of equity to engage in a program of selective investments for the future, emphasizing the transmission of those assets capable of yielding amenity services that will be in relatively short supply."

Stiglitz argues once again (1979, p. 61) that "the appropriate instruments to use for obtaining more equitable distribution of welfare (if one believes the present distribution is not equitable) are general instruments, for example, monetary policy directed at changing the market rate of interest." The profession's emphasis on the relationship between discounting and future generations took a new twist in a recent reanalysis of discount rates and public investment by Robert Lind (1990, p. S-24): "if we are to avoid the type of paradox that in some cases can lead to total neglect of the interests of generations in the distant future, we need to look to new welfare foundations for *our theory of discounting*" (our emphasis).

In short, decisions with respect to the future have been treated as investment decisions yielding returns to this generation, the possibility of transfers is acknowledged but not pursued, and equity questions are treated as matters of the appropriate rate of discount. At the same time, economists have warned against distorting discount rates to try to benefit future generations on the grounds that this would cause inefficient distortions in capital markets that would hurt future generations and would certainly be an unnecessary sacrifice for the current generation (Markandya and Pearce 1988). While it is relatively easy to understand why economists at the World Bank might find it difficult to switch from an investment mode of analysis, it is difficult to understand why academic economists have put so much emphasis on discounting.

About twenty years ago the international agencies began to acknowledge that "trickle down" provided too few benefits for the poor. Development agencies assumed the responsibility of trying to help countries redistribute wealth towards the poor and gave priority to projects which met "Basic Needs." Just as earlier project evaluation methods only benefitted poor people through "trickle down" at best, current project evaluation methods only benefit future generations through "trickle ahead" at best. We need to develop ways of thinking about "Future Needs" systematically.

During the past decade the international agencies have put greater emphasis on how getting prices right is critical to development. Historically, developing countries tried to address the problems of the poor through keeping the prices of food low, the major item in the family budgets of the poor. But the poor were also farmers, and low food prices hurt them. Trying to help future generations through policies to lower discount rates is analogous to trying to help the poor through low food price policies.

INTERGENERATIONAL TRANSFERS AND THE INTEREST RATE

The distinction between efficiency and equity across generations can be demonstrated with a simple overlapping generations model in which transfers between generations are explicitly considered. We demonstrate two important points through such a model. First, intergenerational transfers should not be evaluated by efficiency criteria. Second, intergenerational transfers affect the rate of interest and thereby efficiency criteria. The first point, the distinction between decisions to transfer wealth to future generations from decisions to use this generation's resources most efficiently, resolves both the conservationist's dilemma and the long standing contradictions that have appeared in the literature with respect to equity and the discount rate.

Consider an economy with an arbitrarily large finite number of overlapping generations $g = 1, 2, ..., G$, each of which lives for two periods. The first generation is born in period 1 and a new generation is born in each successive period so that the gth generation lives in periods g and $g+1$. For simplicity, assume that each generation consists of homogeneous individuals who can be represented as a single agent. There is a single consumption good C and the consumption levels of the gth generation in periods g and $g+1$ are C_{gg} and C_{gg+1}. Each generation has a utility function $U_g = U_g(C_{gg}, C_{gg+1})$ which is concave, monotonically increasing, and differentiable on the nonnegative orthant..

A single representative firm produces a homogeneous output in periods $t = 1$, 2, ..., $G+1$ using labor (L_t), capital (K_t), and a nonreproducible natural resource (R_t) according to a constant returns to scale production function $f_t = f_t(L_t, K_t, R_t)$ that is monotonically increasing, differentiable, and concave on the nonnegative orthant. Assume further that production is zero when the levels of all inputs are zero so that $f_t(0,0,0) = 0$. Note that the time subscript (t) allows for exogenous technological improvement through changes in the parameters or functional form of f_t over time.

Output is distributed between consumption and net capital investment $(K_{t+1} - K_t)$. Capital may be freely converted into consumption so that there may be net consumption of the capital stock. Each generation is endowed with a single unit of labor in each period that it supplies inelastically to the firm. The initial stocks of resources (S_1) and capital (K_1) are owned by the first generation and take on strictly positive values. Each successive generation receives an income transfer T_g from its predecessor during the first period of its existence. While, in reality, intergenerational transfers are effected by both private individuals and by public agencies, here we assume that the transfers are selected and enforced by the government and are thus taken as exogenous to individual decision making.

Intergenerational competitive equilibrium is achieved via the temporary equilibria established through trading between the generations alive in each period subject to their expectations concerning future prices and economic conditions. The conditions that describe the competitive equilibria that arise under alternative income transfer regimes may thus be derived by evaluating the maximization problems faced by each agent. Consider first the profit maximization problem confronting the firm. For the sake of simplicity, assume that the firm is myopic and that there are no futures markets so that trading is limited to goods that are made available during the period in which they are purchased. Defining Q_t as the firm's profit in period t, w_t as the wage rate, r_t as the interest rate or price of capital services, p_t as the price of the resource, and taking the firms output as numeraire, the firm's problem is to:

$$\text{Max } P_t = f_t(L_t, K_t, R_t) - w_t L_t - r_t K_t - p_t R_t \tag{1}$$

subject to $L_t, K_t, R_t \geq 0$. Since the firm behaves competitively, this problem generates the first order conditions:

$$w_t = \frac{\partial f_t}{\partial L_t} \tag{2}$$

$$r_t = \frac{\partial f_t}{\partial K_t} \tag{3}$$

$$p_t = \frac{\partial f_t}{\partial R_t} \tag{4}$$

that are necessary and sufficient for the attainment of an interior solution to the profit maximization problem under the restrictions on f_t. The assumption that the production function exhibits constant returns to scale implies that profits are zero in each period.

Now consider the utility maximization problem confronting the gth generation in period g. It must choose its period g consumption and net investments in capital and resources based on the prices it observes in period g and the prices it expects for period g + 1 so as to maximize its expected intertemporal utility $U_g(C_{gg}, C_{gg+1})$. For simplicity, we assume that each generation has perfect foresight so that its expectations regarding future prices are borne out in reality.

The budget constraints may be derived by noting that, since no two generations overlap for more than one period, there are no opportunities for loans, and income must equal expenditure in each period. In period 1, generation 1 makes a payment of $C_{11} + K_2 - K_1$ for consumption and net capital investment while its income from sales of labor, capital services, and resources is $w_1 + r_1 K_1 + p_1(S_1 - S_2)$. Note that:

$$S_t = S_1 - \sum_{\tau=1}^{t-1} R_\tau$$

(5)

is the resource stock remaining at the beginning of period t. In period 2, its expected expenditure is C_{12} while its expected income is $w_2 + (1 + r_2)K_2 + p_2 S_2 - T_2$. This holds because, at the end of its lifetime, each generation consumes the remainder of its non-transferred capital stock; therefore, periods 1 and 2 budget constraints for generation 1 reduce to:

$$w_1 + (1 + r_1)K_1 + p_1(S_1 - S_2) - C_{11} - K_2 = 0$$

(6)

$$w_2 + (1 + r_2)K_2 + p_2 S_2 - T_2 - C_{12} = 0$$

(7)

The budget constraints for generation $g > 1$ are somewhat different in form. Generation g must purchase its stocks of capital and resources for use in period $g+1$ from the preceding generation and/or the firm, so its period g expenditure on consumption, capital, and resources is $C_{gg} + K_{g+1} + p_g S_{g+1}$ while its income from labor sales and the transfer it receives from its predecessor is $w_g + T_g$. In period g + 1, its expected expenditure is C_{gg+1} while its expected income is $w_{g+1} + (1 + r_{g+1})K_{g+1} + p_{g+1}S_{g+1} - T_{g+1}$. Hence its period g and period g +1 budget constraints are:

$$w_g - p_g S_{g+1} + T_g - C_{gg} - K_{g+1} = 0$$

(8)

$$w_{g+1} + (1 + r_{g+1})K_{g+1} + p_{g+1}S_{g+1} - T_{g+1} - C_{gg+1} = 0$$

(9)

The problem confronting generation g is to maximize $U_g(C_{gg}, C_{gg+1})$ subject to the budget constraints and the nonnegativity constraints $C_{gg}, C_{gg+1}, K_{g+1}, S_{g+1} \geq 0$. In the case of an interior solution, this yields the first order conditions:

$$\frac{\partial U_g/\partial C_{gg}}{\partial U_g/\partial C_{gg+1}} = \frac{P_{g+1}}{P_g} = 1 + r_{g+1}$$

$$(10)$$

for $g = 1, 2, ..., G$ that are both necessary and sufficient given the assumptions imposed on the utility functions.

A competitive equilibrium will exist for this model if we can find a set of prices and quantities that simultaneously satisfies the conditions of utility and profit maximization. Howarth (1989) has shown the existence and Pareto efficiency of equilibria provided that the set of income transfers is technically feasible. The possibility of corner solutions implies that the equilibrium conditions derived above are not completely general, although this technicality need not concern us here.

These conditions yield some interesting if familiar interpretations. Along an equilibrium path, the marginal rate of time preference or discount rate of each successive generation must equal the interest rate or return on capital, and the discount rate is always greater than zero provided that the marginal productivity of capital is positive. Moreover, the resource price must rise at the rate of interest over time, confirming the Hotelling (1931) rule. The competitive equilibrium and hence the discount rate, however, depend on income distribution across generations. In a numerical example, (Howarth and Norgaard 1990b), we show how prices, including the discount rate, vary from period to period under different intergenerational distributions of income. Thus intergenerational distribution determines not only the efficient allocation of natural resources but also the discount rate and all other prices and quantities relevant to applying conventional benefit-cost analysis.

SPECIFIC CONCLUSIONS

First, the conservationist's dilemma is effectively resolved by returning to the basic framework of economics. Each distribution of resources or income between people, in our case generations, defines an efficient allocation of resources between end uses and users. Conservation or sustainability cannot be addressed simply through efficiency. While efficiency is important, intergenerational distribution is also important. Both concerns must be addressed, and when they are, the conservationist's dilemma is resolved.

Second, prices, including the rate of interest or discount, equilibrate resource allocations at the margin. With different distributions and efficient allocations, new prices arise. One can no more speak of "the" rate of interest when societies are giving major consideration to the sustainability of development than one can speak of "the" price of timber when deciding whether to conserve forests. Redistributions change equilibrium prices. The rate of interest is undoubtedly distorted by market failures just as is the price of timber, and adjustments are thereby in order. But it is inefficient to adjust either the rate of interest or the price of timber in order to achieve distributional goals.

This second conclusion points to the pitfalls of evaluation techniques when they are used without due consideration to the distribution of welfare across generations. Since all of the variables that go into the calculation of net present value of a project or policy proposal depend on the intergenerational distribution of income, what might appear to be a good social investment under one income distribution may be marginal or worse under another. One practical difficulty is that, while the sustainability of economic systems can only be addressed through the integration of policies in virtually every area, resource and environmental policies must generally be reviewed and assessed one by one using partial equilibrium analysis. In this sense, the idea of sustainable development presents a broad challenge to piece-wise analysis and action.

Third, in our model and perhaps in reality, there is nothing intrinsic about economies that ensures that living standards will continue to improve over time or even remain at current levels. The future will unfold from the choices, including sacrifices, made by our ancestors and those we make ourselves. The ongoing discussion within the profession of economics and international development agencies as to whether sustainable development and intergenerational equity can be addressed through ad hoc manipulations of the discount rate are rooted in an inappropriate theoretical framing of the choices before us. Questions which are fundamentally matters of equity should be treated as such. If we are concerned about the distribution of welfare across generations, then we should transfer wealth, not engage in inefficient investments. Transfer mechanisms might include setting aside natural resources and protecting environments, educating the young, and developing technologies for the sustainable management of renewable resources. Some of these might be viewed as worthwhile investments on the part of this generation, but if their intent is to function as transfers, then they should not be evaluated as investments. The benefits from transfers, in short, should not be discounted.

Limiting how we evaluate choices with respect to the future to merely investment criteria could be a tragic mistake. Such a mistake may well be perpetrated through current economic practice, but it is not intrinsic to economic theory.

BROADER INTERPRETATIONS

This reconstruction of how economics frames decisions with respect to the future has several important broader interpretations which are critical to the evolution of an ecological economics.

We have framed the issue of sustainability as an equity decision that needs to be resolved politically rather than imposed through sustainability constraints on the set of options people have before them. This framing is compatible with the economic model and how most economists think. While it may not be an ideal fit for those holding to an environmental world view, it might better bridge the gap between the two secular religions of the West than an imposition of environmental limits. Economists understand technical tradeoffs and choices and bristle when confronted with constraints. The relationships between economic and environmental systems must be understood to make effective public choices about how income should be distributed across generations. By putting these environmental considerations into the arena of public choice, our framing assumes politically functional societies. We would rather further the development of functioning political societies than presume this possibility away through the derivation and imposition of environmental constraints imposed directly at the bureaucratic level.

Both economics and ecology build on multiple, simple logical patterns of thought which are intrinsically useful and intrinsically limited (Norgaard 1989). The elaboration of these arguments in practice, however, has been influenced by social organization and the nature of particular problems. This leads to the second broader interpretation. In the case of economics, the distinction between efficiency and equity is clear in economic theory. Economists are well aware that their theory assumes a distribution of income and that they have almost no basis for deciding between alternative distributions of income. Though every student learns at the outset that the efficient allocation of resources in static general equilibrium models depends on the initial assignment of property rights, this principle is quickly forgotten as students learn ever more sophisticated ways of thinking about efficiency alone. In practice, furthermore, politicians have instructed agencies to make decisions with equity implications on economic grounds. And economists have accepted the distributive tasks which politicians have dodged. The subdiscipline of "welfare economics," the history of which parallels the incorporation of economic arguments and economists into public decision making, has been a search for arguments rooted in efficiency reasoning which justify equity choices. Thus economic thinking has been coevolving with our political irresponsibility, the rise of bureaucratic decision making, and the increasingly complex world into which our technologies and this form of social organization have led.

Ecological economists need to carefully distinguish between basic theories and how both the elaboration and the use of those theories have coevolved in association with politics, bureaucratic organization, and social and environmental problems. It is easy to question the current state of economics. Differentiating between what economics is and what it could be, and thereby constructively redirecting the profession and social organization, requires a solid understanding of basic theory as well as knowledge of the factors affecting the historic evolution of thought. Though ecological thinking has not coevolved so tightly with social organization, we expect similarly sophisticated distinctions are important. Redirecting development will entail redirecting the coevolution of theory, social organization, and practice. A better comprehension of political, bureaucratic, and science histories is needed as a part of a new understanding of ecological histories (Martinez-Alier, 1990).

REFERENCES

Arrow, K. J. and A. C. Fisher. 1974. Environmental Preservation, Uncertainty, and Irreversibility. *Quarterly Journal of Economics* 55:313-19.

Arrow, K. and R. C. Lind. 1970. Uncertainty and the Evaluation of Public Investment. *American Economic Review* 60:364-78.

Barnett, H. J. and C. Morse. 1963. *Scarcity and Growth: The Economics of Natural Resource Availability*. Baltimore: Johns Hopkins University Press.

Batie, S. S. 1989. Sustainable Development: Challenges to the Profession of Agricultural Economics. *American Journal of Agricultural Economics* 71:1083-1101.

Baumol, W. J. 1968. On the Social Rate of Discount. *American Economic Review* 58:788-802.

Bromley, D. W. 1989. *Economic Interests and Institutions: The Conceptual Foundations of Public Policy*. New York: Basil Blackwell.

Castle, E. N. 1965. The Market Mechanism, Externalities, and Land Economics. *Journal of Farm Economics* 47:542-556.

Castle, E. N., M. M. Kelso, J. B. Stevens, and H. H. Stoevner. 1981. Natural Resource Economics: 1946-75. In Lee R. Martin, ed. *A Survey of Agricultural Economic Literature*, vol 3. Minneapolis: University of Minnesota Press.

Ciriacy-Wantrup, S. von. 1952. *Resource Conservation: Economics and Policies*. Agricultural Experiment Station. Berkeley: University of California Press.

Daly, H. E. 1990. Personal conversations during March and April. Dr. Daly was referring to the direction conversations are taking in the Environment Division of the World Bank, not to World Bank policy.

Dasgupta, P. S. and G. M. Heal. 1979. *Economic Theory and Exhaustible Resources*. Cambridge: Cambridge University Press.

Ely, R. T. and G. S. Wehrwein. 1940. *Land Economics*. Reprinted 1964. Madison: University of Wisconsin Press.

Fisher, A. C. 1981. *Resource and Environmental Economics*. New York: Cambridge University Press.

Fisher, A. C. 1982. "Economic Analysis and the Extinction of Species." Paper presented at AAAS Symposium on Estimating the Value of Endangered Plants. Washington, D.C., January 3-8.

Gray, L. C. 1913. Economic Possibilities of Conservation. *Quarterly Journal of Economics* 27:497-519.

Hall, D. C. and J. V. Hall 1984. Concepts and Measures of Natural Resource Scarcity with a Summary of Recent Trends. *Journal of Environmental Economics and Management* 11:363-379.

Hirshleifer, J., J. C. De Haven, and J. W. Milliman. 1960. *Water Supply: Economics, Technology, and Policy*. Chicago: University of Chicago Press.

Hotelling, H. 1931. The Economics of Exhaustible Resources. *Journal of Political Economy* 39:137-75.

Howarth, R. B. 1989. "Intertemporal Equilibria and Exhaustible Resources: An Overlapping Generations Approach." Paper delivered at the Annual Meeting of the American Economics Association joint session with the Association of Environmental and Resource Economists. Atlanta, December.

Howarth, R. B. 1990. Economic Theory, Natural Resources, and Intergenerational Equity. Ph.D. dissertation in Energy and Resources. University of California.

Howarth, R. B. and R. B. Norgaard. 1990a. Intergenerational Resource Rights, Efficiency, and Social Optimality. *Land Economics* 66:1-11.

Howarth, R. B. and R. B. Norgaard. 1990b. Intergenerational Transfers and the Social Discount Rate. Under review.

Kelso, M. M. 1977. Natural Resource Economics: The Upsetting Discipline. *American Journal of Agricultural Economics* 59:814-823.

Krutilla, J. V. and O. Eckstein. 1958. *Multiple Purpose River Development: Studies in Applied Economic Analysis*. Baltimore: Johns Hopkins University Press.

Krutilla, J. V. and A. C. Fisher. 1975. *The Economics of Natural Environments: Studies in the Valuation and of Commodity and Amenity Resources*. 2d. ed. 1985. Baltimore: Johns Hopkins University Press.

Lin, S. A. Y., 1976. *The Theory and Measurement of Economic Externalities*. New York: Academic Press.

Lind, R. C. 1982. *Discounting for Time and Risk in Energy Policy*. Baltimore: Johns Hopkins University Press.

Lind, R. C. 1990. Reassessing the Government's Discount Rate Policy in Light of New Theory and Data in a World Economy with a High Degree of Capital Mobility. *Journal of Environmental Economics and Management* 18(2, part 2):S-8-28).

Little, I. M. D. and J. A. Mirrlees 1974. *Project Appraisal and Planning for Developing Countries*. London: Heinmann.

Marglin, S. A. 1963. The Social Rate of Discount and the Optimal Rate of Investment. *Quarterly Journal of Economics* 77:95-112.

Markandya, A. and D. Pearce. 1988. Environmental Considerations and the Choice of the Discount Rate in Developing Countries. Environment Department Working Paper no. 3. Washington, D.C: The World Bank. 79pp.

Martinez-Alier, J. 1990. "Ecological Perception, Environmental Policy, and Distributional Conflicts: Some Lessons from History." Paper presented at the European Association of Environmental Economics, Venice, April 17-20, 1990, and International Society for Ecological Economics, Washington, D.C., May 21-23, 1990.

Maxwell, J. and A. Randall. 1989. Ecological Modeling in a Pluralistic Society. *Ecological Economics* 1(3):233-250.

McCloskey, D. N. 1985. *The Rhetoric of Economics*. Madison: University of Wisconsin.

Nelson, R. H. 1987. The Economics Profession and Public Policy. *Journal of Economic Literature* 25 (1):49-91.

Norgaard, R. B. 1985. Environmental Economics: An Evolutionary Critique and a Plea for Pluralism. *Journal of Environmental Economics and Management* 12:382-394.

Norgaard, R. B. 1988. Sustainable Development: A Co-evolutionary View. *Futures* (December):606-620.

Norgaard, R. B. 1989. The Case for Methodological Pluralism. *Ecological Economics* 1:37-57.

Norgaard, R. B. 1990. Economic Indicators of Resource Scarcity: A Critical Essay. *Journal of Environmental Economics and Management* 19:19-25.

Norgaard, R. B. and J. A. Dixon 1986. Pluralistic Project Design: An Argument for Combining Economic and Coevolutionary Methodologies. *Policy Sciences* 19:297-317.

Norton, B. G., Ed. 1986. *The Preservation of Species: The Value of Biological Diversity*. Princeton: Princeton University Press.

Partridge, E., Ed. 1981. *Responsibilities to Future Generations: Environmental Ethics*. Buffalo, N.Y.: Prometheus Books.

Pearce, D. W. and R. K. Turner. 1990. *Economics of Natural Resources and the Environment*. Baltimore: Johns Hopkins University Press.

Porter, R. C. 1982. The New Approach to Wilderness Preservation through Benefit-Cost Analysis. *Journal of Environmental Economics and Management* 9:59-80.

Reisner, M. P. 1986. *Cadillac Desert: The American West and Its Disappearing Water*. New York: Viking.

Schmid, A. A. 1987. *Property, Power, and Public Choice: An Inquiry into Law and Economics*. New York: Praeger.

Scott, A. 1955. *Natural Resources: The Economics of Conservation*. Toronto: University of Toronto Press.

Slade, M. E. 1982. Trends in Natural Resource Commodity Prices: An Analysis of the Time Domain. *Journal of Environmental Economics and Management* 9:122-137.

Smith, V. K. 1972. The Effect of Technological Change on Different Uses of Environmental Resources. In J. V. Krutilla, Ed., *Natural Environments: Studies in Theoretical and Applied Analysis*. Baltimore: Johns Hopkins University Press.

Solow, R. M. 1974. The Economics of Resources or the Resources of Economics. *American Economic Review* 64:1-21.

Stern, N. 1977. The Marginal Valuation of Income. In M. J. Artis and A. R. Nobay, eds., *Studies in Modern Economic Analysis*. Oxford, Blackwell, 244pp.

Stiglitz, J. E. 1974. Growth with Exhaustible Natural Resources: The Competitive Economy. *Review of Economic Studies* (Symposium Issue): 123-138.

Stiglitz, J. E. 1979. A Neoclassical Analysis of the Economics of Natural Resources. In V. K. Smith, Ed., *Scarcity and Growth Reconsidered*. Baltimore: Johns Hopkins University Press.

Stone, C. D. 1987. *Earth and Other Ethics: The Case for Moral Pluralism*. New York: Harper & Row.

Tietenberg, T. 1988. *Environmental and Natural Resource Economics*. 2d ed. Glenview, Ill.: Scott, Foresman.

World Commission on Environment and Development. 1987. *Our Common Future*. Oxford: Oxford University Press.

8

ECOLOGICAL HEALTH AND SUSTAINABLE RESOURCE MANAGEMENT

Bryan G. Norton
Georgia Institute of Technology
School of Social Sciences
Atlanta, GA 30332 USA

ABSTRACT

A contextual approach to environmental management requires a distinction between Resource Management, the management of a resource-producing cell such as a field or fishery, and Environmental Management, which involves concern for the larger systems which environ those cells. Mainstream natural resource economics defines sustainability mainly by reference to undiminished outputs of economically marketable products and emphasizes productivity criteria in judging management regimens. This approach is appropriate in many cases for guiding Resource Management but, taken alone, it provides no guidance regarding the protection of the larger, environmental context of resource-producing activities.

Aldo Leopold, and most environmentalists following him, have applied a contextual approach in which resource-producing cells are understood as subsystems of larger, slower-changing (but still dynamic) ecological systems. According to this approach, Resource Management should be limited when resource-producing activities approach a threshold beyond which they alter larger systems and instigate rapid change in environing systems. Metaphorically, this result is referred to as "illness" in the ecological community, but the rules and criteria for describing these limits have never been stated precisely in ecological terms. A definition of "ecosystem health," based on biologically formulated criteria for judging larger ecological systems, will be proposed and integrated into a hierarchical approach to environmental management.

INTRODUCTION

Conservation biologists and environmental managers say, with depressing frequency today, "This system is degraded and requires restoration." Every time they say this, they imply that they know what that system *should* look like, and that they have some idea about how to improve on the present situation. Do we know these things?

Throughout its history the environmental movement has been mainly reactive, usually defining its positions in negative terms, opposing dams on free-flowing rivers, urban sprawl, fighting wasteful use of renewable but fragile resources, and generally trying with varied success to stall some of the more distasteful effects of human industrial progress. If environmentalists wish to seize the initiative, to go beyond opposition, and propose a positive plan whereby human communities can develop in harmony with their landscapes, they must propose, in clear terms, what standards they use when they say that a system is being "restored." This cluster of terms, "ecosystem health," "restoration," and "illness," or "degradation" are all normative terms. To define and explain them properly would go a long way toward clarifying the political objectives of environmentalism.

Both the practice of environmental management and the politics of environmental protection, therefore, demand a clear, concise, and understandable standard of ecosystem health. It would be nice, also, if the definition were in terms quantifiable and measurable; but here I will limit my remarks to the explaining in general terms what I think environmental managers mean, or at least *should* mean, when they say or imply that they know how to define a healthy ecological system.

HEALTH AND MEDICINE ANALOGIES

Environmentalists and environmental managers, following Aldo Leopold, have often availed themselves of analogies from medicine to explain, at least metaphorically, a general standard that we should manage so as to protect the health and integrity of ecological systems. Let us inquire how far such analogies carry us toward an adequate and useful definition of ecosystem health, and toward adequate standards for environmental management.

The health analogy and its associated organicist metaphors, are unquestionably illuminating, in several ways. First, the health analogy suggests that the object of management, analogous to the human patient, is a dynamic, changing system rather than a static, unchanging machine. Environmentalists are opposed to economic reductionism because they doubt that any reductionistic system can adequately model management problems. The land ethic is an affirmation that land must be seen systematically, not atomistically as Pinchot did (Norton, forthcoming).

Second, these analogies and metaphors emphasize that good management is inherently relational—it is understood as relating activities in parts of the "body" to the larger whole of which those parts are functioning elements. This lesson is the one Leopold learned when he removed predators from game ranges. The mountain and its vegetative cover can be made ill by atomistic management which manipulates populations without care for the impact of these manipulations on the larger, ecological system. Deer and wolf management must, he concluded, be limited by concern for the larger system, the "mountain," (Leopold 1949, pp. 129-133; Norton, forthcoming, ch. 3). This relational model, which I call contextualism, mainly focuses on the scale of economic activities and their impacts on larger ecological systems. This emphasis on scale is often expressed informally as an endorsement of organicism; nature resembles an organism, with organs organized and coordinated to function as a larger whole. More formally, it is expressed as an endorsement of an essentially systematic approach to management. To the extent that the health analogy promotes this more systematic thinking, it is surely helpful.

Moreover, the health analogy and its associated organic metaphor lead us to another, equally fruitful analogy: because ecological systems are more like organisms than they are like machines, conservation biologists and environmental managers are more like physicians than they are like mechanics. And this new analogy spins out, in turn, a series of useful insights regarding the practice of environmental management. Environmental management, like medicine, must be frankly and unapologetically normative (Norton 1988). It acts to achieve a goal, recognized as a worthy one, to protect the organic systems of nature from "illness."

I think the analogy is useful, also, in understanding the role of environmental managers and conservation biologists in the larger society. On the one hand, like physicians, they act to protect and enhance the well-being of their "patients," the living systems which are the object of their therapeutic practices. In this sense, they react to "complaints" of specific patients—as when they apply their skills to reduce erosion on hillsides or to mitigate effects of fish kills in streams shocked by industrial wastes. But on the other hand, like physicians, managers also have a broader public responsibility. Because medical practice affects larger social goals and because the exact nature of human health is not a sharply defined concept, physicians have an obligation to participate in a public debate regarding the meaning of "health" and the goals of medical practice and health policy.

Similarly, conservation biologists and environmental managers have a professional obligation to participate in a public debate regarding the meaning to be attached by our society to ecosystem health and regarding the goals of conservation practice. Like physicians, that obligation is to educate the public, not to enforce from on high definitions and standards which the public does not understand and cannot relate to. Rather, that obligation is to participate as professionals with technical expertise in a broader, democratically understood search for common goals and acceptable standards of conservation practice. One can call this the "educative function" of normative sciences within democratic societies, provided this is not understood to imply that ecologists know the answers and need only communicate them to a passive public. The goal must be to promote education in the richer sense of helping the public to think more clearly about the social goals of environmental management.

Carrying this analogy of environmental managers with physicians further, it is possible to borrow concepts from the theory of medicine, in order to propose and illuminate important choices in the practice of environmental management. For example, the discussion of "holistic" medicine as an alternative or supplement to the usual reactive practices of treating specific illnesses as they arise may prove a useful formulation of important differences in management approaches. Just as a physician would not treat an organ, such as a kidney, without attention to the effects of that treatment on the larger system, intelligent and contextually oriented management will not manage productive subsystems without attention to the impacts of those manipulations on the larger ecological system in which they are embedded.

Another analogy, one that seems to have guided Leopold's thinking in a number of situations, is an application of the idea of preventive medicine to management[1]. Just as physicians counsel their patients to take precautions against illness, especially those for which a cure can be difficult, painful, or even impossible, environmental managers can often counsel that the best treatment may in many cases be to avoid degradation because restoration will be too costly or ineffective.

So far, then, we have seen that medical analogies and organic metaphors can guide us to a variety of useful insights about standards for ecological management. They encapsulate certain hard-learned principles, such as the recognition of the dynamism of natural systems and the need to emphasize relationships among parts of the system and the relationship of the functioning parts to the larger whole. Further, they encourage useful rules of thumb in the area of management practice. In the absence of a "unified field theory" for ecology, which still lacks general and mathematically representable laws to predict events in naturally organized systems, these analogies and metaphors provide useful, if informal, guidance in the form of a unifying image to guide contextual management.

It is tempting, then, to press forward with the analogies, and to examine substantive norms of human health as a guide to providing a definition of ecosystem health. But here, I think, we have reached the limits of the usefulness of the medical analogy. In order to see the crucial disanalogy, focus for a moment on the analogy between holistic medicine and holistic environmental management. While in vague terms this analogy may seem comfortable to environmentalists who favor more holistic management approaches, there is an important difference: whereas the medical profession's tradition has been anchored, from the beginning, by a firm commitment to treating the health of the individual patient as the object of overriding value,[2] environmental managers have not agreed regarding the proper object of management. Thus, while the object of holistic medicine is unquestionable—the object is the *person-patient*, and all decisions will be made from the perspective of that patient and his/her health—the object of holistic therapies in environmental health remains highly controversial and confused. To recall the organicist analogy, environmental managers have not agreed what to consider an organ and what to consider an organism.

What, exactly, *is* the larger system that contextualists such as Leopold attempted to keep healthy? Metaphorically, Leopold had no doubts. It was the mountain—a large and autonomous system, held in a sort of dynamic equilibrium of processes which, more or less mysteriously to us, hold back change and allow species to carve out a stable niche, even as the system continuously changes around them. He was saying essentially that his choice of the deer-wolf-hunter complex as the management system was shortsighted. Since the vegetation holding the soil onto the mountain, the mountain's skeleton, was an

---

[1] See Hargrove 1989 for a discussion of "therapeutic nihilism." While I disagree with some elements of Hargrove's interpretation of Leopold, this discussion raises a number of useful issues regarding the medical analogy.

[2] It is worth noting in passing that, while medical doctors never question their obligation to the individual patient, public health officials and social psychologists sometime focus attention on the family as the operative unit. While the unit of analysis and "management" can in these situations be controversial, the goal remains that of helping individual patients, and there exists a strong tradition establishing this as the central goal of medical practice.

essential part of that system, he had to take into account the welfare of a larger system than deer-wolf-hunter. A more holistic and farsighted management plan would have foreseen the value of wolves and would have maintained populations of predators, at least in remote areas.

But this metaphor is no positive and operational characterization of health; Leopold suggested this metaphor to explain a mistake. Nor does it provide us much guidance in choosing the object, or scale, of management. Is the "mountain" literally that? No, it would be more likely a range of mountains, some fairly self-contained, large, geographical unit—an ecosystem, perhaps. But how does one define the boundaries of the "mountain" in this sense? Environmental management is bedeviled not just by lack of a 2,500-year tradition of good medical practice; it lacks even a clear conception of what is to be managed, and which of many possible perspectives from which natural systems should be viewed when setting management goals. Environmental management has not, as yet, resolved the problem of "parts and wholes."

Before examining the problem of parts and wholes in detail, it will be worthwhile to note the tremendous advantages that physicians, who have resolved their problem of parts and wholes decisively, have over environmental managers, who have not. The organicist conception of medicine, central to any holistic approach, guides the overall therapeutic strategy, and the organism to be protected is unquestionably the human person as a biological organism. But this organism also has values, and places highest value on his/her quality of life. If a given therapy is not serving the health and overall welfare of the individual patient, it should be revised or terminated. Discussions of therapeutic options are not bedeviled by attributions of rights of appendices or interests of tumors.

Because patients ultimately pay the bills, they are the ultimate judges of whether a given course of therapy, and a particular physician's care, shall be continued. If the physician cannot convince his/her patient that the proposed therapy is effective, the patient votes with his/her feet. Given this focus on a single patient, the medical profession can get by with a "privative" definition of human health.

While bodybuilders, health-food faddists, and the United Nations Health Organization will offer very different accounts of health as a positive state of affairs, the medical profession operates, despite these differing visions of health, within an adequate margin of shared comprehension about treatment goals because they share a *privative* conception of health. Although they cannot agree on what health is, they generally agree when some event or condition represents *loss of health*. In many cases, therapy's goal is obvious once the physician has explained test outcomes—as when a patient's blood sugar levels are fluctuating wildly or when the x-rays show a clear fracture—the patient's definition of effective therapy will be close enough to the physician's to allow communication.

Now, one might suggest that conservation biology and environmental management can follow the lead of the medical community in this respect, as well.[3] Why not emphasize only the negative aspects of a definition of ecosystem health, concentrating on developing a consensual criterion for determining "illness" of ecological communities? There are many cases in which the privative concept of ecological illness is sufficiently clear to conclude that degradation has occurred. There is seldom significant disagreement

---

[3] See Sagoff 1988 for a proposal in this direction.

that a river with high levels of toxic chemicals is worse off than it would be without them. Nevertheless, there are real disagreements regarding the scale and object of management. Industrial companies, who have an economic interest in continued dumping, will argue that the river should be considered a part of the economic system and clean-up should only be insisted upon up to the point that economic interests are maximized. Environmentalists insist on different boundaries of the object of system management—they think the boundaries should correspond to reasonable ecological boundaries, such as the watershed, and that goals should be formulated in ecological, not economic, terms.

Note that values will have much to do with which system is chosen—conservation biology must be normative—so we cannot separate the "scientific" question of "correct" boundaries for the management system from the inherently value-laden questions regarding the goals of management. Environmental managers, unlike medical personnel, lack agreement regarding the scale of management and the perspective from which it will be carried out. Consequently, important questions, including value questions regarding the proper goals of management, remain unresolved.

Lacking the realistic test of the individual patient's judgment, I doubt that a privative definition of ecosystem health will provide sufficient guidance to judge whether conservation efforts are really "restorative."[4] To judge this, it is helpful to have a positive image of what the ecosystem *should* be moving toward. Lacking that goal, we are in danger of treating symptoms, while making underlying conditions worse. Medicine, of course, faces similar problems, and physicians and the public bemoan the high incidence of "iatrogeny"—illness caused or compounded by the misguided application of medical practices. Nevertheless, medicine which has achieved a consensus regarding the scale and perspective of therapy, has an important advantage over environmental management in the search for consensually acceptable and scientifically adequate standards.

What unit should be the "patient" in ecological therapy? What level of organization of the complex systems and processes of nature should be chosen as the therapeutic unit, analogous to the individual person in medicine? I call this the problem of "parts and wholes"—the problem of how to choose the proper unit of management. Or, to put the point slightly differently, it is a question of the proper "place" or "perspective" from which to interpret a management problem. The devilish difficulty of problems in environmental management derives from the fact that neither the goals nor the scale of environmental management are given. A solution to the problem of parts and wholes would carry us a long way toward a better understanding of ecosystem health and ecosystem management.

PARTS AND WHOLES

The concept of whole ecosystem management has come under serious criticism recently, precisely because of failures by its advocates to resolve the problem of parts and wholes.

---

[4] Consider, for example, the case of the Exxon oil spill in Prince William Sound. Nobody doubts the spill caused a deterioration in the health of the Sound. Determining whether clean-up efforts are successful, however, is more difficult; lacking a clear, positive conception of health, many experts have denounced the clean-up effort as causing more harm than good.

For example, Tom Regan, reacting to Callicott's suggestion that the land ethic implies that the ecological value of all individuals is to be found in their contribution to the functioning of the larger ecosystem, has charged land ethicists with embracing "environmental fascism" (Regan 1983).

Alston Chase, in a criticism of Park Service policies for the management of large mammal populations, has argued that "the vaunted self-regulating ecosystem, the supreme assumption that guided management of wildlife in Yellowstone was just a fictitious axiom, conceived to satisfy philosophical and mathematical conceptions of symmetry," and that whole ecosystem management is based on "California cosmology" rather than science (Chase 1986, p. 319).

Mark Sagoff, using arguments based in the literature of ecology, has argued that systems theory, which he equates with a search for mathematical models of great generality, has not fulfilled its promise as a guide to management. Following Levins, he says that no model can achieve generality and realistic description as well as mathematical precision, and concludes that "both population and systems ecology have failed to deliver the facts, understanding, and predictions that were anticipated in environmental law [in the writing of systematic management into legal statutes, for example]" (Sagoff 1988, p. 115). He quotes ecologist Levandowsky: "It's unlikely that there are general quantitative principles, comparable to Newton's laws, on which to base a general mathematical theory, and there seems little prospect of a physics of organisms" (Sagoff 1988, p. 161).

Sagoff embraces the health and medical analogy, arguing that ecology must be an eclectic science of the particular, as medicine is, rather than a deductive science based on a grand and unified theory of systems. He exhorts ecologists to stop waiting for their Newton, and get on with solving specific problems caused by modern technologies and lifestyles. He argues, in effect, that ecologists should follow physicians in employing a privative conception of health; they should treat individual ecological "patients" and give up the search for a general, mathematically precise theory of natural systems.

It seems to me that all of these criticisms of whole ecosystem management can be answered, though each of them has an important point to make, by recognizing and explaining some crucial ambiguities in the concept of wholeness. I note at least four meanings—one metaphysical, one contextual, one boundary-drawing, and one pyramidal—given to the term "whole" in theoretical and practical discussions of the phrase "*whole* ecosystem management."

Metaphysical Holism

It is the metaphysical concept that Chase uses when he castigates environmentalists for pursuing California Cosmology, rather than sound ecological science. He first notes that any ecosystem, no matter how large, will have inputs from outside it. "Could it be," Chase asks, "that only the universe taken as a whole was a complete ecosystem? Coincidentally, pursuing this scientific idea to its logical conclusion ran head-on into the new theology. The universe, pantheists were saying, was indivisible. Everything was interconnected. And now, increasingly, scientists agreed" (Chase 1986, p. 318).

Leopold's dream of an ecologically informed management was dashed, Chase thinks, because ecological theory deals with abstractions not directly applicable to real systems:

"while scientists moved ever more narrowly into the abstract realms of mathematical manipulation, environmentalists traveled more widely in search of a spiritual cosmology of nature" (Chase 1986, p. 325).

I agree with Chase that mystical concepts of holism are not very helpful in management decisions.[5] How does knowing that nature is sacred-as-a-whole tell us where to put the garbage? Worse, mystical holism is often associated with a dangerous "top-down" thinking. If a supraorganismic intelligence, with a good of its own, directs natural events, it is tempting to conclude that the part has value insofar as it serves the ends of a larger, supraorganismic whole. It is this tendency that leads, with some justification, to Regan's charges of environmental fascism.

But environmentalists have been careful to dissociate themselves from a view of ecosystem health that elevates ecosystem functioning above the basic rights of human individuals. For example, while environmentalists often emphasize the disastrous ecological effects of overpopulation, they have argued for population control through birth rate reduction, and have rejected extreme methods that fail to protect basic human rights of children, once born. Any hint that human rights should be sacrificed for the good of a mystical, supraorganismic being have been rejected. The metaphysical interpretation of holism seems to me to have caused enough confusion and mischief. Let us keep it out of public policy discussions, altogether.

Contextual Holism

Fortunately, environmental management theory need not be based on such a virulent form of metaphysical holism. Contextualism, the view that parts must be understood within larger systems, but that no "whole" will be reified as absolute, represents a limited form of holism (see Norton 1990). The part, from human economic actor to exploitable subsystems like forests and farms, must be considered as wholes as well as parts.[6] While this raises all sorts of paradoxes, and causes conflicting priorities in practical management situations, contextualism is committed to managing ecological systems, as well as the smaller, resource-producing units which are embedded in them. But this is not supraorganismic holism, managing the parts for the sake of some mystical whole; it is simply a commitment to see the world from a systematic viewpoint, to see individual actions as creating effects also in larger systems, effects which might compound to the point that individual actions may boomerang, destroying the ecological context that sustains those very activities through time. Contextual management is committed to seeing management problems from a local viewpoint or perspective, but recognizes also that any local perspective is limited, and that the problem must be understood also in its larger, systematic context, as a part of many larger and larger "wholes." No particular conceptualization, like a mystical supraorganism is given priority; no moral imperatives

---

[5] This is not to say they are not useful in educational or aesthetic contexts.

[6] As attention turns from local to more global problems such as the greenhouse effect, the "wholes" in question will be larger, and may be more social and climatological than ecological. But these broader concerns can still be understood contextually, as impacts of local activities throughout the world on the larger, climatological system.

are invoked from on high, applied to particular cases. Generalizations are considered "rules-of-thumb," not unquestionable axioms determining the moral acceptability of choices in concrete local situations.

My claim here is not that contextualism solves the problems of perspective and scale in ecological management. I argue only that a generally contextualist approach provides a framework of analysis within which these questions can be understood, discussed, and eventually resolved rationally. This framework recognizes that individuals see the world from their own perspective, but that that perspective includes placement of the individual within many nested systems. The problem, as it is posed in contextualism, is to choose the right one, or more, of these systems as the unit(s) of management.

The Boundary Sense of Holism

If the contextualist synthesis can help us to slip between the horns of the dilemma between atomism and metaphysical holism, we can focus attention on more concrete questions of the proper perspective from which, and the proper scale on which, to approach particular management problems. Note that the perception of a situation as a management problem already implies a normative judgment. I propose the following hypothesis: determination of the proper perspective and scale for addressing environmental problems cannot be separated from the determination of goals for environmental management. The "whole ecosystem" to be managed will be a function of the goals chosen; but the goals, as chosen, must embody a choice regarding the proper scale, and the proper "boundaries" for a management plan.

Take the example of Chesapeake Bay management. The Chesapeake Bay Regional Council has decided to manage the water quality of the Bay system, instituting policies to reduce nutrient loading of the Bay by 40% of 1987 levels by the year 2000 (Krupnick 1987, p. 1). The setting of that goal requires, given hydrological and ecological principles, that the management unit be the whole watershed. See Figure 8.1.

Advocates of management of the whole watershed of the Chesapeake region need not commit themselves to the view that the Bay system receives no impacts from outside the basin as Chase suggests; nor need they advocate mystical holism in order to emphasize spillover effects of economic and leisure activities on larger systems which we also value. Whole ecosystem management in the Chesapeake region represents a simple pragmatic decision. Given the shared public policy goal of improving Bay water quality, the boundary of the whole system must be the watershed. The management unit should be, given our best scientific understanding of the system, tailored to the management goals chosen. We draw the boundaries where we do because smaller boundaries would create situations in which inputs from the agricultural sectors in Pennsylvania would overwhelm the manager's ability to manipulate nutrient levels, swamping management efforts. Again, there is nothing metaphysical in this pragmatic decision. Just as ecologists choose their unit of study, their ecosystem, managers likewise should draw their conventional boundaries so as to give them maximal understanding of environmental problems and adequate management units so that they can suggest and test various therapeutic strategies for specific problems.

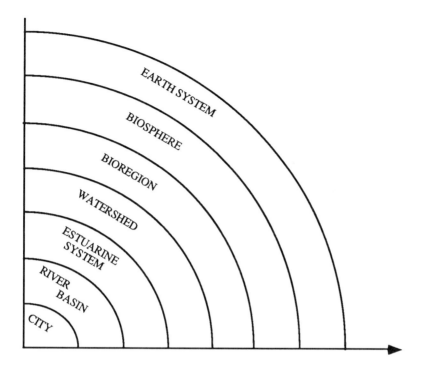

FIGURE 8.1 Whole Ecosystem Management. The scale of the management unit "whole ecosystem" varies according to management goals. Here "whole" is used in the boundary sense.

Sagoff's recommendation that ecologists forsake the search for a unified field theory for ecology can be respected by ecologists using the term "whole" in this sense. There is no claim that management of the whole must be according to a unified set of principles applicable to all ecological systems. The principles invoked can be tailored to the perspective and scale of the management problem at hand, and they can be related to specific, not universal, management goals. Further, they can be justified in pragmatic terms, not as necessary to satisfy ecologists' subconscious yearnings for systematicity—the physics envy that causes some ecologists to seek shortcuts that would be available only after ecology's Newton has worked his theoretical and mathematical magic—but as systems that are practically useful to conceive management goals.

The contribution of systems thinking to ecology will not be, it becomes more and more obvious, the provision of a simple and general mathematical model from which effects of actions such as nutrient loading can be deduced, and prescriptions derived. Systems thinking, however, fulfills another, equally important, if more modest, function: it helps to explain, to conceptualize, the incredibly complex and difficult problems of scale and perspective which face any attempt to work out a comprehensive and integrated theory of environmental management. It is in this sense that ecologists are far behind the medical profession; the scale and perspective from which management will take place is open to political negotiation. Before environmental managers could seriously attack the problem of Bay water quality, it was necessary to establish the Chesapeake Bay Regional Council.

Advocates of whole ecosystem management in the Yellowstone are currently trying to encourage inter-agency cooperation with the political goal of establishing a management unit or cooperative plan for management of the Greater Yellowstone Ecosystem. Until these questions of scale and perspective—management boundaries—are resolved in a way that is both scientifically respectable and politically saleable, calls for the protection of nature will remain far more ambiguous than public lobbying for policies to improve social health.

Environmentalists, with the support of conservation biologists and academic ecologists, have generally agreed that questions of environmental management must be viewed from a systematic perspective, as well as from the perspective of individual economic atoms. But that agreement leaves plenty of room for confusion, until general rules are accepted for determining the appropriate scale of the management unit. And it is in this political arena in which environmentalists of different stripes and economic determinists alike compete to determine the proper scale on which to conceive management problems.

As Sagoff points out, public goals formulated in purely economic terms will not adequately express the concern of most Americans to save our natural heritage (Sagoff 1984). And the role of ecologists in public policy will be determined by the perspective adopted. Environmentalists and academic biologists should exert their influence to determine boundaries of managerial decisions, because that aspect of the public debate is what most directly affects policy formation. Determining boundaries determines scale and perspective; and that is why environmentalists and conservation biologists fight an uphill battle. Economic forces provide reasons on a different scale, the scale of production and consumption of human individuals, not on the larger scale of dust bowls and species extinctions, which involve long-term trends relating processes in a larger system. Environmentalists believe we need to set boundaries in a way that makes sense, ecologically, and that the boundaries of a system should correspond mainly to ecological geography rather than human demography. In this sense, a broad public consensus is evolving to manage contextually, by considering effects of individuals on larger *ecological* systems, and to choose an appropriate management unit. Chesapeake water quality concerns lead to insistence on whole (whole watershed) management plans. Attempts to manage Yellowstone to maximize wildness and protect populations of large mammals implies managing a system large enough to encompass most of the important migration routes of those targeted species. But these are pragmatic arguments, arguments that are normatively loaded, based on the specific, local goals, plus enlightened scientific advice regarding the proper management unit, given those emerging policy goals (see figures 8.1 and 8.2).

Pyramidal Holism

When environmentalists and scientists advocate "whole ecosystem management" for the Greater Yellowstone Ecosystem, GYE, they are using yet a fourth concept of holism, although one that embodies the boundary sense as one element (Agee and Johnson 1988;

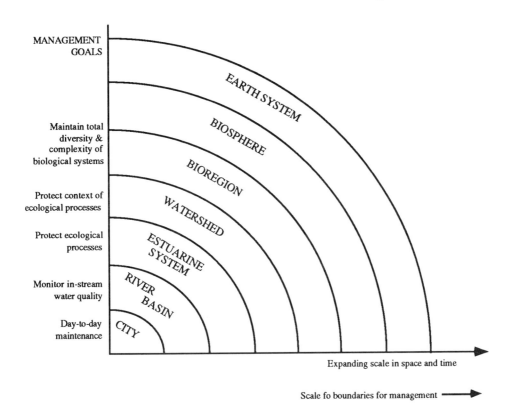

FIGURE 8.2   The scale of whole ecosystem management

McNamee 1986). When those who advocate whole ecosystem management advocate pro
tecting the wild processes, wildness, autonomy, and so forth, of an ecological system,
they are going beyond the boundary sense. Besides suggesting that the National Parks and
National Forests in the Greater Yellowstone area be designated a management unit
("whole" in the boundary sense), and that public policy goals be formulated from that
larger perspective, they are also proposing that this large unit be managed with an overall
purpose of protecting its wildness, the multi-leveled complexity which constitutes total
diversity in a patchy landscape. The hope here is that in that one area, by protecting the
fabric of interlocking habitats, protecting islands of wildness, we can protect a system
complete with stable populations of predators.

When Leopold returned from a hunting trip to the Sierra Madre, he was ready to propose an ideal of ecosystem health. After seeing unspoiled game ranges with stable populations of ungulates coexisting with predators, he said, "All my life I had seen only sick
land, whereas here was a biota still in perfect aboriginal health" (Meine 1988, p. 368). I
call this the pyramidal conception of environmental wholeness, and it is based on
Leopold's conception of an energy pyramid as an interlocking system of species occupying the various levels of a complete food web. (see figure 8.3.)

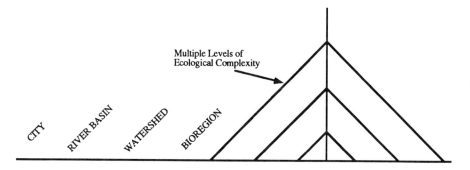

FIGURE 8.3   Whole Ecosystem Management. Management to protect all levels of the biotic community—its complexity and total diversity. Here "whole" is used in the pyramidal sense.

Advocating this approach to ecosystem management requires a stronger concept of ecosystem health, a positive management ideal. That ideal is to save as much as possible of the multi-levelled complexity of dynamic systems; in the paradigmatic case of management to protect wildness at Yellowstone, this ideal implies saving and reintroducing predators. The goal is to perpetuate the natural evolutionary forces that embody biological creativity. Assuredly, we cannot expect this ideal to support the same goals—such as predator reintroduction—in all management situations. Nevertheless, the positive ideal has implications even in "civilized" areas such as the Chesapeake region; the ideal of managing for naturalness and autonomy of natural processes in larger systems must, in these cases, apply by degrees.[7] To manage an ecosystem as a whole, in both the boundary sense and in the pyramidal sense, is to manage it to save as many as possible of the levels of its complexity, thereby protecting its autonomy across time.

The proposed ideal represents the daring hypothesis of a positive conception of ecosystem health—not a unified, single, mathematically measurable criterion that must apply to every natural system, but a consensus opinion that it is valuable and important to protect, to the degree possible, whole ecosystems. These are understood as large, ecologically organized units, such as the Chesapeake Bay and the large mammal populations of the Northern Rockies, including the total diversity they embody. The main practical application of this idea is that the landscape, to provide for the economic goals of the region *and* the setting aside of large enough areas to harbor wild populations of wide-ranging species, will have to be a patchy landscape, one that is not homogenized by slavish commitment to the *economic* concept of comparative advantage.

TOWARD A PRAGMATIC THEORY OF NATURAL SYSTEMS

I have argued that Leopold's land ethic amounts to a sort of poor man's systems theory. It considers resource-producing activities as embedded in larger systems, as elements in a

---

[7] I have discussed the problem of formulating a sliding scale of "naturalness" elsewhere. See Norton, forthcoming, ch. 9.

larger landscape. Economics normally guides management of the elements, but the field or forest is also part of a larger ecological context. It must be considered both as a part and as a whole, a system embedded within systems. I have asserted that it is this contextualist/systems thinking which marks environmentalists off from their opponents, who favor reductionistic, economic approaches (Norton, in preparation, ch. 8). Recognizing that our activities have an impact in a larger ecological context, the environing systems which give context and meaning to our culture, environmentalists are committed to protecting the "health and autonomy" of larger systems.

Systems approaches to ecological theory and practice have been roundly criticized, by a number of recent commentators, for failing to provide scientific guidance in solving environmental problems (Sagoff 1988; McIntosh 1985). In such criticisms, it is assumed that a systems approach to ecology will describe one system, characterized by unifying mathematical principles, which can model all natural systems simply by plugging local variables into one, universally applicable mathematical model. Perhaps systems approaches have failed because too much, or rather the wrong things, has been asked of them. The search for a single, unified system of ecological analysis is, I believe, closely related to metaphysical holism. Both rest on two assumptions:

1. A single system-model should (must) be developed for understanding and discussing all ecological systems, and
2. This system corresponds, in some philosophical sense, to the "Real World." Good philosophical arguments can be presented for questioning the "realism" of any particular model for representing reality[8]; I prefer to assume:
1'. There will be many systems of differing scales that can be used to illuminate any given management situation.
2'. We should choose as our management model a system of appropriate scale, *given the management goals chosen in particular situations.*

We saw, for example, that the goal of improving water quality in Chesapeake Bay indicates a certain management perspective and a certain context, or scale, on which management should be understood, and in terms of which management goals should be formulated. In the Yellowstone Area, quite different public goals are pursued; the boundary of the management unit must be, given the goal of managing the area to protect healthy populations of all the large mammals indigenous to the area, an area bounded by the outer limits of the migration patterns of those animal populations.

Conservation biology cannot be a value-free science and rely on "objectively" true scientific models of reality. It must concurrently formulate the goals which guide management even while choosing and developing the model/system which will be considered the system of concern. The role of ecology as a publicly responsible science is therefore to stimulate dialogue regarding the goals of management strategies, and criteria for success. The particular, distinctive role of professional ecologists in this process will be to help the public, with scientific expertise, to formulate goals and management plans. Science can aid management mainly in developing models that characterize the context as systems constructed on a scale that provides understanding of specific and evolving public goals.

---

8 See, for example, Willard Van Orman Quine 1960, ch. 1.

The process of developing management goals must be a local process which will take place within larger and larger contexts.[9]

Ecologists should act as midwives to the public's conception of management goals, largely by providing systems-models for understanding management problems. But if these models are likely to have a positive effect on policy discourse, they must be communicatable and communicated to the public.

What then of the definition of ecosystem health, the helpful, but limited analogy discussed above. Does this poor-man's systems approach—little more than a style of thinking, a tendency to look at events as elements also in a larger context—point us toward a definition of ecosystem health? If the hope is for some simple and mathematically precise definition applicable in every context, the answer is of course, no. But it may be possible to create a set of rules of thumb, which can be thought of analogically, for analyzing what is going wrong when environing systems undergo rapid change, a sort of owner's manual for thinking about environmental problems (Slobodkin 1986). With this more modest goal in mind, a general definition of health, or perhaps even more modestly, a set of general "standards of conservation" to apply to systems, may be developed (Leopold, forthcoming; Callicott, forthcoming). The undefined variable in the definition is one of scale. The choice of a proper scale on which to set boundaries, and the democratic choice of conservation goals and standards must be determined, I conclude, in the rough and tumble world of politics.

And, if I may hazard a guess, I think these standards will emphasize the protection of the "integrity" of ecological systems. Integrity will then be understood as a function of two more concrete factors: the complexity of the "whole" system under management, and the autonomy of that larger system. Complexity, I think, can be simply understood as the multi-layered complexity represented in the total diversity of the dynamic system (Norton 1987). Autonomy is more difficult to explain, but I would follow Leopold in seeing the context—the mountain—as a larger system that would normally change on a slower scale. Accordingly, we can define autonomy in terms of complexity or total diversity as well. The autonomy of a large, environing system is that system's ability to sustain its total diversity through time, to maintain its complex structure, even through the changes that are inevitable in any dynamic system.[10]

It will be said, of course, that my approach is a throwback to the naive notions of twenty years ago, and that systems theory failed to pay off on its pledges. But I am suggesting a more modest goal for systems ecology and a more modest role for systems thinking in environmental management. If ecology could throw off positivism and abandon the search for a single, universally applicable, mathematically precise model, the discipline could serve the public by helping to conceive management problems in a

---

[9] See Daly and Cobb 1989 for a more detailed explanation of how we might strive for a "community of communities."

[10] Personally, I am optimistic that a general conceptualization—though not necessarily a mathematically precise one—will be provided eventually, perhaps soon, by information-theoretic approaches. See, for example, Ulanowicz 1986. According to my argument, however, the urgent need in public policy is not for a mathematically precise and general model, but for multiple models, a kit of tools applicable in specific situations, situations varying both according to local physical features, but also according to management goals.

proper scale of time and space. But this goal can be understood modestly, and pragmatically, as choosing scales that help the public to formulate goals for public policies. This is a scientific process in that it requires sophistication in constructing models, but it is also a political process. If ecological systems are not given in nature, but are pragmatic tools of the understanding, ecologists must accept a modest role, not as aspiring Newtons, but as designers of publicly understandable models useful in understanding the health and illness of locally evolving ecological systems.

REFERENCES

Agee, J. K. and D. R. Johnson, eds. 1988. *Ecosystem Management for Parks and Wilderness.* Seattle, Wash.: University of Washington Press.
Callicott, J. B. forthcoming. Standards of Conservation: Then and Now. In J.B. Callicott and S. L. Flader, eds. *The River of the Mother of God and Other Essays by Aldo Leopold.* Madison, Wisc.: University of Wisconsin Press.
Chase, A. 1986. *Playing God in Yellowstone.* Boston, Mass./New York: Atlantic Monthly Press.
Daly, H. E. and J. B. Cobb, Jr. 1989. *For The Common Good: Redirecting the Economy Toward Community, the Environment, and a Sustainable Future.* Boston: Beacon.
Hargrove, E. 1989. *Foundations of Environmental Ethics.* Englewood Cliffs, N.J.: Prentice Hall.
Krupnick, A. J. 1987. Reducing Bay Nutrients: An Economic Perspective. Resources for the Future, Discussion Paper QE87-12, Washington, D.C.
Leopold, A. 1949. *A Sand County Almanac.* Oxford: Oxford University Press.
Leopold, A. forthcoming. Standards of Conservation. In J. B. Callicott and S. L. Flader, eds. *The River of the Mother of God.*
McIntosh, R. P. 1985. *The Background of Ecology: Concept and Theory.* Cambridge: Cambridge University Press.
McNamee, T. M. 1986. Putting Nature First: A Proposal for Whole Ecosystem Management. *Orion Nature Quarterly* 5: 4-15.
Meine, C. 1988. *Aldo Leopold: His Life and Work.* Madison, Wisc.: University of Wisconsin Press.
Norton, B. G. 1987. *Why Preserve Natural Variety?* Princeton, N.J.: Princeton University Press.
Norton, B. G. 1988. What is a Conservation Biologist? *Conservation Biology* 2:237-238.
Norton, B. G. 1990. Context and Hierarchy in Aldo Leopold's Theory of Environmental Management. *Ecological Economics* 2:119-127.
Norton, B. G. forthcoming. *The Unity of Environmentalism.* New York: Oxford University Press.
Quine, W. V. O. 1960. *Word and Object.* Cambridge, Mass.: Massachusetts Institute of Technology Press.
Regan, T. 1983. *The Case for Animal Rights.* Berkeley, Calif.: University of California Press.
Sagoff, M. 1984. Ethics and Economics in Environmental Law. In T. Regan, ed., *Earthbound.* New York: Random House.
Sagoff, M. 1988. Ethics, Ecology, and the Environment: Integrating Science and Law. *Tennessee Law Review* 56: 77-229.
Slobodkin, L. B. 1986. On the Susceptibility of Different Species to Extinction: Elementary Instructions for Owners of a World. In B. G. Norton, ed., *The Preservation of Species.* Princeton: Princeton University Press.
Ulanowicz, R. E., 1986. A Phenomenological Perspective of Ecological Development. In T. M. Poston and R. Purdy, eds., *Aquatic Toxicology and Environmental Fate: Ninth Volume,* ASTM STP 921. Philadelphia: American Society for Testing and Materials.

9

ECOLOGICAL PERCEPTION, ENVIRONMENTAL POLICY AND DISTRIBUTIONAL CONFLICTS: SOME LESSONS FROM HISTORY

J. Martinez-Alier
Department of Economics and Economic History
Universitat Autonoma,
Bellaterra, Barcelona 08193 Spain

ABSTRACT

Ecological awareness is socially molded. In the social sciences, we know that neither history nor geography have paid much attention to ecology, but there have been some critiques of economics from the ecological point of view. Such critiques have been directed against both mainstream and Marxist economics. Economic values assigned in the conventional way to externalities (such as exhaustion of nonrenewable resources, global warming, or radioactive pollution) would be so arbitrary that they cannot serve as a base for rational environmental policies. Lack of economic commensurability exists not only in market economies but also in centrally planned economies (as Otto Neurath pointed out in the 1920s). Externalities, defined as uncertain social costs transferred to other social groups, or to future generations, must be perceived before they are valued. The notion of "positional goods" is also discussed. The paper argues, therefore, against environmental policies based upon a presumed economic rationality. On the other hand, it also argues against the new attempts by international eco-managerialism to base policies only on an ecological rationality (in terms of carrying capacity norms or "sustainability"). Such attempts are blatantly ideological because they forget that the differences in the consumption of energy and materials by the human species, and the territorial distribution of the human species in the globe, are not fully explained by ecology.

ENERGY AND THE ECONOMY: A HISTORICAL VIEW

In the 1970s, an attempt to fit *some* environmental problems into the framework of externality analysis in welfare economics became known as "environmental and resource economics." In this paper, I argue against the virtues of this economic theory as a basis for environmental policy.

Because "environmental and resource economics" has too narrow a focus, while biology and ecology do not deal with political economy, there is a no man's field of study, which ought to be called Human Ecology. Unfortunately, "Human Ecology" as used in the 1920s meant urban sociology (of the Chicago school). In present usage, it has come to mean, more or less, the same thing as ecological anthropology—the discipline dealing with the ecological adaptations of primitive, assumedly isolated peoples (e.g., the type of research usually published in the journal *Human Ecology*). The term must be redefined, as the study of the relations between humankind and the environment in all historical and geographical contexts. One instrument for such a study, though certainly not the only one, is the analysis of the energy flow.

Human relations with the environment have a history, and the perception of such relations has also changed over time. Ecological awareness is now increasing everywhere; however, ecological historiography is still in its infancy. Ecology should not always be seen as a Braudelian *longue durée*, a slowly-moving geographical backcloth to economic changes and political events. The irreversible destruction of fossil fuels is proceeding at a rapid pace. Perhaps the increased "greenhouse effect" is being felt already, even though most people in the world have an exosomatic (outside the body) energy consumption level more typical of the period before the Industrial Revolution than of advanced capitalism. The thinning of the ozone layer is taking place over an even shorter span. Fisheries were destroyed in Peru between the 1960s and early 1970s, even quicker than guano deposits had been exported in 1840-80. Earlier, the European invasion of America (and other overseas territories) became an environmental disaster; the native populations swiftly suffered a demographic collapse (Crosby 1986). Ecological history deals therefore with many subjects, some of them slow-moving and majestic, some quick and irregular. The study of the use of energy in the economy is the easiest type of ecological history. Wilhelm Ostwald believed that the history of humanity could be understood as a history of two great regularities: a greater and greater use of energy per person and also an increasing efficiency in its use. Henry Adams' "law of acceleration" of energy use had no influence; it was characteristically dismissed by Karl Popper in a footnote to *The Poverty of Historicism*. Time and again there were modest attempts to provide some figures on the use of energy. Nevertheless, the first academically successful work in this line was Carlo Cipolla's in 1962, representing a delay of almost one hundred years. The European Society for Environmental History didn't have its first meeting until 1988!

The discussion about energy and the economy has included two mistaken views, and one constructive view. One mistaken view is the "energy theory of value." (See Punti 1988, for an argument against this, based on the fact that similar amounts of energy from different sources have different "production times".) Another mistaken view is based on the isomorphism between the equations of mechanics and the equations of economic equilibrium of neoclassical economics after 1870. It was believed that in economic exchange there was an exchange of psychic energy. Winiarski, at the turn of the century, was a spokesman for this absurd view. Leslie White, the ecological anthropologist, complained: "He discusses social systems in terms of the First and Second Laws of Thermodynamics, and uses differential equations to describe certain social processes. But for all this, he seems merely to present social systems as analogous to physical systems, to describe them in the language of physics, rather than to apply physical concepts to gain new insights and understandings of socio-cultural systems" (White 1954; repr. 1987, p. 219).

The third view has been accepted in our time by a few economists, Nicholas Georgescu-Roegen, Kenneth Boulding, William Kapp, Herman Daly, and earlier by Frederick Soddy, Patrick Geddes, Josef Popper-Lynkeus, and Sergei Podolinsky. The economy should not be seen as a circular or spiral flow of exchange value, a merry-go-round between producers and consumers, but rather as the one-way entropic throughput of energy, and materials. This field of study has a long unacknowledged lineage (Martinez-Alier and Schluepmann 1987).

To see the economy as entropic does not imply, in the least, ignorance of the anti-entropic properties of life (or, in general, of open systems). This point must be made explicitly because of the growth of "social-Prigoginism," i.e., the doctrine that human societies (for instance, Japan, or the European Common Market, or New York City) self-organize themselves in such a way as to make worries about depletion of resources and pollution of the environment redundant (Proops 1989, p. 62). If one goes beyond the title of Georgescu-Roegen's book, *The Entropy Law and the Economic Process*, it is clear that Georgescu-Roegen's ecological economics would not give support to what I have called "social-Prigoninism" but would not oppose the view that systems which receive energy from outside (such as the Earth) may exhibit a steady growth of organization and complexity over time (cf. Grinevald 1987). Vernadsky (1863-1945) explained (in a section of his book *La Geochimie* explicitly entitled *Energie de la matière vivante el le principe de Carnot*), that the energetics of life were contrary to the energetics *de la matière brute*. This had been pointed out by authors such as the Irish geologist John Joly and the German physicist Felix Auerbach (with his notion of *Ektropismus*), and one could find the idea already in J. R. Mayer, Helmholtz, and William Thomson (Kelvin). Vernadsky added:

The history of ideas concerning the energetics of life presents an almost unbroken series of thinkers, scientists, and philosophers arriving at the same ideas more or less independently. . . . A Ukrainian scientist who died young, S. Podolinsky, understood all the significance of such ideas and he tried to apply them to the study of economic phenomena." (Vernadsky 1924, pp. 334-335)

Given the importance of Vernadsky in the science of ecology and also in the current ecological revival in the Soviet Union, this endorsement of Podolinsky's ecological economics is likely to become famous, in retrospect (cf. Martinez-Alier and Schluepmann 1987, ch. 3). Podolinsky (1850-1891), though he was a Darwinist, was not a social-Darwinist. He attributed differences in the use of energy within and between nations not to any evolutionary superiority, but rather to the inequality bred by capitalism. This was contrary to the social-Darwinists who, a few years later, applied to human groups Boltzmann's dictum of 1886, "the struggle for life is a struggle for available energy." Nowadays, as was the case one hundred years ago, the ecological point of view is not politically univocal. It leads some towards social-Darwinism (Hardin's "lifeboat ethics" is a notorious example), and leads others (the German Greens, for instance, and many scholars and activists in the Third World) towards international egalitarianism. (I see the new international managerial ecologism nearer to social-Darwinism than to egalitarianism.)

RAUBWIRTSCHAFT—A CONCEPT OF ECOLOGICAL GEOGRAPHY

The social sciences, including mainstream economics, have not shown much interest in the ecology of humans. Such interest could have come from disciplines such as geography and history. However, ecological history is new in academia. Geography has studied some questions of human ecology but it has not focused on the study of other issues such as the flow of energy and materials in human ecosystems (which ecological anthropologists have studied since the 1960s). The so-called geography of energy was merely a description of the location of energy sources and the transport of some forms of energy, it was not an analysis of the energy systems of humanity.

Geography could have become much more ecological at least since the turn of the century, if the lead of Bernard and Jean Brunhes had been followed. One of the chapters of Jean Brunhes' *La Géographie Humaine* developed the notion of *Raubwirtschaft* introduced by the German geographer Ernst Friedrich (b. 1867, professor at Koenisberg): "it seems particularly strange that characteristic devastation with all its grave consequences should especially accompany civilization, while primitive folk know only milder forms of it" (Jean Brunhes 1920, ed. 1978, p. 331). An ecological geography could have been born in French universities out of these reflections by such a prominent geographer as Jean Brunhes. A well known American geographer of German origins, Carl Sauer, who did not explicitly use the concept of *Raubwirtschaft*, was nevertheless influenced by George Perkins Marsh and this led him to ask: "Must we not admit that much of what we call *production* is *extraction*?" (Sauer 1956). Geographers had nothing to lose and much to gain professionally by becoming human ecologists and environmental managers, but a notion like *Raubwirtschaft* would not be politically popular in colonialist Europe, particularly in colonialist France.

After giving examples of *Raubwirtschaft* (see also Raumolin 1984), Jean Brunhes mentioned the book by his brother Bernard Brunhes, *La Dégradation de l'Energie*. Bernard Brunhes was the director of the meteorological observatory at Puy de Dome in central France, and he died young. He studied the flow of energy and also land erosion. He blamed deforestation on the privatization of common lands: he quoted Proudhon's views on private property, which were strong anarchist stuff. The idea of *Raubwirtschaft* could have been linked therefore by geographers to a notion of "tragedy of the enclosures" rather than "tragedy of the commons" because, although private owners carry the full short-term costs of land degradation (compared to users of communal lands), as far as long-term costs are concerned (and this is a relevant consideration for deforestation and land erosion), their time-horizons might well be shorter, and their implicit discount rates higher, than those of communal managers. Today, in the Amazon, we see one of the most massive processes of privatization of land in human history. Such "enclosures" are not only a social tragedy in the form of loss of access to common lands and proletarization, they also become an ecological tragedy.

A CASE STUDY IN ENVIRONMENTAL PERCEPTION: GLOBAL WARMING AS AN INVALUABLE EXTERNALITY

Let us take as established that most historians and social scientists, including economists, did not show much ecological awareness, or at least they were not able to integrate their private worries about the conservation of nature into their professional work. One immediate excuse, or explanation, would be that ecological awareness was not present in the population at large. This does not seem to me a convincing explanation, particularly in low-income countries. I agree that the political philosophies predominant on the side of the poor of the world have not been ecological. Thus, there has been no ecological Marxism, no ecological anarchism, not even an ecological narodnism, or an explicitly ecological Gandhian philosophy. However, it is my contention that, if we look again, we shall find ecological roots and ecological contents in social movements by poor populations, in history and at present, and that ecologism is potentially a stronger force in the South than in the North. I call Third World egalitarian ecologism "ecological narodnism," but it also has been called "ecological socialism" (Ruha 1988). Although its constituency has been and is extremely large, it has unequal opportunities to set the international environmental agenda. For instance, the Supreme Court's recent decision in India on damages for Bhopal's "silent spring" has been quietly accepted by international environmental managerialism, and the complaints in India are not heard in the North Atlantic. The reopening of the case, on the grounds that indemnities are far too low, and because of India's scandalous colonial inability to bring penal charges against Union Carbide officials, would raise the question of how the valuation of externalities depends on geography and social class.

Surely, we would find many spontaneous ecological movements in social history, which have not been researched, trying to keep natural resources away from the inroads of the generalized market system (Ruha and Gadgil 1989; Ruha 1989). The struggle of *seringueiros* in Acre, and Chico Mendes' death in 1988, are relevant examples. Such social movements have tried to keep resources in the sphere of what E. P. Thompson called "the moral economy." Environmental history should add the ecological dimension to an institutional socioeconomic history, rather than introducing the apparatus of formalized environmental economics into a history disembodied of social and political institutions, where environmental history would be explained on the basis of methodological individualism, where no attention would be paid to the study of the social origins of moral values regarding future generations, and where no attempt would be made to explain the social roots of science and technology.

My favorite introduction to the issue of ecological awareness is Gunnar Myrdal's statement at a conference in 1968: "I have no doubt that within the next five or ten years we are going to have a popular movement within the rich countries which is going to press Congress and the Administration to do many things for solving environmental problems. But the same will not be true in most, if not all, underdeveloped countries" (Farvar and Milton 1972, p. 960). It would appear that Myrdal was right: ecological awareness seems to be stronger in the North than in the South, and Washington, D.C., is becoming the capital of a new North Atlantic ecological bureaucracy backed by political power and economic leverage which captures headlines, pays for conferences, and tries to establish a suitable world environmental agenda, impartially recommending ecological "adjustment" programs to all countries and citizens: "the IMF of Ecology." There is a political attempt to move the ecological agenda away from the issue of *Raubwirtschaft* by the wealthy. Thus, in the wake of the Brundtland Report, the study of poverty as a cause of environmental degradation has become more fashionable (and richly funded) than the study of wealth as the main human threat to the environment.

Some ecological damages are novel only in the sense that they have not been socially acknowledged. Hence, for instance, today's German usage of *Treibhauseffekt*, retranslated from "greenhouse effect," instead of the original and excellent German word, *Glashauswirkung*. Such ecological impacts as the gradual exhaustion of some fossil fuels caused by the demand in a few countries, species extinction because of tropical deforestation, acid rain, the carbon dioxide build-up and its effects on climate change, accidents in nuclear power plants and the lack of a technical solution for the disposal of radioactive waste, were discussed at least fifty years ago, and some, one hundred years ago. Their novelty is "socially constructed ignorance" (Ravetz 1989). Other environmental effects might be genuinely surprising. For instance, the effects of CFC on the ozone layer were unknown until the 1970s, and the alarm at the possible ecological effects of genetically-engineered organisms, is also new.

A major constraint to global environmental policy is, therefore, not only the uneven distribution of current emissions (and current consumptions of exhaustible resources), but also the uneven distribution of past emissions and consumptions. Svante Arrhenius (1903, p. 171) explained in his textbook on global ecology that the *Glashauswirkung* which helped to keep the Earth warm would increase with the increase in carbon dioxide in the atmosphere. In 1937, it was estimated that fuel combustion had added around 150 billion tons of carbon dioxide to the air in the past fifty years, three quarters of which had remained in the atmosphere. The rate of increase in mean temperature was estimated at 0.005 degrees centigrade per year: "the combustion of fossil fuel . . . is likely to prove beneficial to mankind in several ways, besides the provision of heat and power. For instance, the above mentioned small increase of mean temperature would be important at the northern margin of cultivation" (Callendar 1938, p. 236). The author was, by his own description, "steam technologist to the British Electrical and Allied Industries Research Association," but his paper was received amiably by disinterested, objective scientists belonging to the Royal Meteorological Society of Great Britain. These scientists questioned Callendar's temperature statistics but did not question the view that increased carbon dioxide would be a positive externality. Their reaction showed that people living in northern latitudes and with high standards of living are not intrinsically more environmentally perceptive. Research on the sociointellectual history of climatic change (Budyko 1980) through the scare in the United States in the summer of 1988 has now become an interesting subject, and perhaps it will appear that some scientists soon took a pessimistic view about the global effects of the increase of CO_2 in the atmosphere.

If international environmental policies based on CO_2 budgets are established (either setting compulsory upper limits, or taxing emissions over a stated limit), they should include in each country's budget the accumulated past emissions—if not from the beginning of the Industrial Revolution, at least since 1900. (It could also be argued that such CO_2 budgets should be set not on a country basis, but on a per capita basis). Other ideas (clean slate, country-based CO_2 budgets) have *already* been proposed by the North Atlantic ecological establishment. Such environmental policy agenda tries to defuse the distributional impact of the ecological critique of economics (so clearly put forward by authors such as Commoner and Georgescu-Roegen). It also assumes ignorance of the history of ecological damages, unaware that the social construction of scientific facts is seldom socially neutral. Thus, the "adjustment" programs to be recommended by the "IMF of Ecology" are, for some, reducing CO_2 emissions by increasing car mileage; for some, by burning less wood in the kitchen through improved stoves; for some, the very poor, presumably by exhaling less CO_2 through breathing slowly or not at all (or, at least, reduced methane emissions—also a greenhouse gas—by having fewer rice paddies).

The history of global warming shows that the critique ecologists make against mainstream economics is based on the question of unknown future agents' preferences and their inability to bid in today's market, and therefore on the arbitrariness of the values given at present to exhaustible resources or to external effects. The ecological critique is also based (as David Pearce has written) on the uncertainty about the workings of environmental systems which prevents the application of externality analysis we do not recognize many externalities, and of the externalities which we *do* recognize, we do not even know whether some are negative or positive, let alone being able to assign a present value in monetary terms.

THE DOUBTFUL ECONOMICS OF NUCLEAR POWER

Global warming is now being used as an argument for nuclear power. But nuclear power also provides good examples of invaluable externalities: present values must be given to the costs of dismantling power stations in a few decades, and to the costs of keeping radioactive waste under control for thousands of years. Such values depend on the rate of discount chosen. Moreover, there are possible by-products of nuclear power, such as plutonium, which we do not know whether to classify as positive or negative externalities, let alone attribute a monetary value to. Since the plutonium produced as a by-product of the nuclear civil program may have a military use, it can be given a positive value, thus improving the economics of nuclear power (in the chrematistic sense). This "plutonium credit" was factored into the accounts of the initial British nuclear power stations (Jeffery 1988). However, plutonium might come to be seen in future as a negative externality, especially if owned by unpopular foreign governments or activist groups. In fact, Frederick Soddy, who was a well qualified nuclear scientist, warned against the "peaceful" use of nuclear energy in 1947 because of "the virtual impossibility of preventing the use of non-fission products of the pile, such as plutonium, for war purposes" (Soddy 1947, p. 12). This worrisome fact did not reach public opinion in the West until the 1970s because of the propaganda barrage in favor of "atoms for peace," which started under Eisenhower's administration. Awareness of the environmental dangers of "peaceful" nuclear power, before the accident in Three Mile Island in 1979, was found in only a few scientists, a few citizens' groups in areas directly threatened by nuclear power stations, and a socially powerless "lunatic fringe" comprising the leftovers of 1968 and a few younger recruits. Awareness of the environmental dangers of nuclear arms (even if they are not used) is also increasing.

Conventional environmental economics, we conclude, is rather useless as an instrument of environmental management, because the concept of "externalities" merely hides the inability to value uncertain social costs which we shifted to other social groups or to future generations. The World Bank, which is becoming one of the main institutions in defining the agenda of international managerial ecologism, did not consider loans for nuclear power stations. However, last year it seemed that the bank was about to make an exception for a new nuclear power station at Angra dos Reis (near Rio de Janeiro) (New York Times Mar. 11, 1989): it would be interesting to see which costs and benefits are factored in, at which rates of discount, but the World Bank, if it gave money for nuclear power, would certainly, for political reasons, disguise this loan inside a large loan for the whole electrical sector, with no specific cost-benefit analysis for Angra dos Reis. In Brazil, the discussion on debt-for-nature swaps reveals the difficulties in the valuation of other externalities. Thus, a "generous" proposal for buying 4 billion dollars worth of the Brazilian debt in order to save the Amazonian rainforest was tentatively launched some time ago (New York Times, Feb. 3, 1989). The total external debt is around $115 billion, so that the proposal, although far from negligible, is of limited importance in terms of finance. Moreover, to offer less than one dollar per inhabitant of the world as the price for the preservation of the Amazonian rainforest, looks cheap. An annual value of, say, $50 billion could be assigned to the positive externalities provided to the rest of humankind by the conservation of the Amazonian rainforest which at present is being privatized. After all, nobody knows how to give present values to the future benefits of preserving tropical biodiversity. (This ecologically conservationist, financially aggressive policy, is now part of the platform of the Brazilian Labor Party).

ECOLOGICAL ECONOMICS AND MARXISM

Since the economy is entropic, resources are exhausted and waste is produced. But ecologists question the ability of the market to accurately value such effects. Ecological economics is not necessarily pessimistic on economic growth, it merely points out that growth cannot be predicted by purely economic models from which the flow of energy and materials is excluded. The ecological critique points out that due to the temporal dimension, the economy involves allocations (of waste, of diminished resources) to future generations, without such allocations arising from transactions with them. Therefore, the economy cannot be explained on the basis of individual choices and preferences. Methodological individualism encounters the insuperable ontological difficulty of coping with future generations. Waste and diminished resources are allocated to them but they do not arise from transactions with them. Because of this, ecological economics is a main enemy of orthodox economics. It belongs with political economy, or institutionalist economics (cf. Martinez-Alier and Schluepmann 1987). Could ecological economics develop close links with Marxian economics? There has been no Marxian ecological history. However, since Marx and Engels were skeptical about benefits of the market's invisible hand, they should have had no *parti pris* against ecological economics. I now consider this question, going beyond the story of Engels' dismissing Podolinsky's ecological economics in 1882, which was surely a missed chance for the birth of eco-Marxism and which is (still?) of more than academic interest in India and Latin America.

Ecological Marxism would comprise both the theory of economic crises, and the history of social movements. Marxian economics has traditionally seen a contradiction between the overproduction of capital in the capitalist, metropolitan countries, and the deficiency in buying power from their own domestic exploited working class, or from the external, exploited economies. In ecological Marxism, one would focus not on the overproduction of capital (and on its consequences in the form of a falling profit rate and periodically increasing class struggle) but on the impairment or destruction of the conditions for the reproduction of capital. Up to now, Marxian economics, to the extent that it has dealt with natural resources, has taken a Ricardian view. Thus, in the 1970s, it was argued that the increase in the price of oil could be analyzed in terms similar to the increase in agricultural prices required to cover costs in marginal land in the Ricardian theory of differential rent, plus costs due to an element of monopoly (as in Marx's "absolute rent"). It was further argued that the resulting increase in rents relative to profits would alter the pattern between consumption and savings (and investment) and so slow down the accumulation of capital. We know, however, that oil prices came down in the 1980s, and nevertheless there is less oil left in nature in the 1980s than in the 1970s. The point is that in Ricardo's theory of rent, the "production price" of agricultural produce in marginal land must cover the cost of production (including profit, without rent) while the corresponding "production price" of an exhaustible resource must simply cover the cost of extraction (plus profit) at the margin. Oil is not produced, it is extracted.

Although Marx agreed with Liebig's argument in favor of small scale agriculture because it would be more conducive to the recycling of nutrients, and although he shared Liebig's enthusiasm for the new chemical fertilizers, he did not discuss whether agricultural prices should not only pay for current production costs but also secure the long-term fertility of the land. In any case, while soil conservation means to use it without erosion, oil conservation means not to use it at all. The reproduction or replacement of fossil fuels is not assured by high prices (although conservation might be helped by high prices). Marxian (or Sraffian) schemes of "simple reproduction" have not yet taken into account the exhaustibility of resources, and other irreversible environmental effects (Christensen 1989, p. 34). If, in Marxian economics "ecological costs" need to be transformed into increased prices in order to have a negative influence on capital accumulation (as argued by an ecological Marxist theory of crisis, cf. Leff 1986; O'Connor 1988), then the ecological critique is also valid against such ecological Marxism, precisely because social costs, and the needs of future generations, are usually not reflected in prices. Those needs remain external to the market. There can be increasing ecological destruction for a long time, without it being reflected in a capitalist crisis.

Now, however, it has been argued that new social movements are the agencies which increase private monetary capitalist costs bringing them nearer to social costs (Leff 1986; O'Connor 1988). This is a nice argument, because it brings together "objective" and "subjective" factors, in a very Marxist way. This is also the line taken by the new socio-ecological history in India. Thus, Ramachandra Ruha's work (1989) on the ecological reasons for agrarian protests provides an explanation of the remote origins of the Chipko movement and other similar movements from the time of British domination to the 1970s. Of course, ecological perceptions in history will not be expressed by the actors themselves in terms familiar to ecologists, of flows of energy and materials and of exhaustible resources and pollution. This is the language of scientists, and of some ecological movements (such as *part* of the German Greens) but is certainly not the language used, by other, as yet undiscovered, ecological movements which have tried to keep natural resources out of the generalized market system, which have set a "moral economy," and therefore an ecological economy, in opposition to the market economy.

If the development of capitalism is understood in terms of increased *Raubwirtschaft* in order to support the living standards of the rich, then many social movements of the poor against the rich will be seen to have an ecological contents and even perhaps an ecological idiom. Social movements, even if they cannot keep natural resources out of the chrematistic economy and under communal control, will at least force capital to internalize some externalities (O'Connor 1988) by struggling over workplace health and safety, toxic waste disposal, water availability in urban areas, conservation of forests by native peoples against paper factories or hydroelectric dams or cattle ranches, or higher prices for exhaustible resources from the Third World. Nevertheless, one may remain unconvinced whether such social movements, even when allowed to exist by state authorities, would really give a voice to future generations. An environmental policy which would rely on the impetus provided by popular ecological movements, might sometimes place the temporal horizon too near. On the other hand, a market-oriented economy will be ecologically even more shortsighted, because those not yet born cannot come to today's market.

ECOLOGY AND THE DEBATE ON ECONOMIC CALCULUS IN SOCIALIST ECONOMIES

It looks as if centrally planned economies will disappear before the issue of ecological externalities in economic planning is realized. Therefore, this section should not be seen so much as a contribution to socialist economics as a further historical case study in ecological perception, and as a contribution to the history of ecological economics.

The current splendid antibureaucratic, democratic crusade in Eastern Europe and the Soviet Union should not lead to a glorification of the market solution to ecological problems. The market cannot count long-term ecological damages. This was clearly stated by William Kapp, who started his career with a doctoral thesis in Geneva on the valuation of externalities (Leipert and Steppacher 1987). This thesis was meant as a contribution to the debate of the 1920s and 1930s on economic rationality in a socialist economy. Towards the end of his life, Kapp wrote:

The fact of the matter is that both, disruption and improvement of our environment, involve us in decisions which have the most heterogeneous long-term effects and which, moreover, are decisions made by one generation with consequences to be borne by the next. To place a monetary value on and apply a discount rate to future utilities or disutilities in order to express their present capitalized value may give us a precise monetary calculation, but it does not get us out of the dilemma of a choice and the fact that we take a risk with human health and survival. For this reason, I am inclined to consider the attempt at measuring social costs and social benefits simply in terms of monetary or market values as doomed to failure. Social costs and social benefits have to be considered as extra-market phenomena; they are borne and accrue to society as a whole; they are heterogeneous and cannot be compared quantitatively among themselves and with each other, not even in principle. (Kapp 1983, p. 49)

This very same view on the lack of economic commensurability had been expressed by Otto Neurath's concept of a *Naturalrechnung*. Neurath's idea was received by market economists as could be predicted: Hayek wrote that Neurath's proposal that all calculations of the central planning authorities should and could be carried out *in natura*, showed that Neurath was quite oblivious of the insuperable difficulties which the absence of value calculations would put in the way of any rational economic use of the resources (Hayek 1935, pp. 30-31). Hayek, on his part, as almost all participants in the debate on economic rationality under socialism (on both sides of the divide), was quite oblivious of problems of resource depletion and pollution. Hayek's glorification of the market principle and of individualism led him to dismiss authors who developed a critique of economics from the ecological point of view—such as Frederick Soddy, Lancelot Hogben, Lewis Mumford, and also Otto Neurath—as totalitarian "social engineers" (Hayek 1952). Hundreds of teachers of "comparative economic systems" have taught the debate on economic calculus in a socialist economy, perhaps praising Lange's and Taylor's "market socialist" solution to Max Weber's, Ludwig von Mises' and Hayek's objections, without realizing that the debate should have included a discussion on the intergenerational allocation of exhaustible resources. The intergenerational point is a different matter from discussing whether coal or oil should be priced according to marginal cost of extraction instead of average cost, as if this would ensure an optimal intergenerational allocation The debate should also have included a discussion on the allocation of waste.

Neurath, inspired by Popper-Lynkeus and by Ballod-Atlanticus, was aware that the market could not give values to intergenerational effects. In his writings on a socialist economy, starting in 1919, he gave the following example: two capitalist factories, achieving the same production, one with two hundred workers and one hundred tons of coal, the second one with three hundred workers and only forty tons of coal, would compete in the market, and that using a more "economic" process would achieve an advantage (where "economic" is used in its chrematistic sense, and not in its "human-ecological" sense). In a socialist economy, in order to compare two economic plans, both of them achieving the same result, one using less coal and more human labor, the other using more coal and less human labor, we would have to give a present value to future needs for coal. We must therefore decide politically, not only on a rate of discount and on the time horizon, but we must also guess the evolution of technology (including estimates for global warming, acid rain, radioactive pollution, which Neurath could have mentioned). Because of this heterogeneity, a decision on which plan to implement could not be reached on the basis of a common unit of measurement. Elements of the economy were not commensurable, hence the need for a *Naturalrechnung*. One can see why Neurath became Hayek's bête noire; but from the opposite political trench to Hayek's, Neurath got no praise. A critic remarked that Neurath's skepticism about economic planning led him to think *auf so primitive chiliastische Weise* that he was *im Utopismus stecken geblieben!* (Weil 1926, p. 457).

Since environmental concerns show the weakness of the market economy, they should have figured in the debates on economic planning in a Central European context of the late 19th century and until the 1930s. However, ecological issues were absent from the article by Enrico Barone (of Padua) on the Ministry of Production in a Collectivist State and in subsequent contributions. Some remarkable exceptions were Popper-Lynkeus (1838-1921) and Ballod-Atlanticus (1864-1933) (see Martinez-Alier and Schluepmann 1987, ch. 13), and also, as we have seen, Otto Neurath (1882-1945) and William Kapp (1910-1976). Only very recently was the conflict between ecology and economics discussed in Eastern Europe (Graf 1984, provides a good analysis and bibliography). Otto Neurath was not only a dissident economist, and a political radical (active in the revolution in Munich in 1919), but a major analytical philosopher of the Vienna Circle, the manifesto which he wrote himself. While most of Neurath's writings on socialist economics are available only in German (there are bibliographies in Weissel 1976 and in Stadler 1982), and while William Kapp's thesis written in the mid-1930s has been practically unknown, the same cannot be said of Kapp's later discussion in plain English, of the social costs and social benefits of economic development: "we are dealing with essentially heterogeneous magnitudes and quantities for which there can be no common denominator . . . a commensurability which simply does not exist" (1983, p. 37). Here, environmental economics is not seen as a minor complement to welfare economics, dealing with sporadic, exceptional cases of "market failure." On the contrary, we reopen one of the major polemics of our age, by pointing out that the market economy cannot provide a guide by itself for a rational intertemporal allocation of resources and waste. This does not imply, however, that the Minister of Production of a Collectivist State would be able to rely on an ecological rationality. The question is rather, who should decide environmental and economic policies and how should they do it?

POSITIONAL GOODS AND FORDISM IN THE PERIPHERY

This section explores whether Fred Hirsch's concept of positional goods has value for our deliberations. In his influential book, *Social Limits to Growth* (1976), Fred Hirsch dismissed an ecological approach to the economy while trying to explain the persistence of strong distributive conflicts in high-income countries. The title was polemical against the sudden fashion for "ecological limits to growth" after the oil price increase in 1973. Hirsch argued that, as wages rose in proportion to productivity, mass-consumption goods produced through mass-production methods became available to everybody (in a Fordist pattern, to use the terminology of another school of political economy). In Western countries, despite the growth in consumption, there was dissatisfaction as manifested in the wave of labor unrest at the end of the 1960s and at the beginning of the 1970s. One of the roots of such dissatisfaction was, precisely, according to Hirsch, the "positional" character of some goods and services. Veblen's conspicuous consumption comprises one class of "positional goods," the "exclusive" goods bought by the snobs. But Hirsch's concept goes beyond this, the satisfaction drawn from positional goods diminishes when other people have them because they impose social costs. His examples were as follows: if everybody has a car, or if everybody strives after a good education which qualifies him or her for a job at a good wage, or if everybody has a country cottage or a yacht, the satisfaction of these wants remains unfulfilled because of traffic congestion and lack of clean air, because there will be not enough jobs for all qualified people, and because the agglomeration of country cottages and yachts makes them unattractive. Hirsch's emphasis was more on the congestion of European cities, roads, and beaches than on exhaustion of resources or on world pollution effects (and nobody living in Catalonia at the peak of summer holidays would deny that he had a point).

According to Hirsch, the "material economy" was defined as "output amenable to continued increases in productivity per unit of labour input," while the "positional economy" could not grow without limit because of increasing social costs. This distinction was in parallel to Harrod's distinction between "democratic wealth" and "oligarchic wealth." However, from an ecological point of view, it appears (and it could already have appeared in 1976) that a "material economy" is also a "positional economy" which shifts costs inside the present generation or shifts costs to future generations. The increase in productivity per unit of labor which in some parts of the world has allowed the generalization of "democratic wealth" in the form of mass-consumption goods, has been partly achieved at the expense of exhaustion of resources and pollution of the environment. That is, unless the economy were delinked or decoupled from the use of energy and materials and the production of waste, certain forms of wealth will never become universal. Also, some forms of wealth are causes of poverty, now or in the future.

However, in Hirsch's view, the limits to growth were "social," not ecological, hence statements such as: "An acre of land used for the satisfaction of hunger can, in principle, be expanded two-, ten-, or a thousand-fold by technological advances. By contrast, an acre of land used as a pleasure garden for the enjoyment of a single family can never rise about its initial productivity in that use" (Hirsch 1976). While the second part of this statement is true, the first part is metaphysical since Hirsch provided no analysis of the meaning of "technological advances" in terms of the flow of energy and materials in the economy. Modern agriculture has a lower ratio of production to fossil fuel input than traditional agriculture, and its "higher" productivity is a consequence of low prices for extracted resources and inserted pollutants. Therefore, the relevance of Hirsch's concept of positional goods is greater than he himself supposed. For instance, a world with a stable population of ten billion people, and with a North Atlantic car density, would have about four billion cars, and this is ten times the present number of cars in the world. A pattern of industrial development without cars would be a novelty in the second half of the 20th century: the economies of the successful newly industrialized countries (Italy, Japan, Spain, South Korea) were or are still led by the car sector. Mexico, Brazil, Eastern Europe and the USSR, India, and indeed China, would like to follow suit. Suitable information on the energetic and material side of production and consumption (which was certainly available by 1975) would have led Hirsch to think not only of traffic jams. Cars will not become mass consumption goods because of their thirst for fossil fuels and also because of their environmental impact in terms of CO_2 and NO_x. Fordism in the periphery will be in any case Fordism without Fords, even perhaps Fordism without meat (or at least, not with a North Atlantic meat consumption of over 50 kilograms per person/year). Therefore distributional conflicts cannot be solved by a universal Fordist pattern of economic growth, not only because of the social limits emphasized by Hirsch but also because of ecological limits. The number and distribution of cars in the world ought to be part of the world environmental policy agenda. Here, as always, the technological optimists may use the fact that the future is uncertain in order to argue against an ecological orientation of economic policy. It is not known whether, for instance, a technology based on photovoltaics (with sun energy) and hydrogen as fuel (taken from water by electrolysis) will soon become available. In the meantime, Fordism with Fords is not a realistic prospect for the "periphery" of the world, where most people live. It is a reality in the metropolitan countries only because there is no competition from oil coming from the poor, peripheral countries, where people lack even the oil they would need (as kerosene or butane gas) to substitute for scarce cooking fuelwood. If you run a car, not only are you preventing another family from having a car (at least in the future, if not already now), you are also increasing the "other energy crisis," the lack of fuelwood.

The Limits of Ecological Rationality

The ecological critique of economics began over one hundred years ago (cf. Martinez-Alier and Schluepmann 1987). Such "ecological economics" is represented today by

Georgescu-Roegen and a few other authors. I do not know why "ecological economics," which is a fundamental challenge to economics, did not take roots in universities. The main reason was probably the separation between the natural and the social sciences, but this amounts to saying that human ecology has not been a prestigious subject among natural scientists. Why? I do not know. Human ecology is different from the ecology of other animals because of the lack of genetic instructions on human exosomatic consumption (and waste) of energy and material resources, and because of the peculiar political, social, territorial human arrangements. Human ecology is a type of study which cannot be reduced to the natural sciences.

Ecological economics, although it has a long history, had no impact on mainstream economics. On the other hand, environmental and resource economics attempted in the 1970s to treat some ecological issues in terms of applied welfare economics (as in the *Journal of Environmental Economics and Management*). This attempt would also lead to the conclusion that there is no economic commensurability, if questions of uncertainty, time-horizon, and discount rates were honestly addressed. For economists, ecological awareness threatens to swamp economic values in a sea of invaluable externalities.

Throughout this paper, I have argued against an environmental policy based on the conceptual apparatus of economics, but in this section I would like to point out some limitations of a purely ecological approach. Specifically, I shall consider, as an example, the failure of the notion of carrying capacity as an instrument of ecological and population policy. "Carrying capacity" refers in ecology to the maximum population of a given species which can be supported indefinitely in a given territory, without a degradation of the resource base that would diminish the maximum population in the future. Here, ecology and economics come again into conflict in the definition of "degradation of the resource base." Economists would claim that use of resources, even if they are not produced but merely extracted, is not necessarily economic degradation because, before they are exhausted, they will be substituted by new resources. A strict conservationist posture, which would give equal values to future and present consumption, would perhaps lead to the conclusion that some resources would be left unused when techniques change. Economists would also point out that, although there is no guarantee of such technical substitutions, nevertheless resources should be used now, because the (assumed) growth of the economy makes future consumption at the margin less valuable than today's consumption. The ecologists may point out, with reason, that the economists have no strong arguments in order to impose a particular rate of discount, and ecologists could even argue for a negative rate of discount. Nevertheless, because of uncertainties about future technical changes, a so-called ecological rationality is not an indisputably better base for policy than the usual economic rationality.

Attempts at using the notion of carrying capacity (for poor countries only) as a basis for policies of "sustainable development," are made by respectable international agencies and the multilateral lending banks. However, the area of cropland per person in Europe is low compared to the world average. (In the Netherlands, Belgium, the Federal Republic of Germany, and the United Kingdom, that ratio is lower than in Haiti). The European population draws upon exhaustible energy and material resources, not only for industry but even for agriculture. Why not ask whether the EC and Japan have exceeded their carrying capacity, and whether their patterns of development are "sustainable"? Moreover, in several European countries there is a policy which promotes increasing or at least maintaining the present population, not through migration but on the contrary, through a higher birth rate in order to produce babies of good European stock. (This is a mistake if the ozone layer thins out, since white people are more likely to have skin cancer). In pushing for a higher population, there is an implicit assumption either of decoupling growth from the use of energy and materials (by increased efficiencies and recycling), or of a continuing ability to extract energy and materials at a cheap price from overseas countries in the characteristic European pattern of *Raubwirtschaft*. We have explored the first assumption in the previous section. The second assumption may prove realistic, if military power is applied when needed.

On March 10, 1989, there was in the Mediterranean an accident similar to those which sometimes happen between Santo Domingo (and Haiti) and the United States. Ten Moroccan would-be immigrants died at sea while attempting to reach Spain. The right to choose one's place of habitation on earth remains the most elusive of human rights. On the same day, by chance, it was announced that Spain, in keeping with the notion of a "fortress Europe" (perhaps a "lifeboat Europe," in Hardin's sense), will require visas for all Moroccan, Algerian, and Tunisian travellers (and also for Latin Americans), after March 1990. A government official (in his role as a sort of Maxwell's Demon) explained that Spain has a long coast near *paises con problemas demograficos* (El Pais Semanal, March 13, 1989, p. 14). Thus, Morocco has *problemas demograficos*, and Europe has no *problemas demograficos*. Migration, and the prohibition of migration, are not seen as a function of the difference in standard of living but as the consequence of the pressure of population on resources in the South. Nevertheless, when Italy, Spain, Portugal, and Greece were countries of outmigration, not so long ago, their population densities (perhaps with the exception of Portugal) were lower than today. Earlier, Germany and Britain were sending large numbers of migrants overseas when their population densities were lower than they are today. Migration usually is a result of "pull" factors, and in any case carrying capacity can be increased, if not from domestic resources, then by energy and materials subsidies from outside. Inside state frontiers, where there is usually freedom to migrate, migrants leave regions which are far from reaching the limit of carrying capacity (such as Western Andalusia in the 1960s, a region of great agricultural surplus). Across states, frontier police stop migrants who come from territories where they are not necessarily starving, but where there is a comparatively low level of consumption of energy and materials. Migration will stop only by the threat of violence, or by a greater equality in living standards. It is doubtful that equalization may be reached by overall economic growth rather than global redistribution.

States, frontiers, and policemen are social, historical products. Hence, the Maxwell's demons analogy, since Maxwell's demons were unnatural beings, who were supposed to be able to maintain, or even increase the difference in temperature between communicating gases by sorting out high-speed and low-speed molecules. Ecologists are quite good at explaining the movements of birds and fish, but today they are unable to explain the geographical distribution of the human population. The territorial-political units where environmental policy is made and applied have no ecological logic, and they are apt at shifting social costs out of their borders. Thus, arguments based on "carrying capacities" and the sustainability of development are blatantly ideological in their selective application.

A POLITICAL CONCLUSION

This paper has presented some old and new questions in ecological economics. "Externalities" is a word that describes the shifting of uncertain social costs (or possibly benefits) to other social groups (whether "foreigners" or not), or to future generations. The conclusion has been reached that because of big, diachronic, invaluable "externalities," economic commensurability does not exist separately from a social distribution of moral values regarding the rights of other social groups. This includes future generations. Economic commensurability (the ability to compare apples and oranges via some intermediary like price) is also not separate from social views (whether pessimistic or optimistic) regarding future technical changes. Such moral values, and views on technical change, perhaps are not class-specific, or gender-specific, or age-specific, but they are not distributed in the world at random, and they are historical; they change. On the other hand, attempts to base policy decisions not on economics but on an ecological rationality are bound to fail because trade-offs require values to be assigned to alternative results and costs, and ecology cannot provide such a system of valuation. A lack of commensurability surfaces again.

My conclusion is that the impossibility of an economic rationality (either based on the market or on central planning) which takes into account ecological side effects and uncertainties, and the impossibility, also, of deciding human affairs according to purely ecological planning, lead towards the politization of the economy. In other words, my conclusion is that the economy and the ecology of humans are embedded in politics. This, in turn, raises the question as to which are to be the territorial units and the procedures for decision making. Thus, many conferences have tried recently to define environmental agendas, which precedes making environmental decisions. Such conferences have unequal representation, as they lack representation from future generations, and probably, also, from large groups of our own generation, such as the bottom three or four billion poor. Because of the shortcomings of both ecological and economic rationalities, decision making in environmental policy is placed squarely back into the political field, away from the defensive screens provided by conventional environmental economics or by ecological planning. This political conclusion remains to be developed. It leads towards questions such as: Which are the territorial-political units which will decide environmental policy, and how will this affect the shifting of social costs to the poor, to "foreigners," and to future generations? How does politics determine not only environmental policy but also the environmental agenda, and even (the lack of) environmental education and environmental perception? For instance, how should we explain the current "global warming" scare, and how should we explain that it did not begin in 1900? How should we explain that the notions of *Raubwirtschaft* and of ecologically "unequal exchange" are not much used by international agencies? Will the resurrected faith in the market prove stronger than the new ecological awareness? Will ecological perspectives in rich countries lead once again to social-Darwinist views, while ecological socialism tends to grow in poor countries? To sum up: What are the politics of environmental policy?

REFERENCES

Arrhenius, Svante. 1903. *Lehrbuch der kosmischen Physik.* Leipzig: Hirzel.

Brunhes, Bernard. 1912. *La Degradation de l'Energie.* 2d. ed. Paris: Flammarion.

Brunhes, Jean. 1920. *Human Geography* Reprinted 1978. Chicago/New York: Rand McNally.

Brunhes, Jean. 1925. *La Geographie Humaine*, 3d. ed. Paris: Alcan.

Budyko, M. I. 1980. *Global Ecology* Moscow: Progress Publishers.

Callendar, G. S. 1938. The Artificial Production of Carbon Dioxide and its Influence on Temperature. *Quarterly Journal of the Royal Meteorological Society* 64:223-237.

Christensen, P. P. 1989. Historical Roots for Ecological Economics: Biophysical Versus Allocative Approaches. *Ecological Economics* 1(1):17-36.

Cipolla, Carlo. 1972. *The Economic History of World Population.* 6th ed. Penguin.

Crosby, A. 1986. *Ecological Imperialism: The Biological Expansion of Europe 900-1900.* Cambridge Cambridge University Press.

El Pais; Panorama Semanal. March 13, 1989. Madrid.

Farvar, M. Taghi and John P. Milton, eds. 1972. *The Careless Technology. Ecology and International Development.* Garden City, N.Y.: The Natural History Press. (Proceedings of a conference chaired by Barry Commoner and Kenneth Boulding.)

Georgescu-Roegen, N. 1971. *The Entropy Law and the Economic Process.* Cambridge, Mass.: Harvard University Press.

Georgescu-Roegen, N. 1986. The Entropy Law and the Economic Process in Retrospect. *Eastern Economic Journal* 12(1): 3-25.

Graf, Dieter (Hrsg.). 1984. *Oekonomie und Oekologie der Naturnutzung.* Jena: Gustav Fischer.

Grinevald, Jacques. 1987. Vernadsky and Lotka as Sources for Georgescu-Roegen's Bioeconomics. Draft Paper. 2nd Vienna Centre Conference on Economics and Ecology, Barcelona.

Hayek, F. A. von, ed. 1935. *Collectivist Economic Planning.* London: Routledge.

Hayek, F. A. von. 1952. *The Counter-Revolution of Science.* Glencoe, Ill.: Free Press. New ed. 1979. Indianapolis: Liberty Press.

Hirsch, F. 1976. *Social Limits to Growth.* Cambridge, Mass.: Harvard University Press.

Jeffery, J. W., 1988. The Collapse of Nuclear Economics. *The Ecologist* 18(1):9-13.

Kapp, K. W. 1983. *Social Costs, Economic Development, and Environmental Disruption.* Introduction by John E. Ullmann, ed. Lanham, Md.: University Press of America.

Kapp, K. W. 1987. *Fuer eine oekosoziale Oekonomie. Entwuerfe und Ideen.* Christian Leipert and Rolf Steppacher, eds. Frankfurt: Fischer (includes Kapp's biography and bibliography).

Leff, Enrique. 1986. *Ecologia y Capital.* Mexico: UNAM.

Lutz, Ernst and Salah El Serafy, eds., 1989. *Environmental and Natural Resource Accounting and their Relevance to the Measurement of Sustainable Development.* Washington D.C.: World Bank.

Martinez-Alier, J. with Klaus Schluepmann. 1987. *Ecological Economics* Oxford/New York: Blackwell.

Neurath, Otto. 1925. *Wirtschaftsplan und Naturalrechnung.* Berlin: Laub.

O'Connor, James. 1988. Introduction. *Capitalism, Nature, Socialism: A Journal of Socialist Ecology* 1(1):Fall.

Proops, John L. R. 1989. Ecological Economics: Rationale and Problem Areas, *Ecological Economics* 1(1):59-76.

Punti, Albert. 1988. Energy Accounting: Some New Proposals. *Human Ecology* 16(1):79-86.

Raumolin, J. 1984. L'homme et la Destruction des Ressources Naturelles: La Raubwirtschaf au Tournant du Siecle *Annales. E.S.C.* 39(4).

Ravetz, J. R. 1989. *The Merger of Knowledge with Power*. London: Mansell.

Ruha, R. 1988. Ideological Trends in Indian Environmentalism, *Economic and Political Weekly* 23(49), Dec. 3.

Ruha, R. 1990. *The Unquiet Woods: The Chipko Movement*. Berkeley: University of California Press.

Ruha, R. and M. Gadgil. 1989. State Forestry and Social Conflict in British India. *Past and Present* 123:141-177.

Sauer, Carl. 1956. The Agency of Man on Earth. In William L. Thomas, Jr., ed., *Man's Role in Changing the Face of the Earth*. Chicago: University of Chicago Press. (Proceedings of a conference in 1952, co-organized by Lewis Mumford.)

Soddy, Frederick. 1947. *Atomic Energy for the Future*. London: Constitutional Research Association.

Stadler, Friedrich. 1982. *Vom Positivismus zur "Wissenschaftlichen Weltauffassung.* Vienna/Munich: Loecker.

Tsuru, Shigeto. 1972. In Place of GNP. In Ignacy Sachs, ed., *Political Economy of Environment: Problems of Method*. Paris/The Hague: Mouton.

Vernadsky, W. 1924. *La Geochimie*. Paris: Felix Alcan.

Weil, Felix. 1926. Review of Otto Neurath, Wirtschaftsplan und Naturalrechnung. In *Archiv fuer Geschichte des Sozialismus*, hrsg. von Carl Gruenberg, 12 (Reprinted Graz, Syndikat, 1979).

Weissel, Erwin. 1976. *Die Ohnmacht des Sieges*. Vienna: Europaverlag.

White, Leslie. 1954. The Energy Theory of Cultural Development. In Beth Dillingham and Robert Carneiro, eds., *Ethnological Essays*. 1987 Albuquerque: University of New Mexico Press.

10

A NEW SCIENTIFIC METHODOLOGY FOR GLOBAL ENVIRONMENTAL ISSUES

Silvio O. Funtowicz
Institute for Systems Engineering and Informatics
Joint Research Centre
Commission of the European Communities
I-21020 Ispra (Va), Italy

Jerome R. Ravetz
The Research Methods Consultancy
13 Temple Gardens
London NW11 OLP, England

ABSTRACT

The extreme uncertainty of the methods used to address the disturbed global environment limits the application of traditional scientific methodologies to current problems. The use of computer models, which are inherently untestable but still are the best tools available, illustrates our dilemma with modern methods.

We use a simple diagram based on two attributes, "decisions stakes" and "systems uncertainties," to illustrate a threefold classification of kinds of science. First is the Applied Science reminiscent of Kuhnian puzzle-solving; second is Professional Consultancy; and third is Second Order Science, characteristic of the new sciences of cleanup and survival. For Second Order Science, facts are uncertain, values in dispute, stakes high and decisions urgent. Such sciences are important when, paradoxically, "hard" policy decisions depend on "soft" scientific inputs.

A new methodology for Second Order Science will require "extended peer communities" because quality assurance requires participants outside the classic peer communities of experts, including investigative journalists and laypersons. Similarly, "extended facts" will be relevant, such as evidence that is initially anecdotal or information that is restricted to the public. By these extensions, Second Order Science can lead to greater democracy in the scientific endeavor, complementary to the diffusion of science by traditional popularization.

INTRODUCTION

Few will still doubt that our modern technological culture has reached a turning point and that it must change drastically if we are to manage our environmental problems. It may not yet be as widely appreciated that science, hitherto the mainspring of technological progress, must also change. From now on, its central task must be to address the pathologies of our industrial system by reacting to new problems and devising new methods. These new problems and methods are the subject our of study.

The fundamental achievements of science, like those of all creative activities, have a timeless quality. The social activity of science, like any other, evolves in response to its changing circumstances. In the high Middle Ages, secular learning was removed from the monasteries and established in the universities; the boundary between the sacred and private on the one hand, and the secular and public on the other, was thus set for European culture. The Scientific Revolution of the seventeenth century was one of the great intellectual mutations of mankind, and reinforced the growing hegemony of European civilization in the world. The nineteenth century saw the replacement of "natural philosophy" by science, the growth of subject specialties, the institution of a value-free Scientific Method, and the first career opportunities for scientists. With this came the consolidation of the science-based professions, with their own institutions and formalized social contracts. In the recent postwar period we have experienced the industrialization of science, with the growth in scale and capital-intensity of research and its intimate connection with technology and political power. Paradoxically, as science prospered materially, it was losing its ideological function as the unique bearer of the True and therefore the Good.

Now global environmental issues present new tasks for science; instead of discovery and application of facts, the new fundamental achievements for science must be in meeting these challenges. Because of the very rapid changes in environment, society and science itself, and in their interactions, a general awareness of the new state of science has yet to be achieved. In this essay, we make the first articulation of a new scientific method, which does not pretend to be either value-free or ethically neutral. The product of such a method, applied to this new enterprise, is what we call "post-normal science."

We adopt the term "post-normal" to mark the passing of an age when the norm for effective scientific practice could be a process of puzzle-solving in ignorance of the wider methodological, societal, and ethical issues raised by the activity and its results. The scientific problems which are addressed can no longer be chosen on the basis of abstract scientific curiosity or industrial imperatives. Instead, scientists now tackle problems introduced through policy issues where, typically, facts are uncertain, values in dispute, stakes high, and decisions urgent. When research is called for, the problem must first be defined, and this will depend on which aspects of the issue are most salient. Hence, political considerations constrain which results are produced and thereby which policy implications are supported. In general, the post-normal situation is one where the traditional opposition of "hard" facts and "soft" values is inverted. Here we find decisions that are 'hard' in every sense, for which the scientific inputs are irremediably "soft."

UNCERTAINTIES IN RESEARCH RELATED TO POLICY

The concept of uncertainty is at the core of the new conception of science, for hitherto it has been kept at the margin of the understanding of science, for laypersons and scientists alike. Whereas science was previously understood as steadily advancing the certainty of our knowledge and control of the natural world, now science is seen as coping with many uncertainties in urgent technological and environmental decisions on a global scale. A new role for scientists will involve the management of these crucial uncertainties; therein lies the task of quality assurance of the scientific information provided for policy.

The new global environmental issues have common features that distinguish them from traditional scientific problems. They are global in scale and long term in their impact. Data on their effects and even data for baselines of "undisturbed" systems are radically inadequate. The phenomena being novel, complex and variable are themselves not well understood. Science cannot always provide well-founded theories based on experiments for explanation and prediction, but can frequently only achieve at best mathematical models and computer simulations, which are essentially untestable. On the basis of such uncertain inputs, decisions must be made, under somewhat urgent conditions. Therefore, science cannot proceed on the basis of factual predictions, but only on policy forecasts.

Computer models are the most widely used method for producing statements about the future based on data of the past and present. For many, there is still a magical quality about computers, since they are believed to perform reasoning operations faultlessly and rapidly. But what comes out at the end of a program is not necessarily a scientific prediction; and it may not even be a particularly good policy forecast. The numerical data used for inputs may not come from experimental or field studies; the best numbers available, as in many studies of industrial risks, may simply be guesses collected from experts. (And who has expertise in choosing experts?) Instead of theories which give some deeper representation of the natural processes in question, standard software packages may be applied with the best-fitting numerical parameters. And instead of experimental, field or historical evidence, which would normally form the basis for scientific theories, calculated outputs may merely be compared with those produced by other equally untestable computer models. Thus, in the practice of simulation, computer power allows articulation and flexibility to substitute for verisimilitude and testing against an external reality.

In spite of the enormous effort and resources that have gone into developing and applying such methods, there has been little concerted attempt to see whether they contribute significantly to our knowledge or to the quality of our decisions. In research related to policy on risks and the environment, apparently so crucial for our wellbeing, there has been very little of the sort of quality assurance that the traditional experimental sciences take for granted in ordinary practice. Whereas computers could in principle be used to enhance human skill and creativity by doing all the routine work swiftly and effortlessly, they have tended to become substitutes for thought and scientific rigor. Indeed, some distinguished scientists have questioned whether computer models should be used at all in the study of the global environmental problems. Thus, the American mathematician S. Mac Lane describes "systems analysis" as:

the construction of massive imaginary future "scenarios" with elaborate equations for quantitative "models" which combine to provide predictions or projections (gloomy or otherwise), but which cannot be verified by checking against objective facts. Instead, [such] studies often proceed by combining in series a number of such unverified models, feeding the output of one such model as input into another equally unverified model. . . . Such studies as these are speculations without empirical check and so cannot count as science. (1988, p. 1144)

In his defense of the field, N. Keyfitz reminded us that "many of the most difficult problems we have to face cannot even be precisely formulated in the present state of knowledge, let alone solved by existing techniques of science. . . . Such models, although unsatisfying to many scientists, are still the best guide to policy that we have." (Keyfitz 1988, p. 496).

In his reply, Mac Lane continued to doubt that the global problems should be tackled by making models "that in the first instance are not verifiable," and he added, "problems are not solved and science is not helped by unfounded speculation about unverifiable models." His concluding comment on quality assurance was a criticism of a certain research institution that he said, "does not appear to have an adequate critical mechanism, by discipline or by report review" (Mac Lane 1988, p. 1624).

To believe that the calculated outputs of untestable computer simulations should determine policies, is to indulge in the purest rationalistic fantasies, reminiscent of Leibniz or better of Ramon Lull. Indeed, we may speak of a new sort of pseudo-science, depending not on magic but on computers, which can be called GIGO ("Garbage In, Garbage Out"). This can be defined as a computational field where the uncertainties in the inputs must be systematically suppressed, lest the outputs become completely indeterminate. How much of our present social and environmental science belongs to this category, is an interesting and urgent question. Parallel to these computer-based pseudo-sciences are the computer-based pseudo-technologies. These base their appeal on a confusion between adequate computer graphics of an excellent technological system and excellent computer graphics of an imaginary technological system.

It is clear that the dilemmas of computer modeling in research related to policy cannot be resolved at the technical level alone. No one claims that computer models are adequate tools; and yet nothing better can be provided by traditional science. The critics basically judge them by the standards of mathematical-experimental science, and of course in those terms they are nearly vacuous. Their defenders advocate them on the grounds that they are the best possible, without appreciating how very different these new sciences of cleanup and survival are with respect to their complex uncertainties, new criteria for quality, and sociopolitical involvements. Exceptionally dedicated efforts are needed for the management of uncertainty, the assurance of quality, and the fostering of the skills necessary for these. Such skills will not be easily developed within the old framework of assumptions about the methods, social functions and qualified participants in the scientific enterprise.

The uncertainties in research related to policy are not restricted to computer models. Even the empirical data that serve as direct inputs to the policy process may be of doubtful quality. Their uncertainties frequently cannot be managed by traditional statistical techniques. As J. C. Bailar puts it:

All the statistical algebra and all the statistical computations are of value only to the extent that they add to the process of inference. Often they do not aid in making sound inferences;

indeed they may work the other way, and in my experience that is because the kinds of random variability we see in the big problems of the day tend to be small relative to other uncertainties. This is true, for example, for data on poverty or unemployment; international trade; agricultural production; and basic measures of human health and survival. Closer to home, random variability—the stuff of p-values and confidence limits, is simply swamped by other kinds of uncertainties in assessing the health risks of chemicals exposures, or tracking the movement of an environmental contaminant, or predicting the effects of human activities on global temperature or the ozone layer. (Bailar 1988).

Thus, in every respect the scientific status of research on these policy-related problems is dubious at best. The tasks of uncertainty management and quality assurance, managed in traditional science by individual skill and communal practice, are left in confusion in this new area. New methods must be developed for making our ignorance usable (Ravetz 1990d). The path lies in a radical departure from the total reliance on techniques, to the exclusion of methodological, societal or ethical considerations, that has hitherto characterized traditional science. This is the challenge that has led us to develop the idea of post-normal science, as the science that is appropriate to this post-industrial civilization.

UNCERTAINTY, QUALITY AND VALUES IN SCIENCE FOR POLICY

Any policy decision on global environmental issues will need to be made in the context of uncertainty, dependent on inputs of variable or even unknown quality. There is a growing concern among experts, politicians and the public about the uncertainties affecting data for major environmental issues, such as global warming. There seems to be no systematic solution to this problem; instead, uncertainty is manipulated politically, for accelerating or deferring major initiatives, depending on the outlook of the advocate. By contrast, the problem of quality assurance of information has been almost universally ignored. One reason for this neglect may be in the confusion between uncertainty and quality, and the naive belief that there is a straightforward relationship between them, high quality being equivalent to low uncertainty.

Hitherto the handling of these problems has oscillated between two extremes. At one end, there are perceptive philosophical analyses about the relation of knowledge and ignorance (Ravetz 1990d; Smithson 1989), and about the general phenomenon of quality criteria as employed in the policy process (Clark and Majone 1985). These provide a reflective understanding, but they cannot easily be translated into practical tools for quality evaluation of uncertain information. At the other extreme are the technical uncertainty analyses (see, for example, for nuclear power, Rivard et al. 1984; for climate model projections Hall 1985; and for an insightful general study, Beck 1987) and simple quality taxonomies (see, for example, Health and Safety Executive 1978, and United Nations Environmental Programme 1984). These combine classifications of sources of uncertainty, specific to each field, with mathematical formalisms that treat uncertainty as if it were an additional physical variable. It is small wonder that those who must cope with uncertainty in their work will generally ignore the whole subject in practice.

Whereas uncertainty is an attribute of knowledge, quality is a pragmatic relation between a product, or process, and its intended users. It can be defined as "the totality of characteristics of a product that bear on its ability to satisfy an established use" (see The British Standards for Quality Assurance, British Standards Institution 1979). Uncertainty and quality are two distinct attributes, for information of lesser certainty may yet be of good quality for its intended function. An extreme case of this is provided by deforestation in the Himalayas; although the estimates of the per capita fuel wood consumption vary through a factor of almost a hundred, all serious studies agree that their numerical predictions imply that the problem exists and that its solution is urgent (Thompson and Warburton 1985). An example of high certainty and very poor quality is provided by a prediction of a rise in the average temperature of the earth of 0 to 100°C over the next forty years due to the greenhouse effect. On a common sense basis, we may say that the true value is almost certain to lie within that range; but the climatic consequences in this range vary from the trivial to the nearly catastrophic. The prediction is nearly true by definition; its quality decreases accordingly because the statement approaches being analytical rather than synthetic—in other words, it tells us very little about the real world .

Thus, there are inherent limitations to the reduction of uncertainty in this kind of research. There is no point in ecological modeling (for example) trying to emulate experimental physics in its control of uncertainty. Each field of practice has a characteristic grade of information (rather like hotels or restaurants in grading schemes) appropriate to its needs; within that grade, information may vary in quality (as with hotels), depending on how well its uncertainties are managed and hence how well the information fits its function as an input to a decision process.

In ordinary scientific practice, considerations of values are largely implicit; even if they are operative in the choice of problems, once the research is underway they are put in the background. However, they are always present as part of the framework of the research; the myth of "value-free" science can be sustained only by ignoring the routinely used statistical methods. In any genuine statistical exercise, the design must take into account the error-costs of the possible alternatives. Thus, no single test can optimize both selectivity and sensitivity (avoiding the errors of false positives and false negatives). The choice, as expressed in numerical confidence levels, reflects the background of values, realized as costs and benefits, which condition every experimental program.

When ordinary scientific practice does not provide conclusive solutions for its problems, the values become explicit in the way inferences are made. The growing use of scientific expertise in the courts frequently reveals a mismatch between the traditional value-implicit rules of scientific inference and those appropriate in tribunals. Thus in the law courts, various special principles for controlling error-costs are invoked, including "balance of probabilities" and "burden of proof." Thus, in the latter case, the error-costs of convicting an innocent person are deemed to be higher than those of acquitting the guilty, at least in the Anglo-Saxon tradition. Tribunals of inquiry provide an illuminating case of bridging between the two approaches and their appropriate conceptions of value and error-cost. In the Black enquiry on the excess child leukemia cases in the neighborhood of the Sellafield nuclear reprocessing plant, the Scottish concept of a "not proven" verdict was explicitly applied for the possible cause of the excess leukemias (Macgill 1987).

In problems of risks and the environment, value considerations in scientific practice may be quite explicit. For example, we may consider the statistical design of a program for testing defective items in a large shipment. If it is of apples, then a bad one spoils only its barrel; but if it is of land mines, a premature explosion can destroy the neighborhood. The relative costs of false positives and false negatives are very different in the two cases. This example also illustrates the factor of "dread"—an important dimension of public perception of risks like nuclear power and genetically engineered organisms.

An integrated approach to the problems of uncertainty, quality and values has been provided by the NUSAP system (NUSAP stands for Numeral, Units, Spread, Assessment and Pedigree; for an introduction, see Funtowicz and Ravetz 1987). In its terms different sorts of uncertainty can be expressed and used for an evaluation of quality of scientific information. NUAP enables us to make the distinction between the sources and the sorts of uncertainty. Classification by sources is normally done by experts in a field when they try to comprehend the uncertainties affecting their particular practice. But for a general understanding, we have to distinguish among the technical, methodological and epistemological levels of uncertainty; these correspond to inexactness, unreliability and "bordering on ignorance," respectively (see Funtowicz and Ravetz 1990).

Uncertainty is managed at the technical level when standard routines are adequate; these will usually be derived from statistics (which themselves are essentially symbolic manipulations) which are supplemented by techniques and conventions developed for particular fields. The methodological level is involved when more complex aspects of the information, as values or reliability, are considered. Then personal judgements depending on higher-level skills are required and the practice in question is a professional consultancy, a "learned art" like medicine or engineering. Finally, the epistemological level is involved when irremediable uncertainty is at the core of the problem, as when modelers recognize "completeness uncertainties" which can vitiate the whole exercise, or when "ignorance of ignorance" (or "ignorance squared") is relevant to any possible solution of the problem. In NUSAP, these levels of uncertainty are conveyed by the categories of spread, assessment and pedigree, respectively.

There is no strict correspondence between these conceptual sorts of uncertainty and the sources we mentioned that are derived from practice. All data are affected by inexactness, and all computer models by ignorance. But all data exist within the framework of structures of concepts and procedures for their production, and of theories for their interpretation; hence, the higher levels of uncertainty are relevant to their evaluation and use. Therefore, we may say that ignorance is a part of data uncertainties. Similarly, the lowest level of uncertainty, inexactness, occurs in computer models, through the use of numerical analysis techniques which unavoidably involve rounding off and other approximation methods. Hence, the two approaches to the classification of uncertainty are quite distinct. A taxonomy based on sorts of uncertainty, like that of NUSAP, enables the construction of a general tool for the explicit communication of quality and values of the kind that appear in global environmental issues and policy related research.

THREE TYPES OF PROBLEM-SOLVING STRATEGIES

The inherent limitations of traditional problem-solving strategies are revealed by a structural feature of the new global environmental issues. For in these, decisions depend on evaluations of future states of the natural environment, resources, and human society, all of which are unknown and unknowable. The powers of science have not only produced irremediable uncertainties in knowledge, now we also find moral uncertainties, resulting from the invasion of the domains of the sacred and private. The most notable cases here are reproductive technology and also scientific research that requires the inflicting of pain on aware beings. Under these circumstances of radical uncertainty, a new type of problem-solving strategy is emerging. In post-normal science, the traditional description, "the art of the soluble" is no longer appropriate. For in this work, it is issues rather than problems that are examined, despite the amount of special scientific research and professional consulting conducted. Instead of the traditional images of conquering or managing, now it is better to think of coping and ameliorating. This is a far cry from the old excitement of scientific discovery or engineering creativity; but now we must cope with the consequences of those traditional activities that were conducted for so long in innocence of their effects.

We can compare the different sorts of problem-solving strategies that are now employed, by using a biaxial diagram which shows them in terms of the two attributes of "systems uncertainties" and "decision stakes," ranging from low to high, as in figure 10.1 (Funtowicz and Ravetz 1985; Rayner 1988). For systems uncertainties, the three intervals along the axis correspond nicely with the distinctions we have already made among the different sorts of uncertainty, namely technical, methodological and epistemological. It is easy to see how the different types of practice correspond to these different sorts of uncertainty. The other axis of the diagram relates practice to the world of policy; and the zones in two dimensions provide a full specification for any issue. To define decision stakes, we include the costs, benefits, and commitments of any kind by the parities involved. The diagram shows three divisions, corresponding to the three types of scientific practice that we have discussed. In the case of applied science, decision stakes and systems uncertainty are minimal; it is rare for a policy decision to depend on a single research result. For professional consultancy, the stakes and uncertainty range from moderate to severe: the medical doctor normally cares for the health or life of a single patient, but may also protect a wider community as with epidemiological problems. The engineer must consider the welfare of a client, and in connection with safety, that of a wider community. In post-normal science, when global environmental issues are involved, the stakes can become the survival of civilization as we know it or even of life on the planet. Although these distinctions are real, there is no pretence of quantifying either of the factors. The intervals, and the zones they define, provide a rough gauge which forms a part of an heuristic tool for distinguishing the three types of problem-solving strategies.

Looking at the diagram, we see that applied science is performed when both factors are low; in this case, puzzle solving in the Kuhnian sense is adequate (Kuhn 1962). But when either factor is in the moderate range, something extra must be brought into the

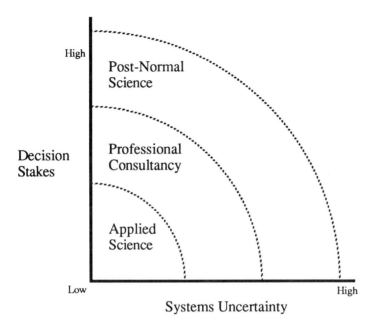

FIGURE 10.1: Three types of problem-solving strategies

work, which we can call consultant's skill or judgment. One very useful feature of the diagram is the way it shows that even when uncertainties are low, if decision stakes are high then puzzle-solving alone will not be effective in a decision process. For no scientific argument can be logically conclusive; even the accepted views of the philosophy of science acknowledge this. Scientific arguments evolve in a continuous dialogue which is incapable of reduction to logic. What makes scientists "rationally" change their opinions is a matter of ongoing debate among philosophers and sociologists (Chalmers 1990). Applying this lesson to policy debates, we can appreciate that when a party finds its interests threatened it can always find some methodological issue on which to challenge results. This is particularly easy in the case of research on risks or the environment. Thus the forum for decision becomes enlarged from that of the technical experts, to include those with a strong stake in the outcome.

All these tendencies to debate appear still more strongly in the case of post-normal science. Although there is still an essential place for professional consultancy and even for applied science, the extremes of decision stakes or of systems uncertainties render them inadequate for the whole work. Research work and the deployment of skills have a central role to play, but this must be done in the epistemological framework in which the narrowly defined problems are integrated into larger issues. In this way, they are provided with direction, quality assurance, and also the means for a consensual solution of policy problems in spite of their inherent uncertainties.

Examples of issues with combined high decision stakes and high systems uncertainties are familiar from the current crop of global environmental problems. Indeed, any of the problems of major technological hazards or large-scale pollution belong here. The paradigm case for post-normal science could be the design of a repository for long-lived nuclear wastes, to be secure for the next ten thousand years. The strength of our diagrammatic scheme can be illustrated by considering cases that fall close to either of the axes.

Examples of problems with low systems uncertainties are among the major disasters that have afflicted our modern industrial societies in recent years. Subsequent inquiries have in many cases established that the disaster had been "waiting to happen" through a combination of physical predisposing causes and management practices which had been well known in advance (e.g., Bhopal, Challenger, Exxon Valdez). Yet the processes of preventing a recurrence through improved regulations, or even of giving redress to the victims or punishing the culpable, can drag on for years or even decades.

A problem with low decision stakes will look very different; let us take for example the field of cosmology. There the data are so sparse, theories so weakly testable, and public interest so lively, that the field is as much "natural philosophy" as science, and experts must share the platform with amateurs, popularizers, philosophers and even theologians. In this example, we see an historical continuity between the science that was practiced before the establishment of authoritarian paradigms, and the post-normal science of the present. This can help us appreciate the methodological continuity between applied science, professional consultancy and post-normal science. Post-normal science is a development from and extension of traditional science, appropriate to the conditions of the present age. Its essential principle is that uncertainty and ignorance can no longer be expected to be conquered; instead, they must be managed for the common good. Programs of reform of technology or lifestyle which ignore this aspect of knowledge are likely to remain part of the problem rather than contribute to the solution.

The dominant historical experience within the lifetimes of those here now, is that science has created effective ignorance, because of our inabilities to cope with the consequences of progress. Paradoxical as it may appear (and such apparent paradoxes may reveal the leading contradictions of an age) each advance in technique now opens up new areas of ignorance. These are not merely stimuli to curiosity-driven research, but they can threaten to vitiate the practice itself, unless they are appreciated as part of an enriched conception of knowledge. Merely to see ignorance as negative and threatening, is to remain in the old scientific paradigm. In the philosophy that we are now articulating, ignorance is a vital complementary aspect of knowledge, and in many cases becomes the driving force of progress. In this way, ignorance can perform the same functions for scientific methodology as the infinite for mathematics. By definition, the infinite cannot be known completely, but through its fruitful contradictions it has created, not only powerful mathematical tools and also beautiful structures, but even new conceptions of mathematics itself. In such a context, we can genuinely speak of "usable ignorance," with the understanding that this is very different from "usable knowledge." For ignorance is usable when it is an object of awareness, and shows its dynamic interaction with knowledge.

By the use of the diagram, we can better understand the different aspects of complex projects in which all three sorts of practice may be involved. For this we may take an example of a dam, that was discussed previously (Ravetz 1971) in connection with an analogous classification of problems as scientific, technical and practical. First, in dam construction much basic accepted scientific knowledge is used; certain research projects with an "applied science" character will be used to describe the relevant features of the local environment and determine the details of its construction. But making the dam is, in the first place, a design exercise where the shape and structure is not determined by the scientific inputs. A compromise design will be chosen from the various possible functions of the completed dam, which may include water retention, hydroelectric power, flood control, irrigation and recreation opportunities. Achieving the optimum balance, given both the uncertainties in scientific inputs and value conflicts among interests, is a task for a professional. But the matter does not stop there. Some people may find their homes, farms and religious monuments drowned by the artificial lake; can they possibly be adequately recompensed? There may be a possibility of long-term deterioration of the hydrological cycle in the district, and perhaps even local earthquakes. Dams, once seen as a completely benign instrument of human control over raw Nature, have come to be seen as a sort of predatory centralism, practiced by vast impersonal bureaucracies against local communities and the natural environment. When such issues come into play, we are beyond professional consultancy and in the realm of post-normal science.

We can also use the diagram to illustrate how a problem can evolve so that it is tamed and brought towards manageability. When, for example, a risk or pollution problem is first announced, it will almost always be in a condition of considerable uncertainty. Since the problem had not been appreciated previously, it is unlikely that there will be substantial evidence about it. Hence the information will tend to be anecdotal on the experimental side and speculative on the theoretical side. But the strength of the decision stakes will ensure that all interests, aided by the independent media, will offer their opinions apparently with complete certainty. The first phase of the discussion will therefore resemble ordinary political debate, but of a particularly confused kind. Each side will attempt to define the problem in the terms most favorable to its interests; typically proponents will present it as applied science and opponents will stress its uncertainties and ethical aspects. It is a new phenomenon for such debates to be effective. Hitherto, commercial viability or state security was the overriding consideration for industrial development, subject to a natural concern for health and safety. Indeed, in recent decades scientists and engineers have been bewildered and dismayed when confronted with those who try to block progress on the basis of such intangible and nonscientific arguments. One of the last debates of the old sort was that over recombinant-DNA research in the 1970s, when the evolving problem was kept firmly under the control of the scientists (Ravetz 1990b); in the sequel conflict about genetic engineering, the critics have scored some signal successes, and those critics are now generally accepted as legitimate participants to the debate (Fincham and Ravetz 1991).

If such issues remained in the realm of pure power politics, the outlook for our policies for science, technology and the environment would be grim. But the pattern of evolution of issues, with different leading strategies coming to prominence, gives hope that the science may yet have an important role in such debates. As the debate develops from its initial confused phase, positions are clarified and new research is stimulated. Although the definition of problems is (as we have seen) never free of politics, an open dialogue ensures that such considerations are neither one-sided nor covert. As the discussion on the technical aspects develops, no advocates need admit they were wrong; a tacit shift in the terms of the dialogue is sufficient. And as new research eventually brings in new facts, the issue becomes more amenable to the approach of professional consultancy. A good example of this evolutionary pattern is the debate over lead in gasoline. In spite of the absence of conclusive environmental or epidemiological information, a consensus was eventually reached that the hazards of leaded gasoline were not acceptable.

Thus, the simple diagram of the three strategies for problem-solving enables us to see where traditional scientific practice is not effective, and why new dimensions need to be added to the problem-solving process. The diagram allows us to make a dynamic analysis of the evolution of an issue involving science and policy. Post-normal science is thereby given its place as a complement to the other more traditional problem-solving strategies.

It is important to appreciate that post-normal science is complementary to applied science and professional consultancy. It is not a challenge to the traditional practice of science, nor does it dispute that science provides reliable knowledge or exclusive expertise in its legitimate contexts. Recent critiques of science that examined scientific knowledge alienated from its social context, have created the view that "anything goes" in science. It is as if any charlatan and crank should have equal standing with qualified scientists or professionals (see, notably, Feyerabend 1975, and comment in Ravetz 1990a).

Another basis for criticism is that of quality assurance. The technical expertise of qualified scientists and professionals in accepted spheres of work is not being contested. What can be questioned is the quality of that work, especially with respect to its environmental, societal and ethical aspects. Previously, the ruling assumption was that these were somehow "external" to the work of science itself and that such problems could be managed by some appropriate societal mechanism. Now the task is to see what sorts of changes in the practice of science, and in its institutions, will be called for by the extensions of problems that are relevant to the quality of scientific and professional work. We have introduced these new aspects through the three strategies of problem solving.

"EXTENDED PEER COMMUNITIES" IN POST-NORMAL SCIENCE

In what we might now call "pre-normal" science, nearly all the practitioners were amateurs. They could and did debate vigorously on all aspects of the work, from data to methodology, but there was no in-group of established practitioners in conflict with an out-group of critics. In normal science, any outsiders were effectively excluded from dialogue; only in a Kuhnian "pre-revolutionary" situation, when the ruling paradigm (cognitive and social) could not deliver the goods in steady progress, would outsiders get the chance to be heard. In post-normal science there is still a distinction between insiders and outsiders, based (on the side of knowledge) on certified expertise and (on the social side) on occupation. But since the insiders are manifestly incapable of providing effective conclusive answers to many of the problems they confront, the outsiders are capable of forcing their way into a dialogue. When the debate is conducted before a lay public, the outsiders (including community activists, lawyers, legislators and journalists) may on occasion even set the agenda.

Because of these human aspects of the issues giving rise to post-normal science, all the elements of the scientific enterprise must be extended. First, experts whose roots and affiliations lie outside that of those involved in creating or officially regulating the issue must be brought in. These new participants, enriching the traditional peer communities and creating "extended peer communities" are necessary for the transmission of skills and for quality assurance of results. For in the case of the new sort of science, who are the "peers"? In Kuhn's normal science, they are colleagues on the job, engaged in that "strenuous and devoted effort to force Nature into the conceptual boxes provided by professional education." Such peers are still there, as scientists and experts, and they exercise quality control within the technical paradigm of their expertise. But the problems of the new sort of science are not ones of purely knowing-that within stable paradigms; they include knowing-how, incorporating broad and complex issues of environment, society and ethics. Hence it is necessary and appropriate for quality assurance in these cases to be enriched at the very least by the contribution of other scientists and experts, technically competent but representing interests outside the social paradigm of the official expertise.

It is important to realize that the need for enriched input is not merely the result of the external political pressures on science that occur when the general public is concerned about some issue. Rather, in the conditions of post-normal science, the essential function of quality assurance can no longer be performed by a restricted corps of insiders. When problems do not have neat solutions, when the phenomena themselves are ambiguous, when all mathematical techniques are open to methodological criticism, then the debates on quality are not enhanced by the exclusion of all but the academic or official experts. Knowledge of local conditions may not merely shape the policy problems, it can also determine which data is strong and relevant. Such knowledge cannot be the exclusive property of experts whose training and employment inclines them to abstract, generalized conceptions. Those whose lives and livelihood depend on the solution of the problems will have a keen awareness of how general principles are realized in their "back yards." It may be argued that they lack theoretical knowledge and are biased by self-interest; but it can equally well be argued that the experts lack practical knowledge and have their own forms of bias.

A study examining how local knowledge can be used to solve scientific and technological problems is only now getting underway. Some authors have recognized this as the key to genuinely sustainable development. Arnold Pacey gives examples to show how a truly successful technology is the outcome of a "dialogue" between what is an apparently more advanced innovative culture, and the apparently traditionalist receiving culture. Thus in African agriculture, the previous dominance of colonially introduced temperate-zone concepts is being replaced by the integration of tree and field crops (incomprehensible to Western experts), together with irrigation and minimal engineering (Pacey 1990).

In Europe, a recent survey by Brian Wynne of the University of Lancaster has shown how the sheep farmers of Cumbria in England sometimes have a better understanding of the ecology of radioactive deposition than the official scientists (Wynne 1990). The farmers would not have made the assumption that radioactive contaminants would drain away through their thin cover of moorland soil at the same rapid rate as through lowland pastures. Also, they would have recognized that high ground lying directly downwind of a major reprocessing plant (the nearby Sellafield plant of British Nuclear Fuels Ltd.) is likely to have a different deposition pattern from remote fields. Although they could not criticize the technically esoteric measurements made by the official scientists, they were fully competent to evaluate their methods and interpretations at every stage.

Along with the enrichment of the traditional scientific peer-communities we have a parallel enrichment of the cognitive basis of post-normal science; we speak of "extended facts." This is the material which is effectively introduced into a scientific debate on policy issues. It is now widely appreciated that the beliefs and feelings of local people, whatever their source and validity, must be recognized and respected, lest the people become totally alienated and mistrustful. But extended facts go beyond a purely subjective base. Anecdotes will also be circulated verbally, and the edited collections of such materials will then be prepared for public use by citizens' groups and the media. This information will not usually be in a traditional scientific form, but may be essential for establishing a prima facie case for the existence of a problem, and therefore the urgency of systematic research. When such testimonies are introduced into scientific debate, and subject to some degree of peer review before reporting or acceptance, they approach the status of scientific facts. Similarly, the experiences of persons with a deep knowledge of a particular environment and its problems, like the hill farmers of Cumbria reported by Wynne, can provide equally strong evidence. We should also consider material discovered by investigative journalism. Finally, the category of extended facts can also be applied to information which is quite orthodox in its production, but which for political or bureaucratic reasons is officially secret in some way or other. The information can then function covertly, forming a background for loaded public questions. This last sort of "fact" may seem very strange to those whose idea of science is derived from textbooks and academic research laboratories. But for those who are familiar with science in the policy context, such extended facts may be quite crucial for quality assurance of results on which health and safety depend.

Since post-normal science depends so critically on data which are frequently inadequate in quantity and quality, the pitfalls in its production and interpretation are particularly severe. Scientists who are engaged in an academic exercise, or those working for a bureaucracy with a vested interest in the issue, will not normally be inclined to check for all the possible hidden traps that could vitiate their results. It is entirely natural and appropriate for those with a personal interest in the issue, and a personal knowledge of the phenomena, to engage in a dialogue on quality assurance. As yet this has happened only sporadically, and in a context of conflict and polarization of interests. The task is to create the conceptual structures, along with the political institutions, whereby a creative dialogue may be developed. For this, post-normal science is a foundational element.

CONCLUSION

We have now reached the point where a narrow scientific tradition is no longer appropriate to our needs. Unless we find a way of enriching our science to include practice, we will fail to create methods for coping with the environmental challenges, in all their complexity, variability and uncertainty. Fortunately, the conditions are ripe in the changing social distribution of knowledge and skills for making these changes. For now the liberal arts, as rhetoric, are no longer restricted to a tiny privileged elite in society, and the manual arts have lost the stigma of belonging to the oppressed majority. The improvement of manners and morals, from the Enlightenment through industrialized society, has been real. In modern societies there are now large constituencies of ordinary people who can read, write, vote and debate. The democratization of political life is now commonplace; its hazards are accepted as a small price to pay. Now it becomes possible to achieve a parallel democratization of knowledge, not merely in mass education but in enhanced participation in decision making for common problems.

The democratization of science in this respect is therefore not a matter of benevolence by the established groups. Rather it is (as in the sphere of politics) the creation of a system which in spite of its inefficiencies is the most effective means for avoiding the disasters that in our environmental affairs, as much as in society, result from the prolonged stifling of criticism. Let us be quite clear on this: we are not calling for the democratization of science out of some generalized wish for the greatest possible extension of democracy in society. The epistemological analysis of post-normal science, rooted in the practical tasks of quality assurance, shows that such an extension of peer communities, with the corresponding extension of facts, is necessary for the effectiveness of this new sort of science in meeting the challenges of global environmental issues.

REFERENCES

Bailar, J. C. 1988. *Scientific Inferences and Environmental Problems: The Uses of Statistical Thinking*. Institute for Environmental Studies. University of North Carolina, Chapel Hill.

Beck, M. B. 1987. Water Quality Modelling: A Review of the Analysis of Uncertainty. *Water Resources Research* 23:1393-1442.

British Standards Institution. 1979. BS 4778, London.

Chalmers, A. 1990. *Science and Its Fabrications*. Milton Keynes: Open University Press.

Clark, W. C. and G. Majone. 1985. The Critical Appraisal of Scientific Inquiries with Policy Implications. *Science, Technology and Human Values* 10(3): 6-19.

Feyerabend, P. K. 1975. *Against Method*. London: New Left Books.

Fincham, J. R. S. and J. R. Ravetz. 1991. *Risks and Benefits of Genetically Engineered Organisms*. Milton Keynes: Open University Press.

Funtowicz, S. O. and J. R. Ravetz. 1985. Three Types of Risk Assessment: A Methodological Analysis. In C. Whipple and V. Covello, eds., *Risk Analysis in the Private Sector*. New York: Plenum.

Funtowicz, S. O. and J. R. Ravetz. 1987. The Arithmetic of Scientific Uncertainty. *Physics Bulletin* 38: 412-414.

Funtowicz, S. O. and J. R. Ravetz. 1990. *Uncertainty and Quality in Science for Policy*. Dordrecht: Kluwer.

Hall, M. C. G. 1985. Estimating the Reliability of Climate Model Projections—Steps towards a Solution. In M. C. MacCraken and F. M. Luther, eds., *The Potential Climate Effects of Increasing Carbon Dioxide*. DOE/ER-0237, Washington D.C.: U.S. Department of Energy.

Health and Safety Executive. 1978. Canvey: An Investigation of Potential Hazards from Operations in the Canvey Island/Thurrock Area. London, HMSO, 48.

Keyfitz, N. 1988. Letters. *Science* 242:496.

Kuhn, T. S. 1962. *The Structure of Scientific Revolutions*. Chicago: University of Chicago Press.

Macgill, S. M. 1987. *The Politics of Anxiety*. Chapter 8. London: Pion.

Mac Lane, S. 1988. Letters. *Science* 241:1144; 242:1623-1624.

Pacey, A. 1990. *Technology in World Civilization*. Oxford/Cambridge, Mass. Blackwell.

Ravetz, J. R. 1971. *Scientific Knowledge and Its Social Problems*. Oxford: Clarendon.

Ravetz, J. R. 1989a. Ideological Commitments in the Philosophy of Science. In J. R. Ravetz, 1989c.

Ravetz, J. R. 1989b. Recombinant DNA Research: Whose Risks? In J. R. Ravetz, 1990c.

Ravetz, J. R. 1989c. *The Merger of Knowledge with Power*. London: Mansell

Ravetz, J. R. 1989d. *Usable Knowledge, Usable Ignorance: Incomplete Science with Policy Implications*. In J. R. Ravetz, 1990c.

Rayner, S. 1988. Risk Communication in the Search for a Global Climatic Management Strategy. In H. Jungermann, R. E. Kasperson and P. M. Wiedemann, eds., *Risk Communication*. Julich, KFA, 169-176.

Rivard, J. V. et al. 1984. Identification of Severe Accident Uncertainties. NUREG/CR-3440, SAND 83-1689, USNRC, Washington D.C.: U.S. Government Printing Office.

Smithson, M. 1989. *Ignorance and Uncertainty*. New York: Springer-Verlag.

Thompson, M. and M. Warburton. 1985. Decision Making under Contradictory Certainties. *Journal of Applied Systems Analysis* 12:3-34.

United Nations Environment Programme. 1984. State of the Environment, Addendum. GC.12/11/Add.2.

Wynne, B. 1990. Personal communication.

11

RESERVED RATIONALITY AND THE PRECAUTIONARY PRINCIPLE: TECHNOLOGICAL CHANGE, TIME AND UNCERTAINTY IN ENVIRONMENTAL DECISION MAKING

Charles Perrings
University of California
Department of Economics
Riverside, CA 92521 USA

ABSTRACT

Many of the most intractable environmental problems are those in which the use of environmental resources in novel ways has effects that are highly uncertain in both their spread and duration. The greater the uncertainty of the effects of technologically innovative use of environmental resources, the greater is the difficulty in evaluating associated environmental damage or the marginal social costs. The wider and more durable the environmental effects of economic activities are, the less is the scope for a market solution involving the allocation of property rights. Problems of uncertain environmental effects that may be global in spread or may endure for generations require responses that go beyond existing evaluation techniques. This paper considers the theoretical content of the precautionary principle n connection with problems of this sort. It discusses an approach to sequential decision making under uncertainty in which decision makers may choose to reserve their position on certain data by taking an initially cautious approach that may be relaxed as the data set is enriched by experience. The main case addressed here is one that existing economic approaches are the least capable of addressing—when the probability of distant but potentially catastrophic environmental damage is admitted, but thought to be very low.

INTRODUCTION

In recent debates over the problem of global climate change there has been strong support for the application of a "precautionary principle."[1] There is, however, no consensus on

---

[1] This was, for example, the main recommendation of the conference on Sustainable Development, Science and Policy, Bergen, May 1990.

what the principle means for decision making under uncertainty, either generally or in the context of specific issues such as global climate change. While it is clear that a precautionary principle implies the commitment of resources now to safeguard against the potentially adverse future outcomes of some decision, it is not clear how this should be related to the uncertainty associated with those outcomes. The sort of decisions in which the precautionary principle is currently being invoked are those for which the probability distribution of future outcomes cannot be known with confidence. Indeed, all that may be known is that the probability of distant but potentially catastrophic outcomes is positive—even if there is no information on the precise nature, timing or incidence of those outcomes. They are outcomes against which it is impossible to insure commercially because there are insufficient data on which to estimate an expected value for future losses within acceptable limits of confidence. This implies two sources of difficulty. First, it is not clear what relationship should exist between "safeguard" resources committed now and the expected value of future losses. Second, since the principle implies the commitment of public resources, it concerns collective decision making, and so raises the familiar difficulties associated with collective responsibility for future generations.

This paper's main concern is with the precautionary element in decision making under uncertainty, and with the relationship between decision making processes reflecting a precautionary principle and more conventional minimax (minimizing the potential for the worst case) and stochastic (probabalistic) approaches. Specifically, the paper considers estimating future environmental costs where there is a high degree of uncertainty about the nature, incidence and timing of those costs, and where the possible effects may be catastrophic for future generations. It is argued that the application of a precautionary principle in these circumstances implies a reservation on the expected value of those costs. Put another way, decision makers applying a precautionary principle will reserve their position on data for which the expected value (taken with whatever risk aversion coefficient applies) is judged to be an inappropriate measure. By analogy with the concept of bounded rationality developed for decision making under uncertainty in the industrial context (cf. Williamson 1985), we may refer to this phenomenon as reserved rationality.

The paper is in five sections. The following section addresses the second source of difficulty identified: the ethical issues raised by the responsibility borne by present generations for the welfare of future generations. The next section returns to the main concern of the paper and introduces a formal framework within which to consider the problem of decision making under incomplete information. A fourth section considers the decision making process under the precautionary principle, and a final section discusses the link between reserved rationality in decision making under uncertainty and the commitment of safeguard-resources under the precautionary principle.

DISCOUNTING, VALUATION AND TREATMENT OF FUTURE GENERATIONS

One of the strongest arguments in favor of the precautionary principle is that environmental effects that are both distant in time (are visited only on future generations), and thought to have a low probability of occurring, necessarily receive little attention in decision models focussing on the discounted expected value of future effects. The prospect of catastrophe in such cases is necessarily assigned a weight close to zero. The ethical questions this raises concern the right of present generations to put at risk not just the marginal benefits to future generations of access to a particular resource or ecosystem, but the very survival of those generations. There are two main questions at issue: the first concerns the rate of discount, the second concerns the valuation of resources under the current structure of property rights.

The rate of discount is an issue for the simple reason that discounting implicitly involves ethical judgements both about intertemporal or intergenerational equity, and about the appropriate treatment of uncertainty. The social discount rate is a measure of the rate at which it is considered socially desirable to substitute consumption in the present for consumption in the future. It accordingly involves a judgement about the responsibility that the present generation should bear for future generations. At positive discount rates, an unsustainable development strategy may well be judged to be optimal—implying, at the very least, indifference to the welfare of future generations. Indeed, as has been pointed out elsewhere, the impoverishment of future generations as a result of the profligacy of present generations in such cases is not just an incidental, but a desirable outcome of the decision making process (Dasgupta and Heal 1979). At zero discount rates, on the other hand, a development strategy will only be judged to be optimal if it yields a constant income stream in all periods. The optimality criterion of intertemporally egalitarian social welfare functions, and the maximization of the minimum period-income, implies a strategy that delivers maximum constant income.

Moreover, a positive discount rate implies a judgement about the responsibility of present generations for potentially catastrophic impacts about which there is a high level of uncertainty. Since uncertainty tends to be greater for longer time-horizons of the decision maker, positive discount rates act to screen uncertainty out of the information relevant to the decision making process (Perrings 1987). The more the future is discounted, the fewer the uncertain future effects that are relevant to the optimization problem, and the fewer the data that will be sought to solve the problem. Decisions taken on the basis of data relating to a short time horizon will accordingly tend to yield more unexpected effects than decisions taken on the basis of data relating to a longer time horizon. This is partly because in any given resource-based activity the rate of depletion of both renewable and nonrenewable resources, and/or the rate of emission of pollutants, are increasing functions of the discount rate. In the depletion problem, a low discount rate implies that more of an exhaustible resource will be left in the environment for future extraction than under a high discount rate. In the pollution problem, a low discount rate implies that less of the pollutant will be emitted in the present than under a high discount rate. The build-up of pollutants will proceed at a slower pace. The effect of discounting in this case is thus both to increase the potential for unexpected future costs, and to eliminate those costs from present consideration. We shall return to this momentarily.

Historically, economists have taken very different positions on the ethics of discounting. Ramsey, Pigou, and Harrod—representing the mainstream view between the 1920s and the 1950s—were all strongly critical of the ethical judgements involved in positive discount rates. Subsequently, however, the sovereignty of the present generation of consumers has been invoked both to deny any role for the state in securing the welfare of future generations, and to assert the propriety of discounting (Marglin 1963). Indeed, those economists who continue to hold that sustainable development implies discount rates at or close to zero are in a small minority. The arguments of Randers and Meadows (1973) and Myrdal (1975) against the rights of individuals to inflict harm on future generations through myopic behavior are still reflected in the work of Daly (Foy and Daly 1989), though the latter acknowledges that sustainability cannot be guaranteed by adjustments to the discount rate alone. Leaving the impatience of sovereign consumers to one side, the main factor in this shift in perception on the ethics of discounting is the recognition that a positive rate of growth of services in the system "authorizes" the discounting of the future at that rate. The argument against the rates of discount embodied in current real rates of interest is then an argument about the actual or potential growth rate of the system, and the position taken by those seeking to integrate the environment in the analysis of economic systems is that the marginal efficiency of capital assessed over the current capital base (which excludes natural capital) is a highly misleading proxy for the growth potential of the whole system (which includes natural capital) (Daly 1973; Perrings 1987).

It should be said that this is not necessarily an argument for zero rates. Although it has long been recognized that discount rates greater than the natural rate of regeneration of an environmental resource will lead to the depletion of that resource, it is now also argued that discount rates that are too low may result in unsustainable levels of investment in environmental resource-using activities (Pearce, Markandya and Barbier 1989). There is, in fact, a strong argument that the discount rate should not be manipulated to achieve intergenerational equity goals, and that these goals should be satisfied directly through intergenerational transfers (Norgaard and Howarth 1990). The argument is well-founded in the sense that the welfare of future generations may be compromised by the adoption of unsustainable discount rates. But it should not be taken to mean that we may ignore the intergenerational equity implications of specific rates. We should not, for example, be blind to the potential costs imposed on future generations by the adoption of real interest rates that bear no relation to the productive potential of the global system—that are intertemporally inefficient. Real interest rates that are excessive in these terms not only invite the depletion of stocks of natural capital, they also ensure that no matter how catastrophic the future environmental effects of present activities are thought to be, they will be irrelevant if they lie beyond a generation in the future.

The second set of ethical questions raised by this class of problems concerns the impact of the structure of property rights on the valuation of resources. Under existing property rights environmental resources are not allocated on the basis of their social cost: their full intertemporal opportunity cost. Property rights tend to be such that resource users are confronted by the direct costs of resource use only, and are able to ignore the other important components of intertemporal opportunity cost: intersectoral costs (intratemporal external costs) and user costs (intertemporal external costs) (see Pearce and Markandya 1989). The fact that an activity imposes uncompensated costs on future generations poses as much of an ethical problem as the fact that it imposes uncompensated costs on other members of the present generation. Certainly, there is scope for both intergenerational and intragenerational transfers to effect compensation in the two cases, but whereas it is possible to assign property rights in a way that ensures that intratemporal external costs are accounted for, the same is not true for all intertemporal external costs.

While the existence of overlapping generations means that continuous costs known with certainty can be compensated in intergenerational transactions, the same is not true of discontinuous and uncertain costs which might be borne in the distant future. Firstly, there exist no means for estimating such costs directly. Much effort has gone into the estimation of the value placed by the present generation on future access to a resource (its option and existence value), but the class of problems requires the estimation of costs actually to be born by future generations, not the future costs to be borne by the present generation(s). It is not possible to apply hedonic price or contingent valuation methods in such cases. Secondly, there exists no means of allocating property rights to unborn generations in any meaningful way. The net result is that such user costs will be taken into account only in so far as the present generation(s) feel an ethical responsibility for the generations to come.

DECISION MAKING UNDER INCOMPLETE INFORMATION

The problem of determining the optimal use of resources under uncertainty belongs to the general class of problems of decision making under incomplete information. How the problem is framed in any given circumstance will depend on the sense in which information is incomplete. If information is incomplete only in the sense that it is not known which of a well defined set of outcomes will eventuate, but that the probability of each outcome is known, then the problem is one of decision making under risk. If information is incomplete in the sense that it is not known in advance whether the outcome will belong to a particular set, nor is it known in advance what the probability of each outcome in that set is, then the problem is one of decision making under uncertainty.[2] It is a characteristic of decision making under risk that no new relevant information is acquired in the course of time. By contrast, it is a characteristic of decision making under uncertainty, that new information is acquired over time.

Uncertainty will typically exist where some activity, or a constituent part of some activity, is without historical precedent. It will not then be possible to predict the set of outcomes of such activities, let alone the probability of each outcome in that set. That is, the future states of nature associated with such activities and the probability of occurrence of the future states of nature will be unknown at the moment the activities are

---

[2] The distinction between risk and uncertainty in this paper is Knightian, in the sense that it emphasizes the difference between those cases in which the set of possible outcomes and the probability of those outcomes is known in advance, and those cases in which it is not.

undertaken. However, as the constituent parts of an economic activity are undertaken a historical record of their outcomes will be built up, thereby changing the boundary between the known and unknown outcomes of the activity as a whole. Repeated enumeration of events makes possible the development of increasingly robust predictions. Symmetrically, as the limits of what is "known" in a probabilistic sense change, so too does the list of outcomes of the remaining aspects of the activity. Hence the decision making process will itself evolve sequentially in response to the changing information available to the decision maker, and to the decision maker's changing perceptions of that information.

To provide a framework for the discussion of decision making under the precautionary principle, this section specifies a model of sequential decision making under incomplete information in two variants: stochastic and minimax. Each variant requires sufficient initial information to initiate an iterative decision making process, although the nature of such information requirements differs in the two cases. Each also allows the acquisition of new information at successive stages of the process. Formally, the two variants are very similar.

Let the state of knowledge of the system at time t be denoted W_t and assume perfect recall of all observations before time t. Hence W_t gives the history of all observations on the system up to time t. The evolutionary nature of the system is such that new unanticipated (and unanticipatable) observations are added in each period. Hence, W_t may be said to be observable at W_{t+1}, but W_{t+1} is not observable at W_t. This is known as the monotonic property of knowledge: the future is not knowable on the basis of the past history of the system. Further, let all actions depend on the state of knowledge at the time they are made, and define an action in time t as $u_t = u(W_t)$. A rule specifying u_t for all t is called a policy, and may be denoted $\upsilon = \{u_t\}$. From the monotonic property of knowledge, u_t may be said to be observable given W_{t+1}, but not to be observable given W_t. That is, W_t specifies the history of observations on the system *before* u_t is undertaken. The problem is to determine an optimal policy in circumstances where each action has implications that are not observable at the time the action is undertaken.

To specialize this general problem, let the social objective be to minimize the environmental costs, C, of a policy, υ, for the use of natural resources, given the developing history of the ecosystem, $\{W_t\}$. Incomplete information in this case follows from the fact that C is a function of quantities that are not observable at the moment each action is undertaken.

In the stochastic approach to the problem, the initial estimate of these environmental costs is the expected value of those costs conditional on the observed history of the system at time zero 0 and on the adoption of policy υ, denoted $E\upsilon(C|W_0)$. The initial informational requirements of this approach are accordingly quite severe. It requires a stochastic model of the environmental damage function, which assumes knowledge of the probability distribution of future environmental costs. The optimal policy in such a case is the one that minimizes average environmental costs over variation in the unobserved part of the history of the system.

In the minimax approach to the problem, the initial estimate of environmental costs is the "maximum" of C, conditional on the observed history of the system at time 0 and adoption of policy υ, which may be written $M\upsilon(C|W_0)$. Uncertainty about environmental costs in this case is rendered tractable by assigning a "worst case" value to C, given W_0. The optimal policy in the minimax case is the one that minimizes the maximum environmental costs over variation in the unobserved part of the history of the system.

Mathematically, the stochastic and minimax variants are similar, and the results carry over readily from one to the other. Taking the minimax example and assuming a finite horizon, T, the "worst case" costs of policy υ from time t are defined as

$$Hm(u, W_t) = Mu(C|W_t) \tag{1}$$

From the fact that u_t is observable at W_{t+1}, this may be written in the form of the recursion

$$Hm(u, W_t) = Mu[Hm(u, W_{t+1})|W_t] \tag{2}$$

The implication of the finite horizon, T, is that the costs of the policy u at time T are independent of u_T, so that

$$H_M(\upsilon, W_T) = H_M(W_T) \tag{3}$$

Defining $G_M(W_t)$ to be the minimum of $H_M(\upsilon, W_t)$ over all possible u_t, this yields the optimality equation

$$G_M(W_t) = \inf_{u_t} M[G_M(W_{t+1})|W_t, u_t] \tag{4}$$

with terminal condition

$$G_M(W_T) = H_M(W_T) \tag{5}$$

The values of u_t that minimize the maximum environmental costs for all t in (4), conditional on the state of knowledge in each period, define the optimal policy, υ^*. It can be shown[3] that with such a policy $H_M(\upsilon^*, W_t) \pounds H_M(\upsilon, W_t)$ for all υ, all t, and all W_t.

The stochastic version of this problem is entirely analogous, with the optimality equation identical to (4) except that the maximizing operator, M, is replaced by the expectation operator, E. That is, we have:

$$G_E(W_t) = \inf_{u_t} E[G_E(W_{t+1})|W_t, u_t] \tag{6}$$

with terminal condition

$$G_E(W_T) = H_E(W_T) \tag{7}$$

The minimizing value of u_t in (6) is optimal, and defines the minimum expected environmental costs over all policies, conditional on the state of knowledge at each point in

---

[3] By reasoning analogous to that in the steps before equation (16).

time. This completes our description of a sequential decision making framework under incomplete information.

DECISION MAKING UNDER THE PRECAUTIONARY PRINCIPLE

The class of problems for which the precautionary principle is advocated includes those in which both the level of fundamental uncertainty and the potential costs (or stakes) are high. This places such problems in the realm of what Funtowicz and Ravetz (1990) have referred to as "second order science", where "traditional" science is argued to be "inadequate" and ethical judgements are argued to be "ubiquitous". The precautionary principle does indeed involve a highly normative judgement about the responsibility borne by present generations towards future generations. Nor can this be captured in existing models of rational decision making. Nevertheless, the gulf between such "second order" problems as global climate change where the precautionary principle is invoked and the more mundane sequential decision making problems for which dynamic programming techniques were devised is not so wide that the techniques themselves are irrelevant. As shown in section above, neither of the stochastic and the minimax variants of the decision making model is formally privileged over the other. Yet the minimax variant does, as we shall see, imply a normative judgement about the appropriate attitude to uncertainty. Indeed, there is a strong sense among some economists that a minimax approach is, for this reason, less rational than an expected value approach (Maler, 1989a, 1989b).

The link between the precautionary principle and the "worst case" orientation of the minimax approach is quite intuitive. The precautionary principle requires the commitment of resources now to safeguard against the potentially catastrophic future effects of current activity. It is those effects which provide a point of reference for the principle, and it is natural to think of them as the "worst case" effects. But it is important to clarify how the "worst case" is arrived at in conditions of incomplete information, where the range and distribution of outcomes is not in fact known in advance. The "worst case" identified for decision making purposes in minimax approach under incomplete information of the sort discussed here cannot be the most extreme of a known range of outcomes, since the range of outcomes is not known. Nor can it be the worst imaginable case. It is always possible to construct a story in which the worst case environmental costs approach infinity for any policy, but such a construction would not only paralyze all activity, it would fail utterly to discriminate between different policies. Since it has to be believable enough to command the attention of decision makers, an operational "worst case" must be something else.

Recent work in the theory of decision making under uncertainty by Katzner (1986, 1989) suggests that what is currently interpreted as the "worst case" may better be represented by Shackle's concept of the focus loss of a decision.[4] In the Shackle approach, it is argued that decision makers conjecture a set of future states associated with any given action, and form an opinion on nonprobabilistic grounds about the degree of disbelief

---

[4] See also Perrings, 1989.

they would have in the occurrence of each state (Shackle, 1955, 1969). More particularly, to each future state they attach a measure of the potential surprise that they imagine they would experience if that state actually occurred. Since these measures can be mapped into the closed interval [0,1] they are obviously akin to a subjective probability distribution. The set of choice options—in this case the set of policies available to the decision maker—is ordered by an "attractiveness function" that registers the power of each outcome to command the attention of the decision maker. In this paper, the outcomes of interest are the environmental costs of each option, and the cost which becomes the reference point for the decision maker is the focus loss of the option. Outcomes will typically attract greater attention, the smaller the potential surprise they involve and the greater the extent of the damage they imply. Potentially catastrophic effects will, however, only seize the decision maker's attention if the prospect of their occurrence excites minimal disbelief. Thus, catastrophic but scarcely believable outcomes of vanishingly small probability will be ignored. On the other hand, outcomes such as Chernobyl may become a focus of decision making despite an estimated probability of occurrence of 0.0001 simply because they are believable events.

A point of central importance here is that the focus loss of a decision is adopted as the point of reference in the decision making process whenever the data are judged to be insufficient to support a decision on expected values. This will happen wherever the decision maker is unable to accept the data, whether because it is objectively scarce or because of the overriding importance attached to protecting some population placed at risk by the activity. The scarcity of data is closely associated with the evolutionary nature of the global system, and with technologically innovative uses of environmental resources. It follows that in almost any activity data will be scarce with respect to certain outcomes. However, whether any data set is regarded with confidence by the decision maker will depend on ethical judgements they may make about the weight attached to the welfare of those affected. Inadequate data will often be accepted as the basis of decisions where the welfare implications of a bad decision are trivial. This is the notion of bounded rationality. On the other hand, data that look very solid to scientists may be treated as suspect by decision makers if the welfare costs associated with adverse outcomes are unacceptable, regardless of the objective probability of such adverse outcomes. We shall return to this point later. In both cases, decision makers will tend to reserve their position by taking an initially cautious stance—using the "worst case" or focus loss as their point of reference. We may refer to this as reserved rationality.

The main implication of this discussion is that decisions involving uncertain environmental effects on future generations will tend to contain elements of both approaches, the balance reflecting both the degree to which the activity has historical precedents and the ethical environment within which it is made. The estimate of environmental costs of activities will tend to comprise two broad parts: the expected value of the environmental costs of those constituent actions for which the data set is sufficient to identify the set of all possible outcomes, and the focus loss of those constituent actions for which this is not true. The present value of such costs is obtained by discounting both parts at the appropriate rate. The stochastic case addresses the first data set, providing a rule for assigning probabilities to each of the set of outcomes for which the mean and variance is supplied by the decision maker. The focus loss case addresses the second data set, identifying the focus environmental damage associated with each policy available to the decision maker (analogous to the subjective expected value of the environmental damage function). Choice of the most appropriate criterion for decision making accordingly rests both on the nature of the data base, and on the institutional and ethical conditions under which the policy is devised.

To see the implications of this for the theory of decision making under incomplete information, let us first relate the focus loss of a policy to the expected cost of the same policy. If decision makers choose to employ the focus loss of a policy as their point of reference, then there must exist some variation in cost that would compensate them if they were constrained to adopt the expected cost of the same policy. We may denote this compensating variation in cost $V < 0$, and define the focus loss of a policy $\{u_t\}$ to be:

$$F\upsilon(C|W_t) \equiv E\upsilon(C-V|W_t) \tag{8}$$

where $E\upsilon(-V|W_t)$ is the reduction in expected environmental costs that would be necessary to compensate decision makers for adopting those expected costs, $E\upsilon(C|W_t)$, as their point of reference. The expected costs are defined by the expected value of *available* cost estimates: scientific predictions as the effects of global warming based on model simulations, say. This expression merely asserts that it is possible to define a relation between the expected value of available cost estimates associated with a given policy and the focus loss of that policy.

Now, in general, we would not expect the decision maker to reserve judgement on all available data. There exist a range of outcomes that have sufficient historical precedent that that their distribution is known with certainty. Suppose, therefore, that a subset only of the current state of knowledge on the global system, U_t, is treated with reservations. The rest of the current state of knowledge, the complement of U_t in W_t, denoted $\hat{U}_t$, is treated "at face value." We can then define the expected augmented costs, C^+, of the policy u at time t given W_t as

$$E\upsilon(C^+|W_t) \equiv E\upsilon[(C-V|U_t) + (C|\hat{U}_t)] \tag{9}$$

These augmented costs are simply the expected costs of the policy, plus the compensating variation in cost required for outcomes associated with data in respect of which decision makers wish to reserve their position. Let $H_E^+(\upsilon,U_t,\hat{U}_t)$ denote such expected augmented costs. We then have:

$$H_E^+(\upsilon,U_t,\hat{U}_t) \equiv E\upsilon[H_E^+(\upsilon,U_{t+1},\hat{U}_{t+1})|U_t,\hat{U}_t] \tag{10}$$

The optimal policy will be the one that minimizes this. To see what this policy is, we need to be more specific about costs. Define augmented costs from time t to time T as C^{+t}, implying that $C^+ = C^{+0}$, and let these be generated recursively by the function:

$$C^{+t} = \phi[U_{t+1},\hat{U}_{t+1},C^{+t+1}] \tag{11}$$

with terminal condition

$$C^{+T} = \phi[U_T, \hat{U}_T] \tag{12}$$

If the minimum expected augmented cost is given by

$$G_E^+(U_T, \hat{U}_T) = \inf_{u_t} E[C^{+t}|U_t, \hat{U}_t, u_t] \tag{13}$$

then the optimality equation takes the form

$$G_E^+(U_T, \hat{U}_T) = \inf_{u_t} E[\phi(U_{t+1}, \hat{U}_{t+1}, G_E^+(U_{t+1}, \hat{U}_{t+1}))|U_t, \hat{U}_t, u_t] \tag{14}$$

with terminal condition

$$G_E^{+T}(U_T, \hat{U}_T) = \phi(U_T, \hat{U}_T) \tag{15}$$

The minimizing u_t in (14) for all $t < T$, U_t, and $\hat{U}_t$, defines the optimal policy, υ^*, under which $H_E^+(\upsilon^*, U_t, \hat{U}_t) \leq H_E^+(\upsilon, U_t, \hat{U}_t)$ for all $t \leq T$, U_t, $\hat{U}_t$ and υ.

This follows directly, since environmental costs are independent of policy at time T,

$$H_E^+(\upsilon^*, U_T, \hat{U}_T) = H_E^+(\upsilon, U_T, \hat{U}_T). \tag{16}$$

For time less than time T,

$$
\begin{aligned}
H_E^+(\upsilon, U_t, \hat{U}_t) &= E\upsilon[H_E^+(\upsilon, U_{t+1}, \hat{U}_{t+1})|U_t, \hat{U}_t] \\
&\geq \inf_{u_t} E[H_E^+(\upsilon^*, U_{t+1}, \hat{U}_{t+1})|U_t, \hat{U}_t, u_t] \\
&\geq \inf_{u_t} E[H_E^+(\upsilon^*, U_{t+1}, \hat{U}_{t+1})|U_t, \hat{U}_t, u_t^*] \\
&= H_E^+(\upsilon^*, U_t, \hat{U}_t) \tag{17}
\end{aligned}
$$

Moreover

$$
\begin{aligned}
H_E^+(\upsilon^*, U_t, \hat{U}_t) &= \inf_{u_t} E[\phi(U_{t+1}, \hat{U}_{t+1}, G_E^+(U_{t+1}, \hat{U}_{t+1}))|U_t, \hat{U}_t, u_t^*] \\
&= G_E^+(U_T, \hat{U}_T) \tag{18}
\end{aligned}
$$

The model is similar to both the stochastic and minimax variants discussed earlier. Formally, the only difference is that the state of knowledge is now partitioned between knowledge which is accorded provisional or reserved status, U_t, and knowledge accorded confirmed status, $\hat{U}_t$. Yet this involves a substantially different way of thinking about the decision making process. Within the sequence $\{W_t\}$ there exists an ever-widening range of outcomes, each of which may be subject to a Bayesian learning process as more and more observations on it are recorded. However, where the decision maker happens to be in the Bayesian learning process with respect to any outcome or outcomes determines how existing observations are regarded. For example, if there is no past history of observations on an outcome, and the decision maker has only a set of priors provided by the scientific community, it is reasonable to expect those priors to be treated with some reservation. It is tempting to think of this as a problem of cascading risk—the risk of an outcome might be interpreted as its risk under the prior distribution combined with the risk that the prior distribution is biased in some way. But the point is that there exists no basis on which to assess risks of the latter sort.

The optimal policy will be that which minimizes a) the focus losses of activities for which the distribution of outcomes has reserved status in the decision making process, and b) the expected losses of all other activities. The identification of only two such categories of knowledge is, of course, an abstraction from the complexity of the real world, where their exist a continuum of subtly distinct qualities of knowledge, but it serves our purpose. Knowledge accorded provisional status in this case is that on which the decision maker chooses to reserve judgement. Knowledge accorded confirmed status is that which the decision maker is content to evaluate on an expected value basis. Neither subset of the global state of knowledge will be constant over time. New observations on the state of the system may be expected to induce decision makers to vary the proportion of the data set on which they wish to reserve their position. Indeed, if new observations on the system are not surprising they will lead decision makers to revise upwards the judgement they make about their understanding of the system, and U_t will tend to be reduced relative to $\hat{U}_t$. On the other hand, if new observations on the state of the system are surprising, they will tend to have the opposite effect. Decision makers will become less confident of their understanding of the global system. Notice, though, that there is no reason to believe that the acquisition of new knowledge by itself necessarily improves understanding of the global system—except in the negative sense of persuading us that we understand less than we thought.

It is a characteristic of the increasing flow of observations on the state of the global system that it contains surprises. The emission of greenhouse gases is not new, but the notion that they may damage global life support systems is. Destruction of rainforest is not new, but the idea that it may contribute to climate change is. As our knowledge of the global system increases, so does our uncertainty about the long term implications of present economic activity. Combined with the uncertainty caused by the rapid pace of change in resource use technology, this suggests that the increasing flow of information does not in fact give more complete information. The problem for decision makers does not get easier. Not only is the perceived range and severity of the possible environmental effects of economic activity expanding, so is the gestation period.

RESERVED RATIONALITY AND THE PRECAUTIONARY PRINCIPLE

The notion of reserved rationality describes those decision making processes where ignorance as to the probability distribution of outcomes, and so ignorance as to the magnitude of potential losses, makes it natural to proceed cautiously—to safeguard initially against the possibility of unexpectedly severe future costs. It seems quite intuitive that where policies have the potential to destroy crucial life-support systems, it is prudent to leave some margin for error as one learns the outcomes of the policy. But there is mounting evidence that decision making reflects this property in a much wider set of circumstances. In economic experiments, for example, it has been found that where subjects are given a substantial sum initially, and where they have no personal experience from which to construct a probability distribution of outcomes during experiments involving repeated trials, their initial strategies will be designed to minimize maximum loss—even where those subjects have been advised of the expected value of the outcome in advance. However, as the subjects build up personal experience over a sequence of trials, their strategies tend to move towards the maximization of expected value.[5] All that is required is that a (subjectively) valued asset be subject to a threat of unknown dimensions.

The relation between the notion of reserved rationality and the precautionary principle is equally intuitive. The principle requires that allowance be made for the potential, though uncertain, future losses associated with the use of environmental resources. Consider that component of the expected augmented costs of policy u associated with the uncertain knowledge in W_t, $E\upsilon(C-V|U_t)$. We have defined this to be the focus loss of actions informed by U_t. The difference between this and the expected or reference costs of the same policy represents the decision maker's allowance for error in the expected costs. In the language of Shackle, $E\upsilon(-V|U_t)$ is a measure of those costs of the policy, over and above the expected costs, which would cause the decision maker no surprise. An efficient intertemporal allocation of resources in these circumstances requires that the expected benefits of the policy exceed the expected costs by $E\upsilon(-V|U_t)$. This then defines the upper bound on the value of resources that may be committed as preventive expenditures under the precautionary principle.[6]

It is, finally, worth repeating that there is a very strong normative, ethical, content to these expenditures. Recall that there are two elements to the focus loss of a policy: the first is a set of conjectured outcomes which would cause greater or lesser disbelief to the decision maker. The second is a set of weights on those outcomes reflecting the ethical judgement of the decision maker as to the importance of the incidence and timing of those outcomes. These weights may not be explicit. They may be implicit in the "attractiveness function" which draws the decision maker's attention to some outcomes rather than others. But whether implicit or explicit, they indicate the importance of equity issues in fixing the value of the safeguard allowance under a precautionary principle. A common thread in the various interpretations of sustainable development in the wake of the Brundtland report is the necessity to preserve the options available to future generations. Intergenerational equity, in this view, will be satisfied if the activities of the present generation do not impose irreversible costs on future generations. This principle can be interpreted as saying that if it is known that an action may cause profound and irreversible environmental damage which permanently reduces the welfare of future

---

[5] W. Schultze, personal communication.

[6] It should be clear, therefore, that it is the precautionary principle and the reserved rationality that underpins it, which lies behind the environmental bonds recommended elsewhere for innovative activities with uncertain future environmental effects [cf. Perrings 1989; Costanza and Perrings 1990].

generations, but the probability of such damage is not known, then it is inequitable to act as if the probability is known. The decision on whether to accept the expected or reference costs of a policy involving fundamental uncertainty is, in this sense, a function of the ethics underpinning an intertemporal social welfare function.

REFERENCES

Costanza, R. and C. Perrings. 1990. A Flexible Assurance Bonding System for Improved Environmental Management. *Ecological Economics* 2(1): 57-76.

Daly, H. E. 1973. The Steady State Economy: Toward a Political Economy of Biophysical Equilibrium and Moral Growth. In: H. Daly, ed., *Toward A Steady State Economy*, pp. 149-174. San Francisco: W. H. Freeman.

Dasgupta, P. S. and G. M. Heal. 1979. *Economic Theory and Exhaustible Resources*. Cambridge: Cambridge University Press.

Foy, G. and H. E. Daly. 1989. Allocation, Distribution and Scale as Determinants of Environmental Degradation: Case Studies of Haiti, El Salvador and Costa Rica. World Bank, Environment Department Working Paper 19.

Funtowicz, S. O. and J. R. Ravetz. 1990. Global Environmental Issues and the Emergence of Second Order Science. Paper presented at the ISEE Conference, The Ecological Economics of Sustainability, Washington D.C., May.

Katzner, D. W. 1986. Potential Surprise, Potential Confirmation and Probability. *Journal of Post Keynesian Economics* 9:58-78.

Katzner, D. W. 1989. The Comparative Statics of the Shackle-Vickers Approach to Decision making in Ignorance. In T. B. Fomby and T. K. Seo, eds., *Studies in the Economics of Uncertainty: In Honour of Joseph Hadar*. Berlin: Springer-Verlag.

Maler, K.-G. 1989a. Risk and the Environment: An Attempt to a Theory. Stockholm School of Economics, Research Paper 6390.

Maler, K.-G., 1989b. Environmental Resources, Risk and Bayesian Decision Rules. Stockholm School of Economics, Research Paper 6391.

Marglin, S. A. 1963. The Social Rate of Discount and the Optimal Rate of Investment. *Quarterly Journal of Economics* 77:95-112.

Myrdal, G. 1975. *Against the Stream*. New York: Vintage Books.

Norgaard, R. B. and R. B. Howarth. 1990. Sustainability and the Rate of Discount. Paper presented at the ISEE Conference, The Ecological Economics of Sustainability, Washington D.C., May.

Pearce, D. W. and A. Markandya. 1989. Marginal Opportunity Cost as a Planning Concept in Natural Resource Management. In G. Schramm and J. J. Warford, eds., *Environmental Management and Economic Development*. Baltimore: Johns Hopkins University Press for World Bank.

Pearce, D. W., A. Markandya and E. B. Barbier. 1989. *Blueprint for a Green Economy*. London: Earthscan.

Perrings, C. A. 1987. *Economy and Environment*. New York: Cambridge University Press.

Perrings, C. A. 1989. Environmental Bonds and Environmental Research in Innovative Activities. *Ecological Economics* 1:95-110.

Randers, J. and D. Meadows. 1973. The Carrying Capacity of Our Global Environment: A Look at the Ethical Alternatives. In H. E. Daly, ed., *Toward a Steady State Economy*. San Francisco: W.H. Freeman.

Shackle, G. L. S. 1955. *Uncertainty in Economics*. Cambridge: Cambridge University Press.

Shackle, G. L. S. 1969. *Decision, Order and Time in Human Affairs*. Cambridge: Cambridge University Press.

Williamson, O. E. 1985. *The Economic Institutions of Capitalism*. New York: Free Press.

Part II

Accounting, Modeling and Analysis

12

THE ENVIRONMENT AS CAPITAL

*Salah El Serafy**
The World Bank
1818 H Street, N.W.
Washington, DC 20433 USA

ABSTRACT

Inasmuch as the environment contributes to the productive process, even when it is not appropriable, it should be considered as a factor of production. This paper will consider the contribution the environment makes to production, and examine the substitutability between environmental elements and factors of production, notably capital. The paper will emphasize the necessity of keeping environmental capital intact, for proper national income measurement, while distinguishing between renewable and nonrenewable resources. To keep renewable environmental capital intact, provision should be made for its depreciation. Depreciation, however, is inappropriate for depletable resources, and this paper will explain why. The paper will end with a recommended approach to integrating these capital conservation concerns in environmental accounting. The recommendations stress that we should proceed without delay to incorporate ascertainable environmental degradation into national accounting, however imprecisely, fully realizing that such an approach will remain partial, but is bound to be expanded gradually as our knowledge of the facts improves, and as we bring more environmental concerns "into relation with the measuring rod of money."

INTRODUCTION

The capital of an economy is its stock of real goods, with power of producing further goods (or utilities) in the future. Such a definition of capital will probably be acceptable to most economists (Hicks 1974). Viewed as such, capital would comprise land, consid-

---

* The views expressed in this paper are those of the author and should not be construed as necessarily reflecting those of the World Bank.

ered in classical economic thinking as a separate factor of production. Land would qualify as part of the stock of real goods, capable of producing further goods. And it is only a short step to extend such a definition to Nature, both as a source of raw materials and as a receptor of wastes generated in the course of economic activity. Alfred Marshall viewed the distinction between land and capital in their capacity as factors of production as rather artificial, just as he viewed the distinction between rent and profits.[1]

Marshall, who may appropriately be regarded as the father of neoclassical economics, was so conscious of the contribution of Nature to production, that he inscribed the adage, "Natura non facit saltum" (which can be translated as "Nature does not make a leap" or that it only proceeds slowly) on the frontispiece of his *Principles*. Although he subscribed to the Ricardian theory of rent, which ascribes rent to the "inherent" and "indestructible" properties of the soil, and accepted the distinction between these original properties of land and the "artificial properties which it owes to human action," he expressed his reservations thus: "provided we remember that the first include the space-relations of the plot in question, and the annuity that nature has given it of sunlight and air and rain; and that in many cases these are the chief of the inherent properties of the soil" (Marshall 1947, p. 147).

Stressing the *capital* quality of land, Marshall (1947) wrote:

All that lies just below the surface has in it a large element of capital, the produce of man's past labour. Those free gifts of nature which Ricardo classed as the "inherent" and "indestructible" properties of the soil, have been largely modified; partly impoverished and partly enriched by the work of many generations of men.

We have only to think of the "impoverishment" of land in the passage just cited, as disinvestment, and of land as a proxy for Nature to conclude that the father of neoclassical economics himself was not unmindful of the contribution of nature to the production of goods and services and would not be averse to the notion that the degradation of the environment should be charged against production as capital depreciation. It is not therefore in any sense revolutionary to think of Nature as a factor of production. The acid test of what makes a factor of production:

is that it should make a contribution to production, in the sense that if it were removed, production (or output) would be diminished. Or more usefully, if a part of it were to be removed, production would be diminished. Which comes to the same thing as saying that the factor must have a marginal product (Hicks 1983, in his essay, "Is interest the price of a factor of production?").

That Nature has a marginal product is not difficult to demonstrate and therefore can be acceptable in the neo-classical economic framework as a factor of production. Even when Nature is held in common, such as international waters, the ozone layer or the air we breathe, it can still be viewed as a factor of production: even if it is not tradable and does

---

[1] Marshall 1947. See, in particular, p. 432, where he draws attention to the views in this regard of Senior and J. S. Mill.

not carry a market price. In the words of Hicks, "In order that a thing should have a price, it must be appropriable, but it is not necessary that a thing should be appropriable for it to be a factor of production."

Which factor of production should Nature come under in order for it to be included in the economic calculus? The answer is that, for *theoretical* reasons, it does not matter. We may recall that Marshall applied the concept of rent to machines, inventing the category of "quasi-rent" to denote income derived from capital that is in short supply in the near term. Also, Hicks questioned the traditional demarcation of factors: "The factors of production are Land, Labour and Capital; or just Labour and Land; or just Labour and Capital; or just Labour" (Hicks 1983, pp. 121-122), and questioned their supposed independence of each other. For practical reasons, however, there are compelling reasons why Nature should be treated as capital. That it contributes to economic activity is not beyond dispute. To subsume it under land, would trivialize its contribution, recalling Herman Daly's strictures against economics, having reduced land's contribution to production to rent, and rent to a surplus that is excluded from the determination of prices (Daly and Cobb 1989). After all, the most fecund form of the production function is provided by the Cobb-Douglas model which most commonly reduces the factors of production to only two: capital and labor. Under capital, one could bring in all kinds of things—including land and technology.

THE POWER OF TECHNOLOGY

In the debate between the environmentalists and traditional economists, a great deal has been made of the notion as to whether capital and natural resources are substitutes or complements. This issue appears important not theoretically, but empirically, as the environmentalists, often viewed as doomsayers—very much as Thomas Robert Malthus has been viewed on population—insist that our planet's capacity to absorb our wastes and to provide raw materials and energy is limited, and that this limitedness cannot be assumed away in the belief that advances in technology are bound to ease out the constraints. Technology has been remarkably successful in that it substituted synthetic rubber for natural rubber, plastics for copper, and may well replace renewable fossil fuels by renewable energy. It has recycled some of our wastes and will probably recycle others. The electronic revolution is allowing much work to be carried out at home rather than on business premises, and is thus bound to reduce road congestion and economize on transport.

In all this, technology seems to have consistently economized on material inputs as well as labor. And yet, overoptimistic views about the power of technology, however, have failed us over the population problem—a major source of environmental stress—and are bound, if they continue to prevail, to fail us again in regard to the environment. Enough has been said in the first part of this paper to show that theoretically all factors are substitutable for each other, but it is empirical substitution that is in question, as reflected in, say, the elasticity of substitution between energy and capital in the production process. It would be foolhardy to argue that such an elasticity is zero, even for the shortest of all runs. That it is bound to rise with time, in common with all elasticity measurements, is true but trite, but this is not the issue.

The issue, in my judgment, is whether in practical terms technology is developing rapidly enough to solve our environmental degradation, and the answer is clearly no. And it is the duty of the economist to play the role of the pessimist. If things are left to go the way they are going, repairing the damage would be far more costly than attempting to avert it before it has taken place. As all can see, too much damage has taken place already for complacency to continue to prevail.

NATIONAL ACCOUNTING FOR NATURE

The foregoing suffices to show that there is much in modern economics that is in harmony with environmental thinking. This, as I have shown, is particularly evident in the writings of John Hicks, a Nobel Laureate in economics who has had a profound influence on the profession. Not only has he supplied an excellent definition of sustainability by defining income, but he has also striven to give economic meaning to business practices, including accounting.[2] Besides, his numerous works on capital theory contain fertile notions about time's contribution to production from which environmental economics will eventually benefit. For me, it was his reflections on income from wasting assets in his magnum opus, *Value and Capital*, that led me to develop a model for calculating income from depletable natural resources (El Serafy 1981). It is not, however, the contribution to production of natural resources that I wish to address here, but simply how to reflect, in national accounting measurements, the changes in the stock of available natural resources brought about by economic activity.

Quite clearly, if what is conventionally measured as income ignores the deterioration of the environment, either as a source of materials or as a sink into which we pour emissions that result in environmental degradation, then such an income is overstated. It is now pretty well established that national accounting should reflect such environmental degradation. But there are still a number of controversies on how to make the required adjustments. Before addressing these controversies, however, it is necessary to point out that accounting has, on the whole, a limited function; it merely assesses the implications of past behavior for profits or income, thus providing a measurement of performance, and therefore indicating net worth, in that the accountants usually also produce balance sheets of assets and liabilities for the entity concerned, whether an individual, a corporation or a nation. On the basis of the accounts, entrepreneurs can make decisions about the future, relying also on many other factors, including their expectations. Where the accounts are wrong, in the sense that the accountant's income contains elements of capital

---

[2] Hicks was a product of the English neoclassical school of economics, who was also greatly influenced by Pareto and more profoundly by the Austrians. His great interest in accounting is reflected in many contributions, including *The Social Framework: An Introduction to Economics*, (Oxford: Clarendon Press, 1942). An article on him published by Arjo Klamer after his death was titled "An Accountant among Economists: Conversations with Sir John R. Hicks," *The Journal of Economic Perspectives*, (November 4, 1989), vol. 3.

(representing running down stocks of natural resources or polluted air or water), such an "income" exaggerates true income, and if consumed, could lead to inevitable ruin. In other words, accounting would be encouraging behavior that cannot be sustained. Meanwhile, since macroeconomic policy makes use of income as a "touchstone" against which various economic aggregates are tested (money supply; savings and investment, fiscal and current account deficits, etc.), false income measurements lead to faulty economic policies, besides failing to gauge true economic performance. Thus a country may be presumed to be achieving a high economic growth rate on the strength of the accountant's measurements, whereas in reality its true growth would be slower, nonexistent, or even negative if the accounts were properly to reflect the diminution of the natural resource stocks and the deterioration and degradation of the environment.

PARTIAL ADJUSTMENTS TO NATIONAL ACCOUNTS

I find it interesting that most people who see the necessity of adjusting standard national income calculations to reflect environmental concerns, wish to do so under two constraints. First, they wish to have a totally integrated system starting with a complete inventory of environmental assets, and setting money values on these in order to construct a balance sheet of all assets, whether nature- or man-made. Changes in such a balance sheet from year to year, as a result of degradation, renovation, locating new deposits, as well as economic exploitation, would be reflected in the end-period balance sheet. The impact on the flow of income would simply be derived from the change in wealth from one balance sheet to the next. The second constraint I see, which again seems to be quite popular, is the view that the gross domestic product as conventionally calculated needs no adjustment at all. All that is needed is to reflect environmental degradation only in net income, by deducting from the gross values calculated, a magnitude for "depreciation." This course would leave the GDP and GNP series, as previously calculated, without change, and adjust only NDP and NNP.

The first constraint is self-imposed and unnecessary; it is also so constrictive that it is likely to impede progress on the adjustment of national accounts. It should be obvious that no balance sheet can be constructed which would not only cover the totality of natural assets in quantity and quality, but also put a money value on all these assets. Moreover, to attempt to reflect the year-to-year changes in the value of environmental assets in the flow accounts would introduce large adjustments which can dwarf the annual economic activities which should be the legitimate basis for income calculation. Re-estimation of mineral deposits, either as a result of new discoveries or a reassessment of reserves (remembering also that reserves are often many many times annual extraction), can thus play havoc with solidly-based income calculations without, in my view, being at all essential.

I have made two suggestions in this regard (El Serafy, forthcoming) First this "holistic approach" should not be attempted at all, not even as an eventual goal, because it is impossible to attain and its adoption is bound to impede progress on adjusting national income. All that is required is to take advantage of the Satellite Accounts, which have now been agreed to be developed under the United Nations System of National Accounts, so that *partial* adjustments to income can be made. On this, I have taken my cue from Pigou (1924, pp. 10-11) who, in *The Economics of Welfare* was facing a similar problem. Pigou perceived that economic welfare was only part of human welfare, but saw that human welfare was such a vast and complex subject—just like the environment—that it could not be profitably studied by economists. Choosing a partial approach, he proposed that economists focused on those aspects of human welfare that can be "brought into relation with the measuring rod of money."

I see the road very clearly ahead: we must not be too ambitious and aim for a comprehensiveness that will forever remain elusive. Let us adjust income gradually for degradation of petroleum, forestry and fisheries, water quality, soil erosion, one at a time and additionally as our methodologies firm up and the physical basis of our calculations improves, leaving economic valuation of thorny areas such as biodiversity to the last. We must also bear in mind the fact that accounting has a limited function, in that it should be complemented by sound environmental policies including proper incentives for conservation, disincentives against pollution and eventually regulation, if neither a carrot nor a stick is applicable.

Apart from the holistic approach, the other major concern I want to stress is depreciation. I find no fault in applying the accounting convention of depreciation of assets that wear out in the process of production to those environmental assets which are renewable. For renewability gets such assets very close to buildings and machines that can be renovated or replaced. In respect of resources such as forests and fish, sustainable yields can be calculated, and exploitation over and above such yields may be considered as comparable to depreciation. "Positive depreciation" may be possible if replanting or restocking exceeds exploitation, but this should more appropriately be treated as capital formation. Where I think depreciation is not applicable is in the case of nonrenewable natural resources such as fossil fuels that cannot be recycled or reused once they have been combusted. I have argued elsewhere (El Serafy 1989) that in their case, we need to adjust gross income itself and not just net income. It is wrong to reckon receipts from the sale of such nonrenewable assets as value added, to be included in GDP. And taking the whole part, representing asset diminution, out of gross income would wipe out from net income, the income effect of mineral exploitation. Sustainability, which is the hallmark of income, compels us to adjust GDP itself along the lines I proposed.

FUNDISTS VERSUS MATERIALISTS

The controversy on this issue, if I may call it thus, recalls two different views of capital which have been lucidly demarcated by Hicks (1974). He distinguished between two schools of thought, the "fundists" and the "materialists" The former view capital as a fund, a sum of money that can be embodied in physical goods, which is revolved over time in the process of production, whereas the latter, the materialists, think of capital in physical terms as machinery and equipment, goods in process, and stocks of raw materials and finished products. Among the fundists, he listed Adam Smith, Marx, Jevons, as well as the accountants who, up to the present, regard capital as a sum of values which may be embodied in physical goods in different ways. Hicks believed that fundist thinking derived from the pattern of economic (largely mercantile) activities that preceded the Industrial Revolution. Among the materialists, he classed Cannan, Marshall, Pigou and J. B. Clark. Keynes, he thought, was brought up as a materialist, a post-industrial school of thought to which practically all neoclassical economists belonged, but his writings led to a revival of fundism. According to Hicks: "If the Production Function is a hallmark of materialism, the capital-output ratio is the hallmark of modern Fundism."

As accountants have remained fundists, they naturally have had difficulty dealing with material capital, such as machines, which do not circulate like trading stocks and last considerably longer than one accounting period. The imputation for depreciation of machinery and equipment is done on simple assumptions with which the economist feels uncomfortable. This probably explains, at least in part, why economists prefer to use GNP rather than NNP for macroeconomic analysis. The accountant, anxious to keep capital intact, uses rules of thumb, approximations and—above all—the assumption that technology remains constant over the life of the asset he is depreciating. He resorts to approximations and shortcuts, because his primary concern is to indicate a level of income that can be safely consumed, leaning always to the side of caution, underestimating income if in doubt, in order to preserve the intactness of capital. The accountant's income is primarily a level that indicates prudent and sustainable behavior, though it often lacks the precision the economist is usually seeking.

When it comes to the treatment of exhaustible resources in national accounts I find myself using the fundist approach to capital. The method I devised which converts receipts from mineral exploitation into a permanent stream representing true income, I did unconsciously, using the methods of the accountant. Petroleum reserves are part of a stock. They can be sold *in toto* or in part, and their proceeds can be sunk in other assets. I asked the accountant's question: what proportion of the total stock does the annual sale represent? In the light of the answer, and with the aid of a discount rate, I could convert the proceeds into a permanent income stream.[3] That Keynes (1949), according to Hicks, became a fundist is clearly evident from his treatment of User Cost in *The General Theory of Employment, Interest and Money* (appendix to ch. 6). Very clearly, I treated sales of exhaustible natural assets as sales out of stock, not as production, creating value added which they are not. It was later, that I called the capital element I calculated a "user cost" to correspond to Keynes's approach which, interestingly—as he says—he brought over from the world of using up "copper stocks," seeking to apply it to the utilization of machinery.

---

[3] The discount rate has also relevance to the ex ante optimization of extraction, for it should guide the owner of a resource as to whether he should leave his resource in the ground to appreciate per Hotelling or reinvest the proceeds in alternative assets. See El Serafy 1989 and Hotelling 1931.

CONCLUSIONS

Finally, I want to end on a practical note. Accounting has never found a totally satis-factory way of treating capital consumption for the purposes of net income estimation. Problems of inflation and changed technology have remained difficult to handle. To the economist, who is by nature forward looking, the value of capital lies in its ability to generate future production, and this may have little relation to the historic cost of making the capital in question. This latter cost is recorded in the books, and is the stock-in-trade of the accountant. The accountant, also by nature, is backward looking, and it is wrong, as frequently happens, for people to confuse his functions with those of the economist. But the accountant performs a most useful function, and his results are at once approxi-mate and cautious. If we follow his reckoning, we strive to keep capital intact so that we can make use of it to give us a "sustainable" income in the future. All we need to do now is to try to apply accounting to environmental capital, without waiting, focusing on in-come flows and leaving aside the valuation of the total environment. Our approach should be gradual, attempting to bring measurable elements into the process as our knowledge improves. But to wait until everything falls properly into place will mean that we shall have to wait forever.

REFERENCES

Daly, Herman E. and John B. Cobb, Jr. 1989. *For the Common Good*. Boston: Beacon Press.

El Serafy, Salah. 1981. Absorptive Capacity, the Demand for Revenue, and the Supply of Petroleum. *The Journal of Energy and Development* 7(1)

El Serafy, Salah.1989. The Proper Calculation of Income from Depletable Natural Resources. In Ahmad, El Serafy and Lutz, eds., *Environmental Accounting for Sustainable Development: a UNDP - World Bank Symposium*. Washington, D.C.: The World Bank.

El Serafy, Salah. (forthcoming). Natural Resource Accounting—An Overview. Paper read to a conference organized by the Overseas Development Institute, London, March 1990.

Hicks, John. 1942. *The Social Framework: An Introduction to Economics*. Oxford: Clarendon.

Hicks, John. 1974. Capital Controversies: Ancient and Modern. *American Economic Review* (May) 64:301-316..

Hicks, John.1983. *Classics and Moderns: Collected Essays on Economic Theory*. Cambridge, Mass.: Harvard University Press.

Hotelling, Harold. 1931. The Economics of Exhaustible Resources. *Journal of Political Economy* (April).

Keynes, J. M. 1949. *The General Theory of Employment, Interest and Money*. London: MacMillan.

Marshall, Alfred. 1947. *Principles of Economics*. 8th ed. London: Macmillan.

Pigou, A. C. 1924. *The Economics of Welfare*. 2d ed. London: MacMillan.

13

ALTERNATIVE ENVIRONMENTAL AND RESOURCE ACCOUNTING APPROACHES

Henry M. Peskin
Edgevale Associates, Inc.
1210 Edgevale Road
Silver Spring, MD 20910 U.S.A.

ABSTRACT

Models of the interrelationships between economic and environment systems are usefully supported by accounting systems that provide relevant measures of economic performance and consistent data bases. Since the conventional System of National Accounts (or SNA) neglects the services of environmental assets and their deterioration, it is deficient in both respects.

This paper discusses an extension to the conventional economic accounts. This extension treats the services and depreciation of environmental, natural resource, and marketed assets in a manner consistent with neoclassical economic principles. Modified income aggregates drawn from this system have been used to describe the macroeconomic effects of environmental policy in the United States. In addition, the system's data bases have supported a number of environmental policy analyses.

The paper compares the system with other environmental accounting approaches that are being implemented or discussed in other countries.

INTRODUCTION

The national economic accounts summarize the flows of services, materials, and products which characterize a nation's economic activity. The system accounts for those flows that are generally reflected in monetary transactions.[1] Economists have emphasized the

---

[1] The major exception is the accounting for the depreciation of capital, where the relevant "transaction" is a bookkeeping entry. Most national accounts cover a few "implicit" transactions for which money does not change hands. For example, the accounts usually include an entry for the implicit rent earned by the owners of owner-occupied housing and the

limitations of using monetary transactions to measure total economic activity, let alone to measure total societal well-being. Yet, the national accounts—and especially certain subtotals drawn from these accounts such as the Gross National Product—have gained popular status, as a key measure, if not *the* key measure, of a nation's economic and social performance.

Members of the environmental community have joined those economists that are unhappy with the status given to the national accounts as a barometer of societal performance. Environmentalists are especially concerned that the accounts fail to reflect pollution and general environmental deterioration. Moreover, since the conventional national economic accounts also ignore deterioration of the nation's environmental and economic resource base, they paint a falsely optimistic picture of a nation's prospects for sustainable economic growth.[2]

These environmental issues and concerns about the national accounts are attaining worldwide prominence, as evidenced by the concern about global warming and the transnational damage caused by acid rain. Yet, before there can be any serious effort to address the deficiencies in the standard national accounts, there needs to be a clear understanding of just how these accounts react to environmental and natural resource change; how the accounts might be altered to reflect these changes more adequately; and what the practical implementation problems are of effecting such alterations.

DEFICIENCIES IN THE CONVENTIONAL NATIONAL ACCOUNTS

One can identify three deficiencies in the standard national economic accounts resulting from inadequate treatment of the environment and natural resources. The conventional accounts provide a poor measure of social and economic performance, treat different forms of national economic wealth inconsistently, and ignore important variables explaining economic activity. It is the last of these deficiencies that is most related to the question of economic-environmental linkage—the subject of this paper.

Inadequacies as a Measure of Environmental Performance

One frequently heard criticism of the conventional national accounts is that they respond poorly (some would say "perversely") to changes in environmental and resource conditions. Pollution, congestion, and the depletion of natural resources are often unfortunate side effects of economic growth. Given these conditions, it is disturbing to much of the public that economic data point in a positive direction. In addition, the conventional economic indicators can poorly reflect efforts to defend against

---

implicit cost of own-produced food consumed by farm families. However, specific accounting practices may deviate from the norm, especially in less developed countries, because of data limitations and special needs.

[2] See, in particular, Repetto et al. 1989.

environmental insult when, for example, increased expenditures on medical services or household cleaning that are incurred because of pollution, lead to an increase in the GNP. Also, efforts to clean up the environment could lead to a decrease in real GNP to the extent that these efforts divert resources from the production of ordinary output.

Inconsistent Treatment of Income and Wealth

The conventional accounts fail to treat different forms of wealth consistently, even though these different forms may be highly substitutable as contributors to income and well-being. The inconsistency has to do with how we define "income."

Conventionally, income is defined as the sum of consumption expenditures plus investment. Yet, the conventional definition further distinguishes between gross investment and investment less depreciation, or net investment. It further distinguishes between gross income and net income, where the latter is defined as consumption plus net investment.

Many, if not most economists follow J. R. Hicks in the belief that net income is the more relevant indicator of the economic well-being of society, since it better represents the amount society can consume *after* allowing for the production of resources necessary to maintain society's stock of capital.[3] Gross income, in contrast, may not be *sustainable* to the extent that its level is supported by diminishing capital stock. Consequently, one of the most important entries in standard economic accounts is "depreciation," which allows the translation of gross income (or product) to net income (or product).

The conventional national accounts inconsistently measure the depreciation of certain forms of capital, such as plant and machinery, but neglect to account for the depreciation of other forms of capital such as natural resources and environmental wealth. As both environmental and natural resource capital are crucial to the production of goods and services, neglecting to value their depletion necessarily means that net or sustainable income is overstated.[4]

Neglect of Important Determinants of Economic Activity

An important function of a system of national accounts is that it serves as an information system containing those statistics that determine and define the nation's

---

[3] This concept of income has been well defended by Hicks 1946. However, net income is hardly a perfect indicator of a nation's well-being. For example, two countries can have the same net income; but where one country has a high savings or investment rate and the other has a high consumption rate, the long-term prospects of the former could be far better than that of the latter.

[4] Other forms of capital depreciation are also neglected in the standard accounts. Of particular importance is the neglect of the depreciation of (as well as investment in) human capital, even though the services of this capital (or "labor") account for most of a nation's income.

economic activity. In a sense, the economic accounts provide a snapshot of the economy's "production function": an instantaneous picture of the transformation of factors of production into product and services. Neglecting environmental and natural resources in the accounts distorts the picture of production in two ways. The oversight ignores the production of some undesirable outputs (e.g., pollution) and leaves out a number of crucial inputs to the production of both desirable and undesirable product.

The desire for improved understanding of the linkages between environmental activity and economic activity is equivalent to the search for improved understanding of the nation's production function.

SUGGESTED APPROACHES FOR IMPROVING THE NATIONAL ACCOUNTS

Most suggested approaches for modifying standard national accounting practices involve an expansion of the conventional accounts either by direct modification of these accounts or by the construction of separate "satellite" accounts. There are, in fact, a broad spectrum of options for environmental accounting. These will be discussed briefly, starting, more or less, from those measures that make modest or minor demands on existing national accounting systems to those that would entail major changes in the existing structure.

Identification of Environmental Expenditures

Both expenditures to clean up the environment and expenditures to "defend" against environmental insult are already covered in the conventional accounts, but they may not be identified as such. For example, industrial purchases of pollution control equipment, such as scrubbers and filters, and the labor and materials needed to operate such equipment are usually co-mingled with other expenditures associated with the company's ordinary business activity. Pollution expenditures are also difficult to identify since much pollution control products and materials are not the exclusive output of a single "pollution-control" industry. The data situation is slightly better with respect to non-military governmental pollution-control activities. Municipal sewage treatment, for example, is usually a separately defined sector in most national accounting tables.

The standard accounts are of little help in identifying defensive outlays, such as for home air cleaners or for the purchase of protective coats of house paint. While such outlays are covered (within the consumption expenditure totals), they are not distinguished from ordinary outlays for goods and services that serve nondefensive purposes. Thus, for example, while some portion of total expenditures for air conditioning is to "defend" against polluted air, another portion is simply for the air conditioner's major purpose of providing cooler air. The accounts provide no information to allow one to determine which portion is which.

The mere collection of data identifying defensive outlays and pollution control expenditures would appear to be worthwhile and would not necessitate any modification to the conventional accounts. In both West Germany and the United States efforts have been made in this direction. For over 15 years, the U.S. Bureau of Economic Analysis (BEA) has been publishing pollution-control expenditure data.[5] These data have been used to analyze the macroeconomic effects of environmental policy of the U.S. economy.[6] Similar analyses have been conducted in West Germany.[7]

Physical Resource and Environmental Accounting Systems

A frequent and practical suggestion for rectifying the deficiencies with the conventional economic accounts is to develop separate or "satellite" accounts that describe the flows of resources, materials (including pollutants), and energy that underlie any economic activity. These accounts display input-output balances that are necessary consequences of physical conservation laws. The accounts show an initial stock (or "opening balance") of a resource, its diminution through use and degradation, its augmentation through discovery or, in the case of renewable resources, through natural growth, and, finally, the total stock at the end of the accounting period (or "closing balance"). Thus, in principle, such accounts show the depletion of natural resources but also their transformation into goods and materials, some of which may find their way back to the environment in the form of pollutants. The material or energy accounts can be linked to the conventional economic accounts through the use of ratios (or input-output coefficients) that express units of energy or material use per unit of production or sales.

On a more or less "official" governmental level, this approach is being tried in France and Norway. However, it appears most fully implemented in Norway, where a number of resource accounting tables have been published.[8]

Depreciation of Marketed Natural Resources

Another approach to modifying the standard economic accounts is to focus on their failure to depreciate natural resource and environmental assets. This particular strategy has received recent popular attention through the work of Robert Repetto and his colleagues at the World Resources Institute.[9] Concentrating on the depreciation problem may be

---

[5] These are published regularly in the *Survey of Current Business*.

[6] Generally the effects have been small. See Peskin et al. (1981). For a contrary view, see Jorgenson and Wilcoxen (1989).

[7] See, for example, Schäfer and Stahmer (1988) and Leipert and Simonis (1989).

[8] This is not to imply that the Norwegian and French approaches are the same. Indeed, the French approach is more ambitious in that it attempts to cover a broader range of environmental consequences of economic activity. Further discussion may be found for Norway in Alfsen et al. (1987) and for France in Theys (1989).

[9] See Repetto (1989).

especially important in resource-based developing countries where resource problems may be quantitatively more important than environmental problems. Thus, Repetto's adjustments have been implemented in Indonesia and further studies are planned or are currently in progress in the Philippines, Costa Rica and China.

The depreciation calculations have depended on estimates of changes in the physical stock of the natural resource and on the market values of commodities generated by this stock. Thus, the depreciated value of a forest is calculated in terms of the loss of the forest's ability to generate marketed product such as hardwood. Estimates of the loss of the forest's ability to generate "nonmarketed" environmental services (e.g., species protection, species diversity, esthetic services, CO_2 absorption, etc.) have yet to be made. Nevertheless, even though the estimates thus understate the full value to society of the depreciation of certain resources, the calculations do suggest that conventionally estimated net income may be grossly overestimated in resource-based economies. Repetto, for example, estimates that the Indonesian annual income growth rate from 1971-1984 would be reduced by about 3% were the effects of resource depletion accounted for.

Full Environmental and Natural Resource Accounts with Valuation

The most ambitious approach would be to include all the elements of physical resource accounting and natural resource depreciation calculations and to place monetary values on all physical entries. In addition to the data needed for the above three approaches, this approach also requires a substantial amount of data for valuation. A recent example of this approach is the expanded SNA system developed by Bartelmus et al. (1989). A (perhaps less formal) version of such a full system has been suggested by Hueting (1980). See also Hueting, and Hannon, this volume. Finally, there is the Peskin "neoclassical" framework, which is the subject of the next section.

A NEOCLASSICAL ENVIRONMENTAL ACCOUNTING FRAMEWORK

While all of the above approaches display some linkages between pure environmental accounting and conventional economic accounting, they treat conventional economic decision making as taking place somewhat independently of the decision making that directly affects the quality and quantity of natural resources and the quality of the environment. In contrast, the linkage in the Peskin framework is more intimate since it assumes that decision makers recognize that environmental asset services are substitutes in production and consumption for the services of other forms of reproducible wealth.[10]

Essentially, the framework treats the environment as a generator of services which are in demand by production and consumption sectors. As Hueting (1980) notes, these services may compete with each other. Thus, the positive waste disposal input service that a lake provides a factory may compete with the lake's ability to provide recreation services.

---

[10] A detailed discussion of the Peskin framework may be found in Peskin (1989).

The framework treats such losses in final demand services as negative output. The value of the services are theoretically measured by what users would be willing to pay for them. In practice these willingness to pay values are estimated by a number of techniques drawn from the benefit-cost literature. For example, the cost of waste disposal services have been estimated by least-cost pollution control methods that are based on engineering studies. Recreation losses have been estimated by travel demand methods. Since in principle the value of input services need not equal the value of resulting environmental damages (except with Pareto optimality), the framework includes a balancing entry.

One way this framework differs from other approaches is in how it treats environmental control costs. The theory behind the framework suggests that users of an environmental asset may experience two types of costs. First, there are the costs that users may have already incurred if they have had to take measures to protect a resource. (Land restoration after strip mining might be an example.) Secondly, there are the costs faced by users, in terms of foregone benefits, due to the denial of further access to the resource. The former costs are, in principle, already accounted for in the conventional economic accounts. The latter costs are not currently accounted for. Yet estimating these costs is important since it provides an estimate of the value of services generated by the environment.

This distinction between costs already incurred and opportunity costs to the user denied access to the environment are shown schematically in figure 13.1. MB is marginal benefit to environmental asset service users and MD is resulting marginal disbenefits.

Suppose that a user of the resource, because of environmental regulations, is observed consuming environmental assets just to level V, rather than level A which is the level likely to be observed in the absence of any constraints on the user's behavior. The actual

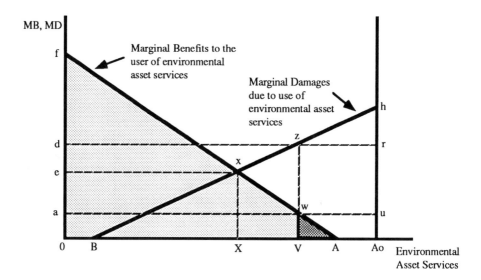

FIGURE 13.1 Marginal benefits and marginal damages of environmental asset services

costs to this user—in terms of direct environmental regulatory costs and any loss in profits—is just equal to the area of the triangle VwA.[11] While these control costs are already accounted for in the standard national accounts, their separate identification is the focus of the ongoing program at the U.S. Bureau of Economic Analysis and of recent West German studies.

What has not been accounted for is the value to the user of being allowed access to the resource to level V. The total amount of this value is the cross-hatched area OVwf. This amount could also be interpreted as a "cost"—in this case, the *prospective* cost to the user of being denied use of the environmental asset. This sort of cost has not been included in the conventional accounts. However, a complete accounting of all sources of economic income would include such costs since they measure the value of a nonmarketed, but valuable, factor input (the environmental asset service) just as, say, the wage bill measures the value of marketed factor input (labor).

Both the Peskin and the Dutch (Hueting) accounting approaches attempt to measure this prospective cost by estimating the cost polluters would incur were they forced to meet some low-polluting standard—although not necessarily to the zero level in the diagram. In the Netherlands, the plan is to define this standard to be consistent with some level that will assure sustainability.

The above diagram can also be used to illustrate a feature that is peculiar to the Peskin framework. As environmental assets are used, there is, as has been noted, the "production" of disbenefit due to any pollution generated and due to any denial of access to the resource by other potential users. This "negative" production is also neglected in conventional accounting systems. At an environmental asset use equal to level V, the total value of this disbenefit equals the triangular area BzV.

It is apparent from the figure that the triangular areas representing polluter benefits and injured party disbenefits are generally unequal in size. In other words, there is no reason why either actual or prospective pollution control costs should measure pollution damages. However, a number of suggested approaches (e.g., Hueting 1980 and Bartelmus et al. 1989) do just that.[12]

There is, however, one situation where the cost and damage valuations are the same. Two conditions must hold. First, suppose that instead of using triangles, values are measured as the product of the value of the *marginal* unit of pollution times the amount of pollution. Second, suppose that current pollution levels are Pareto optimally set at level X instead of level V. In that case, both prospective costs and damages will equal the rectangular area OXxe. Thus, one way of interpreting those systems that measure

---

[11] The proposition that area Vwa equals this cost is a logical consequence of the definition of the marginal benefit curve. If regulatory costs were greater than this area, then the benefit to the polluter by not having to reduce asset use to level V would be greater than area Vwa. But this result is a contradiction since, by definition, this benefit just equals area VwA. Similarly, if regulatory costs were less than area VwA, the benefits of moving from V to A would be less than VwA—again a contradiction.

[12] It should be noted that Hueting recognizes the theoretical problem but he feels that attempts to measure true disbenefit functions are futile.

pollution damages by the costs of reducing those damages is to say that they assume perfect policy performance—at least, according to criteria of economic efficiency. Of course, a system based on such assumptions cannot be used to generate data for the purposes of investigating the economic efficiency of environmental policy—an important concern of the U.S. EPA in recent years.

There are, in sum, two lessons to be learned about valuation of environmental and natural resource services from this theoretical analysis:

- There is generally both a positive (benefit) and a negative (disbenefit) value associated with the use of environmental and natural resource services. In general, these values are not equal.
- The overall social value of environmental and natural resource services depend on the values *all* interested parties place on the services that are being provided.

These two features of the theory are important because they point out the inadequacy of accounting for a particular service value and resulting environmental asset value without looking at all competing uses of the asset. Thus, for example, the value of a forest has to be ascertained by looking at all existing and potential competing uses of the forest simultaneously. One benefit of a national accounting structure is that, in principle, it can picture these competing uses.

While this theoretical framework describes the flows of services from environmental assets and their valuation, we have yet to discuss the treatment of the deterioration of the assets. In the spirit of treating environmental assets in the accounts like other, marketed assets, it is necessary to depreciate them.

One reason that depreciation exists is because of a reduction over time in the physical ability of capital to generate consumable services. This loss in physical ability—physical depreciation—may also lead to a loss in the value of the capital stock—value depreciation. That is, value depreciation may be caused by physical depreciation. However, value depreciation can also arise for other reasons. For example, the value of capital can fall due to a change in tastes for those consumption items produced by the capital or simply because of a change in interest rates. Thus, it is quite possible that an asset that remains physically intact (or even grows) can "depreciate."

Environmental capital has value presumably because it generates a stream of goods and services. Let V_0 represent this value at the beginning of the year and, Q_1, Q_2, etc., represent the services generated (that is, "gross environmental income") at the beginning of the next and subsequent years. Thus, Q_1 is gross environmental income in year 1.

The theory of investment relates V_0 to the Q's as follows:

$$V_0 = \frac{Q_1}{(1+i)} + \frac{Q_2}{(1+i)^2} + \dots + \frac{Q_n}{(1+i)^n} + \dots ,$$

(1)

where i is the rate of interest. Since V_1, the value of V_0 at the end of year 1, is simply

$$\frac{Q_2}{(1+i)} + \frac{Q_3}{(1+i)^2} + \dots + \frac{Q_{n+1}}{(1+i)^{n+1}} + \dots \, ,$$

equation (1) can also be written:

$$V_0 = \frac{Q_1}{(1+i)} + \frac{V_1}{(1+i)} = \frac{Q_1 + V_1}{(1+i)} \, , \tag{2}$$

from which it follows that

$$Q_1 = iV_0 + (V_0 - V_1) \, . \tag{3}$$

The term $(V_0 - V_1)$, representing the loss in value of the initial capital stock, is, by definition, *value* depreciation occurring in year one, or D_1. The relationship between income and depreciation was developed without any reference to the physical destruction of capital. It is true that as capital wears out physically, future Q's may fall, explaining why V_0 may exceed V_1. However, future Q's may fall for other reasons such as a fall in the demand for the capital's services. The Q's may also decline because of an inability to employ the capital fully. That is, if capital is complementary to labor, a decline in labor services would also bring about a decline in future Q's.

Moreover (in contrast to *physical* depreciation), value depreciation, $V_0 - V_1$, need not necessarily be positive, i.e., V_1 could in principle exceed V_0, i.e., there could be a "capital gain." In fact, value depreciation can be decomposed into two components: the portion of the difference between V_0 and V_1 that is due to actual physical depreciation and the portion which is due to other causes. The latter, if positive, is conventionally termed "capital gain" and, if negative, is termed "capital loss." Thus, if physical depreciation is D_p and capital gain (or loss) is G, this decomposition of value depreciation requires that

$$V_1 - V_0 = D_p + G \tag{4}$$

or

Value depreciation = physical depreciation - capital gain or + capital loss

Of course, the physical depreciation must be valued in order to make this computation possible. How the units of physical loss should be valued—using the original price of the capital, its current price, or some other price—is a matter of some controversy. However, if the focus is on *value* depreciation, the implications of choosing among alternative capital prices can be avoided by estimating value depreciation by successive application of equation (1) one year apart rather than by applications of equation (5).

The importance of this formulation of depreciation is that, like any other capital asset, the depreciation in value of an environmental asset such as a forest or other natural resource is due both to physical depreciation and capital loss (or gain). For example, consider a lake. Because of its role as a generator of waste disposal services, its value as a source of other services such as water supply is likely to diminish as the lake deteriorates physically. But what if its primary purpose was to provide recreation services. Clearly, as the lake becomes more and more congested, its physical quality declines and the service value *to each user* declines. However, depending on the rate of increase in the total number of users, the overall service value of the lake for recreational purposes may actually increase.[13] Thus, for reasons of changes in demand, changes in taste, changes in interest rates, it is dangerous to assume that physical declines in environmental and natural resources necessarily imply true economic declines.

THE ACCOUNTING STRUCTURE

The above discussion suggests both a framework for introducing the services of environmental assets into the conventional accounts and the principles for evaluating these services. Basically, the framework assumes that these services can be treated as unpaid inputs to either production or consumption activities. The adverse effects of consuming these services, either due to pollution or to the denial of environmental services to other users, can be treated as negative output or damages. A balancing term is required to assure that the conventional equality between total inputs and outputs is maintained.

The framework assumes a set of accounts that is rather similar to the conventional input-output structure underlying most national economic accounting systems including the United Nations System of National Accounts. However, in addition to the three conventional sectoring groups—Industries, Governments, and Households—we add a fourth: the environment (Nature). Nature is shown as the primary source of all environmental asset services and as the final "consumer" of environmental damages. Nature also must be included for practical reasons because it generates a substantial portion of environmental damage. For example, a large portion of both dissolved solids and suspended sediment in water have a natural origin and, on average, naturally generated particulates and nitrogen oxides (other than NO_2) greatly exceed the manmade production of these air pollutants. Available estimates of damages due to air and water pollutants cannot distinguish between damages from those offensive residuals that have a human origin and damages from residuals with a natural origin. Rather than attribute all the damage to nonnatural causes, it is more accurate to prorate the damage total between the two sources. The Natural account has the form shown in Table 13.1.

The accounting framework "assigns," as inputs, all damages to the Natural sector even though the actual damaged parties may be industries, households, and governments.

---

[13] Suppose each of n user's utility is $u = u(z/n)$ where z is meters of lake shoreline. Total utility is $nu(z/n)$ and the change in total utility as n increases is $u(z/n) - u(z/n)$, which can be positive or negative.

TABLE 13.1 Natural Account

Input	Output
1. Environmental damages (Including those naturally generated)	2. Environmental services
a. Air	3. Net environmental benefit
b. Water	
c. Land	
NATURAL SECTOR INPUT	NATURAL SECTOR OUTPUT

(Knowledge of actual damaged parties is required for empirical estimates of damage.) The outputs of this sector are the various environmental asset services enjoyed by industries, households and governments. The "Net environmental benefit" term is simply the difference between damages and services and is required to assure equality of input and output totals since, as was argued above, damage values are generally unequal to service values.

The principal features of the accounting framework are illustrated in the consolidated gross product accounts (see Table 13.2). All intra-sector flows have been eliminated. Furthermore, households and nature are not included in the consolidation. Following conventional national accounting practice, neither sector is viewed as undertaking production activities.

The unconventional environmental entries are shown in such a way as to preserve the conventional account entries, enabling those who may not be interested in the modifications to simply ignore them. Thus, for example, while the modified accounts show environmental depreciation as a negative adjustment to Net National Product (NNP), it is added back in order to leave the conventional measure of Gross National Product (GNP) unchanged. By arranging the entries in this way, an effort is made to alleviate the fears of those who object to modifying the conventional accounts on the grounds that such modifications destroy the "integrity" of the existing system.

Inspection of the modified consolidated account indicates that modified GNP equals conventional GNP less environmental damage. Actually, this relationship is an identity: it is necessarily true because of the way we chose to arrange the entries in our accounting structure. However, a number of other arrangements are possible, each leading to its own formula relating the conventional GNP to a "modified" GNP.

Using the following notation:

VA: Charges against conventional GNP or value added
GNP: Conventional GNP
GNP_i: Modified GNP, definition i (i = 1, 2, 3, and 4)
ES: Environmental services
NEB: Net environmental benefit
ED: Environmental damage,

all of the following "GNP" definitions are consistent with the accounting structure:

$$GNP_1 = GNP - ED$$
$$GNP_2 = GNP + ES$$
$$GNP_3 = GNP + NEB$$

and

$$GNP_4 = GNP.$$

(Corresponding "net" concepts can be defined by subtracting off ordinary and natural resource depreciation.)

TABLE 13.2 Consolidated National Income and Product Account

Input	Output
Compensation of employees and proprietors (incl. rental income)	Personal consumption
Profits after inventory valuation & capital consumption adjustment	Gross private domestic investment
	Exports
	Imports (-)
	Governmental goods and services
Net interest	
NATIONAL INCOME	
Transfer payments)	
Indirect Taxes	
Subsidies (-)	
Statistical discrepancy	
NET NATIONAL PRODUCT	
Environmental depreciation (-)	
MODIFIED NET NATIONAL PRODUCT	
Capital consumption	
Environmental depreciation (+)	
CHARGES AGAINST GROSS NATIONAL PRODUCT	GROSS NATIONAL PRODUCT
Environmental services (-)	Environmental damages (-)
a. Air	a. Air
b. Water	b. Water
c. Land	c. Land
Net Environmental Benefit	
CHARGES AGAINST MODIFIED GROSS NATIONAL PRODUCT	MODIFIED GROSS NATIONAL PRODUCT

Each of these definitions has its own welfare interpretation and thus may serve different policy objectives. The point is that they are all consistent with the same accounting framework. A similar situation holds with other environmental accounting approaches. Although the individual authors of the systems may promote particular accounting aggregates, alternatives are easily constructed within the same accounting structure.

This "neoclassical" framework was partially implemented with U.S. data for the years 1972 and 1978 (Peskin 1989). More importantly, data from the implementation exercise was used for a number of policy studies by the U.S. EPA and the U.S. Department of Agriculture.[14] Several features of the system explain its usefulness as a policy data system: its complete coverage of all economic activity, the fact that it relies on micro-data bases that permit detailed geographical aggregations, and the fact that the different valuations of environmental service inputs as opposed to environmental damages are useful for benefit-cost comparisons.

Because of its policy orientation, the EPA has recently decided to use the framework as the basis of developing resource and environmental accounts for the Chesapeake Bay region.

DIFFICULTIES WITH THE ABOVE FRAMEWORK

While the above system may serve U.S. purposes, it is not necessarily superior to approaches developed in other countries. In the first place, the uniqueness of policy needs and different data availabilities suggest that no single framework will serve all countries equally well. Moreover, there are a number of problems with the Peskin framework, some of which are also shared by other suggested approaches. Four problems are particularly troublesome, as noted below.

Appropriate Units of Measurement

The above system assumes that all entities—values of environmental services and depreciation—will be measured in money terms. This approach has been attacked as unrealistic and arbitrary by those who question the validity of the recommended benefit-cost valuation techniques and the ability to implement those techniques in countries with very poor data bases. Instead, they recommend that all environmental entities be measured in physical units, recognizing that full integration with the economic accounts will not be possible.

However, adopting a physical accounting system has its own costs. In the first place, it makes it impossible to obtain an objective measure of the importance of

---

[14] Most of these are in unpublished reports. However, some findings have found their way to journal articles. For example, see Gianessi, Peskin, and Wolff (1977, 1979), Gianessi and Peskin (1980), and Peskin (1986).

environmental services relative to ordinary marketed goods and services. Thus, analysis of policies that may require a trade-off between the development of marketed capital at the expense of environmental capital become very difficult. In the second place, a lack of a common monetary measure makes it impossible to adjust conventional GNP for reductions in environmental quality and conventional NNP for environmental degradation.

A third, more practical problem of measuring environmental assets and services in physical units is that it is not clear what units should be used. Because most environmental assets generate a number of very different services, no single physical unit of measure seems appropriate for all of them. While the number of hectares may describe something about a forest, it says little about its ability to generate timber, to provide a wilderness experience or to protect endangered species.

Disagreements over the Appropriateness of Discounting

As was discussed earlier, one way to measure the monetary value (and depreciation) of an environmental asset is to estimate the monetary value of the future "income" stream generated by the asset and then discounting this stream using an appropriate rate of interest. It is not clear, however, what interest rate, if any, is "appropriate."

There are, of course, vast amounts of literature on this subject, much of which has been aptly summarized by Robert Lind (see Lind et al. 1982). One crucial issue is not so much the appropriate rate of interest but rather the discounting process itself. A feature of a neoclassical economic framework is that environmental asset values are being determined according to the preferences of today's generation and not future generations. As Page (1977) and others have shown, it is quite possible that a society could choose for itself an "optimal" allocation of capital that, under certain circumstances, could bring the economy to a halt in the future. That is, the "optimal" growth path, from the point of view of the present generation, may not be sustainable.

Whether this result occurs depends on such factors as the degree of capital substitutability and the techniques of production. Thus, discounting is not necessarily inappropriate for all countries. However, for certain countries, some other methods might have to be found for valuing the stock of environmental and natural resource capital.

Overdependence on the Classical Economic Model

The theoretical structure and accounting framework presented in this paper is consistent with the neoclassical economic model commonly taught in most Western universities. It is not self-evident that this model is acceptable to or relevant for all societies.

For example, the individualistic aspect of the theory may be suspect. The system addresses competing users of environmental assets and how the value of the asset services depend on the values as perceived by these users. The system does not address the possible independent value society might place on the same asset, since in the Western model, society's valuation is the consolidation of the valuation of each society member.

These Western economic concepts, which also underlie the benefit-cost techniques recommended for obtaining monetary valuation, may not be acceptable to countries with very different cultural traditions. If so, some other evaluation and accounting scheme might have to be found.

Unacceptable Demands on the Availability of Data and Skills

Finally, the implemention of the theoretical and accounting framework presented here may make unrealistic demands on the available data. It is important to realize, however, that questions of data and skill adequacy are ultimately empirical matters. As argued elsewhere (Peskin 1988), they can only be addressed through experimental case studies. Indeed, all four problem areas identified above need to be addressed through a program of research and experiment.

FINAL THOUGHTS

This paper identified some problems with conventional economic accounting and suggests an approach that may provide a better link with environmental activity. Some comparisons were made with other approaches although a much longer document would be required to give them the discussion they deserve.

A recent survey of these alternatives (Peskin, 1990) has convinced the author that none of the systems will satisfy all the critics of the conventional economic accounts. Moreover, there is no "best" system as each reflects individual country's needs and data availabilities. However, the author also notes that much of the needed data for the various alternatives are common to all the systems. As a result, it is not necessary to make a firm commitment to any particular framework before initiating a program in environmental accounting. Also, the author's experience suggests that "learning-by-doing" is the best approach to building any data system.

The survey also found little sentiment for abolishing the conventional national accounting system. Rather, the intent is to modify or otherwise supplement the conventional system with "satellite" accounts. While this sentiment may reflect political reality or simply a desire to preserve tradition, it also reflects the view that economic valuations are important in measuring the well-being of a society.

It is understandable that one can become frustrated with the inability of the conventional, neoclassical model to reflect important environmental concerns. And some, therefore, have called for replacing economic valuations with some other valuation approach. However, it is easy to overlook a fundamental feature of the economic model. In well-functioning markets, economic valuations reflect the sovereignty of the consumer—not the sovereignty of the government, of religious institutions, or of well-intentioned individuals. In our search for a compatible relationship between the economic system and the ecologic system, let us not forget this essential democratic feature of the economic system. The challenge is to develop approaches that preserve the nondictatorial features represented by the concept of consumer sovereignty and at the same time address the environmental issues neglected by the economic paradigm.

Acknowledgments

An earlier version of this paper was presented to the Conference on Environmental Cooperation and Policy in the Single European Market, Venice, Italy, April 17-20, 1990. References to related work in other countries are drawn from a study prepared for the World Bank. The World Bank, of course, does not necessarily endorse any of the conclusions in this paper.

REFERENCES

Alfsen, Knut H., Torstein Bye, and Lorents Lorentsen. 1987. *Natural Resource Accounting and Analysis: The Norwegian Experience, 1977-1986.* Oslo: Central Bureau of Statistics of Norway.

Bartelmus, Peter, Carsten Stahmer, and Jan van Tongeren. 1989. *SNA Framework for Integrated Environmental and Economic Accounting.* Paper presented to the 21st Conference of the International Association for Research in Income and Wealth. Lahnstein, Germany (August 21).

Gianessi, Leonard P. and Henry M. Peskin. 1980. The Distribution of the Costs of Federal Water Pollution Control Policy. *Land Economics* 56(1):83-102.

Gianessi, Leonard P., Henry M. Peskin, and Edward Wolff. 1977. The Distributional Implications of National Air Pollution Damage Estimates. In F. Thomas Juster, ed., *The Distribution of Economic Well-Being.* Cambridge, Mass.: Ballinger (for the National Bureau of Economic Research).

Gianessi, Leonard P., Henry M. Peskin and Edward Wolff. 1979. The Distributional Effects of Uniform Air Pollution Policy in the United States. *Quarterly Journal of Economics* 93(2): 281-302.

Hicks, J. R. 1946. *Value and Capital.* 2d ed. Oxford: Oxford University Press.

Hueting, Roefie. 1980. *New Scarcity and Economic Growth.* Amsterdam: North Holland.

Jorgenson, Dale W. and Peter J. Wilcoxen. 1989. Environmental Regulation and U.S. Economic Growth. Paper presented for discussion at the M. I. T. Workshop in Energy and Environmental Modeling and Policy Analysis. Cambridge, Mass. (July 31-August 1).

Leipert, Christian and Udo E. Simonis. 1989. Environmental Protection Expenditures: The German Example. *Rivista Internazionale di Scienze Economiche e Commerciali,* 36(3):255-270.

Lind, Robert C., Kenneth J. Arrow, Gordon R. Corey, Partha Dasgupta, Amartya K. Sen, Thomas Stauffer, Joseph E. Stiglitz, J. A. Stockfisch, and Robert Wilson. 1982. *Discounting for Time and Risk in Energy Policy.* Washington, D.C.: Resources for the Future.

Page, Talbot. 1977. *Conservation and Economic Efficiency.* Baltimore: Johns Hopkins University Press (for Resources for the Future).

Peskin, Henry M. 1986. Cropland Sources of Water Pollution. *Environment* 28(4):30-35.

Peskin, Henry M. 1988. A Program of Research in Support of the Development of Integrated Environmental-Economic Accounts. Unpublished draft prepared for the World Bank (October 20).

Peskin, Henry M. 1989. *Accounting for Natural Resource Depletion and Degradation in Developing Countries.* Environmental Department Working Paper no. 13. Washington, D.C.: The World Bank (January).

Peskin, Henry M., with Ernst Lutz. 1990. *A Survey of Resource and Economic Accounting in Industrialized Countries*. Environment Working Paper no. 37. Washington, D.C.: The World Bank (August).

Peskin, Henry M., Paul R. Portney, and Allen V. Kneese, eds. 1981. *Environmental Regulation and the U.S. Economy*. Baltimore: Johns Hopkins University Press (for Resources for the Future).

Repetto, Robert, William Magrath, Michael Wells, Christine Beer, and Fabrizio Rossini. 1989. *Wasting Assets: Natural Resources in the National Income Accounts*. Washington, D.C.: World Resources Institute.

Schäfer, Dieter and Carsten Stahmer. 1988. Input-Output Model for the Analysis of Environmental Protection Activities. Prepared for the 2d International Meeting on Compilation of Input-Output Tables. Baden, Austria (March 13-19).

Theys, J. 1989. Environmental Accounting in Development Policy: The French Experience. In Yusuf J. Ahmad, Salah El Serafy, and Ernst Lutz, eds., *Environmental Accounting for Sustainable Development*. Washington, D.C.: The World Bank.

14

CORRECTING NATIONAL INCOME FOR ENVIRONMENTAL LOSSES: A PRACTICAL SOLUTION FOR A THEORETICAL DILEMMA

Roefie Hueting
Netherlands Central Bureau of Statistics
Prinses Beatrixlaan 428
2273X2 Voorburg, Netherlands

ABSTRACT

Increase in production as measured in national income is generally called economic growth, identified with an increase in welfare and conceived as *the* indicator for economic success. All countries of the world give it the highest priority in their economic policy. At the same time, the present increases in national income are being accompanied by the destruction of the most fundamentally scarce resource at man's disposal and, consequently, the resource with the most economic good, namely, the environment. This paper will 1) Outline information which should be included when National Accounts are published in order to avoid misinterpretation of the changes in the level of national income by politicians and the public; 2) Examine whether it is possible to correct national income for environmental losses; 3) Come to the conclusion that such a correction runs up against the impossibility of constructing shadow prices for environmental functions that are directly comparable to the market prices of goods and services produced by man (the valuation problem); 4) Propose a practical and defensible solution for this problem, namely, estimating the costs of the measures (including a direct decrease in activities) that are necessary to meet standards for sustainable use of the functions of the environment; 5) Sum up the advantages and disadvantages of this solution.

THE INTERACTION BETWEEN ENVIRONMENT AND PRODUCTION: THE URGENCY OF CORRECTION

In economic policy, the news media and, alas, also in some economic literature, the increase in production as measured in national income (or Gross National Product, GNP) is called economic growth. This growth is identified with an increase in welfare and is viewed as *the* indicator for economic success. Defining production growth as economic growth means defining economics as production Such a definition excludes, among other things, scarce environmental resources from economics. Economic growth, defined in this manner, obtains the highest priority in the economic policy in all countries of the world.[1] At the same time, we see across the world the growth of national income accompanied by the destruction of the most fundamental, scarce, and economically beneficial resource at man's disposal, namely the environment.

From this simple observation, three conclusions can be drawn: 1) society is sailing by the wrong compass, at the expense of the environment; 2) the error is covered up by using terms incorrectly; 3) the belief in continuous exponential growth in production, as measured in national income, is the heart of the environmental problem.

The current terminology regarding growth and welfare is an expression of the strong belief that things are going well, economically speaking, solely when production, as measured in GNP, increases. The notion that production should be increased in order to create support for financing the conservation of the environment reflects this belief. This notion is widespread and highly popular in official economic and environmental policy. The proposition is disputable, because environmental deterioration is to a large extent precisely a *consequence* of production growth. The production growth attained in the North is largely the result of increases in productivity[2] and has required the loss of scarce environmental goods; this loss has not been taken into account.[3] Few people seem to be aware of the following. One quarter to one third of the activities making up national income (notably state consumption) do not contribute to its growth, because by definition no increase in productivity can result from them.[4] Other activities result only

---

[1] Even though in most countries there are politicians and bureaucrats who have understood that continuing growth which conforms to the current pattern of production and consumption for more and more people is accompanied by the destruction of the environment, the official policy still gives the highest priority to increasing production. The population growth is, with a time lag, also reflected in GNP.

[2] Production growth results from growth in the labor force and increase in labor productivity. The labor force growth in the North has been small during the last few decades. The production growth in the North is consequently mainly the result of an increase in labor productivity. See also note 4.

[3] Polluting, degrading and depleting environment and resources is free of charge. Preventing this process, by levies or by regulations, means that, given the existing technology, more labor input is required for the production of a given number of goods. This reduces labor productivity and consequently checks production growth. Saving the environment without checking production growth (corrected for double counting, such as treatment plants), is only possible if a technology is invented which is sufficiently clean, reduces the use of space sufficiently, leaves the soil intact, does not deplete energy and resources (i.e. uses energy derived from the sun and recycling) and is cheaper (or at least not more expensive) than current technology. This is hardly imaginable for our whole range of current activities.

[4] GNP growth results from two factors: increase in productivity and increase in labor volume (LV). In the 15-year period examined (see note 5), the growth of GNP amounted to 72%, of which 5% was caused by LV increase. In principle the allocation of this 5% forms an unsolvable allocation problem. But allocating this LV increase to the various sectors in

in slight improvements in productivity. The minimal growth rate of 3% (a doubling of production in 23 years) promoted through official policy across the world, and advocated in the Brundtland report (WCED, 1987), must therefore be achieved by much higher growth in remaining activities. Unfortunately, these activities which, by their use of space, soil and resources or by the pollution they generate, in production or consumption, most harm the environment. Some notable examples are oil and petrochemical processing and manufacturing, agriculture, public utilities, road construction and mining.[5]

A shift in human activities to reduce the burden on the environment and resources can be achieved in two ways: first by dictating environment-saving measures for production and consumption and secondly by directly changing production and consumption patterns.

The first method, e.g., applying end-of-line provisions or changing processes, mostly results in higher real prices of the products and thus in a decrease in the growth of national income.[6] (Note that when process changes do not result in higher prices or even result in lower prices, no environmental problem exists: the market forces will bring about these changes "automatically."[7] Of course, the price increases resulting from the environment-saving measures cause a shift towards more environment-friendly activities.

Technical measures often do not really solve environmental problems, because growth of an activity overrides the effect of the measure, or because, owing to the persistent and cumulative character of the burden, the measure only slows down the rate of deterioration. In these cases, in addition to the technical measures, a direct shift in behavior patterns must ensue, forced by do's and don'ts and levies. Thus, it is estimated that to stop its contribution to the acidification of forests and lakes, apart from applying all available technical means, the Netherlands must reduce the number of car miles and farm livestock by 50%. A direct shift in production and consumption patterns (the second method) will usually also check the growth of GNP, as follows from the above mentioned analysis of the National Accounts; (the environmentally most burdensome activities contribute most to GNP growth). Bicycling contributes less to GNP growth than does the use of private

---

proportion to their value added would, first of all, hardly change their relative contribution to growth of GNP, and secondly, would have a negligible influence on this contribution in absolute figures. However, a slight growth of state consumption (governmental services) and the like would of course result.

[5] This emerges from an analysis of the basic material for the Dutch National Accounts, of which the sectorial composition does not differ appreciably from that of the UK nor probably from that of most other Northern countries (see Hueting 1981).

[6] A disturbing factor is that provisions made by those outside of private firms are entered in the System of National Accounts (SNA) as final deliveries instead of as intermediate (costs). This is generally considered as double counting. When the text says: growth is checked, it always means growth, corrected for double counting. The double counting, however, is marginal to the losses, because by far the greatest part is not repaired.

[7] Of course the market is not perfect, and it is possible that introduction of processes that are both clean and cheaper is delayed for some time by inertia or lack of knowledge. It is, however, very unlikely that clean processes that are cheaper than the polluting ones are not introduced after some time. As soon as one firm applies such processes, competition forces other firms to follow suit.

cars. Saving energy and materials checks growth insofar as it either anticipates the rise in prices or it pays as a result of price increases. (Again: when this is not the case there is no environmental problem.) However, such a shift would increase our welfare (satisfaction of wants evoked by dealing with scarce means) and economic growth in the true sense (namely, increase in welfare) if we value bicycling, forests, and safeguarding the future, more than we value traffic jams, species extinction, and increasing the risks of depleting resources. Unfortunately, there is no method to state whether or not this is the case, as we shall see in the next section.

Two conclusions can be drawn from the above. First, it is unlikely that stimulating a GNP increase in industrialized countries will solve the problems of the developing countries. Such increases will most likely be possible only by accelerating encroachment on the limited energy stocks and the limited carrying capacity of the environment, which would be at the expense of developing countries. (If we try to avoid this encroachment, the growth would be checked). Secondly, GNP growth and safeguarding the environment and resources are two conflicting ends. Sustainable use of our planet's resources requires a shift in priority from increasing GNP to saving the environment. This certainly does not mean "stop production growth," but rather it requires a shift in production and consumer activities in an environmentally acceptable direction in order to arrive at sustainable economic development. We would then wait to see what the increase in production would be. Those who advocate both ends are apparently either blind to present-day reality or are speculating on as yet uninvented technologies while putting at risk the basis of our existence. Such advice will likely do more harm than good to the environment, because it strengthens the forces behind the increase of national income, which are already much stronger than those defending the environment.

The recommended shift in priority in economic policy would avoid both risks and future financial losses. Restoration after the event is usually much, often very much, more expensive than prevention and a number of environmental losses are irreversible. This shift would also stimulate the search for and application of environment-friendly technologies much more strongly than current policy.

THE UNSOLVABLE PROBLEM OF SHADOW PRICES FOR ENVIRONMENTAL FUNCTIONS

It follows from the above that the environment constantly risks falling victim to the misconceptions of economic growth and welfare. Economic policy has a one-sided stress on increasing production, as measured in national income. Therefore, a correction of national income for environmental losses seems highly recommendable, provided that it is made clear in the presentation of the results that the figures do not constitute a complete indicator for society's welfare in the course of time (see next section). In view of the severe criticism and the pressure for carrying out such a correction, that have been going on for decades, one might be surprised that this work has not been completed, or even started. The main reason for this is that it is impossible to find a theoretically sound solution for one of the two problems involved in such correction.

First of all, "the environment" has to be defined in a manageable way and the link between environment and economics must be made. This problem can be solved with the aid of the concept of environmental functions.[8] Very briefly, the reasoning is as follows.

For an economic approach the environment can best be interpreted as the physical surroundings of humans, on which they are completely dependent for all activities. Within the environment, a number of possible uses can be distinguished. These are called environmental functions or, for short, functions. When the use of an environmental function conflicts with the use of another or the same function, either in the present or future, loss of function occurs. We call this competition between functions and make a distinction among qualitative, spatial and quantitative competition.

When competition of functions occurs, the environment takes on an economic aspect. Economics boils down to the study of the problems of choice entailed by the use of scarce means for the satisfaction of wants. A good is scarce if the demand for it exceeds its availability, or, when something else we would like to have (an alternative) has to be sacrificed to acquire it. Environmental functions meet this definition fully as soon as they compete. Competing functions are scarce goods. Losses of function are costs, irrespective of whether they are expressed in monetary terms; the terms "money" and "market" do not occur in the definition of the subject matter of economics.

Qualitative competition occurs when the use of the environmental function "dumping ground for waste" (or: "withdrawal or addition of species and matter") is at the expense of other functions. There is, however, an intermediate step. An activity introduces or withdraws an agent into or from the environment, thereby changing the quality. This may make other uses more difficult or impossible. An agent is a constituent or amount of energy (in whatsoever form) which may cause loss of function by its addition or withdrawal from the environment by humans. An agent could be a chemical, plant, animal, heat, noise, radioactivity, etc. In spatial and quantitative competition, the amount of space or matter is insufficient to meet the existing wants for it. Note that the use of a function also includes the passive use of the function "natural environment." This conserves the actual and potential utilities of ecosystems, now and in the future, and retains the diversity of species of our planet. Competition between functions can take all sort of forms. But in most cases by far, it is a question of the environment being used for current production and consumption activities at the expense of other desired uses or of future possible uses, including production and consumption. A well-known example of the latter is the loss of topsoil resulting from deforestation.[8]

The second problem pertains to the construction of shadow prices. National income is recorded in market terms. For confronting environmental losses with this figure it is necessary to construct shadow prices for environmental functions that are directly comparable with market prices. For this, demand and supply curves have to be constructed. In the period 1969-1974, the Netherlands Central Bureau of Statistics attempted to do this (CBS 1972, 1973, 1975b; Hueting 1980), with, briefly, the following results.

---

[8] For a detailed description and elaboration of this approach see Hueting (1980), which is based on Hueting (1970).

The supply curve can, in principle, always be constructed by estimating the costs of measures needed to prevent loss of function. The measures will often be a mix of technical provisions such as add-on technology (e.g., treatment plants) and changes in processes, and reductions or stoppage of the burdening activities (which also can be expressed in monetary terms). The supply curve is called the elimination cost curve.

Constructing a complete demand curve, however, is mostly not possible because only in exceptional cases can the intensity of the *individual preferences* for environmental functions be entirely expressed in market behavior or other behavior that can be translated into market terms (money).[9] Loss of function can sometimes be partly compensated by provisions which substitute for the original function but in other cases loss of function causes financial damage. When, for instance, water is polluted by chemicals, compensation of the function "drinking water" or "water for agriculture" is possible to a certain degree by purifying the intake of the polluted ground or surface water. In the long run, however, elimination of the pollution is necessary, because of the cumulative effect. An example of financial damage is flooding of cropland and property resulting from loss of the forest function "regulator of the water management".

Both compensation and financial damage can be interpreted as revealed preferences for a given function. In compensating for a function, provisions are made to replace the function that was originally present. Amounts of damage can also be conceived of as revealed preferences, since they are losses suffered as a result of the disappearance of the function. As stated above, preferences can seldom be manifested entirely via the market. It is clear that only a very small proportion of the losses of environmental functions are compensated, and they are not always reflected in financial damage. Often, too, the possibility of compensation does not exist. Thus, a compensatory measure like moving to a clean area is feasible only for the happy few. Moreover, it evokes new traffic streams causing new losses of function. Financial damage through noise nuisance and air pollution is very incompletely reflected in the decline of a house's value because prices are also affected by tightening of the housing market and people's immobility caused by ties to work and the neighborhood (Jansen and Opschoor 1972). The construction of new forests and lakes is pointless as long as the process of acidification is not halted. The loss of soil by erosion cannot be compensated. Most important of all, much of the damage caused by losses of function will occur in the future, such as the damage caused by loss of the stability of the climate, by loss of the functions of tropical forests ("gene reserve," "regulator of the water management," "preventor of erosion," "supplier of wood," "buffer for CO_2 and heat," "regulator of the climate" and the like), and by the disruption of ecosystems resulting from the extinction of species. The risks of future damage and the resulting poor prospects for the future cannot manifest themselves via the market of today.[10] Yet there is obviously a great need for unvitiated nature and a safe future.

---

[9] The possibilities for this are limited. An example is the costs of travel involved in visiting a nature area (see Hueting 1980).

[10] Calculating the net present value (NPV) of future damages, the current extent of which can be established via the market (e.g., damage by flooding resulting from loss of the function

Because of the limited way in which preferences for environmental functions can be manifested in market behavior, efforts have been made to trace these preferences by asking people how much they would be prepared to pay to wholly or partially restore functions and to conserve them. Quite a lot of research is going on in the field of willingness to pay for the environment and willingness to accept environmental losses (an overview of the methods used , including quite a few results, can be found in Johansson 1987, p. 32; Kneese 1984; Pearce et al. 1989). It is questionable, however, whether this method is suitable to arrive at the construction of a complete demand curve, certainly on a macro scale and certainly for functions on which current and future life depends. Insofar as people are directly affected by environmental losses, the approach might be justified. Many environmental losses, however, constitute part of a process which may lead to the disruption of the life-support functions of our planet and endanger the living conditions of generations to come, and therefore cannot be considered separately. In cases where this is true, asking people how much a function is worth is pointless. Arguments for this view are listed in Hueting (1989).[11] See also Kapp (1972, pp. 17-28).

From the above, it follows that the construction of shadow prices that are directly comparable with market prices, a prerequisite for a theoretically sound correction of national income, is mostly not possible.

WHAT CAN BE DONE IMMEDIATELY?

In the last section a practical and defensible proposal will be put forward to overcome the problem of the impossibility of constructing shadow prices for environmental functions, comparable with market prices. However, elaboration of the method proposed will probably take at least three years for a country with a relatively well-developed system of environmental statistics like the Netherlands. For some other countries it might take longer. In expectation of the results, one thing can be done immediately: make clear when the National Accounts are published, what the changes in the level of national income do *not* mean, in order to prevent these changes from giving incorrect signals to society about the economic success of its activities. In addition, the costs of compensating, restoring and preventing environmental losses that are wrongly entered as final delivery (see below) can be made visible with relatively little effort.

'regulator of the water household'), breaks down on the unsolvable problem of the level of the discount rate in environmental costs and benefits (see Hueting 1991).

[11] To these arguments the following can be added: 1) A number of people will probably have their doubts about the participation of others (the Prisoner's Dilemma from game theory) or prefer to wait and see (the Free Rider Principle from the theory of collective goods). Thus in developing countries (where the tropical forests are) the view is widespread that people from the rich countries should pay, because these countries a) have much more money to spend, b) nevertheless are destroying their own environment, e.g., what is left of their forests, by acidification, c) contribute considerably to global effects such as the greenhouse effect, and d) have a clear interest in saving natural resources in the third world. 2) In cases where the whole community is involved, the willingness to accept or to pay approach is pointless, for who is paying whom?

Proposal for a Text to Avoid Misinterpretation of National Income Figures

Economics boils down to the study of the problems of choice entailed by the use of scarce means to satisfy wants. Welfare is defined as the satisfaction of wants evoked by dealing with scarce means. So welfare, or satisfaction of wants, is a psychical category, an aspect of one's personal experience. Economic theory assumes that, when dealing with scarce means, we try to maximize our welfare (the opposite is nonsensical). Besides maximization of welfare with given means, the desire to raise the level of satisfaction of wants (welfare) in the course of time is also regarded as a motive of economic action.

It follows from this brief description of the subject matter of economics that economic growth and economic success can mean nothing other than an increase in the level of welfare. Our economic actions have scored success when our satisfaction of wants has increased. Since satisfaction of wants is not directly observable "from the outside" and thus not in itself a cardinal measurable quantity, it seems logical to look at factors that *are* measurable and that can arguably be supposed to determine the level of welfare. In this procedure there are a number of objections to using the production of goods and services, as measured in national income, as *the* indicator for welfare, economic success, and economic growth. These objections could be classed in three categories, which will be summarized very briefly.

The *first category* is of a more or less "technical" nature. It encompasses five points:

1. The consumer surplus, which relates to the difference between the total utility of a good and the product of price (as the criterion of marginal utility) and quantity, is not expressed in national income.
2. In national income, the value added by production is calculated at market prices. This means that different people's (marginal) utilities of goods are added. This is already unallowable with equal income distribution because it is impossible to compare utilities between individuals: some people attach greater value to goods than others. With the existing inequality of incomes this is not allowed a fortiori on account of the diminishing marginal utility of money as income grows. The impossibility of comparing utilities between individuals further implies that if part of the population of a country regresses and the rest progresses, no pronouncement can be made on the final result. (This situation certainly occurs.)
3. The law of diminishing marginal utility applies to individuals. However, the same tendency is noticeable for the whole economy, as has been shown by recent research (Van Praag and Spit 1982). It appears from this research that ever more extra goods are necessary for the attainment of the same increase in welfare as income rises. This makes the importance of an ever-growing production relative for welfare.
4. Real national income is obtained by expressing income in current prices in constant prices using a composite price index. This can only be done correctly for a constant package of goods. Because of the constantly changing package of goods the calculated value of the price index varies, depending on the solution chosen (Kuznets 1948; Hicks 1948; Pigou 1949, p. 11). This problem worsens over long time periods.

5. Not all production takes place in business enterprises or in government agencies. This may not only influence the level of national income but also the changes in it, if, for instance, the work of former housewives is taken over by paid domestic help, dishwashers and restaurants.

The *second category* of objections to identifying increases in national income with economic growth and economic success relates to the intermediate character of some elements of national income. Under present conventions for calculating national income, a number of activities which have a cost character, and therefore ought to be entered as intermediate deliveries, are designated as final consumption. S. Kuznets (1947, 1948), one of the great theoreticians of national income, emphasizes this. Kuznets divides these activities into three classes.

1. The first class is invoked by the fact that in industrial countries the dominant modes of production impose an urban pattern of living. Urbanization brings in its wake numerous services whose major purpose is to offset disadvantages of urbanization. Kuznets gives as examples the expenditure necessary for bridging the greater distance between home and work, and the money spent on compensation for the inconveniences entailed in living in dense agglomerations.
2. The second class distinguished by Kuznets considers expenditure that is inherent in participation in the technically and monetarily complex civilization of industrial countries. Payments to banks, employment agencies, unions, brokerage houses, etc., including payments for such things as technical education, are not, according to Kuznets, payments for final consumer goods. They are activities necessary to eliminate the frictions of a complicated production system and not net contributions to ultimate consumption.
3. The third class is a major part of government activity. The legislative, judicial, administrative, police and military functions of the state, according to Kuznets, are designed in order to create the conditions under which the economy can function. These services do not provide goods to ultimate consumers. It is wrong to count the whole of government activity as a net contribution to national income.

To this a fourth class can be added. This class would include expenditures on measures that compensate or restore the losses of environmental functions (see the second section) or prevent losses of environmental functions from occurring. These expenditures are entered as intermediate deliveries insofar as the measures are taken and directly paid for by private firms. They are entered as final consumption when the measures are paid for by private households or the government and when they are taken by private firms but financed via levies imposed by the government. All these outlays should be entered as intermediate when a long time series such as that for national income is composed. The losses of environmental functions are not entered as costs at the moment they originate; at that moment the environment is excluded from the System of National Accounts (SNA). When these losses are eliminated or compensated, the environment is included—

although only partially—in the SNA. This is generally considered as double counting. This procedure also makes figures of different years incomparable, at least when they are used as a measure for economic growth and welfare.

The *third category* of objections to identifying increase in national income with economic growth (increase in welfare) relates to the fact that production is only one of the factors that determine the level of welfare. At least seven factors play a role:

1. The package of goods and services produced by humans.
2. The scarce environmental goods in the broad sense (i.e., including space, energy, natural resources, plant and animal species.
3. (Leisure) time.
4. Income distribution.
5. Working conditions.
6. Employment.
7. The safety of the future, insofar as this depends on how we deal with scarce goods (e.g. the life support functions of the environment).

All these seven factors play a part in economic action. They constantly have to be weighed against each other whenever the desired quantity or quality of a given factor is at the expense of one or more other factors. (Note that when this is not the case, no economic aspect is at stake, so no choice has to be made). Seen from the point of view of those who choose, citizens or politicians, there is thus an unbreakable link between all the factors influencing welfare.

From this, it follows that no judgment can be given on the development of welfare over time because, among other things, the factors influencing welfare cannot be linked by a common denominator. To give an example, the effect an increase in production has on our welfare depends on the importance we attach to a greater quantity of produced goods on the one hand, and to the resultant loss of environmental functions on the other. Depending on how we weight these factors, an increase in production may lead to economic progress, a neutral effect or economic decline (loss of welfare). If we rate at the margin, the environment higher than production of large quantities of goods, and the government implements measures that lead to fewer available goods and services[12] and an improvement in the environment, then the overall satisfaction of wants obtained from economic goods is enhanced. In this case, less production leads to greater welfare.

Correcting National Income for Double Counting

A correction for expenditure on compensatory, restoratory and preventive measures would be feasible without theoretical difficulties. This expenditure, which only reestablishes or maintains environmental functions that would remain available without the negative

---

[12] This is very likely to occur; see the first section.

impact of our activities on the environment, is wrongly entered as value added. This leads to an overestimation of the increase in national income and conceals what is going on in the environment: loss is not written off, restoration is written up. On the one hand, such a correction would be a step forward, as it would partly solve the well-known problem of double counting (see above) and provide more information about the relationship between production and environment. For example, it appears from a study by Christian Leipert (1989) that between 1970 and 1985 the defensive outlays in the Federal Republic of Germany increased from 5% to nearly 10% of GNP. The expenses include expenditures induced by manmade environmental deterioration, such as restorative and compensatory outlays, but also defensive expenditure in the fields of traffic, housing, security and health. This means that, in this period, about one-fifth of the GNP growth consisted of an increase in additional costs caused by the same growth. On the other hand, a correction for double counting would express the environmental losses very incompletely and introduce the pars pro toto problem: part of the information is conceived as the total environmental effect. For, as is well known, most environmental losses are not restored or compensated.

A PRACTICAL SOLUTION

Since the period 1969-1974, in which the Netherlands Bureau of Statistics made an attempt to construct shadow prices for environmental functions with the intention of correcting national income, the call for national income to be corrected to include environmental losses has been steadily growing. In the course of a working visit to Indonesia in 1986, I was provoked by the following remark made by the Indonesian minister for Population and Environment: "In my policy-making I need an indicator in money terms for losses in environment and resources, as a counterweight to the indicator for production, namely national income. If a theoretically sound indicator is not possible, then think up one that is rather less theoretically sound."

The answer to the minister's dilemma is obvious; an estimate based on standards. The setting of standards was also discussed at the time (Hueting 1980), but the point was not elaborated because the question "What standards are to be set and by whom?" could not be answered. This situation has now changed, partly due to the publication of the report of The World Commission on Environment and Development *Our Common Future* (1987) (the so-called Brundtland Report). Politicians and organizations across the world have declared themselves in favor of sustainable development. This can be conceived as a preference voiced by society which opens up the possibility of basing a calculation on standards for a sustainable use of environmental functions, instead of on (unknown) individual preferences.

I proposed (in 1986 and 1989) the following procedure (the feasibility of which will be investigated at the Netherlands Central Bureau of Statistics, starting in February, 1991): define physical standards for environmental functions, based on their sustainable use; formulate the measures necessary to meet these standards; and finally, estimate the amounts of money involved in putting the measures into practice. In technical terms, this means that in the familiar diagram of the supply and demand curve for environmental functions we have to determine a point on the abscissa which represents the standard for sustainability (figure 14.1). A perpendicular on this point intersects the supply curve; the perpendicular replaces the (unknown) demand curve. The point of intersection indicates the volume of activities, measured in terms of money, involved in attaining sustainable use of the function. The volume will often be a mix of necessary technical measures and reductions in activities (expressed in money terms) which, even after application of the measures, will be required to attain sustainable use of the function (see figure 14.1).

The reduction of national income (Y) by the volumes thus derived gives a first approximation of the level of activities which, in line with the standards applied, is sustainable. Needless to say the correction for double counting, mentioned previously, must also be made. This amount is also deducted from national income. We call the sustainable level Y'. The difference between Y and Y' indicates, in money terms, how far society has drifted away from the desired sustainable development. In other words, this difference indicates the sacrifice involved in attaining the desired sustainable use of the environment.

To be absolutely clear, it should be pointed out that this is a partial equilibrium and static approach. Effects on other sectors of the economy as a result of taking measures and reducing activities are not considered. Neither are future developments that might be expected involved in the approach. These are not taken into consideration, firstly because the exercise is aimed at a correction of the figure of national income and not at the

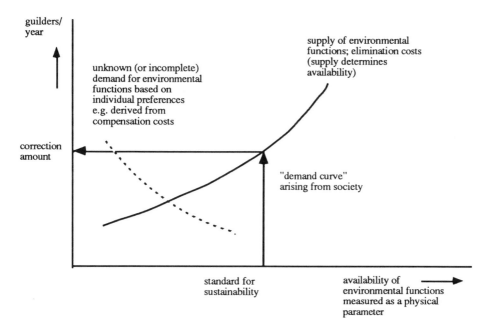

FIGURE 14.1 Supply and demand curve for environmental functions

development of a vision for the future, and secondly because in the model to be used, a large number of assumptions would have to be incorporated. The following may be regarded as an exception.

Reducing Y to account for double counting, for the costs of the measures needed to attain a sustainable society, and for the added value of the activities which have to be reduced, provides a first approach to Y'. The activities burdening the environment which are to be discontinued will in most cases be replaced by environmentally acceptable alternatives. The problem of the environment is a problem of allocation. It is a question of shifting the patterns of activities in an ecologically sound direction. This shift will be brought about both by price increases which will result from taking the discussed measures and by directly reducing activities which place a heavy burden on the environment.

The first estimate of Y' must therefore be raised by the added value of activities that will come into being after reduction of the harmful activities and through the disappearance of the compensation measures. An iterative calculation becomes necessary here, for there are very few activities that do not adversely affect the environment. The replacement activities may lead to the standards being exceeded and thus to the need for further reallocation. I propose postponing this more dynamic approach until the second phase of the research and limiting the first phase to an initial estimate of Y'.

The familiar objection to entering restoration measures as final delivery instead of as an intermediate term (costs) does not apply to Y'. The objection is that the environment is considered outside the System of National Accounts when environmental losses occur (loss is not written off), but is included, albeit partially, in the SNA when the loss is restored (restoration is written up). This means that a comparison cannot be made between various years. Hence the continual pressure for correction of double counting which has been going on for decades now. Because environmental loss at Y' is written off, it is only logical that restoration should be written up. Naturally, an increase in Y' will also occur as clean technology and flow energy become cheaper, for then the deduction is reduced.

Standards and the measures that are based on them may relate to the occupation of space, the use of soil, the availability of stocks of natural resources, the composition of products, the consumption of raw materials and energy, the emission by activities, and the concentration of chemical and other agents.

The standards can be related to environmental functions. Thus it is possible to formulate the way in which a forest should be exploited in order to attain a sustainable use of its functions such as "supplier of wood," "regulator of the water management," "object of study for ecological research," "supplier of natural products for the local population," and "source of income from tourism." The estimated expenditure on the measures required to meet those standards then tells us in monetary terms how far a nation has drifted away from its (supposed) end or standard of sustainable use of its forest resources. Likewise it is possible to formulate the way in which surface and groundwater should be exploited in order to arrive at a sustainable use of its functions such as "waste-dumping site," "water as raw material for drinking water," "water for agriculture," "cooling water," "water for flushing and transport," "process water," "water for recreation," "water for navigation" and "water allowing the existence of natural ecosystems." And the estimated expenditures on the measures required to meet those standards tell us in monetary terms how far a nation has drifted away from a sustainable use of its water resources. The same holds true for the use of air, soil and space.

The measures used to insure sustainability may range from selective cutting of trees, reforestation, building terraces, draining roads, maintaining buffers in the landscape, selective use of pesticides and fertilizers to building treatment plants, recirculation of materials, introducing flow energy, altering industrial processes, making more use of public transport and bicycles instead of private cars, and making use of space that leaves sufficient room for the survival of plant and animal species.

Of course, no measures can be formulated for irreversible losses. If plant and animal species become extinct, no restoration measures are possible. The same probably holds for the total loss of the topsoil of a mountainous area. An arbitrary value then has to be assigned to these losses, of which only one thing can be said for certain: the value is higher than zero.

A similar problem seems to arise in the case of nonrenewable resources. When oil has been exploited and exported or used, it is gone forever. This loss constitutes a cost for the country, for oil reserves are finite. Thus, if no alternatives are developed, the generation which experiences the depletion (or a sharp rise in operating costs) of the resource will suffer a severe economic loss. And since the prospect of a safe and prosperous future for one's children and one's children's children is a normal human need, the diminution of the reserves of nonrenewable resources such as oil constitutes a cost now.

The valuation of the depreciation of nonrenewable resources can be done by estimating the costs involved in the development and practical introduction of alternatives like solar energy, substitutes for minerals and recycling methods.

Comments on the application of the concept of sustainability can be found in ecological literature. E. P. Odum (1971) states that through human activities a development is increasingly taking place which results in mature, stable ecosystems being replaced by more recent, less stable stages. This is the opposite of natural development. As fewer stable stages remain, restoration of impaired systems becomes increasingly difficult and takes longer and longer and the number of potential and actual possible uses falls steadily. An irreversible situation can be created when harm is done on a large scale to predators, when substantial numbers of species are lost or when general biological activity is suppressed. This is a disruption of food chains that may lead inter alia to disruption of the life-support functions of our Earth. The process of the decline and disappearance of species can be seen as an indicator of the extent to which we are already on the way to disrupting life-support functions. The chance of severe disruption can be minimized if human activities, through the use of recycling processes, (again) become part of the biological cycle and the level of activity is limited by the condition that the degree of stability of this cycle does not decrease. A sustainable activity pattern will amount to recycling natural resources, switching to non-polluting sources of flow energy and using land so that sufficient room is left for natural ecosystems to function.

Sustainability is linked to environmental functions (possible uses of the environment) as described and listed in Hueting (1980). Sustainability means that functions must remain intact so that all present and future uses remain available. As for renewable resources such as forests, water, soil and air, it holds that as long as the regenerative capacity remains intact, the functions remain intact, e.g., the function "supplier of wood" of forests, the function "drinking water" of water, the function "soil for raising crops" of soil and the function "air for physiological functioning" of air. This means for instance, emissions of accumulating matters (PCBs, heavy metals, nitrates and CO_2) may not exceed the natural buffering capacity of the environment and that the erosion rate may not exceed the regenerative power of the soil. As for nonrenewable resources, such as oil and copper, "regeneration" takes the form of research and implementation of flow resources such as energy derived from the sun and from recycling materials.

At the request of Dr. Emil Salim, Minister of Population and Environment of Indonesia (see above), I elaborated on the method of setting costs to meet the standards for sustainable use of environmental functions for the case of erosion. I designed a coordinated set of tables which would handle the information needed to arrive at the required figures. The reasoning is as follows.

Suppose the standard for erosion, necessary to arrive at a sustainable economic development is equal to the natural rate of change of the topsoil. This means that the erosion rate in all regions of Indonesia has to be measured. (Quite a bit of measurement has been done already.) For places where the erosion rate exceeds the natural rate of increase, measures to bring the erosion back to the natural rate must be formulated. These measures may differ from place to place. But let us suppose that in a certain region the following measures have to be taken to meet the erosion standard and to maintain the standard in the course of time:[13]

1. Reforestation of the mountain above the agricultural zone or, when farm land has expanded too far uphill, above a certain, not too high, contour line.
2. Building terraces.
3. Setting up a drainage and irrigation systems to prevent gully formation.
4. Draining roads and, if necessary, rebuilding them as far as possible along the contour lines of the mountain.
5. Disseminating information on the necessity of the measures.
6. Installing officials, chosen from the local population, to ensure that rules necessary to meet the standard are followed.

The expenditure necessary to carry out these measures (and other similar measures) can be estimated. In the case of erosion, the following steps have to be taken to arrive at the figure with which the national income figures over a certain period have to be confronted. (Which period this should be is not discussed here, because it is not essential to get an idea of the method.)

---

[13] As a result of the measures, the area of arable land and the employment in agriculture might decrease. This necessitates, for instance, investments for local industry or rehabilitation of waste land. Furthermore, measures such as making tenancies more secure and other land reform measures may be necessary to maintain a stable situation. This kind of secondary measure is not taken into account because correction of national income allows only for the direct measures that are necessary to meet the standard for a sustainable use of the environmental functions.

1. Reviewing the erosion rate for each province. (Use table 14.1.)

TABLE 14.1 Erosion Rate by Region (per slope class) (in year x)

Erosion rate (times the natural rate of increase)	Region (in km^2) 1	2	3	..	n	Total km^2 in the Province with a Given Erosion Rate
1 time	a km^2	b km^2	c km^2			
2 - 3 times	etc.					
3 - 5 times						
5 - 7 times						
etc.						
Total km^2 of region						

2. Establishing the causes of the erosion for each province. Erosion can be caused by 1) Agricultural mismanagement (e.g., using overly steep slopes), 2) Overcutting of wood (e.g., for cooking), 3) Disobeying cutting rules, 4) Natural disaster. More than one cause at the same time can lead to erosion in an area. (The findings should be collected in a table similar to table 14.2.)

TABLE 14.2 Causes of Erosion by Region (in year x)

Main cause	Region (in km^2) 1	2	3	..	n	Total km^2 per Cause in the Province
1	a km^2	b km^2	c km^2	etc.		
2	etc.					
etc.						
Total km^2 per cause in a region						

3. Formulation of the necessary measures for each province to meet the erosion standard. These measures might differ by local geographic situation and by cause. (The findings should be collected in a table similar to table 14.3.)

TABLE 14.3 Measures to Meet the Erosion Standard, by Region (in year x)

Measure	Region (in km^2) 1	2	3	..	n	Total km^2 per Measure in a Province
1	a km^2	b km^2	c km^2			
2	etc.					
3						
etc.						
Total km^2 of region						

4. Estimate of the costs to meet the erosion standard. (Includes costs of forgoing farm land, if necessary. The results of the estimates should be collected in a table similar to table 14.4.)

TABLE 14.4 Costs of the Measures by Province

| Measure | Costs per Province | | | | | Total Costs for Indonesia per Measure |
	1	2	3	..	*n*	
1	a Rp	b Rp	c Rp			
2	etc.					
3						
etc.						
Total costs to meet the erosion standard						Total costs for Indonesia to meet the erosion standard

The necessary calculations of the costs of measures to meet the standards for a sustainable use of environmental functions are not new. As for the Netherlands, the first publications of the Department for Environmental Statistics of the Central Bureau of Statistics resulted in estimates of the costs of measures for various degrees of restoration of function (the supply curve, see above) (CBS 1972, 1973, 1975b; Hueting 1980). Later the CBS undertook such calculations for the Netherlands Scientific Council for Government Policy (1978) and for the scenario studies in the context of broad social discussion on the future energy supply (Hueting 1987). In the chapter on costs and benefits of environmental measures in a recent report by the National Institute of Public Health and Environmental Hygiene (1988), a number of "supply curves" are included; the institute uses these data for scenario studies. Other countries have carried out similar estimates, for instance the Economic Council of Japan (1974). Apart from these and other, more integral, approaches, numerous studies on, most notably, the measures and costs involved in the reduction of emissions of harmful agents have been carried out in quite a number of countries.

Based on the experiences mentioned above, I believe that a correction of national income figures on the basis of standards for sustainable use of the environment is feasible. The method has certain advantages and drawbacks.

The drawbacks (or imperfections) are as follows:

1. The results of the approach do not represent individual valuations in the true sense, as has been extensively explained above. For, among other things, the intensity of the preferences for a sustainable use of the environment cannot be measured. However, this simultaneously implies that the intensity of the preferences for accepting adverse effects and future risks involved in the present growth pattern of production and consumption, and thus for the growth of GNP, is equally unknown. Both of these points should be clearly mentioned in the results presentation of National Accounts.

2. The approach is strictly static. Only first-order effects are taken into account. In reality, two additional movements will take place. The technical measures and the reduction of activities will bring about a contraction or an expansion of related industry, depending on the type of industry. Presumably the net effect will be an increase in the correction amount. In the course of time, however, the economy will adapt itself to the new situation. A second movement takes place as a reaction to the reduction of polluting activities: people will seek environmentally acceptable alternatives. This would reduce the correction amount. Both effects could in principle be estimated with the aid of a dynamic model. However, the reliability of the results of such a model would be questionable, because changing to sustainable use of the environment means a break in the trend, thereby surpassing the predictive capabilities of the current models.

3. The method ignores the loss of welfare suffered by those people who have a strong preference for the survival of plant and animal species apart from their role in the maintenance of the life-support functions of our planet. This preference could be compared with the preference for creating and maintaining art or churches, which might not be considered indispensable for sustainable development and yields, but the loss of which would constitute a decrease in welfare for those who need them. A solution would be to class the diversity of species under the function "gene reserve." Another similar example is noise. Noise does not affect sustainability, but it can be very disturbing. A solution would be to class this under the function "quiet in the living area." Standards for noise levels are already operational in many countries. This too should be mentioned in the presentation of the results.

4. The results of the approach do not indicate the state of the environment. If, for instance, a cheaper anti-pollution technology is invented, the distance between national income (Y) and the estimated sustainable activity level (Y') becomes smaller. But if the technology is not applied or is not generally used, the state of the environment changes very little or not at all. Furthermore, a decrease in costs does not necessarily run parallel with changes in physical parameters. Therefore, environmental statistics in physical units remain indispensable.

5. For irreversible losses no measures can be formulated, of course. This holds true for any method.

6. The method is laborious.

The advantages are as follows:

1. As far as we can see, the method is the only way to confront the national income figures with the losses of environmental functions in monetary terms.

2. The method compels us to explicitly define "sustainable economic development." Without such a content, the term remains vague and not operational in economic policy regarding the environment.

3. The physical data required for comparison with the standards come down to basic environmental statistics which have to be made anyway, if a government is to get a grip on the state of the environment. The formulation of the measures to meet the

standards and the estimates of the expenditure involved are indispensable for policy decisions. Or in other words: the work for supplementing national income figures might be laborious, but it has to be done if one wants to implement a deliberate environmental policy.

On the strength of the arguments mentioned above, I recommend a correction of national income for environmental losses (including resources). The correction should be based on standards for sustainable use of environmental functions in order to arrive at a figure for national income that can be compared with the current one.

REFERENCES

Central Bureau of Statistics. 1972. Waterverontreiniging met Afbreekbaar Organisch en Eutrofierend Materiaal (Water Pollution by Biodegradable Organic and Eutrophicating Matter). 's-Gravenhage.

Central Bureau of Statistics. 1973. Waterverontreiniging ten gevolge van Verzilting, 1950-1970 (Water Pollution Caused by Salinization, 1950-1970). 's-Gravenhage.

Central Bureau of Statistics. 1975a. Luchtverontreiniging door Verbranding van Fossiele Brandstoffen 1960-1972 (Air Pollution Resulting from the Combustion of Fossil Fuels 1960-1972). 's-Gravenhage.

Central Bureau of Statistics. 1975b. Functieverliezen in het Milieu (Losses of Function in the Environment). Statistische en Econometrische Onderzoekingen (18), 's-Gravenhage.

Economic Council of Japan. 1974. Measuring Net National Welfare of Japan. Tokyo: Japanese Bureau of Statistics.

Hicks, J.R. 1948. On the Valuation of Social Income. *Economica* (August).

Hueting, R. 1970. Moet de Natuur Worden Gekwantificeerd? (Should Nature Be Quantified?). *Economisch-Statistische Berichten* (21 January).

Hueting, R. 1980. *New Scarcity and Economic Growth*. Amsterdam/New York: Oxford University Press.

Hueting, R. 1981. Some Comments on the Report "A Low Energy Strategy for the United Kingdom," compiled by Gerald Leach et al. for the International Institute for Environment and Development (IIED). Paper prepared for the Working Party on Integral Energy Scenarios. The Hague, May 20, 1981. Also published under the title: De Relatie Tussen Produktiegroei en Energieverbruik. Maakt Groeifanatisme Blind? (The Relation Between Production Growth and Energy Use. Does Growth Fanaticism Blind?). *Economisch-Statistische Berichten* (June 24, 1981):609-611.

Hueting, R. 1987. An Economic Scenario that Gives Top Priority to Saving the Environment. *Ecological Modelling* 38: 123-140.

Hueting, R. 1989. Correcting National Income for Environmental Losses: Towards a Practical Solution. In Y. Ahmad, S. El Serafy, and E. Lutz, eds., *Environmental Accounting for Sustainable Development*. Washington, D.C.: The World Bank.

Hueting, R. 1991. The Use of the Discount Rate in a Cost-Benefit Analysis for Different Uses of a Humid Tropical Forest Area. *Ecological Economics* 3(1991) (in press).

Jansen, H. M. A. and J. B. Opschoor. 1972. De Invloed van Geluidshinder op de Prijzen van Woningen: Verslag van een Enquàte onder Makelaars in en rond de Haarlemmermeer. (The Influence of Noise-nuisance on the Prices of Houses: Report of a Survey among House-Agents in and around the Haarlemmermeer.) Instituut voor Milieuvraagstukken van de Vrije Universiteit. Werknota (15). Amsterdam.

Johansson, P. 1987. *The Economic Theory and Measurement of Environmental Benefits*. p. 32. Cambridge: Cambridge University Press.

Kapp, K. W. 1972. Social Costs, Neo-Classical Economics, Environmental Planning: A Reply. *Social Science Information*, 11(1):17-28.

Kneese, A. V. 1984. *Measuring the Benefits of Clean Air and Water*. Washington, D.C.: Resources for the Future.

Kuznets, S. 1947. National Income and Industrial Structure. The Econometric Society Meeting. Washington, D.C., September 6-18, 1947. *Proceedings of the International Statistical Conferences*, Calcutta. 5:205.

Kuznets, S. 1948. On the Valuation of Social Income. *Economica* (February/May).

Leipert, C. 1989. Social Costs of the Economic Process and National Accounts: The Example of Defensive Expenditures. *Journal of Interdisciplinary Economics* 3(1): 27-46.

National Institute of Public Health and Environmental Hygiene. 1988. *Concern for Tomorrow*. Samsom H.D. Pjeenk Willink.

Netherlands Scientific Council for Government Policy. 1978. *The Next Twenty Five Years*. The Hague: Staatsuitgeverij.

Odum, E. P. 1971. *Fundamentals of Ecology*, 3d ed. Philadelphia: Saunders.

Pearce, D. W., A. Markandya and E. B. Barbier. 1989. *Blueprint for a Green Economy*. London: Earthscan.

Pigou, A. C. 1949. *Income*, p. 11. London: MacMillan.

Van Praag, B. M. S. and Spit, J. S. 1982. The Social Filter Process and Income Evaluation: an Empirical Study in the Social Reference Mechanism. Rep. 82.08. Centre for Research in Public Economics. Leyden University, Leyden.

World Commission on Environment and Development (WCED). 1987. *Our Common Future*. New York: Oxford University Press.

15

NATIONAL ACCOUNTING, TIME AND THE ENVIRONMENT: A NEO-AUSTRIAN APPROACH

Malte Faber
Alfred Weber Institute
University of Heidelberg
Grabengasse 14, 6900 Heidelberg, Germany

John L. R. Proops
Department of Economics and Management Science
University of Keele
Staffordshire, ST5 5BG, UK

ABSTRACT

National accounting is an area of controversy. In this paper, we explore the conceptual foundations of the difficulties of determining accurate national accounts. The fundamental requirement of a system of national accounts is that national income equal national product. Further, if the economy is composed of several producing sectors, then the value of total output must equal the value of total input for each sector. This requires strong assumptions about the nature of price formation, if the use of ad hoc "balancing" items is to be avoided. In particular, if balance is to be achieved in each sector of the economy, then the prices used must be the shadow prices derived from an optimizing economy. That balancing items are generally necessary therefore reflects the deviation of real economies from the competitive equilibrium.

To achieve accounting balance, the shadow price of capital accumulation must be established. This implies the even stronger condition that for national accounts to be, and to remain, in balance, the quantities and prices used must derive from intertemporal optimization in the economy. The necessity of taking time into consideration in conceptualizing national accounts becomes even more clear when the productive role of the natural environment is considered; in particular, we examine the valuation of natural pollution degradation, and of natural resources. The appropriate valuation of the stocks and flows of the natural world must be an intertemporal valuation, if they are not to add to the problems of national accounting. Our analysis of national accounts is illustrated with input-output tables derived from a neo-Austrian model of intertemporal choice between investment and consumption. This gives an optimal intertemporal price system. Comparing this model with the one used in reality lays bare the reasons why one cannot expect to find a consistent national accounting system in use. From this, it follows that alternative institutions and conventions have to be developed to measure and value the scarcity of the services provided by the ecosystem.

INTRODUCTION

As Norgaard (1989, p. 303) recently pointed out, "Economists during the last two decades have tried to logically extend the system of national accounts to include the value of resources, environmental systems, and their services. Alternative proposals for improving the accounts have been extensively debated, but no agreement has been reached." He goes on to expose three dilemmas of environmental accounting. From these, he concludes: "The development of multiple approaches would be scientifically rigorous and more likely fail-safe."

While we in essence agree with Norgaard, we consider it useful first to go as far as possible in developing a single conceptual approach to national accounts. To this end, we take recourse to two strands in the literature. The first strand was initiated by Fisher (1930), and further developed by Samuelson (1961) and Weitzman (1976). They stress the importance of dynamic considerations in national accounting, particularly with respect to the representation of consumption over time as being the true measure of an economy's "wealth." The second strand is the inclusion of the value of services rendered by the natural world into the national accounting framework (e.g., Peskin 1976). These two strands have already been related to the use of natural resources, by Dasgupta and Heal (1979, pp. 244-246), but in a rather aggregated framework. We intend to integrate these two strands into an accessible form by using an input-output framework and a neo-Austrian model of intertemporal production. We will show that the corresponding intertemporal price system is crucial for the consistency of national accounts.

Our paper is organized as follows. First, we develop the concepts of national accounting in a disaggregated input-output framework. We use this framework to show that national accounting balance depends crucially on the quantities produced in the economy being "supported" by an appropriate set of prices. We use a neo-Austrian production model to show that these appropriate prices can be obtained if we assume the economy to be subject to an optimal intertemporal plan, whether through the operations of a central planning agency, or through the existence of a set of efficient markets. We then examine the problems of valuing the natural environment service of pollution degradation. We note that no fully satisfactory practical solution is available, because of the public good aspect of this service. We suggest a partial solution, based upon the known cost of pollution abatement. We then turn to the inclusion of the services of nonrenewable resources within a national accounting framework. We note in particular the crucial role played by the "royalty element" in the intertemporally efficient resource price, and the strong implications that underestimating this royalty element has in the long-run assessment of national income. We contrast the theoretically consistent intertemporal price system with the one which exists in reality, and conclude that there are inherent discrepancies which cannot be overcome, even at a theoretical level. Finally, we draw our environmental policy conclusion. Because of missing markets, and thus missing prices, for environmental goods, and given that the decision horizon of many economic and political agents is rather short, it is urgently necessary to establish new institutions which formulate appropriate indicators for environmental services, and ensure that they are reflected in the opportunity costs of all relevant political and economic decision makers.

It should be made clear at the outset that this paper is a contribution to the conceptual foundations of national accounts and the environment. As a result, it does not discuss at length the controversy relating to the appropriate valuation in national accounts of natural resource depletion. (For a clear discussion of this debate, see El Serafy 1991). Indeed, in the simple model we present, the depletion of resources is not fully treated. However, the conceptual framework we present can be extended to include the deterioration of capital, both physical and natural.

NATIONAL ACCOUNTS AND THE INPUT-OUTPUT FRAMEWORK

The aim of national accounts is to give a "snapshot" of economic activity in an entire economy. The accounts for successive years give a series of such snapshots. Now, a series of consecutive photographs gives an indication of the dynamic processes involved in a particular scene; e.g., stop-frame movie photography of a flower as it grows in spring and summer allows a better understanding of flower development than any individual photograph. Similarly, understanding the processes in a national economy using national accounts will be much easier (we shall suggest below, even necessary) if we consider a series of national accounts over time rather than examine one single year in isolation.

However, before we move on to this consideration of national accounts over time, it will be useful to examine the structure of national accounts and their interpretation. Economic activity can be considered to be the transformation of inputs into outputs. In particular, we usually make the distinction between the "basic" inputs and the "final" outputs. By basic inputs, we mean productive elements such as land, labor and capital goods (as well as environmental services) which are not produced in the year being considered; i.e., these elements are "endowments" from previous years' economic activity, or from nature. Final output is the provision of good and services in the current period, or the provision of productive elements for use in later periods.

The transformation of the basic inputs to the final outputs is mediated by a complex, interconnected set of production activities. The transactions within this mediating production process are termed "intermediate demand." In our view, the clearest way to represent national accounts is in the input-ouput form, originally devised by Leontief (1953). In this approach, the economy is broken down into several sectors which are simultaneously using basic inputs and intermediate products and producing final outputs and intermediate products. An example of a three-sector model economy is shown in table 15.1.

Element (i,j) [i = 1, 2, 3; j = 1, 2, 3] of the i-th row and the j-th column denotes the output of sector i which is used as an input by sector j. The elements used are in value terms and we follow the usual convention that the total value of inputs into any sector must equal the total value of outputs from that sector.

For the purposes of national accounts, we are only interested in net inputs and net outputs. As the elements of intermediate demand are simultaneously inputs and outputs, they can be ignored for national accounting purposes (though they are extremely useful for many other types of economic analysis).

TABLE 15.1 A Three Sector Input-Output Model. (VA = Value Added, FD = Final Demand, TI = Total Input, TO = Total Output)

	1	2	3	FD	TO
1	5	10	-	5	20
2	-	20	3	12	35
3	6	-	10	9	25
VA	9	5	12		
TI	20	35	25		

We can therefore concentrate upon the value of basic inputs and the value of final outputs. In a sense all economic value derives from basic inputs, so they are often referred to as the "value-added" components. When the value-added terms are summed, they represent the payment made to the productive elements and therefore constitute "national income." The final demand elements when summed give the total value of production by the economy, which is known as the "national product." In table 15.1 the total value-added and national product are both 26 units of value.

Our above assumption, that for each sector the value of inputs must equal the value of outputs, implies that we would expect that national income always equals national product. We turn to this assumed equality in the next section. We find a proper understanding of the relationship between national income and national product depends on an intertemporal view of national accounts and requires making certain strong assumptions about the underlying economic processes of production and social choice. When these intertemporal relationships are established, we will have completed the groundwork necessary to extend our national accounting framework to include services provided by the natural world.

PRICES, QUANTITIES AND OPTIMIZATION

A Static Approach: An Ex Post Procedure

The conventional approach to national accounts uses only values. In this paper, we shall decompose these values into the products of the quantities of goods and their corresponding prices. Taking this approach naturally leads one to an analysis of the production process, and the mechanism by which the corresponding prices are established.

To illustrate the importance of the distinction between values and the corresponding quantities and prices, we begin with a "model economy" where the quantities of inputs and outputs are given. We represent these flows of quantities with an input-output table.

We suppose there are three producing sectors. The first sector produces the consumption good using only labor as an input. The second sector also produces the consumption good, using both labor and capital as inputs. The third sector produces the capital good i.e., it is the investment sector) using only labor. The flows of quantities of factors and produced goods in this model may be represented by table 15.2.

TABLE 15.2 Quantities of Inputs and Outputs in a Three-Sector Model.
(L = Labor, C = Consumption, K = Capital, I = Investment)

	1	2	3	C	I
1	-	-	-	4	-
2	-	-	-	10	-
3	-	-	-	-	5
L	40	30	30		
K	-	10	-		

We note that in this model there are no intermediate flows, i.e., we ignore any inter-industry purchasing that may take place. Such inter-industry trading makes no contribution to gross domestic product, (our main concern), so the blanks in the top left quadrant of table 15.2 are of no concern.

We can now use our normal national accounting conventions to:
1. Find a set of prices which ensure that for each sector the value of the inputs exactly equals the value of the outputs, and
2. Thence derive a set of balanced national accounts.

To solve this problem we introduce the following set of prices. The price of labor is P_L; the price of capital is P_K; the price of the consumption good is P_C; and the price of the investment good is P_I.
For sector 1, the only input is 40 units of labor, of value $40P_L$. The output from sector 1 is 4 units of the consumption good, of value $4P_C$. Thus we require:
$$40P_L = 4P_C \text{ i.e., } P_C/P_L = 10$$
Similarly, for sector 2 we have:
$$30P_L + 10P_K = 10P_C$$
Finally, for sector 3 we have:
$$30P_L = 5P_I$$
Of course, these three equations in four unknowns cannot be solved completely. It is necessary to take one of the prices as a numeraire. We take labor as the numeraire good, and set $P_L=1$. We then get:
$$P_L = 1, P_K = 7, P_C = 10, P_I = 6$$
Multiplying the quantities shown in table 15.2 by the corresponding prices gives the set of national accounts for this model economy shown in table 15.3.

We see that the usual national accounting identity is obeyed.

Total Value of Inputs = Total Value of Outputs = 170

i.e., National Income = National Product = 170

TABLE 15.3 A Balanced Three-Sector Input-Output Model

	1	2	3	C	I
1	-	-	-	40	-
2	-	-	-	100	-
3	-	-	-	-	30
L	40	30	30		
K	-	70	-		

Thus, if input and output quantities are given for the various sectors in our model economy, one can find a set of prices so that the usual national accounting balance can be maintained. However, it is also possible to specify prices which do not give rise to balance automatically. If this is the case, and one wishes to maintain balance, it is necessary to introduce a "balancing items" row or column. For example, suppose the prices used were:

$$P_L = 1, P_K = 6, P_C = 11, P_I = 5$$

The balanced national accounts would now be as shown in table 15.4. There, the row marked "B" contains the balancing items.

TABLE 15.4 An Unbalanced Three-Sector Input-Output Model

	1	2	3	C	I
1	-	-	-	44	-
2	-	-	-	110	-
3	-	-	-	-	25
L	40	30	30		
K	-	60	-		
B	4	20	-5		

If the balancing items are excluded we have:

National Income = 160   National Product = 179

i.e.,   National Income ≠ National Product.

We could now ask, very reasonably, what might be the mechanism which could ensure that the prices observed in an economy were such as to ensure sectoral balance, and thus overall balance? Before demonstrating this mechanism in detail, it is useful to recast the above numerical example into algebraic form. We rewrite table 15.2 as shown in table 15.5.

The X_i terms represent the quantities of the consumption good produced by sectors 1 and 2. I is the quantity of the investment good produced by the third sector. The L_i and K_i terms are the quantities of labor and capital used in the three sectors.

TABLE 15.5 Quantities of Inputs and Outputs in Algebraic Form

	1	2	3	C	I
1	-	-	-	X_1	-
2	-	-	-	X_2	-
3	-	-	-	-	I
L	L_1	L_2	L_3		
K	-	K	-		

The requirement for balance by each of the three sectors is now given by the following three equations, by setting the value of income (value added) equal to the value of production for each sector.

$$L_1 P_L = X_1 P_C$$

$$L_2 P_L + K P_K = X_2 P_C$$

$$L_3 P_L = I P_I$$

To solve this system of equations in a way compatible with the later analysis, we define the following production coefficients:

$$l_1 = L_1/X_1, \quad l_2 = L_2/X_2, \quad l_3 = L_3/I, \quad k_2 = K/X_2$$

If we set $P_L = 1$, as before, then substituting for the above production coefficients in the three balance equations, we get the prices expressed by algebraic coefficients in a general form:

$$P_L = 1, \quad P_K = (l_1 - l_2)/k_2, \quad P_C = l_1, \quad P_I = l_3$$

A NEO-AUSTRIAN PRODUCTION MODEL: AN EX ANTE PROCEDURE

We can now turn to the problem of how a set of prices which give national accounting balance might be established. That is, whereas the prices that give balance were derived above ex post, we now seek a model of economic activity where the prices are developed ex ante (i.e., simultaneously with the quantities). To include the production and use of capital goods within our approach, which requires an intertemporal approach to production, we use a two-period production model. We draw upon a neo-Austrian approach to intertemporal production (Bernholz 1971; Faber 1979; Faber et al. 1990; Faber and Proops 1990, ch. 8-9).

To proceed, we specify a three-process, two-period production model. The production processes used correspond to the three producing sectors in the above simple accounting model. We assume the use of fixed-coefficient production functions. Sector 1 corresponds to process 1, producing the consumption good using only labor, according to:

$$X_1 = L_1/l_1$$

Sector 2 corresponds to process 1, producing the consumption good using both labor and capital, according to:

$$X_2 = L_2/l_2 = K/k_2$$

Sector 3 correspond to process 3, producing extra capital good as investment, using only labor, according to:

$$I = L_3/l_3$$

As this is a two-period model, the capital available for process 2 in the second period is the initial capital stock plus the first period investment, i.e.:

$$K(2) = K(1) + I(1)$$

The number in parenthesis indicates the corresponding period.

We may make the reasonable assumption that, ceteris paribus, in any period more consumption is preferred to less. As our model is for only two periods, and as the opportunity cost of investment in any period is consumption in that period, it is clear that there will be no second period investment. i.e., $I(2) = 0$.

For simplicity, we suppose that the labor force does not vary over time and is always fully employed. We also assume that all capital is fully utilized in each period and that the capital good does not deteriorate over time.[1]

From the above, it is apparent that the entire two-period model hinges upon the choice of first period investment. If first-period investment is zero, then consumption over the two periods will be unchanging, as there will be a constant stock of labor and capital. If there is first-period investment, this has the negative impact of reduced first-period consumption, but the corresponding benefit of increased second-period consumption. Indeed, it is relatively easy to show that total first-period consumption, $X(1)$ $[=X_1(1)+X_2(1)]$ and total second-period consumption, $X(2)$ $[=X_1(2)+X_2(2)]$, are related through the labor force, L, and the initial capital stock, $K(1)$, by the equation which defines the "intertemporal production possibility frontier" of our intertemporal model economy (Faber and Proops 1990, p. 135):

$$X(2) = [L/l_1+(K(1)/k_2)(1-l_2/l_1)][1+(l_1-l_2)/l_3k_2] - [(l_1-l_2)/l_3k_2]X(1)$$

This equation represents the downward sloping relationship between consumption in the first and second periods. Thus it reflects the trade-off between consumption "today" and consumption "tomorrow."

The problem we face is that of choosing the proper point upon this intertemporal production possibility frontier. In principle, we could invoke any mechanism of choosing a point on this frontier. For example, the choice might be determined by habit or at random. In line with the conventional approach, here we adopt a simple utility/welfare maximizing model. We define an intertemporal welfare function as $W(X(1),X(2))$, which we maximize subject to the constraints on production defined by this intertemporal model.

---

[1] We ignore capital deterioration for simplicity of presentation. The corresponding model with capital deterioration is discussed in Faber (1979), Faber and Proops (1990, ch. 9, 10) and Faber et al. (1990).

We may write the full optimization problem as follows:

$$\underset{X(1),\, X(2)}{\text{Max}} \quad W(X(1),X(2))$$

subject to:

$$X_1(1) + X_2(1) = X(1)$$

$$l_1 X_1(1) + l_2 X_2(1) + l_3 I(1) = L$$

$$k_2 X_2(1) = K(1)$$

$$X_1(2) + X_2(2) = X(2)$$

$$l_1 X_1(2) + l_2 X_2(2) = L$$

$$k_2 X_2(2) = K(2)$$

$$K(2) = K(1) + I(1)$$

$$X_1(1), X_2(1), X_1(2), X_2(2), I(1), L, K(1) > 0$$

This problem may be solved by the method of Lagrange, which introduces Lagrange multipliers, a separate multiplier being associated with each constraint (Baumol 1977, ch. 6). The advantage of this approach is twofold. First, the original constrained optimization problem becomes an unconstrained optimization problem which is relatively easy to solve. Second, one can interpret the Lagrange multiplier as the opportunity cost, in welfare units, of the corresponding constraint. This is the valuation that the society should place on the constraint variable, to reflect its scarcity, i.e., of the consumption good, the investment good, labor and the capital good. It is because we wish to obtain these shadow prices that we constrain the optimization with the full model description, rather than with just the production possibility frontier.

Since there are two periods in our model economy, we get two shadow prices for each of the consumption good, labor and the capital good, one for each period. It is important to realize that these shadow prices are evaluations which are determined in the *first* period. Therefore they are called "present prices."

How do these shadow prices relate to the current prices in an economy? Obviously, in the first period, shadow prices and current prices are identical. But the shadow prices for the second period are different from the current prices of the second period. The difference depends on the time preference between the two periods. For the sake of simplicity, we assume that the rate of time preference is measured by the discount rate, d. Then it can be shown that the shadow prices of the second period are equal to the discounted current prices of the second period and vice versa; the current prices of the second period are equal to the shadow prices multiplied by $(1 + d)$ (Malinvaud 1985, ch. 10).

In this context, it is helpful to note that these shadow prices may be used to decentralize an economy (e.g., Malinvaud 1953; Koopmans 1957). In this case, the current prices are equal to the observed market prices.

The full details of the determination of the shadow prices is not presented here (see Faber and Proops 1990, pp. 149-153). We simply summarize the results for the shadow prices as follows, using W_1 as the marginal welfare from first-period consumption, and W_2 as the marginal welfare from second-period consumption (i.e., $W_t = \partial W/\partial X(t)$).

$$P_{X(1)} = W_1 \qquad\qquad P_{X(2)} = W_2$$

$$P_{L(1)} = W_1/l_1 \qquad\qquad P_{L(2)} = W_2/l_1$$

$$P_{K(1)} = W_1(1-l_2/l_1)/k_2 \qquad\qquad P_{K(2)} = W_2(1-l_2/l_1)/k_2$$

$$P_{I(1)} = W_1 l_3/l_1$$

These shadow prices can be interpreted as the present prices that would be found in an efficient market economy whose choice process could be described by the corresponding intertemporal choice mechanism. As before, we adopt the price of labor, in the first period, as the numeraire, so that the relative shadow prices in this intertemporal economy are given by:

$$P_{X(1)} = l_1 \qquad\qquad P_{X(2)} = l_1/Q$$

$$P_{L(1)} = 1 \qquad\qquad P_{L(2)} = 1/Q$$

$$P_{K(1)} = (l_1-l_2)/k_2 \qquad\qquad P_{K(2)} = (l_1-l_2)/k_2 Q$$

$$P_{I(1)} = l_3$$
(We define $Q = W_1/W_2$)

Concentrating on the first-period prices, we see that these are identical with the simple accounting prices we derived as necessary to give sectoral balance in the original national accounting model. This has very strong implications for how we conceptualize national accounts in terms of the values of transactions. We began by enquiring how the prices necessary to give accounting balance might be derived. We now see that a sufficient condition for accounting balance to be achieved is that these prices be derived from an intertemporal optimization process whether through a central planning agency, or through a competitive market.

What are the implications of this capital theoretic approach to national accounting? Are there any obvious constancies that show how this intertemporal model might have something to say about the incorporation of interactions of the economy with the natural world into the national accounting framework?

Before we proceed, we need to note a powerful and useful result. This concerns the relationship between the discount rate, d, the intertemporal marginal rate of substitution, $Q = W_1/W_2$, the rate of "superiority," S, and the ratio of the present prices of capital and investment, P_K/P_I. The only term that has not been already discussed is the rate of superiority, S. This is a measure of the benefit to be derived by using resources to allow investment in capital, and thus "waiting" one period for consumption, rather than using those same resources to allow consumption in the present period. Thus S can also be interpreted as the marginal rate of transformation of X(1) (first-period consumption) into X(2) (second-period consumption). For this three-process, two-period model it is straightforward to show that the degree of superiority is given by (see Faber 1979, pp. 71-73; Faber and Proops 1990, pp. 127-130):

$$S = (l_1 - l_2)/l_3 k_2$$

It is possible to show that, at the optimum, the following result holds (see Bernholz and Faber 1973; Faber 1979, pp. 77-78):

$$(1+d) = W1/W2 = S = PK/PI$$

Returning to our search for constancies in this national accounting framework, we first work in current/market (i.e., undiscounted) prices.

Concentrating on first period production, our balanced table (e.g., table 15.3) reassures us that the value of our net product (i.e., the value of consumption and investment) must equal national income (i.e., total value added). The current prices in the first period are determined in this model solely by the technology, so the value of the net product is determined by the available quantities of capital and labor in the first period. Thus, we have here an initial constancy; in this simple closed model with no government sector, the size of national product does not depend on how it is spent.

Considering now second period national accounts in current/market prices, we first note that as $Q = (1+d)$. The second-period current prices of the consumption good, labor and capital are the same as the first-period current prices. Here we exclude the possibility of capital saturation (i.e., the capital stock being too great to be fully employed, through labor shortage). Also, as we use a fixed labor force, the current value of national income cannot be less in the second-period than in the first-period. They are equal when there is zero first-period investment. If there is positive first-period investment, the current value of national income in the second-period must be greater than in the first-period.

Turning now to the current value of consumption, in the first-period it is variable and depends on the amount of first-period investment. However, while first-period consumption decreases with first-period investment, the effect of such investment on second-period consumption is clearly to increase it.

We now consider national income using present (i.e., discounted) prices. The essence of using present prices is to be able to combine first and second-period values. We therefore consider the total present value of national income over the two periods. First-period national income is given by:

$$Y(1) = K(1)P_K + L$$

Using $1+d = S$, the present value of second-period national income is given by:

$$Y(2) = [K(1)+I(1)]P_K/S + L/S$$
$$= Y(1)/S + I(1)P_K/S$$

Now we have seen that $S = P_K/P_I$, so we can rearrange the above expression to give:

$$Y(2) = Y(1)/S + I(1)P_I$$

Thus the present value of national income over the two periods is:

$$Y(1) + Y(2) = Y(1)(1+1/S) + I(1)P_I$$

We have already seen that first-period national income $Y(1)$ is a constant. The above expression shows therefore that the present value of the two-period income, $Y(1)+Y(2)$, varies with first-period investment.

We now consider the total resent value of consumptionover the two periods. First-period consumption at present prices is given by:

$$C(1) = Y(1) - I(1)P_I$$

That is, national income less investment.
Second-period consumption at present prices is simply the present value of second-period national income, as there is no investment in that period. Thus:

$$C(2) = Y(2) = Y(1)/S + I(1)P_I$$

So the present value of two-period consumption is:

$$C(1) + C(2) = [Y(1)-I(1)P_I] + [Y(1)/S+I(1)P_I]$$
$$= Y(1)(1+1/S)$$

Thus, we have a strong and interesting result: the total present value of consumption over the two periods is a constant; it is independent of investment. The intuition behind this result is that here the appropriate rate of discounting future consumption is precisely the marginal rate of transformation of consumption between the two periods minus unity (i.e., $S-1$). This result is in contrast to that for the present value of the two-period income. (For a discussion of the importance of this "wealth-like" measure of welfare, see Samuelson (1961, pp. 51-52)).

It might be argued that a two-period model, as used above, is rather unsympathetic to the sustainability issue, as an optimal system implies that all natural resources may be depleted by the end of the second-period. However, this can be easily dealt with in a neo-Austrian framework, either by extending the model to n-periods, or by invoking successive reoptimization within a "rolling myopic plan" (Faber and Proops 1990, ch. 9-10).

POLLUTION AND NATIONAL ACCOUNTS

We now turn to consider how services provided by the natural world may be evaluated and included in national accounts. We shall consider two services rendered by the natural world: pollution degradation and resources. As we shall see, both involve an intertemporal approach. We begin with the less complex problem of natural pollution degradation.

A great deal of the pollution that we introduce into the natural environment is degraded to an innocuous state by the natural functioning of the ecosystem. Were this not the case, we should have submerged our civilization in its own wastes long ago. Clearly, this natural pollution degradation constitutes a valuable service rendered by nature. But how can we value it? The particular problem we face is that this natural pollution degradation is not marketed, so no price is associated with this service. In this respect, natural pollution degradation is akin to housework; indeed, we can consider natural pollution degradation as being nature's housekeeping.

As well as receiving nature's service, we also clean up some of our own mess with pollution abatement activities, such as the desulfurization of flue gases, water treatment, and solid waste incineration, etc. For this we pay, as these activities require that productive factors be used for this purpose, instead of for producing consumption goods or investment goods; i.e., pollution abatement has an opportunity cost.

However, the level of pollution abatement is not decided by the market; it cannot even be said that any particular level of pollution abatement is in some sense optimal. This is because, of course, the service rendered by pollution abatement has many aspects of a public good, and the problems associated with assessing the optimal levels of public goods are well known and generally accepted to be intractable.[2] Instead, pollution abatement is set at a level that is deemed "appropriate." Therefore, the total value which we should associate with pollution abatement activities is not that which corresponds to an optimal solution. It therefore follows that the shadow prices corresponding to the pollution abatement activities do not reflect the full social opportunity costs. As a rule, they are smaller than the true shadow prices (social costs) of the natural environmental services of pollution abatement. However, as these incomplete shadow prices are the only ones we have, they are the ones we have to use for the national accounting system.

The shadow price of pollution abatement gives us a means to provide an evaluation of the services provided by natural pollution degradation. The fact that nature degrades a unit of pollution means that there is one less unit of pollution that we may feel the need to abate, at some social cost. Therefore, we should value the service of natural pollution degradation at the same unit shadow price as pollution abatement.

Having established the valuation of natural pollution degradation, we can now enquire how it should be represented in a national accounting input-output framework. Here we may again take our clue from pollution abatement. The production of pollution abatement is a service rendered to final consumers, and therefore should appear in the final demand (national product) section of the accounts. The provision of pollution abatement requires the use of production factors, so that this should also appear in the value added (national income) section. For example, we might expand our original three-sector model to four sectors, with the fourth sector being pollution abatement, as shown in table 15.6.

Table 15.6 is a modified version of table 15.3, with the introduction of sector 4 (pollution abatement) and the addition of pollution abatement as a final good. The value of investment has been reduced from 30 to 20, and the 10 units of production factor in value terms thus released have been used for abating pollution. Thus, net output and national income are unchanged at 170 value units. However, the reduction of first-period investment will reduce the value of second-period national income. The constancy of the present value of two-period consumption (including an exogenously given level of pollution abatement) is unaffected.

---

[2] There have been several attempts to devise preference revealing processes. Perhaps the most famous is by Clarke and Groves (see Mueller 1979, ch. 4). However, even this method is unlikely to be practically implementable, and still retains problems of revelation in principle, stemming from the possibility of the formation of coalitions by the economic actors.

TABLE 15.6 A Four Sector Model with Pollution Abatement.
(PA = Pollution Abatement)

	1	2	3	4	C	I	PA
1	-	-	-	-	40	-	-
2	-	-	-	-	100	-	-
3	-	-	-	-	-	20	-
4	-	-	-	-	-	-	10
L	40	30	20	10			
K	-	70	-	-			

TABLE 15.7 A Four Sector Model with Natural Pollution Degradation

	1	2	3	4	C	I	PA
1	-	-	-	-	40	-	-
2	-	-	-	-	100	-	-
3	-	-	-	-	-	20	-
4	-	-	-	-	-	-	90
L	40	30	20	10			
K	-	70	-	-			
PD	-	-	-	80			

(PD = Pollution Degradation)

Using this extended national accounting framework the evaluation of natural pollution degradation can be achieved. On the output side of the accounts, this can be introduced as a good (PA) delivered to final consumers. On the input side, we treat natural pollution degradation as a production factor (PD), and therefore need to extend the value added section. For example, if the shadow price of pollution abatement were 2 per unit, and we are rendered 40 units of natural pollution degradation by our housekeeping ecosystem, then we should include 80 extra units of value in net output and national income, as shown in table 15.7. Here we see that the first-period national income and net output have both increased by 80 units.

There are two temporal aspects to this extended national accounting model which are of interest. The first is a conceptual matter, while the second is of a more practical interest. The conceptual issue is that this new extended model, including pollution abatement and natural pollution degradation, does not have the firm theoretical underpinning that the original, optimizing, model had. This is because the pollution abatement does not appear as a component of the social objective function, so that the level of pollution abatement and the shadow price cannot be determined simultaneously as part of an overall optimization process. Thus the balance of the extended table is not achieved automatically, but is imposed as a requirement. This lack of endogeneity of pollution abatement in the optimization process also means that the intertemporal evaluation, using present prices derived from the optimization procedure, is no longer appropriate. Thus, the useful and necessary inclusion of pollution abatement and natural degradation within the national accounting framework makes the value of national income to some extent arbitrary.

The practical issue concerns the likely effect of pollution on the value of national income when the valuation of natural pollution degradation is included. It is generally recognized that the ability of an ecosystem naturally to degrade pollution diminishes with the pollution load. Thus a "clean" river can easily degrade low levels of pollution, but if the pollution load is high, this ability is impaired. Now if, over time, an ecosystem is subject to continued high levels of pollution, the pollution degradation that takes place may decrease. In consequence, the value of the natural pollution degradation service rendered to the economy may also decrease. As a consequence, the value of national income including natural services may decline, while the value of national income excluding natural services may continue to rise. We suspect that for many national economies this reduction of "real" national income is already underway. This possibility is illustrated in figure 15.1.

Here "measured" national income is that usually presented in national accounts, ignoring the services provided by the natural environment. "Real" national income includes the services rendered by the natural environment, appropriately valued. It is clear that "real" income will always exceed "measured" income. However, if the natural environment is over-exploited, the rapid fall in the services rendered may lead to negative "real" income growth, while "measured" income continues to grow.

EXTENDING THE MODEL TO RESOURCE DEPLETION

Besides pollution degradation, the environment supplies renewable and nonrenewable resources. For the sake of simplicity, we shall deal here only with nonrenewable resources.

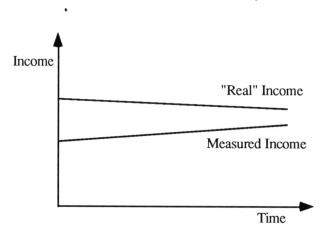

FIGURE 15.1 "Real" and "measured" national income over time

Renewable resources can be handled with a combination of the analysis given above concerning pollution, and that to be given below, concerning nonrenewable resources. Indeed, one could view natural pollution degradation as a renewable resource. For the time being, we shall neglect pollution to make the presentation more tractable. We shall return to a discussion of pollution in the next section.

The resources in the natural environment of an economy constitute a wealth asset. From an economic point of view, they have the same status as labor and capital goods. However, there are obvious differences between the three kinds of inputs to the production process. Thus, unlike capital goods, as discussed in the previous sections, nonrenewable resources cannot be produced. Another important difference is that the scarcity of a nonrenewable resource is necessarily increased with its extraction and consequent use in the production process, or its direct use as a consumption good. A third difference is that, in general, the extraction costs of a nonrenewable resource rise with the amount of the resource already extracted, for a given technology, in line with the Second Law of Thermodynamics (see Faber et al. 1987).

From a theoretical point of view, we now wish to explore how natural resources can be introduced into our intertemporal national accounting approach. In particular, this raises the question of how a unit of resource is to be valued.
This problem has two aspects.

1. The cost of producing the service of the resource, i.e., the extraction costs, has to be determined, and
2. the notional cost of replacing a unit of the resource has to be evaluated so that the wealth base is not eroded; the latter costs are known in the literature as the "royalty," "opportunity cost," or the "resource rent" (e.g., Dasgupta and Heal 1979, p. 159, 169-171). One could extend the model to take account of this negative income increment. This extension is not discussed for the sake of brevity; the various arguments are well presented elsewhere (El Serafy 1991).

It is helpful to note that the price of a capital good is also composed of two components in an analogous way; i.e., the cost of the service of the capital good (P_K) in one period (the "service cost"), and the cost of replacing the deterioration from using the capital good during the corresponding period (P_I) (the "maintenance cost"). It is this opportunity cost of capital replacement (i.e., reinvestment), which corresponds to the royalty charge for a nonrenewable natural resource.

Because of the increasing scarcity of the nonrenewable resource, the royalty element will increase over time, assuming there is no further discovery of resource reserves (i.e., the Hotelling rule; see Dasgupta and Heal 1979, p. 156). It can also be shown that, for a given technology, the longer the time horizon used, the higher the intertemporal optimal royalty will be at any moment in time. This gives an efficient use of the resource (Dasgupta and Heal 1979, p. 171). Therefore, the longer the time horizon employed within the economy for resource extraction decisions, the higher is the intertemporal optimal resource price. However, for a resource which is initially relatively abundant, the royalty element of the price will be small compared with the extraction cost element in the early stages of extraction (Dasgupta and Heal 1979, p. 178). Even if the time horizon used for the assessment of the royalty element is relatively short, this will have little effect in the early stages on the price; that is, the observed current price with a short time horizon will not differ significantly from the price that would be observed in a world with a long time horizon. However, as the resource becomes depleted and its scarcity increases, in a world with a short time horizon the actual price of the resource will fall further and further below the long-run optimal price.

In light of the above intertemporal analysis of resource pricing, how should natural resources be handled within a national accounting framework? Also, what lessons can we draw about the estimation of national income in the light of the likely underpricing of resources, as discussed above?

First, we noted that a natural resource is like labor and capital in that it renders a service to the economy; i.e., natural resource use should be included as a component of value added, and so contribute to national income. This is easily accomplished by adding another row to the national accounts in input-output form.

Second, the price which is used to calculate this component of value added should be the full price, i.e., the extraction cost plus royalty.

Third, the royalty element should be set at a level corresponding to a reasonably long time horizon. It is often felt that the royalty element used in actual economies for establishing the resource part of the national accounts is too low, because the time horizon used is too short. It is more appropriate for an individual firm than an entire society. The consequence of the low royalty element will be an increasing divergence between the apparent national income and that which would be appropriate to a world which is using its resources optimally. This problem of the underestimation of national income is not restricted to the problem of resource pricing. Other problems of the conceptualization in national accounts have led Eisner (1989, p. 6) to comment, "My own estimates for a comprehensive "total incomes system of accounts" put net national product in 1981 at 30% more than its Bureau of Economic Analysis counterpart."

ON INTERTEMPORAL PRICE SYSTEMS: THEORY AND REALITY

In accordance with the two strands of literature (Fisher, 1930; Samuelson, 1961; Weitzman, 1976; and Perkins, 1976; Dasgupta and Heal, 1979), we have tried to develop a theoretically consistent system of national accounts, which includes the services provided by the ecosystem and natural resources. The main tool to develop our framework was the use of an intertemporal price system, based on intertemporal optimization, be it via a planning procedure or via a competitive market system. However, it was already apparent in one of the sections above, on pollution, that it is not always possible to find such intertemporal prices which reflect truly the scarcity conditions. Those conditions depend on the technology, preferences, institutions and nature at large. Hence, we were already compelled to take recourse to some artificial, though reasonable, prices.

It is apparent that the price system in reality deviates in many respects and, to a great extent, from any conceivable optimal price system. This results from inherent market failures, due to the existence of public goods, external effects, interdependencies of preference functions, non-convexities of the technologically feasible set (e.g., because of increasing returns to scale), inconsistent time horizons, and the existence of ignorance and uncertainty about future states of the world.

We have dwelt on the divergence between the optimal price system and the price system in reality so extensively because we wish to stress that we do not believe that it is possible to find a price system which can, in reality, serve as a consistent guide to national accounting. Nevertheless, we consider it also to be important to have established a consistent theoretical framework for national accounting for the following reasons:

1. This theory gives us a guide to how a consistent system of national accounts could be established, at least in principle.
2. It shows us how much the system we use in reality deviates from this ideal, thus giving us clues to the magnitude of the deviation.
3. Last, but not least, we think that perhaps the main meaning of our endeavor is to make us aware of how careful one has to be with conclusions drawn from standard national accounts (cf., the Presidential Address of Eisner 1989).

CONCLUSIONS

What conclusions about methods for accounting for the natural environment are to be drawn from these considerations? We have tried to point out two sorts of problems associated with the use by economists of services rendered by the natural world. First, there is the problem of missing markets for services which have a public good aspect, as in natural pollution degradation. Second, there is the problem of short time-horizons being used by many decision makers, thus leading to royalties which in turn causes market prices for nonrenewable resources to be too low.

A consequence of these two factors is that, generally, current national income is grossly underestimated. This in turn leads to an inappropriate time profile of investment, thereby further distorting economic development from the path that would be intertemporally optimal if there were a long planning horizon and an appropriate evaluation of environmental services. In particular, there will be a tendency to overexploit nonrenewable resources and to overload ecosystems with pollution, thereby eating up the "environmental wealth" available for present and future generations.

It is important to realize that in our society there is really no one to blame for this overexploitation of the natural world. This is because the economic indicators (prices) do not exist in our society to reflect the corresponding intertemporal scarcities. Abelson and Markandya (1985) showed within a dynamic framework how much price estimates can deviate from those which reflect the true effects of environmental variables. Various problems of evaluation of ecosystems, including a brief survey of the literature, are discussed by Costanza et al. (1989).

To give an example, a recent study (Faber et al. 1989) has estimated that the appropriate long-run price of one tonne of waste disposal in the Federal Republic of Germany is five times as high as that presently charged; i.e., instead of DM60 per tonne, one should charge DM300 per tonne. Considering that there are about one hundred million tonnes of waste to be disposed of every year, one sees immediately how great the discrepancy between the measured valuation of this natural service, and its intertemporally optimal valuation is. This DM24 billion underestimate of national income alone is roughly 1% of GDP.

How might this problem of undervaluation be resolved? To this end, we need new institutions and conventions, which have to fulfill the following tasks.

1. The true intertemporal scarcity has to be analyzed, and on the basis of this analysis, appropriate indicators have to be found.
2. These indicators have to be publicized.
3. Institutional frameworks have to be developed to ensure that the measured scarcity is reflected in the economic and political activities of society. This can be carried through by increasing prices, and introducing new prices, for environmental services. For this purpose, one could use—inter alia—licences and charges. In this respect, we can draw upon experience with emission trading in the United States, and charges relating to water quality in Europe.

Acknowledgments

For their helpful comments, we are grateful to Alexander Gerybadze, Charles Perrings, Frank Jöst, Marco Lehman-Waffenschmidt, Armin Schmutzler and Ian Walker.

REFERENCES

Abelson, P. W., and A. Markandya. 1985. The Interpretation of Capitalized Hedonic Prices in a Dynamic Environment. *Journal of Environmental Economics and Management* 12:195-206.

Baumol, W. G. 1977. *Economic Theory and Operations Analysis* 4th ed. Englewood Cliffs, N.J.: Prentice Hall.

Bernholz, P. 1971. Superiority of Roundabout Processes and Positive Rate of Interest: A Simple Model of Capital and Growth. *Kyklos* 24: 687-721.

Bernholz, P. and M. Faber 1973. Technical Productivity of Roundabout Processes and Positive Rate of Interest: A Capital Model with Depreciation and N-Period Horizon. *Zeitschrift für die gesamte Staatswissenschaft* 129: 46-61.

Costanza, R., S. C. Farber and J. Maxwell 1989. Valuation and Management of Wetland Ecosystems. *Ecological Economics* 1:335-361.

Dasgupta, P. S. and Heal, G. M. 1979. *Economic Theory and Exhaustible Resources.* Cambridge: Cambridge University Press.

El Serafy, S. 1991. The Environment as Capital. Chapter 12 this volume.

Eisner, R. 1989. Divergences of Measurement and Theory and Some Implications for Economic Policy. *American Economic Review* 79:1-13.

Faber, M. 1979. *Introduction to Modern Austrian Capital Theory.* Heidelberg: Springer.

Faber, M. and J. L. R. Proops. 1990. *Evolution, Time, Production and the Environment.* Heidelberg: Springer.

Faber, M., H. Niemes and G. Stephan. 1987. *Entropy, Environment and Resources: An Essay in Physico-Economics*. Heidelberg: Springer. First published in German, 1983.

Faber, M., G. Stephan and P. Michaelis. 1989. *Umdenken in der Abfallwirtschaft*. 2d ed. Heidelberg: Springer.

Faber, M., J. L. R. Proops, M. Ruth and P. Michaelis. 1990. Economy-Environment Interactions in the Long-Run: A Neo-Austrian Approach. *Ecological Economics* 2:27-55.

Fisher, I. 1930. *The Theory of Interest*. London: Macmillan.

Koopmans, T. C. 1957. Allocation of Resources and the Price System. In *Three Essays on the State of Economic Science*, pp. 1-126. New York: McGraw-Hill.

Leontief, W. 1953. *Studies in the Structure of the American Economy*. Oxford: Oxford University Press.

Malinvaud, E. 1953. Capital Accumulation and Efficient Allocation of Resources. *Econometrica* 21:233-268.

Malinvaud, E. 1985. *Lectures on Microeconomic Theory*. Amsterdam: North Holland.

Mueller, D. C. 1979. *Public Choice*. Cambridge: Cambridge University Press.

Norgaard, R. B. 1989. Three Dilemmas of Environmental Accounting. *Ecological Economics* 1:303-314.

Peskin, H. M. 1976. A National Accounting Framework for Environmental Assets. *Journal of Environmental Economics and Management*, 2:255-262.

Samuelson, P. A. 1961. The Evaluation of 'Social Income': Capital Formation and Wealth. In F. A. Lutz and D. C. Hague, eds., The Theory of Capital. Proceedings of an IEA Conference. London: Macmillan.

Weitzman, M. 1976. Welfare Significance of National Product in a Dynamic Economy. *Quarterly Journal of Economics* 90:156-162.

16

ACCOUNTING IN ECOLOGICAL SYSTEMS

Bruce Hannon
Department of Geography
Affiliate Scientist, Illinois Natural History Survey
University of Illinois
Urbana, IL 61801 USA

ABSTRACT

Accounting of material and energy flows has long been an important tool in ecosystem ecology. But each material is usually handled separately and independently. The connections *between* materials, energy, plants, animals, etc., have not been incorporated into the accounting framework, and "service" or information flows (such as flower pollination by bees) are usually ignored. A general accounting framework that addresses these deficiencies is developed and a set of systemwide weights for the units of the flows has been developed. In this framework, each connection (both physical and information) can be unambiguously assigned and quantified, and an input–output balance is easily checked and maintained for each product. Costly independent data collections can be integrated into this common framework to amplify their original usefulness and provide the investigator or ecosystem manager with enhanced understanding of the entire ecosystem from which they were taken. The integrated data also allows various ecosystem models to be constructed efficiently, without unnecessary and costly duplication of effort. Nonlinear models can be developed which could be calibrated with existing data and then used to provide estimates of missing or more detailed numbers.

INTRODUCTION

Accounting System Definition

In biology, the assignment of any particular species to a place in the hierarchic system is based on genetic and morphological similarities between organisms. Classifications are often debated and reassignments are occasionally made, but progress in biology without such a system would not have been possible.

In ecology, although functional connections between the many species and their environment have been observed for decades, there is no accepted general accounting system for noting and comparing these connections. An ecological accounting system is a framework in which the quantified connections between organisms (individual species, collections of species) and their abiotic environment can be placed and balanced, without ambiguity, omission or double counting exchanges, at any scale which an investigator chooses. "Connections" means any kind of exchange of product or service (e.g., nectar from a plant, pollination time from an insect) between ecological processes (e.g., insect and plant).

This general ecological accounting system is the result of years of debate among ecological modelers, and it also benefits from years of debate and experience gained in national economic accounting.

Use of an Accounting System

The principal advantage of a universal accounting system in ecology is that it allows the material, energy and service flows between all the parts of an ecosystem to be systematically placed in a common framework. Ecologists have long been involved in material and energy flow accounting (Hannon 1973; Finn 1976), but in the past each type of material (i.e., nitrogen) or energy was accounted for independently of all the rest. To classify as eco*system* accounting, the interconnections between all the material and energy (and service) flows must also be included. This is the purpose of the framework proposed below.

The concept of a common framework can allow extant data bases, gathered for unrelated purposes, to be incorporated into the same framework as current data. Together, old and new data can reside in a framework that provides the format for evaluating whole ecosystem function (e.g., Costanza et. al. 1983; Hannon and Joiris 1989; Baird and Ulanowicz 1989). Models can be based on the accounting system which might be used to estimate the flow and stock changes resulting from the introduction of new species or certain toxicants, for example. The extent and quality of accounting data may also be improved by nonlinear models of all or part of the studied ecosystem. The results of such a model could provide data for the accounting table by allowing estimates of the variation of the exchanges through time. Such a model could also fill in data that are now missing from the exchange table.

The use of an ecological accounting system could therefore assure ecosystem analysts and managers that a research effort on a particular ecosystem was done: 1) with full awareness of the system boundary (in space and time); 2) such that materials, energy and service flows for each compartment of the ecosystem under study were balanced; 3) so that the connection of any particular species to the ecosystem could be quantified; 4) in a way that allows data from different sources and time periods to be compiled into a framework which would have growing utility.

The experience of economists has guided the development of this *ecological* accounting system. However, these accounting systems must consider the consequences of the laws of thermodynamics. With this recognition, the economic approach has been modified to incorporate thermodynamic constraints, in an attempt to create a more comprehensive, less arbitrary accounting system that simultaneously meets the needs of ecologists and economists. This approach also allows for the integration of economic and ecological data for analysis, planning, and management of large regional systems and for global change studies, as so elegantly called for by Roughgarden (1989).

DEVELOPMENT OF ACCOUNTING SYSTEMS IN ECOLOGY

Connecting elements in an overall budget makes it possible to trace and quantify indirect causes of change. However, the number and complexity of such indirect linkages easily overwhelms one's perceptual capacities. Thus, to make any progress in evaluating indirect effects one must resort to systematic analytical techniques. Of course, the question of indirect effects is hardly limited to ecology. It is a significant issue in economics as well. It was an economist (Leontieff 1941) who first successfully applied the techniques of matrix algebra to the task of accounting for indirect effects. Some economists (Daly 1968) have suggested that ecological flows be included in the Leontieff approach.

Hannon (1973) demonstrated how Leontieff's methods could be applied to ecological budgets. Copious efforts to apply "input-output" analysis to various ecosystems and to modify the analytical techniques to better address various issues of ecological concern followed. Patten et al. (1976) combined input-output analysis with general systems theory to extend the notion of an ecological niche to include indirect impacts and effects. Barber (1978) emphasized the probabilistic nature of ecological transfers and spotlighted the Markovian assumptions underlying the Leontieff approach. Finn (1976) studied the indirect effects that many system components exert upon themselves and showed how to estimate the fraction of the total system activity that is devoted to recycling.

Amir (1979) laid out the economist's views on equilibrium in ecological systems, including the interpretations of value, cost and price. Ulanowicz and Kemp (1979) remarked how the algebraic powers of the matrix of normalized transfers (the technical coefficients of economics) provide information on the amount of medium that traverses trophic pathways of various lengths. They used the successive matrix powers to transform arbitrary webs of trophic interactions into chains or pyramids of flows.

Matis and Patten (1981) attempted to incorporate the contents, or stocks associated with the system components into the analysis of indirect effects and sketched out what they called "environ analysis." Unlike the standard input-output analysis of systems that must balance around each component, environ analysis treats systems which are unbalanced and change via an assumed set of linear dynamics. Hannon (1986) examined the stability of such linear systems and evaluated several control strategies (either endogenous or externally applied) that could guide the system towards particular configurations.

Virtually all of the early work in ecosystem accounting focused on the exchange of a single medium. That is, budgets were cast in terms of energy, or carbon, or nitrogen; but seldom was there an attempt to treat systems wherein more than one medium was flowing, or to include connections that involved no physical exchange (e.g., the effect of pheromones, shading, pollination). Implicit in single medium analyses are the assumptions that all inputs to a compartment have equal effect upon the recipient population, and that only one product issues from each compartment (usually its biomass as food to predators). The latter assumption is not valid because organisms produce byproducts such as feces, urine, detritus and other exudates. How does one treat these multiple products that issue from a population or process? But perhaps even more problematical is the likelihood that the various foods consumed by each predator differ in their utilities to the consumer. How does one weight the various inputs to a compartment to reflect their relative utilities? Fortunately, one solution to both these difficulties is suggested by recent work in input-output theory.

In economics, one view of the origin of prices is that they account for the cumulative value added to a particular item at every step in the economic process from raw materials to finished product.[1] In ecology, it is sometimes possible to make analogous calculations (Costanza and Neill 1984). If a system has only one external input of a type not produced in the system, it is then possible to trace back through the network to identify the amounts of that input medium that were necessary to create any given connection. That is, one can estimate the portion of the input that has been "embodied" in each connection. It has been shown (Costanza 1980; Costanza and Herendeen 1984) how all the energies contributing to various economic products are correlated to their market prices. In the special case of a network with only a single medium (e.g., all energy flows), a particular transfer between higher trophic levels may yield value when directly measured (a certain number of calories). However, the de facto value of the flow in a systems context would reflect more the amount of the external input (e.g., sunlight) that went into creating the given flow. The ratio of the latter quantity to the former is called an "intensity" and can be used to weight the various inputs to a population so as to allow a legitimate comparison among these inputs.

In historic applications of the accounting procedure, each of the components in the system was assumed to have only one type of output to other components on that system. This organization allows a relatively straightforward calculation of the intensities of the net input. However, intensities can be calculated for an ecosystem description in which each component has multiple outputs as well as multiple inputs from other components. This procedure is more complex (Costanza and Hannon 1989) and requires the accounting system described below.

Amir (1987) has discussed the use of an ecological accounting system, both static and dynamic, to elaborate on the formal connections between ecology, thermodynamics and economics. He points out that the "intensities" are measured in terms of an external input and that one therefore must choose which external input should form the basis of the intensities. In the development of the general accounting system however, I am not proposing the use of intensities or of any other kind of model. The accounting system is seen as

---

[1] This is the view we take in our paper. Amir (1979) points out the other economic theories on the origin of prices.

a necessary underpinning for any kind of ecosystem model that claims to be consistent with any other kind. The proposed accounting system forms the basis for consistent comparison of model results. Amir (1987) points out that intensities are reminiscent of economic prices which arise from ecological resource allocation problems, just as they do in economics.

A GENERAL ACCOUNTING FRAMEWORK

The minimum desired criteria for an accounting framework capable of handling both ecological and economic systems has been summarized as: 1) generality; 2) comprehensiveness; 3) uniqueness; and 4) quality control and is described in Hannon, Costanza and Ulanowicz (1989).

Choosing Stocks and Flows

As in general systems theory, the delimitation of the system is strictly at the discretion of the observer, i.e., the system boundaries and list of internal elements may be chosen at will. That is not to say that all choices are equally good. Ideally, the physical boundaries are chosen to minimize the amount and diversity of exchanges across them. Then the exogenous transfers can be more easily monitored. Interfaces that inhibit certain classes of transfers, such as air-water and water-land are often convenient for this purpose. Topographic features, such as watershed or airshed boundaries often work nearly as well. The duration and frequency of sampling for data also place hierarchical boundaries on the system description that are every bit as important and restrictive as the demarcation of spatial domain. The temporal limits are thus inexorably connected with the choice of compartments.

What are the important processes and products of the ecosystem? Activity level is a criterion for selection of the stocks. Microbial populations usually have small standing stocks, but engender a disproportionate fraction of system activity. A species of marginal ecological importance will often be included in the accounting by virtue of its rarity or for its economic, aesthetic, recreational or political usefulness. The issue of which components of the system are actually controlling the critical flows is important as well (Hannon, Costanza and Ulanowicz 1989). Once the system boundary in space and time has been identified, and the processes with their input and output products have been chosen, one can begin the accounting procedure.

Quantifying the Flows

There are six distinct types of connections, in addition to the stocks, that must be identified in analyzing any ecosystem: 1) the nonproduced inputs; 2) the net outputs; 3) the product use record; 4) the product production record; 5) the total output, and 6) the waste heat flows. These product flows are shown in figure 16.1.

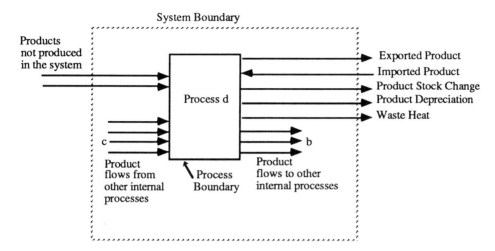

FIGURE 16.1   The product flows into and out of a typical process in an ecosystem. Product flows are mainly distinguished as going to other processes inside the system boundary or, crossing the boundary. Products imported from outside the system are separated depending on whether they are made in the system or not.

1. *The nonproduced inputs*

One must distinguish here between imported/exported products made within the defined ecosystem, and imports of special products which are not made within the system. One thinks of the import of nonproduced products as a constraint on the system's activity level. These are the products which are sometimes deemed as scarce or growth–limiting in some sense. Sunlight is a nonproduced import but not the only possible one. For example, sunlight contains blue, red and infrared radiation, all of which may be used by the ecosystem. While infrared radiation is sometimes thought of as extraneous, it is possible to define this radiation as a product *within* the ecosystem. If the ecologist chooses not to define it in this way, then infrared radiation should be included as a nonproduced input to the system.

2. *The net outputs*

The net output flows are connections from the defined ecosystem across the system boundary to itself at a future time and/or to other ecosystems. There are three kinds of net output flows which must be accommodated in the accounting system.

First, there is the import and export of products made and/or used by the processes in the defined system. For example, in an aquatic ecosystem, one of the processes is likely to be algae, and one of its output products would be "algal biomass." Some of this biomass could be washed from (export) or into (import) the system during a flood.

The second kind of net output flow is the change in the storage level of a particular product which occurs during the designated time period. Since a static picture of the ecosystem is assumed, the changes in stocks explicitly as flows must be included. These flows might be thought of as flows across the time boundary of the chosen period. For example, an increase in storage will be used in the next or later time periods.

Thus, the requirements do not surprise the experienced ecologist. But there remains a third net output flow across the ecosystem boundary that is harder to describe. If, as just assumed, the stock change is a flow across the time boundary of the system, how do we handle the change of stock due to the natural decay and stock replacement processes? These processes are inherent in all products of the ecosystem as required by the second law of thermodynamics. Everyone is aware of the decay processes associated with death. These same processes are at work in the living system, but they are countered by the simultaneous restorative or anabolic processes. Thus, when one observes a quantity of living substance perceived to be of constant stock over the chosen time period, one is really seeing the balanced result of the decay and replacement processes. During the time period of a study, a certain amount of the living product decays into its component parts and some of that lost stock is replaced. This decay-replacement process is known to occur because the products of decomposition and the net absorption of energy are both evident. If, the stock size of the organism does not change during the chosen time period, what is lost or gained by the organism? It is known from thermodynamics that the *order* of the structure of the organism was lost for that stock of product which decayed during the time period. Yet, for the case where the total product stock size did not change, that lost order was exactly made up by what is called the replacement process. One must record the contribution of this replacement process in some manner. It is a product contribution across the time boundary of the system in the same manner as is the net change of a product stock.

Decay rates can be measured directly by isolating a sample of the product in question and observing the loss of order with time. Frequently, this isolation is not possible or practical, however, and an estimate of the magnitude of the decay rate for living products (e.g., biomass) must be substituted. The basal metabolism of living products can be used as an estimate of decay rates, since it represents the replacement flow required to just compensate for decay processes when the organism is at rest.

In a similar way, abiotic products in the ecosystem (e.g., solution ammonia) might also degrade during the time period, and in the process, give off heat. One needs to know what the normal degradation rate would have been for this substance at the system temperature. Such data for most substances can be found in standard chemical reference handbooks. These standard data are accepted as the *replacement* rate of the nonliving substances in question. If there is a need to account for the waste heat associated with the decomposition of these abiotic compounds, one could add another process to the accounting system which describes the input and output of the compound and its degradation (therefore its replacement rate) and record the associated waste heat. Of course, some of the nonliving substances might already be at their lowest energy state, at equilibrium with the system environment (e.g., carbon dioxide), and no decay need be considered.

The assignment of the basal or standard metabolism to the net output of the framework to represent the replacement of the depreciated stock is explicit acknowledgement of the second law of thermodynamics. The remainder of the respiratory flows are not counted as a part of the net output since they are not of the same quality as the flows along that row of the matrix. These flows are recorded only to provide a check on the balance of the total flow of that substance in the system. The assignment process does not double count any of the flows.

3. *The product use record*

The "Use" and "Make" matrices approach (Stone 1966) is modified to meet the special problems involved with ecological flows. The "Use" matrix is the record of which process uses what product. The "Make" matrix is the product production record and is discussed in the next section.

The first accounting step is to record where each product is used and by which process. A "Use" or U matrix is constructed as follows: Each process is placed at the head of a column in the U matrix; each product is placed at the beginning of a row in this matrix. The use of a product (e.g., product "c") by a particular process (e.g., process "d") for the chosen time period (i.e., the flow of "c" into "d"), is the number placed on the cell which lies at the intersection of row "c" and column "d." The quantities of the substances consumed by predators for example, appear in the U matrix (figure 16.2).

The units of measure of product c must be the same wherever product c is used, but these need not be the same as the measure for any of the other products. For example, product c may be measured in grams of carbon per square meter and product d may be

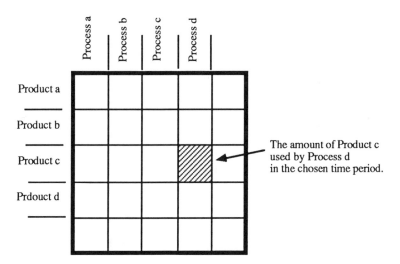

FIGURE 16.2 The "Use" or U matrix. The uses of products by the various processes are tabulated in the Use or U matrix.

measured in kcal per square meter. So long as the units are consistent along the row of U, the products may be measured in any units.

Besides allowing greater convenience in data gathering, the tolerance of the U matrix for any kind of unit of measure lets experimentalists incorporate units of "service" into the matrix. For example, the pollination of flowers by bees may be measured in "pollination bee–seconds" per square meter for the chosen time period. It is known that if some pollutant reduced the bee population by half, flower productivity would decrease. But there is hardly any physical transfer from bees to flowers, except that measurable by the service units of pollination. The "bee–seconds" measure is the collective visitation time spent by bees in flowers. In the U matrix, the process *plant* would have "used" so many bee–seconds of pollination service. So the term "product" as it is used here may mean an actual physical product *or* a service.

4. *The product production record*

In the general accounting system, a second matrix is needed to show where the products are made. Such a matrix is called a "Make" or V matrix and it has the same configuration as the U matrix: the column heads are the processes and the product names and are shown at the beginning of each row. The quantified list of substances produced by prey for example, are listed in the production matrix. The elements of the V matrix are for example, the amount of product b which is produced or made by process d (figure 16.3).

The units of a product must be consistent across the row of the V and U matrix but can vary from product to product.

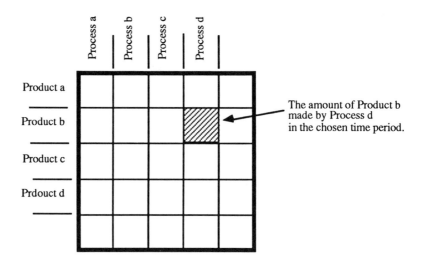

FIGURE 16.3 The "Make or V matrix. The products made by the various processes are tabulated according to where they are made.

5. *The total output flow*

The sum of the entries in a row of U plus the corresponding row of the net outputs is the total output flow for that product. This sum is also equal to the sum of the corresponding row of the V or "Make" matrix. The balance here provides a check on the various product entries.

6. *The waste heat flow*

Respiratory or total waste heat is composed of the basal metabolic heat and the organism's muscle and bone friction during, prey–seeking or predator–avoidance effort, and reproductive effort. To complete the accounting procedure, all of the heat given off by each process is collected into a vector. This heat is considered lost to the system due to its lack of utility to any of the other organisms in the system. If some of this heat were *used* by certain components in the ecosystem, then that quantity of heat would be classed as a product produced and used by the appropriate components in the U and V matrices. If some of this "waste" heat could be used as a measure of reproductive service for example, then that quantity of heat would appear in the U and V matrices as used and made by the same component. All of the "waste" heat *could* be assigned in this manner. Note that the waste heat flow defined in this section is not associated with the basal metabolic processes: those substances representing basal metabolism have been assigned to represent the decay processes and appear in the net output.

To complete the accounting framework, figures 16.2 and 16.3 are augmented with the matrices of product inputs and outputs to the system. Since the non-produced inputs to the system are products used by the system, they are best assembled at the lower edge of the U matrix. The produced import and export products, the changes in product stocks and the depreciation products are all products of the system but they cross the system boundary. Therefore, the matrix of these three types of flows (the net outputs matrix) are associated with the U (Use) matrix on its righthand side for convenience in balancing. The waste heat from the processes of the system are tabulated at the bottom edge of the V (Make) matrix as a notational reminder that they have been accounted for. The complete ecological accounting system therefore looks like figure 16.4.

As a matter of accounting convenience, the U and V matrices are overlaid in tabular form and kept in the same table, with the Make or V matrix entries shown immediately below the Use or U matrix entries. Also note that the U and V matrices will most likely be rectangular and not square, because the number of products used will not necessarily equal the number of products made.

Examples

The following simplified examples will serve to illustrate the accounting system. Simplified examples rather than data from actual systems have been used. A simplified form of the accounting system has been applied to the Southern North Sea ecosystem and published (Hannon and Joiris 1989; Costanza and Hannon 1989). The goal in my paper

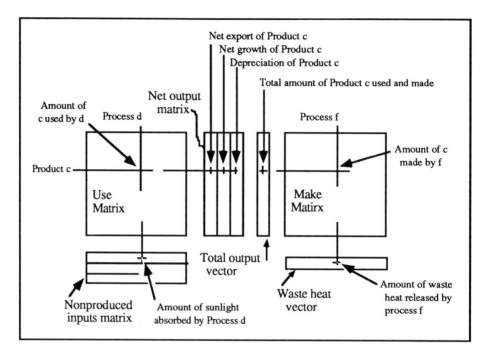

FIGURE 16.4 The general accounting system describing the assignment of all product flows to where they are made and used.

is to demonstrate the full accounting system in the simplest possible form so that each of the principles can be clearly explained.

The first example is a hypothetical ecosystem consisting of three internal processes (abiotic chemical processes, primary producer processes, and consumer processes) and five products (nutrients, plant biomass, animal biomass, waste heat and sunlight). The "use" or inputs, "make" or outputs, and "stock" of each of these products by (or in) each of the internal processes are shown in table 16.1 and figure 16.5. The table and figure are equivalent representations of the system. Figure 16.6 shows the nutrient product flows isolated from the system.

Most past network analysis in ecosystems has concentrated on such a one-product abstraction of the system. The general accounting framework is intended to account for more of the complexity of ecological networks as shown in figure 16.5 and table 16.1.

Table 16.1 represents a compact yet readable representation of the complex flows shown in figure 16.5. For example, one can read from figure 16.5 that abiotic chemical processes use 20 grams per unit time of nutrients, primary producers use 10, and consumers use 0. Nutrients are "made" by all three processes, since plants and animals release nutrients when they die and decompose. In contrast, plant and animal biomass are "made" only by producers and consumers, respectively.

TABLE 16.1 Hypothetical Ecosystem Network Illustrating a General
Accounting Framework

PRODUCTS *Internally Produced* (units)		INTERNAL PROCESSES			EXTERNAL PROCESSES			
		Abiotic Chemical Processes	Biotic Processes producers	consumers	Net Outputs Stock Change	Export/ Import	Depre- ciation	TOTALS
Nutrients	Use	20	10	0	0	5	2	37
(g/t)	Make	20	5	5	-	7	-	37
	Stock	500	0	0	-	-	-	500
Plant Biomass	Use	1	5	6	1	0	1	14
(g/t)	Make	0	14	0	-	0	-	14
	Stock	0	100	0	-	-	-	100
Animal Biomass	Use	1	1	5		0	1	8
(g/t)	Make	0	0	8	-	0	-	8
	Stock	0	0	10	-	-	-	10
Waste Heat	Use	-	-	-	-	-	100	100
(Cal/t)	Make	5	50	45	-	-	-	100
Externally Produced								
Sunlight	Use	0	100	0	-	-	-	100
(Cal/t)	Make	-	-	-	-	100	-	100

A fourth internal product is waste heat. This product is unique in that it is made by all the internal processes but is used by none of them. In addition to the four internal products, one "external" product is noted for this ecosystem. This is sunlight, which is used by producers to drive the system and is not made anywhere in the system. Any product that is not produced anywhere in the system under study but is necessary for the system's survival is called an externally produced product.

There are also three external processes" shown: change in stock, exports/imports, and depreciation replacement, along with their use and make of the system's products. As already mentioned, a change in stock is equivalent to an export to another time period and depreciation is equivalent to export from the system. Thus, taken together, these three external processes represent net outputs from the system of one kind or another.

Note that for this system (and for any system) mass and energy balance are maintained for each product. For example, the total use of nutrients in the system of 37 g/t is equal to the total production of nutrients by internal processes plus imports (also 37 g/t). The total use is measured in quantity per unit time and represents use by internal processes plus use by external processes in the form of change in stock, exports and depreciation. Product use must balance product make for each product in the system.

"Products" in the accounting system can be services or information. An example in which bee pollination services are an important product is shown in table 16.2.

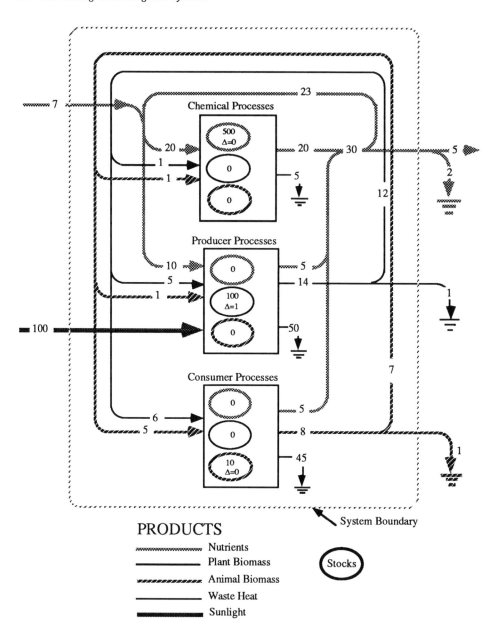

FIGURE 16.5 Flows and stocks of 5 different products in a hypothetical ecosystem with three internal processes.

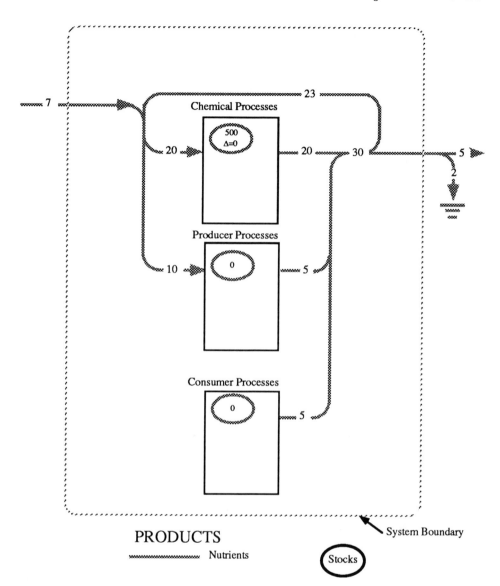

PRODUCTS

〰〰〰〰〰〰〰〰〰 Nutrients

⬤ Stocks (ellipse labeled "Stocks")

System Boundary

FIGURE 16.6 Hypothetical ecosystem flows and stocks of the single product "nutrients."

TABLE 16.2 Hypothetical Ecosystem Network Illustrating the Accounting of Service Type Products

PRODUCTS (units)		INTERNAL PROCESSES			EXTERNAL PROCESSES Net Outputs			
Internally Produced		Bees	Plant 1	Plant 2	Stock Change	Export/ Import	Depre- ciation	TOTAL
1 Bee Biomass	Use	5	0	0	1	0	1	7
(g/t)	Make	7	0	0	-	0	-	7
	Stock	35	0	0	-	-	-	35
2 Honey	Use	30	0	0	0	0	5	35
(g/t)	Make	35	0	0	-	0	-	35
	Stock	0	100	0	0	0	0	100
3 Plant Biomass	Use	5	25	35	3	0	5	73
(g/t)	Make	0	33	40	-	0	-	73
	Stock	0	200	100	-	-	-	300
4 Bee Pollination	Use	0	6	3		0	1	10
Effort	Make	10	0	0	-	0	-	10
(bee sec/t)	Stock	10	0	0	-	-	-	10
Waste Heat	Use	-	-	-	-	-	70	70
(Cal/t)	Make	5	30	35	-	-	-	70
Externally Produced								
Sunlight	Use	0	40	30	-	-	-	70
(Cal/t)	Make	-	-	-	-	70	-	70

In this example the products are bee biomass, honey, plant biomass, and bee pollination effort, in addition to the ubiquitous waste heat and sunlight. The internal processes are between bees and two species of plants that the bees pollinate. Bees in this system produce four different products; 1) bee biomass (7 g/t); 2) honey (35 g/t); 3) pollination effort (10 bee sec/t); and 4) waste heat (5 Cal/t). Note that both bee biomass and honey are used and made exclusively by bees. Bee pollination effort is used more by plant 1 (6 bee sec/t) than by plant 2 (3 bee sec/t). Plant 1 also has twice the biomass stock of plant 2 (200 versus 100) and uses more sunlight (40 versus 30 Cal/t).

USES OF THE ACCOUNTING SYSTEM

The accounting framework described above allows the collection of ecosystem data in a form that is useful for a number of analytical purposes. Some of these uses and their importance are described in detail in a number of recent works (i.e., Hannon and Joiris 1989; Costanza and Hannon 1989; Fasham 1984; Hannon 1986; MacDonald 1983; Ulanowicz 1986; Wulff et al. 1989). Some examples that demonstrate existing and potential uses are briefly described below, but the presence of a consistent accounting framework should allow an explosion of new uses not mentioned or conceived of here.

Most uses of the framework require some assumptions about the functional relationships between process inputs and outputs The examples below are organized in order of increasing sophistication of these assumptions, from static, linear ones, to dynamic, nonlinear ones, to more elaborate optimization functions.

Network Analysis and Energy Intensities

One common use of the framework is to calculate the indirect connection of a particular species to another or to the net input, for example, the connection of the sun to the lion. In general, this idea is captured by the concept of the cost (in terms of the net input) of a unit of output of a particular species or individual.

The "mixed units" problem is a critical one in both ecology and economics. An accounting system which allows multiple products has been elaborated, but most analytical work requires (or, at least, finds more convenient) a definition where only a single product is tracked. One area of research therefore is, the conversion of the full multi-product network into an equivalent single-product network. Work in this area has so far centered on various ways to calculate intensity factors (analogous to prices in economic systems) that allow all the system's products to be converted into a common currency (Hannon, Costanza and Herendeen 1986; Hannon and Costanza 1985; Costanza and Hannon 1989). If the system under study has only one nonproduced input (usually sunlight for ecological systems) and there are an equal number of products and processes, it becomes possible to calculate the input intensities that represent the amount of the non-produced input "embodied" in each of the system's products (Costanza and Hannon 1989). These energy intensities can then be used to convert the system into a single product network (i.e., in "embodied sunlight") that is amenable to further network analysis. The range of possible network analyses that can be performed on single-product ecological networks is given in a recent compendium (Wulff et al. 1989).

The energy intensities are very sensitive to the choice of the split between waste heat and basal metabolic heat as Herendeen (1981) has shown.

The Geographical and Economic Connections

The general accounting framework will also allow analysis of different but connected geographical ecosystems. The matrix of use and make for the first ecosystem would be arranged in the manner stated above. The use matrix of the second, adjoining ecosystem would be appended to the use matrix of the first one, centered on the extended diagonal of the first. Any exchanges between the two ecosystems would be placed in the appropriate cell in the matrix to the right of or below the first matrix. The vectors of net and total output and of net input of the first matrix would be extended in length to accommodate those same vectors from the second system. The combined result could then be analyzed as though it were a single system matrix. For example, the above mentioned energy intensities would then reflect the combination of the two systems with possibly differing intensities for the same product from each system.

In a similar way, the ecosystem framework could be appended to an economic framework (Isard 1951; Isard 1968) with matrices added for the economic system inputs to the ecosystem (including pollution) and another matrix for the services provided to the economy by the ecosystem. With such an arrangement, the economic values of the ecosystem services to the economy could be estimated.

An example of a generic simulation model constructed to match an accounting framework has been reported by Ulanowicz (1989).

Model Building and Data Enhancement Through Modeling

No accounting system of any particular ecosystem is ever complete. One will always long for more detail and for more accounting "snapshots" of the framework at various moments in time (e.g., diurnal, seasonal). Policy makers will continue to ask questions about the impact of pollutant removal or chemical addition to an ecosystem. The only way to improve our ability to answer these questions is to use the mathematical model building process. Nonlinear, dynamic models of the ecosystem under consideration can be calibrated from the framework of existing data. The model must, of course, be constructed to match the data framework. For example, the data in the framework may represent annual averages but we may wish to know the variation in these data through the seasons. A monthly nonlinear dynamic model might be constructed to give reasonable averaged approximations of the average annual data in the framework. The model could then be run to find the seasonal variations in the data. Such a process while not necessarily fully accurate, is the best one can do within constraints of the data. Once this model is verified, it can be used to show which of the system's stocks and flows contribute the most to model parameter changes and which do not disturb the average annual results of the model. The most sensitive parameters point out where the marginal research dollar should be spent to enhance the overall utility of the model. Such a model could also be used to fill in missing data in an accounting table. If, for example, the stocks and some of the flows in part of the ecosystem under study were well known through time, a model of this subsystem could be constructed to allow calculation of the missing flows.

SUMMARY AND CONCLUSIONS

The need for and style of a general accounting system as an ecological research tool has been discussed. When field ecologists investigate one portion of a large ecosystem, they usually design their own research strategy. In so doing, they may miss the opportunity to significantly augment the future work of other scientists by slight changes in their research strategy. They may also leave off some stock or flow measurement which was unimportant to their own needs. By reviewing their research proposal and checking it against the requirements of a general accounting system, these scientists could assure themselves that their work will be usable to others. In this way, their research would contribute to a network of understanding of the larger ecosystem. As research progresses on the pieces of the larger system, the time approaches when scientists will be able to synthesize the collective information into a model of the larger system function.

Acknowledgments

I wish to thank Robert Ulanowicz and Robert Costanza for their many helpful comments.

REFERENCES

Amir, S. 1979. Economic Interpretations of Equilibrium Concepts in Ecological Systems. *Journal of Social and Biological Structures* 2:293-314.

Amir, S. 1987. Energy Pricing, Biomass Accumulation, and Project Appraisal. In G. Pillet and T. Murota, eds., *Environmental Economics: The Analysis of a Major Interface*, 53-103. Geneva: R. Leimbreger.

Baird, D. and R. E. Ulanowicz. 1989. The Seasonal Dynamics of the Chesapeake Bay Ecosystem. *Ecological Monographs* 59:329-364.

Barber, M. 1978. A Retrospective Markovian Model for Ecosystem Resource Flow. *Ecological Modelling* 5:125–35.

Costanza, R. 1980. Embodied Energy and Economic Valuation. *Science* 210:1219-1224.

Costanza, R. and B. Hannon. 1989. Dealing With the "Mixed" Units Problem in Ecosystem Analysis. Wulff et al. 1989:ch. 5.

Costanza, R. and R. Herendeen. 1984. Embodied Energy and Economic Value in the United States Economy: 1963, 1967, and 1972. *Resources and Energy* 6:129-164.

Costanza, R. and C. Neill. 1984. The Energy Embodied in the Products of Ecological Systems: A Linear Programming Approach. *Journal of Theoretical Biology* 106:41–57.

Costanza, R., C. Neill, S. G. Leibowitz, J. R. Fruci, L. M. Bahr and J. W. Day. 1983. Ecological Models of the Mississippi Deltaic Plain Region: Data Collection and Presentation. U.S. Fish and Wildlife Service, Division of Biological Services, Washington, D. C. FWS/OBS-82/68.

Daly, H. 1968. On Economics as Life Science, *Journal of Political Economics.* 76:392–406.

Fasham, M. J. R. 1984. *Flows of Energy and Materials in Marine Ecosystems.* New York: Plenum.

Finn, J. 1976. Measure of Ecosystem Structure and Function Derived from the Analysis of Flows. *Journal of Theoretical Biology* 56:363–80.

Hannon, B. 1973. The Structure of Ecosystems. *Journal of Theoretical Biology* 41:535–46.

Hannon, B. 1986. Ecosystem Control Theory. *Journal of Theoretical Biology* 121:417–37.

Hannon, B. and C. Joiris. 1989. A Seasonal Analysis of the Southern North Sea Ecosystem. *Ecology* 70:1916-1934.

Hannon, B., S. Casler, and T. Blazeck. 1985. Energy Intensities for the U.S. Economy— 1977. Document no. 326. Energy Research Group, Urbana, Ill. University of Illinois.

Hannon, B., R. Costanza, and R. Ulanowicz, 1991. In press. A General Accounting Framework for Ecological Systems. *Theoretical Population Biology*.

Herendeen, R. 1981. Energy Intensities in Ecological and Economic Systems. *Journal of Theoretical Biology* 91:607-620.

Isard, W. 1951. Interregional and Regional Input-Output Analysis: A Model of a Space Economy. *Review of Economics and Statistics* 33:318-328.

Isard, W. 1968. Some Notes on the Linkage of the Ecologic and Economic System. *The Regional Science Association Papers* 22:85-96.

Leontieff, W. 1941. *The Structure of the American Economy, 1919-1939.* New York: Oxford University Press.

MacDonald, N. 1983. *Trees and Networks in Biological Models.* Chichester: John Wiley.

Matis, J. H. and B. C. Patten, 1981. Environ Analysis of Linear Compartmental Systems: The Static, Time Invariant Case. *Bulletin of the International Statistical Institute* 48: 527-565.

Patten, B. C., R. W. Bosserman, J. T. Finn and W. G. Cale 1976. Propagation of Cause in Ecosystems. In B. C. Patten, ed., *Systems Analysis and Simulation in Ecology.* Vol. 4. New York: Academic Press, pp. 457-479.

Roughgarden, J. 1989. The U.S. Needs an Ecological Survey. *Bioscience* 39-1:5.

Stone, R. 1966. *Input–output Relationships 1954–1966; A Programme for Growth,* Cambridge, Mass.: M. I. T. Press.

Ulanowicz, R. and W. Kemp. 1979. Toward Canonical Trophic Aggregation. *American Naturalist* 114:871–83.

Ulanowicz, R. E. 1986. *Growth and Development: Ecosystems Phenomenology.* New York: Springer-Verlag.

Ulanowicz, R. E. 1989. A Generic Simulation Model for Treating Incomplete Sets of Data. In Wulff et. al: 1989:84-89.

Wulff, F., J. G. Field, and K. H. Mann 1989. *Network Analysis in Marine Ecology: Methods and Applications.* Berlin: Springer-Verlag.

17

CONTRIBUTORY VALUES OF ECOSYSTEM RESOURCES

Robert E. Ulanowicz
University of Maryland
Chesapeake Biological Laboratory
Solomons, MD 20688-0038 USA

ABSTRACT

Ecologists and economists use different methods for keeping track of transactions occurring in their respective systems. Until now, those investigators who have sought to analyze indirect influences in multimedia (multiple commodity) ecosystems have employed a supply-side analysis and have advocated that ecologists adopt the style of accounting framework commonly used in economics. Existing ecological bookkeeping practices, however, are more compatible with a complementary demand-side, or input treatment that describes the contributory value of each flow or process towards specified final outputs. The values of these products are determined by an extant market.

The contributory values assigned to each trophic transaction also serve to convert all flows of various media into common dimensions. When combined with an appropriate discounting, or costing scheme, this conversion permits comparison of all the inputs into a compartment, allowing one to identify which nutrient is limiting the production of each node in the network.

INTRODUCTION

In his most recent book, Eugene Odum (1989) repeatedly stresses how urgent a task is the incorporation of ecological support functions into economic theory. Of course, one may attempt to bridge the separation of ecology and economics from either direction. That is, one could try to adopt the "ecological" standpoint and recast economic values in physical or biological terms; or alternatively, one could approach the problem as an economist and endeavor to impart monetary significance to ecological externalities.

To date, most efforts at combining ecology and economics have proceeded from the first viewpoint, i.e., interpreting market goods and services in terms of primary ecological inputs—giving rise to what has become known as the "energy theory of value." That the "ecological" perspective was the first to develop merely reflects the necessity that one first had to perceive the human economy as a threat to the natural order before economists could be informed of the long-term dangers of economic overexpansion. Thus it was that Hannon's (1973) application of Leontief's (1951) Input-Output Theory to the analysis of ecological systems soon was followed by efforts to reinterpret the units of economic transactions in physical terms (Costanza and Neill 1984; Odum 1988; Hannon et al. 1991).

There are also dimensional reasons why more progress has been made to date toward quantifying the ecological viewpoint. In economic communities, the inputs of goods and services are usually reckoned in variegated units, e.g., tons of coal, numbers of televisions, or manhours of sales effort. By assuming that outputs of a single commodity from any process are valued equally by all users of that product, it becomes possible to assign weights (or prices) to each commodity which then permit their intercomparison. These weights allow one to interpret the relative worth of each transaction with respect to some primary input to the system. The most obvious and prevailing input to the combined ecological-economic system is that of solar energy, so that most ecological values for artificial goods have been estimated as "solar equivalents."

Because the converse assumption that a consumer values its different inputs equally is considered economic nonsense, input-side evaluations of transfers are practically never made. (See Augustinovics 1970 for very special exceptions.) Hence, projects to hindcast the monetary contributions of natural subsidies to the functioning of the human economy have languished. The goal of this exercise is to make some modest progress toward rehabilitating the "economic" perspective, so as to lend credibility to the notion that heretofore unassessed natural subsidies (externalities) have made at least nominal cash contributions to the marketplace.

Before proceeding with the technical analysis, it is useful to note several interesting differences between the habits and concerns of economists and those of ecologists. For example, in economics most interest focuses upon exchanges between a given community and the next larger hierarchical unit, i.e., how changes in transfers affect final supplies and demands (whence the name "Input-Output Analysis.") In ecology, more stress is placed upon reckoning how the intermediate transfers affect each other (Szyrmer and Ulanowicz 1987). Concerning preferences for quantitative units, it has already been remarked how economists often deal with idiosyncratic units reflecting the particular nature of the commodity being measured. American ecologists, believing that systems reflect their underlying chemistries, seem less concerned with individual units (organisms) than with the amounts of energy or chemical constituents that accompany feeding transfers. There remains, however, a strong Eltonian tradition in British ecology for quantifying organism sizes and numbers (Ulanowicz 1989).

Differences are marked as well in how economists and ecologists perceive the concept of "value." "Value" in economics has manifest cognitive associations, and for that reason is eschewed by most biologists as inapplicable to nonhuman systems. If they think about value at all, most ecologists gravitate towards physicochemical notions, such as entropy and its derivative property, free energy. However, a few (notably H. T. Odum 1988), feel that expressing magnitudes of trophic exchanges in terms of measured energy or material, without scaling their observations by the "qualities" these media possess in context, will lead to erroneous conclusions. As a measure of quality or value they suggest that I-O supply-side techniques be invoked to calculate how much of the primary inputs are "embodied" downstream in the intermediate flows and final outputs. Embodied flows magnify the otherwise small, higher trophic interactions to portray more accurately their significance in the overall scheme of the ecosystem.

With these distinctions as background, it is most interesting to note that virtually all of the work that led to an energy theory of value (the *ecological* viewpoint) was cast entirely in the lexicon of *economic* discourse. Such orthodoxy has not precluded a host of other ecologists from adapting input-output mathematics for a variety of other purposes (e.g., Patten et al. 1976; Finn 1976; Ulanowicz and Kemp 1979; Levine 1980; Bosserman 1981; Szyrmer and Ulanowicz 1987; Ulanowicz and Puccia 1990), but the larger body of ecologists remains distinctly cool towards the introduction of "foreign" terms and unfamiliar methods of bookkeeping. For their part, economists have shown even less enthusiasm for employing input-output theory to reconcile economics with ecology, feeling instead that I-O methods are academically passe. It would indeed prove ironic should the extension of *economics* into ecology proceed via arguments that are basically *ecological* in form; however, it is precisely this interwoven and mutually dependent nature of the ecological-economic dialogue that I wish to demonstrate.

QUANTITATIVE FORMULATION

Ecologists are wont to create "mass and energy balances" for the ecosystems they study. For example, one might choose the medium carbon, and make an account of the magnitudes of each influx into and efflux out of every element in the system. The bookkeeping on a single element does not constitute a unique description of the flow system, inasmuch as parallel balances of energy, nitrogen, phosphorous or any number of other media are possible. Each account gives rise to a different network description of the system (see also Herendeen 1990). Because the various media are transported together, embodied in the biomass of the component species, the connection topologies of the parallel networks are often very similar. The weightings of the connecting arcs, however, can be quite dissimilar.

To be more precise, it is assumed that the system consists of n distinct compartments and that m different media are considered relevant to the description of the system. The transfer of the kth medium leaving the ith compartment and entering the jth will be denoted by T_{ijk}. In addition to the n specified components, it will be convenient to define three virtual compartments with which to represent the exogenous transfers. All primary inputs to the system will be considered to emanate from compartment zero (0). Similarly, all exports of useful products from any compartment will be regarded as flowing to node $n + 1$, and all egress of degraded, or useless, medium will be denoted as entering $n + 2$ (Hirata and Ulanowicz 1984). Often, the system is considered to be at steady-state, that is, the amount of each medium that enters each compartment balances the amount of the same medium that exits, or

$$\sum_{p=0}^{n} T_{pik} = \sum_{q=1}^{n+2} T_{iqk}$$

(1)

for all i and k. Assumption (1) is not required in what follows.

One now wishes to calculate the relative contribution that one unit of medium k entering compartment i makes towards creating final products (to be denoted by λ_{ik} They are analogous to the "shadow prices" of equilibrium I-O Theory.) A key assumption will be that the per-unit value of k entering i is independent of the source of k. This is the mirror conjugate to the assumption made by Costanza, Hannon, et al. They assume that the per-unit value of a product from a given node is the same to all compartments that utilize that product. In economics, the latter assumption makes good sense. For example, a tele-fax machine usually is sold at the same price, regardless of whether the customer is a law firm, a grocery, a baseball club or whatever. Any attempt at differential pricing would be eroded by market forces. It usually does not make economic sense to break products into their material components and equate the value of, say, the copper in a telefax machine with that of the copper in a personal computer purchased from another vendor. The customer uses the telefax and the personal computer in the forms they were received and (usually) doesn't render them into component parts for reconstitution into some radically different configuration. But such is precisely the nature of trophic transformations! A predator captures a prey organism and catabolizes it (digests it) into elementary chemical forms (not elements, but simple compounds dominated by specific elements) and thereafter anabolizes (assimilates) those simple forms according to a new *bauplan*. The anabolic process incorporates simple elements in nearly fixed ratios, and the residuals are discarded. Glutamine, for example, is required at the same rate by the anabolic system of a fox, regardless of whether it comes from a rabbit or from a mouse. Hence, equivalue of elements as seen from the demand side seems an eminently reasonable assumption to make as a prelude to assigning values to ecological components.

It is useful to compare the values of the inputs with those of the outputs from the same compartment. If the contribution (value) of all the inputs to a node is the same as the contribution its outputs make to their consumers, then

$$\lambda_{ik} \sum_{l=0}^{n} T_{lik} = \sum_{p=1}^{n+2} \lambda_{pk} T_{ipk}$$

(2)

for all components $i = 1, 2, ..., n$ and all media $k = 1, 2, ..., m$. In general, however, the contributory values of inputs and outputs are not equal. The outputs from a compartment often contribute more to the final product than the inputs that sustain them. The difference may be regarded either as the "value-added" by processes that occur within the compartment, or conversely the amount by which subsequent users discount[1] the value of inputs to the given compartment. The fraction by which the users of output from i discount the value of medium k flowing into i will be denoted by D_{ik}. It follows that (2) should read

$$\lambda_{ik} \sum_{l=0}^{n} T_{lik} = (1 - D_{ik}) \sum_{p=1}^{n+2} \lambda_{pk} T_{ipk} \tag{3}$$

Not all of the λ_{ik} are unknown. In particular, one assumes that the degraded products are of no value whatsoever to the sink that receives them, i.e., $\lambda_{n+2} = 0$. By way of contrast, the exports of i to the economic community have values set by the market. Call the market value of exports from compartment i, V_i. It is now assumed that products passing into the market are used in the same forms that they are harvested or otherwise delivered. That is, each of the m elements is necessary to the final product in its natural proportions. Then,

$$V_i = \lambda_{n+1,k} T_{i,n+1,k} \tag{4}$$

for all i and k. (This assumption is *not* crucial to what follows. In the event the various elements in a product are valued differently by the market, one can replace V_i in eqn 4 by V_{ik}, the value of element k in the export from i, and proceed accordingly.)

Substituting (4) into (3) yields

$$\lambda_{ik} \sum_{l=0}^{n} T_{lik} - (1 - D_{ik}) \sum_{p=1}^{n} \lambda_{pk} T_{ipk} = V_i \tag{5}$$

or

$$\sum_{p=1}^{n} K_{pik} \lambda_{pk} = V_i \tag{6}$$

where

$$K_{pik} = \delta_{pi} \sum_{l=0}^{n} T_{lpk} - (1 - D_{ik}) T_{ipk} \tag{7}$$

---

[1] This is not discounting in the traditional sense of the word, which always implies an evaluation over time. Rather, it refers to what might be called "trophic discounting," or that done across trophic levels. Hannon has suggested that "costing" might be the more orthodox terminology, but "discounting" will be retained here in the spirit of transdisciplinary discourse.

and

$$\delta_{pi} = 1 \ \forall \ p = i$$

$$\delta_{pi} = 0 \ \forall \ p \neq i$$

There are $m \times n$ unknown λ_{pk} in the $m \times n$ equations represented by (6). Once the discount rates have been specified, (6) can be solved for the contributions per unit flow, λ_{pk}.

In formulating the discount rates, D_{ik}, it might seem that one should regard the $T_{i,n+2,k}$ as wasted and discount the input flows by the fraction of the input that is dissipated by i. That is,

$$D_{ik} = T_{i,n+2,k} / \sum_{l=0}^{n} T_{lik} \tag{8}$$

However, discounting everything that is dissipated is too radical an assumption. Although the values of the dissipated flows to the sink that receives them is null, it is likely that some of the dissipated energy or material served the purpose of organizing and/or maintaining the internal structure of i. Therefore, some portions of the dissipated media contributed to the values of the outputs from i, and their effects are embodied therein (cf., "basal metabolism" in Costanza and Hannon 1989).

A rational estimate of the discount rates may be achieved by searching for the "limiting element" flowing into each i. If the amount of medium k stored in i is B_{ik}, then the characteristic rate at which element k appears to pass through stock B_{ik} is

$$r_{ik} = (\sum_{l=0}^{n} T_{lik}) / B_{ik} \tag{9}$$

However, population i incorporates media only in fixed proportions[2] (cf., stoichiometry), so the actual rate at which all media pass through biomass i becomes identical to the slowest, or limiting (Liebig 1840) rate of passage–call it $r_{iq} = \min (r_{ik})$ over all $k = 1, 2, ..., m$. Media being received by i faster than this limiting rate are assumed to be wasted in proportion to the relative amounts by which their throughput rates exceed the limiting pace and should be discounted accordingly (cf., Gigantes 1970.) Hence, the discount rate for medium k through component i becomes

$$D_{ik} = (r_{ik} - r_{iq}) / r_{ik} \tag{10}$$

---

[2] Margalef (personal communication), however, warns that molar ratios are prone to shift with suceeding generations.

Nonliving compartments (e.g., detrital pools) are assumed to add no value to the resources passing through them and hence exhibit no discounting. (Of course, microbiota acting on detritus do change the values of the media they process and thus should appear *explicitly* in the network.) In the event that biomass proportions are unknown, the discount rate can be approximated by setting

$$r_{ik} = (\sum_{l=0}^{n} T_{lik}) / (\sum_{p=1}^{n+1} T_{ipk})$$

(9')

and proceeding in similar manner to calculate D_{ik} with (10).

A SIMPLE EXAMPLE

Flows in ecological communities usually are reckoned in terms of a single reference medium. Although budgets of various media appear in the literature, the instances of bookkeeping on more than a single medium for a given system are few. In one such case, Fasham (1985) estimated both carbon and nitrogen flows occurring in the ecosystems of the marine euphotic zone in neritic areas such as the North Sea. Fasham's networks were reported as the flows that would result from a unit of primary input of each element. The carbon network is readily converted into actual mass flows by scaling up the primary production to 90 g carbon $m^{-2}y^{-1}$, as is typical of the North Sea (Steele 1974). The results are shown in figure 17.1.

Nitrogen enters Fasham's system into the dissolved pool and could not be scaled directly. Fortunately, the molar ratio of C:N in the phytoplankton is known rather precisely to be 6.625, so that the nitrogen output from the phytoplankton is obtained accordingly. Back-calculating through the dissolved pool of nitrogen reveals that an input of 3.47 g nitrogen $m^{-2}y^{-1}$ into the system at that point replicates the desired output of nitrogen from the phytoplankton. The network of nitrogen flows appears as figure 17.2.

Fasham gives no values for the standing stocks of carbon and nitrogen in each compartment. Actually, the absolute magnitudes of the stocks are not necessary, as they cancel out in eqn (10). Knowing only the C:N ratios is sufficient to calculate the discount coefficients. The C:N ratios of 5 of Fasham's 7 compartments appear as the ratios of carbon to nitrogen in predatory losses from these nodes. Only the C:N ratios for the fish and the dissolved nutrient pool are not evident from his diagrams, and the latter is moot, because discounting by nonliving compartments is assumed to be zero. As to the carnivores (fish), Jorgensen (1979) reports that the dry weight of fish nominally consists of 16.3% nitrogen and 50% carbon. Thus, the C:N ratios (by weight) are taken to be:

1. Phytoplankton	5.68	5. Protozoa	5.06
2. Planktobacteria	4.54	6. Detritus and Bacteria	2.92
3. Carnivores	3.07	7. Dissolved nutrients	****
4. Omnivorous Zooplankton	5.23		

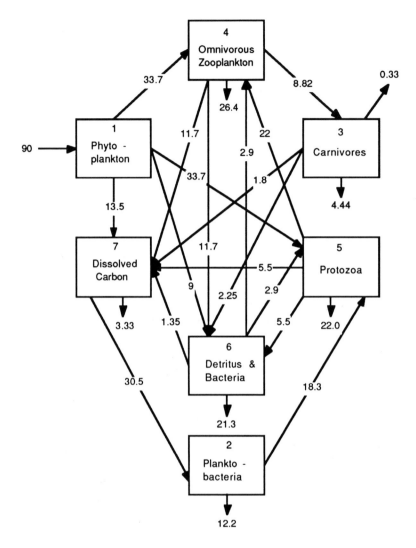

FIGURE 17.1 Flows of carbon in mg m^{-2}y^{-1} among the components of the euphotic zone of the North Sea ecosystem (after Fasham 1985).

The final market product is the carnivores, or fish. Here, it is assumed that all the nitrogen exported from that compartment (0.11 g Nitrogen m^{-2}y^{-1}) appeared in the harvest and was accompanied by a proportionate amount of carbon. The value of the harvest was set arbitrarily to 100 units.

The contributory values per unit flow (the λ_{pk} in eqn. 6) were calculated and multiplied by their corresponding flows (the T_{ipk}) to assign a contributory value to each flow

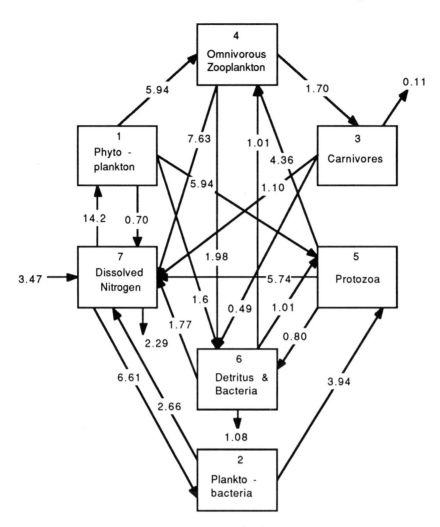

FIGURE 17.2 Exchanges of nitrogen in mg $m^{-2}y^{-1}$ among seven compartments comprising the ecosystem of the North Sea euphotic zone.

in the system. The values corresponding to the flows of carbon are shown in figure 17.3, and those for nitrogen appear in figure 17.4.

One notices immediately that the contributory values of the nitrogen flows are generally higher than their carbon counterparts. The final output from the carnivores (fish) is set equal to 100 for both media, but farther back in the network, the carbon flows diminish rapidly in value. The primary production of carbon contributes only 56 units to the final output, as contrasted with the 95 units contributed by the nitrogen upwelling into the dissolved nutrient compartment. Perusal of the discount coefficients reveals that the

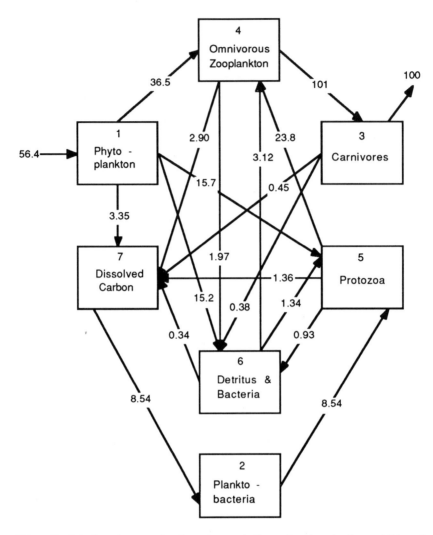

FIGURE 17.3 Relative contributions that each flow of carbon in figure 17.1 makes towards 100 arbitrary units exported from the fish compartment (carnivores).

inputs of carbon and nitrogen are virtually balanced (i.e., near their stoichiometric proportions) coming into the zooplankton and protozoan compartments and limited only slightly by nitrogen going into the phytoplankton (10.5% discounting of carbon) and planktobacteria (2%) compartments. Major nitrogen limitation occurs at the inputs to the detritus-bacteria compartment (50% discounting of carbon) and during carnivore feeding (41%). These two nodes account for most of the excess consumption (i.e., discounting) of carbon by the system.

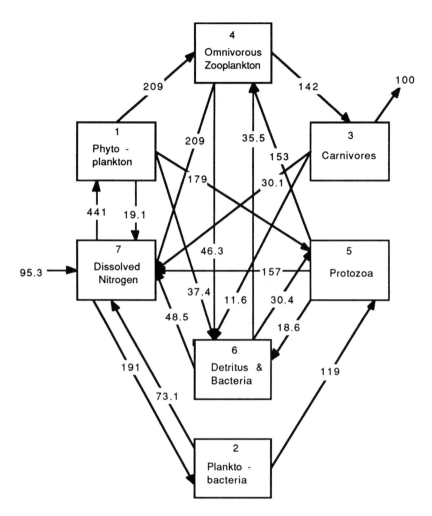

FIGURE 17.4 Relative contributions of nitrogen flows to the 100 units produced by the fish compartment.

The values of many nitrogen flows are high, sometimes even exceeding the value of the exported nitrogen. These high values are due to the large fraction of nitrogen recycling activity within the system (Finn cycling index >87%). In particular, the uptake of nitrogen by phytoplankton contributes 441 units to the final output. This contribution exceeds the 100 unit value of the final output; however, it should be remembered that this nitrogen will spend considerable time recycling within the system. Before it finally leaves the system, it will have contributed not only to the 100 units of fish now being produced, but to several-fold that output during later circuits through the system. It is interesting to note that the major pathways for recycle of valuable nitrogen do not include the detrital stage. (Detritus is prone to settling out of the euphotic zone.) Rather, the planktobacteria, zooplankton and protozoa are the principal agents that keep the scarce nitrogen circulating within the system.

An algorithm exists that is able to separate the cycled components in figure 17.4 from their once-through counterparts (Ulanowicz 1983). Figure 17.5 shows such "circulation" of nitrogen value. As Fasham himself pointed out, the system is strongly nitrogen limited, and one can see in figure 17.5 the emergence of nitrogen as a "proto-currency," which remains in the system for a long time.

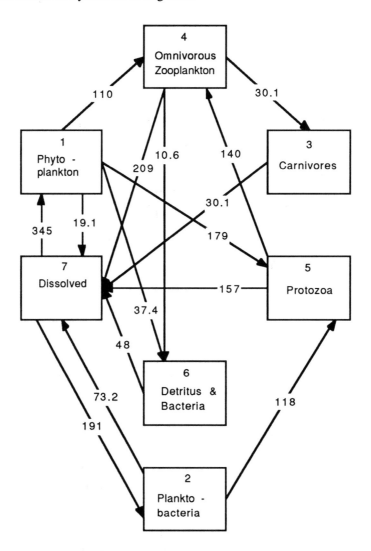

FIGURE 17.5 The circulation of nitrogen value in the North Sea ecosystem.

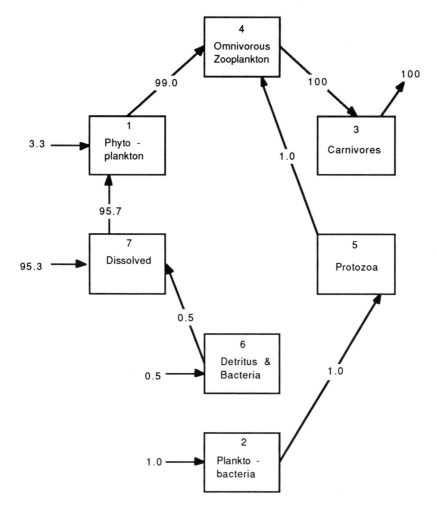

FIGURE 17.6 Residual sources and transfers of nitrogen values.

The actual sources for the nitrogen values appear in figure 17.6, where the residual nitrogen values are depicted. One sees that the "new" nitrogen contained in the upwelling is the primary contributor of *net* nitrogen value to the system. The remaining value-added sources are small by comparison.

DISCUSSION

Like all measures of value, contributory value is a relative concept. It refers specifically to the potential for a particular medium in a given form to contribute towards a designated end product. The magnitude of any contribution is always scaled by the value of the final product, which in turn is set outside the system. For example, the 33.7 g carbon $m^{-2}y^{-1}$ of phytoplankton consumed by the zooplankton potentially contributes 36.5 units to the output of fish valued at 100 arbitrary units. This contributory value does not measure the value of the phytoplankton to the zooplankton, much less any intrinsic worth phytoplankton may have in some other context. This last point deserves emphasis, lest any reader be tempted to use the method described here to assess the absolute worth of an ecological resource per se. Trees in a forest will contribute via the food web to the creation of, say, pelts taken from resident furbearing mammals. If one knows the quantified web of trophic interactions that culminate in the growth of furbearers and the economic value of the harvested pelts, it becomes possible to use the method described above to calculate the contribution of the trees to the harvested pelts via the web of trophic transfers. The same result, however, does not assess the value of the trees in creating habitat for the mammals, nor does it begin to encompass the host of other roles (utilitarian and otherwise, e.g., aesthetic) that trees play in the ecosystem.

These limitations notwithstanding, the calculation of contributory values can prove very useful. Justifiable or not, the perceived value of familiar ecosystems to some segments of the public will be in terms of one or a few items that the natural system provides to these individuals. For example, the primary association that legislators (or many natural resources managers, for that matter) make with the Chesapeake Bay is to a sportsfish, the striped bass (*Morone saxatillis*). Understandably, these individuals wish (for both personal and political reasons) to maintain the value of striped bass provided by the Bay ecosystem at a high level. The same group probably would never give a thought to the condition of ciliate populations (microscopic heterotrophs) in the same water column, until it is pointed out to them that the ciliates contribute 35 units to every 100 units of striped bass taken from the Bay (Baird and Ulanowicz 1989; and eqn. 6 above). At the very least, the contributory values of the supporting members of an ecosystem provide a *conservative* estimate of the values of the ecological foundations underlying a given resource.

In the end, the most important application of contributory values may not be explicitly economic in nature. The index could prove most useful in expanding the utility of the notion of limiting nutrients. Marine and aquatic ecosystems often are characterized as nitrogen-limited or phosphorous-limited, respectively. Some question remains as to what limits the productivity of estuarine systems—it probably would be simplistic to think of estuarine systems as being limited by either one element or the other. It is likely that phosphorous is limiting to some populations at certain times and nitrogen (or even carbon) to the remaining species at other times. Exactly which compartments are controlled by what elements, at which times and places could be addressed by computing the contributory values of each element using quantified networks of carbon, nitrogen and phosphorous flows for the estuarine ecosystem.

Like its supply-side counterpart, the calculation of contributory values converts flows having different sundry physical dimensions into a common currency. Networks with common units are of great advantage to those seeking to characterize ecosystem status in terms of whole-system indices (Kay et al. 1989). Heretofore, such indices have been reckoned in terms of a single medium. However, the magnitudes of flows of other media bear different proportions to each other than do those of the chosen medium, and useful information is lost by neglecting the exchanges of other media (Herendeen 1990). By expressing all flow networks of relevant media in terms of a common currency, one obtains a three-dimensional array (n x n x m) of flows to which one may then attach appropriate three dimensional information measures (see Ulanowicz 1986, p. 146; and Pahl-Wostl 1990). These expanded indices encompass all aspects of multimedia kinetics that comprise the system's organization.

It might further be asked how the demand-side calculations presented here could be related to the results from the supply-side treatment (e.g., Costanza and Hannon 1989). The system topologies used in the two methods are different, and comparing the values of intermediate products could be problematical. The primary inputs and the final products are easier to compare, however, and it would be interesting to calculate the contributory values of sunlight to a number of ecosystem products in order to observe the range in values that would result. Determining a *lower bound* on the economic utility of sunlight and primary production could be a significant step toward attributing value to many ecosystem components whose contributions to human welfare have heretofore languished in the realm of "common property natural resources."

Finally, it bears repeating that this exercise yields a valuable glimpse into the nature of ecological economics. It was mentioned before how progress by others in quantifying an *ecological* description of the natural world and its embedded marketplace was predicated on the methods and conventions of *economic* theory. The work just described points out the complement: How articulation of the *economic* perspective on the same interactions depends crucially on invoking bookkeeping methods peculiar to *ecology*. It may appear trite or self-evident to read that a transdisciplinary endeavor relies on a sufficient knowledge of both separate fields of study, but perhaps nowhere else is the mutually obligatory nature of such a relationship made more starkly evident than in the case of ecological economics!

Acknowledgments

The author would like to thank Dr. Robert Costanza for pointing out to him the existing notion of "contributory value." Drs. John Cumberland, Bruce Hannon, Robert Herendeen, and Ramon Margalef read the draft manuscript and offered many useful suggestions for revision. This work was supported as an element of the National Science Foundation's "Land Margin Ecosystems Research" Program, Grant no. BRS8814272.

REFERENCES

Augustinovics, M. 1970. Methods of International and Intertemporal Comparison of Structure. In A. P. Carter and A. Brody, eds., *Contributions to Input-Output Analysis*, vol. I, pp. 249-269. Amsterdam: North Holland.

Baird, D. and R. E. Ulanowicz. 1989. The Seasonal Dynamics of the Chesapeake Bay Ecosystem. *Ecological Monographs.* 59:329-364.

Bosserman, R. W. 1981. Sensitivity Techniques for Examination of Input-Output Flow Analyses. In W. J. Mitsch and J. M. Klopatek, eds., *Energy and Ecological Modelling*, pp. 653-660. Amsterdam: Elsevier.

Costanza, R. and B. Hannon. 1989. Dealing With the "Mixed Units" Problem in Ecosystem Network Analyses. In F. Wulff, J. G. Field and K. H. Mann, eds., *Network Analysis in Marine Ecology: Methods and Applications*, pp. 90-115. Berlin: Springer-Verlag.

Costanza, R. and C. Neill. 1984. The Energy Embodied in the Products of Ecological Systems: A Linear Programming Approach. *Journal of Theoretical Biology* 106:41-57.

Fasham, M. J. R. 1985. Flow Analysis of Materials in the Marine Euphotic Zone. In R.E. Ulanowicz and T. Platt, eds., *Ecosystem Theory for Biological Oceanography*, pp. 139-175. Canadian Bulletin of Fisheries and Aquatic Sciences, 213.

Finn, J. T. 1976. Measures of Ecosystem Structure and Function Derived from Analysis of Flows. *Journal of Theoretical Biology* 56:363-380.

Gigantes, T. 1970. The Representation of Technology in Input-Output Systems. In A. P. Carter and A. Brody, eds., *Contributions to Input-Output Analysis*, vol. 1, pp. 270-290. Amsterdam: North Holland.

Hannon, B. 1973. The Structure of Ecosystems. *J. theor. Biol.* 41:535-546.

Hannon, B., R. Costanza and R. E. Ulanowicz. 1991. In press. A General Accounting Framework for Ecological Systems: A Functional Taxonomy for Connectionist Ecology. *Theoretical Population Biology*.

Herendeen, R. 1990. System-Level Indicators in Dynamic Ecosystems: Comparison Based on Energy and Nutrient Flows. *Journal of Theoretical Biology* 143:523-553.

Hirata, H. and R. E. Ulanowicz. 1984. Information Theoretical Analysis of Ecological Networks. *International Journal of Systems Science.* 15:261-270.

Jorgensen, S. E., ed. 1979. Handbook of Environmental Data and Ecological Parameters. Copenhagen: International Society for Ecological Modelling.

Kay, J. J., L. A. Graham and R. E. Ulanowicz. 1989. A Detailed Guide to Network Analysis. In F. Wulff, J. G. Field and K. H. Mann, eds., *Network Analysis in Marine Ecology: Methods and Applications*. Berlin: Springer-Verlag, pp. 15-61.

Leontief, W. 1951. *The Structure of the American Economy, 1919-1939*. 2d ed. New York: Oxford University Press.

Levine, S. 1980. Several Measures of Trophic Structure Applicable to Complex Food Webs. *Journal of Theoretical Biology* 83:195-207.

Liebig, J. 1840. *Chemistry in Its Application to Agriculture and Physiology*. London: Taylor and Walton.

Odum, H. T. 1989. Self-Organization, Transformity, and Information. *Science* 242:1132-1139.

Pahl-Wostl, C. 1990. Temporal Organization: A New Perspective on the Ecological Network. *Oikos* 58:293-305.

Patten, B. C., R. W. Bosserman, J. T. Finn and W. G. Cale 1976. Propagation of Cause in Ecosystems. In B. C. Patten, ed., *Systems Analysis and Simulation in Ecology* 4:457-479. New York: Academic Press.

Steele, J. H. 1974. *The Structure of Marine Ecosystems*. Oxford: Blackwell.

Szyrmer, J. and R. E. Ulanowicz. 1987. Total Flows in Ecosystems. *Ecological Modelling* 35:123-136.

Ulanowicz, R. E. 1983. Identifying the Structure of Cycling in Ecosystems. *Mathematical Biosciences* 65:219-237.

Ulanowicz, R. E. 1986. *Growth and Development: Ecosystems Phenomenology*. New York: Springer-Verlag.

Ulanowicz, R. E. 1989. Energy Flows and Productivity in the Oceans. In P. J. Grubb and J. B. Whittaker, eds., *Toward a More Exact Ecology*, pp. 327-351. Oxford: Blackwell.

Ulanowicz, R. E. and W. M. Kemp 1979. Toward Canonical Trophic Aggregations. *American Naturalist* 114:871-883.

Ulanowicz, R. E. and C. J. Puccia 1990. Mixed Trophic Impacts in Ecosystems. *Coenoses*, 5:7-16.

18

ECOLOGICAL-ECONOMIC ANALYSIS FOR REGIONAL SUSTAINABLE DEVELOPMENT

Leon C. Braat and Ineke Steetskamp
Institute for Environmental Studies
Free University
P.O. Box 7161
1007 MC, Amsterdam Netherlands

ABSTRACT

In this paper, we focus on five aspects of open regional system development: the roles of nonrenewable and renewable resources, the importance of trade and exchange flows, the position of environmental and resource policies, and the succession of subsystems in the course of development. We introduce a generic, aggregate economic-ecological model for analysis of regional development processes and trace the dynamics of sustained development in a case study of a region in the Netherlands.

INTRODUCTION

In the last few years, the concept of sustainable development has come to imply conservation. This probably results from the imperative to conserve resources, environmental quality and species diversity as necessary elements in the process of sustainable economic development (WCED 1987). In an attempt to put these conservation issues in a proper perspective, this paper concentrates on the dynamic features of sustainable development, and in particular those at the regional scale.

Natural communities and human economies both are highly dynamic systems, which all show a sequence of growth, decline and replacement of system components. In ecological succession, species colonize an area and may dominate for some time. Eventually, several are outcompeted and, in the long run, many become locally extinct. In the histories of human economies, social groups successively conquer a region, establish a political empire and an economic organization, dominate for decades or centuries and are eventually expelled, absorbed by successors or exterminated (e.g., Sumerians, Romans,

Aztecs, Mayas, the medieval feudal system). On historical time scales, development of local populations of plants, animals and humans may not be sustainable. The species, however, usually survives in more favorable locations. On geological time scales, even the species may not be sustainable, but then new species evolve, and the ecosystem processes are still sustained.

Sustainable development has mostly been addressed at the global scale, thereby considering semiclosed systems on historical time scales. Braat (1990) discusses the structure and dynamics of aggregate generic economic-ecological models for the global system. The usefulness of these models is that the hypothetical futures, generated through computer simulation, provide a view of alternative development patterns and indicate required policies and sacrifices in the component systems.

For semiclosed systems the Growth-Peak-Decline-Stabilize pattern has been suggested as a likely development pattern. The nature of the growth and peak phase has been shown to be related to the stock of nonrenewable resources. Strict and immediate policies regarding use of nonrenewable resources, population growth, capital formation, and pollution generation are indicated in order to minimize the decline phase in the pattern.

The long-term welfare level in the stabilized phase has been shown to depend strongly on the maintenance of a productive capacity of the renewable resources (see Forrester 1971; Meadows et al. 1972; Odum 1983, 1987).

The aggregate models have proven to be useful in the analysis of dynamics of semiclosed systems. They are also extensively used in ecological and economic studies of regional systems where regions are defined as open subsystems of the semiclosed biosphere. Open systems have the added feature of cross-boundary flows, which are often ill-defined but crucial to the dynamics of the regional economic-ecological system (Isard 1972; Zucchetto 1975; Zucchetto and Jansson 1985; Ikeda 1987).

The models of growth and development in semiclosed systems indicate that in planning for sustainability it is useful to focus on nonrenewable resource stocks, the productive capacity of renewable resources, and conservation policies. Thus, in this study of the dynamics of sustainable development we have concentrated on these aspects of the ecological economic systems. As we deal with open regional systems, we, of course, include the cross-boundary flows. We have extended a conceptual semiclosed system model described in Braat (1990) to make it represent an open regional system, selected a region in the Netherlands, and applied the model in a study of the ecological and economic dynamics in the period from about 1800, just before the industrial revolution, until the present.

A CONCEPTUAL MODEL OF REGIONAL SUSTAINABLE DEVELOPMENT

The Concept

Sustainable development has been the key concept in the 1980s in economic planning and environmental management. And, as always happens to new concepts in policy making, it has led to a proliferation of ideas, discussions and publications. Some views have turned out to be politically influential (WCED 1987), others are more educational (Clark and Munn 1986; Costanza and Daly 1987; Pezzey 1989).

In 1987, the World Commission on Environment and Development (WCED) made a widely publicized statement on the necessity of addressing world problems. They viewed sustainable development as combining two basic notions: economic development and ecological sustainability. Ecologically sustainable economic development can then be thought of as changes in economic structure, organization and activity of an economic ecological system that are directed towards maximum welfare and which can be sustained by available resources. We shall refer to this concept as sustainable development.

Most definitions of sustainable development are phrased in general, qualitative terms and include a wide range of phenomena such as economic growth, equitable distribution of wealth within and between generations, supply of resources, environmental quality and the like (see the collection of definitions published in Pezzey 1989). The definition of sustainable development given above refers to a process. It also identifies an aggregate objective: "welfare," and an aggregate constraint: "available resources". This suggests the structure of an optimization problem, i.e., a problem following from the desire or necessity to achieve objectives within a set of constraints.

The objectives included most often in published definitions are: survival (e.g., Daly 1973; WCS 1980; Brown et al. 1986), satisfaction of needs (e.g., WCED 1987) and welfare (Goodland and Ledec 1987; WCED 1987) of the human species. These are clear anthropocentric objectives. Vitality and diversity of natural ecosystems is in some cases an explicit, ecocentric , objective (Opschoor 1989). Most authors, however, consider the role of nature, ecosystems, resources and environment as constraints.

Constraints may be viewed in at least three forms: the physical properties of a particular system, including those in systems of possible trade partners, which ultimately define the welfare level that can be sustained in the long run; self-imposed constraints; and the level of technology. If it has not appeared already as an explicit first objective, "leaving a useful and livable planet" for future generations appears in almost all definitions as a self-imposed (ethically-based) constraint (Goldsmith 1972; Goodland and Ledec 1987; Brown et al. 1987; WCED 1987). The present physico-ecological constraints are referred to in the definitions of, for example, Opschoor (1987) and WCED (1987).

In studies of sustainable development, time is usually dealt with in terms of intertemporal trade-offs, intergenerational equity, long-term planning, or discount rate. The horizon is usually defined arbitrarily at some convenient moment in the future or, in qualitative studies, at infinity. A criterion used often is that more than one generation is considered. The spatial dimensions of sustainable development processes have received more attention. The systems considered range from the biosphere (e.g., Clark and Munn 1986), via the national (see e.g., RIVM 1989) to the regional scale (Turner 1988).

The Open System Model

The semiclosed system model described by Braat (1990) is based on the elements of what is defined in the literature as sustainable development. This model, designed for analysis of global sustainable development has been expanded with cross-boundary flows (resources, goods and services, technology and people) to create an open regional system model. The model is used here to explore sustainable development futures of particular regions (figure 18.1).

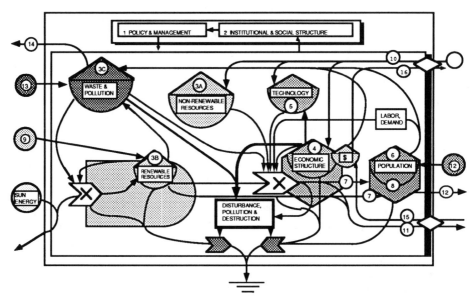

FIGURE 18.1 A generic open system model. Numbers correspond to the numbers in the explanation below.

From left to right, reflecting the basic direction of energy transformations, the diagram shows solar energy entering the system and being captured by the renewable resource systems (3b) (including the natural component of agricultural systems), which are exploited by economic production systems (4) and directly by humans (8) to generate consumable inputs (7). The flow through this major energy transformation chain is amplified by the nonrenewable resources (3a), drawn in and interacting with renewable resources, labor and technology (5) in the production function symbol. The production and depreciation processes generate heat (which ultimately leaves the system) and waste and pollution (3c), which feeds back negatively on stock regeneration and energy transfer processes. Social dynamics in the population and interactions between ecological, economic and human subsystems generate social and institutional structure, which produces policies and management, which modify and control the energy transformation processes. Human needs (6) are implicit in the population stock, but also reflected in the labor flow (willingness to pay for satisfaction of needs by means of work) which influences the production function symbol. Consumable inputs (7) are acquired by the population from the production function or directly from renewable resource stocks.

Stocks of renewable resource populations may be replenished by import of seeds and immigration of animals (9), while trade flows (10, 11) affect stocks of available nonrenewables, technology and economic assets. Numbers of people are influenced by migration between regions (12). Undesirable pollutants are sometimes passively imported (13) and passively or actively exported (14). In most systems, economic trade flows are accompanied by money flows (15, 16) which can be imported or exported as is or as loans.

Welfare, the objective in the optimization philosophy, can be thought of as some function of the per capita total of consumable inputs, environmental quality (the reciprocal of waste and pollution) and natural system amenity. The constraints at any moment in time are the current stocks of economic capital, of resources, level of technology and the effectiveness of policies.

Constraints may in theory be alleviated if there is a possibility to trade. Open systems can "import sustainability" of economic development by importing resources which are in short supply, or goods, services and technology that cannot be produced or can be purchased at lower cost. Open systems can also "export unsustainability" by sending pollution across the border (with or without payments).

This export-import game, if played smartly, could extend the sustained growth of a particular region beyond the point in time where the aggregate world economy is already experiencing decline. Such attempts to deal with uneven distributions of resources and assimilative capacities of natural ecosystems in maximizing welfare may constitute a relatively large factor in determining the sustainability of a region's economic development. It may also involve negative impacts on future generations in other regions.

Open systems are also susceptible to influences from other regions against which they cannot or can only partly protect themselves. For example, cross boundary air pollutants cannot be stopped physically. Regions which depend on specific resource imports to produce products for export are very vulnerable to world market price changes, to whims in the political behavior of suppliers (e.g., OPEC in the oil crises) and of competitors, and to changes of consumer markets. In other words, to open the boundaries for sustainability of economic development is also to open it for economic and environmental risk.

From this brief discussion, it may be obvious that the long-term dynamics of regional open systems are more complicated than those of semiclosed systems. One way to obtain some insight in these complications is to analyze the economic and ecological dynamics in historical development of regional systems. In the next section, we review the results of such a historical study. We used the conceptual open system model to direct our search in historical records for indications of development trends and changes in the economic and ecological structure (see Steetskamp 1989). Because of the degree of detail, the results are presented in a different style of diagrams.

A HISTORY OF SUSTAINED DEVELOPMENT: THE CASE OF NORTH BRABANT

In terms of the open system model, the North Brabant province in The Netherlands (see figure 18.2) during most of the 19th century had a small population, limited technology,

FIGURE 18.2   Location of North Brabant in The Netherlands

simple economic structure, low productivity of resources and low levels of pollution and waste, while the institutional structure had just been modernized in the Napoleonic era. In 1990, the province can be characterized as having a high population density, advanced technology, complex economic structure, small stocks of renewable resources with high productivity, negligible stocks of nonrenewables, high solid waste and pollution levels and a well-developed social and institutional structure. It has become an economically prosperous region, but one which is suffering from a number of serious environmental quality and renewable resource problems.

At the beginning of the 19th century, almost half of this region consisted of so-called wild lands, including dry and wet heathland, drift sands, forests, bogs—and in the east—an almost impenetrable marsh, dominated by peatmoss. The region has sandy soils, poor in nutrients, which form when forest exploitation is followed by grazing. The heath, which thrives on such soils, covered large areas until the beginning of the 20th century. Villages, towns, cities and roads only covered a low percentage of the area. The remainder was pasture and crop land.

A century and a half later, human influence had brought extensive cultivation, drainage of marshland, and peat extraction, all of which created land for agriculture, forestry and urban activities. This left only small pockets of natural systems (see figure 18.3).

Quantitative aspects of ecological and economic development in North Brabant have been traced since 1833, when the first reliable topographic map was produced. We discuss the history of regional development in two ways: first, by means of trends in a series of

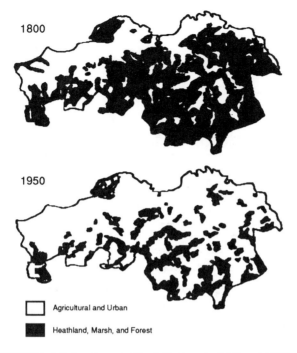

1800

1950

☐ Agricultural and Urban

■ Heathland, Marsh, and Forest

FIGURE 18.3 Land use in North Brabant in 1800 and 1950

economic and ecological indicators, and second, by reviewing changes from 1850 through 1980 in the relationships between the regional resources and socioeconomic user sectors and related import and export flows. The results are summarized in a series of regional system diagrams, in which we concentrated on the role of regional and imported renewable resources. We did not collect data on imports and exports of fossil fuels, minerals and consumer goods and services. Their presence and role, however, is acknowledged in the discussion.

Trends in Ecological and Economic Development Indicators

Human population (see figure 18.4)
In 1850, about 580,000 people lived in North Brabant. Half a century later, the population had slightly decreased, mostly due to emigration, which was probably caused by the national economic depression during the last two decades of the century. After the turn of the century, the population started to grow again, until in the 1980s the population reached 2 million and thus had almost quadrupled since 1850. In 1850, about 40% of the people were included in the labor force. Over the years, this percentage slowly dropped to 36%.

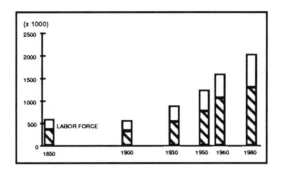

FIGURE 18.4 Growth of human population and the labor force in North Brabant 1850-1980

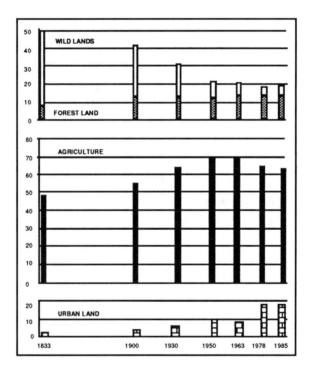

FIGURE 18.5 Land use development in North Brabant 1833-1985

Land use (see figure 18.5)

The map in figure 18.3 shows the aggregate change in land use due to cultivation. The bar graphs in figure 18.5 illustrate that the conversion of wild lands took place rather gradually. From 1850 to 1900, yearly changes in land use were rather small. Most of the wild land was converted (into forest, pasture and crop land) between 1900 and 1950. After that period, urbanization took its toll, partly on the wild lands but mostly on agricultural land. By 1985, about 17% of the region was devoted to urban uses, while seminatural ecosystems had made a minor recovery.

Economic sectors (see figure 18.6)
We derived an indication of the change in economic structure and activity from the changes in percent share of the labor force by each economic sector. We have only included sectors which had consistent data sets available. Figure 18.6 shows the change of total labor force volume and the succession of economic sectors from 1900 to 1980. At the beginning of this century, agriculture was the dominant sector, as it had been in the centuries before, with 41% of the labor force. Industry had developed throughout the second half of the 19th century and reached 36%. This growth was in line with, but lagging slightly behind, other areas in western Europe. Commerce and miscellaneous (services and government) were relatively small sectors at the time of the strong industrial growth. Thirty years later, industry had passed agriculture as the dominant sector; commerce and miscellaneous were increasing their share. Another twenty years did not cause much change in the relative shares, although the labor force grew 55%. From 1950 to 1980, however, the changes were dramatic. Agriculture's share dropped to 6%, while commerce and miscellaneous each attracted a quarter of the labor force and industry remained at about 40%. The real growth of the labor force during this period was almost 50%.

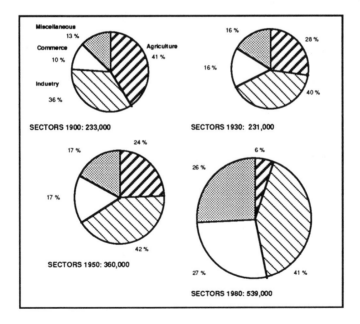

FIGURE 18.6 Change in economic sectors in North Brabant 1900-1980

Livestock (figure 18.7)

Agriculture, by far the most important economic sector throughout the 19th century, developed from an almost closed farm economy, via a market oriented agriculture in the first half of the 20th century, to a so-called bio-industry. Figure 18.7 illustrates this development with the exponential increase of three of the five livestock species, cattle, pigs and poultry, and the simultaneous decline of the other two, horses and sheep. The 1850 populations in North Brabant have been set at index value 1. By 1980, cattle shows a 6-fold increase, pigs an almost 90-fold, and poultry a 170-fold increase. Sheep lost their importance as source of wool with the coming of cheap imported wool, and as source of manure with the advent of artificial fertilizer. The number of horses slightly increased until the 1940s when they were rapidly replaced by automobiles and tractors.

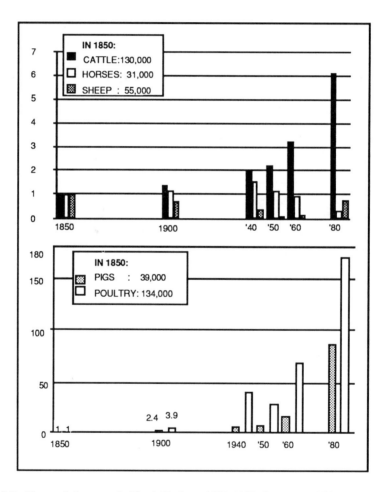

Figure 18.7 Livestock increase in North Brabant 1850-1980 (indexed; 1850 population = 1)

Industry (figure 18.8)

The shift in share of the labor force of each of the major groups of industries in the region provides a view of economic succession. During most of the 19th century, the majority of industries in the region were based fully or partially on one or several of the region's natural resources. In 1930, the original renewable resource based sectors (wood processing, textiles/leather, and food products) together still employed 63% of the labor force, even though much of the raw materials was already imported. In the next few decades, however, these industries were quickly replaced as major industries by electronics (in the metal group) and construction (in 1980 together, 69%).

Changes in the Structure of the Regional Ecological Economic System

The ecological-economic system in 1850 (figure 18.9)

The diagram in figure 18.9 shows the region's major resources (water, agricultural land, forest, wild lands, peat) and the most important components of the regional economy. The closed farm economies derived most of their goods and services from the nearby land.

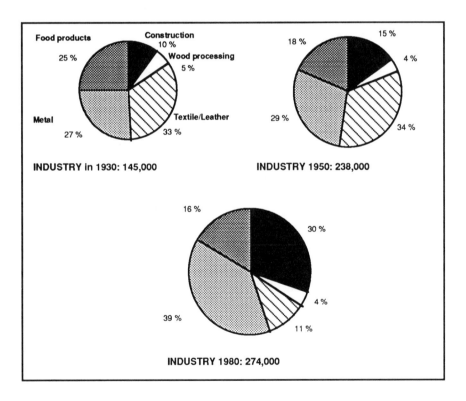

FIGURE 18.8 Change in the industry in North Brabant 1930-1980

The few industries still obtained most of their raw materials and fuels (peat, wood) from the region. The textile and cigar industry relied on imported raw materials. Most of the industrial products were exported. The agricultural system operated only marginally through a market; crops were locally consumed. To produce crops on the poor soils, fertilizer was made of animal manure mixed with heathland top soil and forest litter.

The ecological-economic system in 1900 (figure 18.10)
In fifty years, the role of resources had changed considerably. In 1900 the crops were produced predominantly to feed livestock. Introduction of artificial fertilizer, low grain prices, and relatively high prices for animal products contributed to this change.

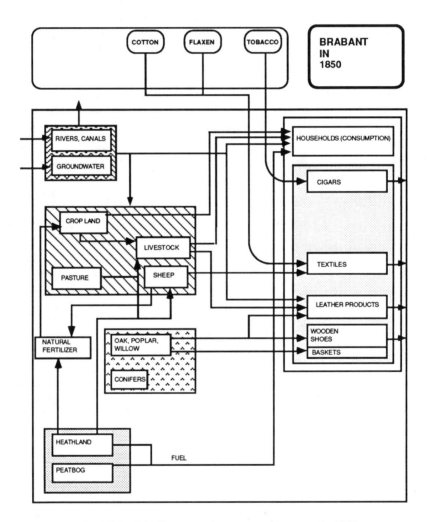

FIGURE 18.9 The ecological-economic system in 1850

Agricultural products were mostly marketed and a major share was exported from the region. After a depression in the last two decades of the 19th century, the economic pace began to pick up again, and the population which had actually decreased (through emigration) began to grow again. Sheep's role as the characteristic livestock species began to diminish, as their wool, meat and manure became less important. Industries were importing more and more of their raw materials and new industries were established that operated for the most part without renewable resources, except for peat, space and water. Peat soon became a nonrenewable resource, as it was being mined in large-scale operations.

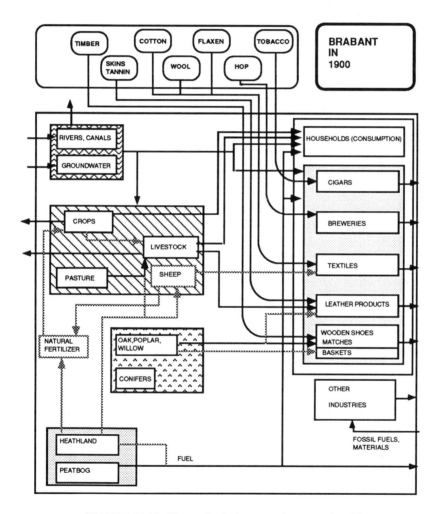

FIGURE 18.10 The ecological-economic system in 1900

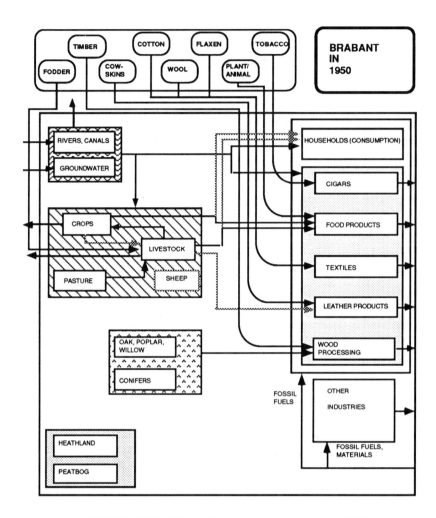

FIGURE 18.11   The ecological-economic system in 1950

The ecological-economic system in 1950 (figure 18.11)
Another 50-year jump through time brings us to 1950. The population has grown to
almost 1.25 million. The natural wild lands have been reduced to 10% of the area.
Forests, mostly tree plantations, cover about 12%. Both agriculture and industry are increasingly more involved in an import-export network. The food products industry is still
largely based on regional agriculture while the other renewable resource based industries
of 1900 have almost disappeared or shifted to imported raw materials. As holdings of cattle, pigs and poultry rapidly increase the livestock sector increasingly turns to imported
fodder. Households acquire food, goods and services almost completely through the

market. Pine trees, which were originally planted to provide poles for nearby coal mines, became available for producing paper and other wood products when the mines started to use steel. With expanding supplies of coal, and decreasing peat stock, mining of peat had almost completely stopped in the 1930s. A few patches of living fens and extensive tracts of drained and mined peatland, ready for agriculture, remained.

The ecological-economic system in 1978 (figure 18.12)
About a quarter of a century later, the face of the region had changed again quite considerably. At the end of the 1970s more than 2 million people live and work in the

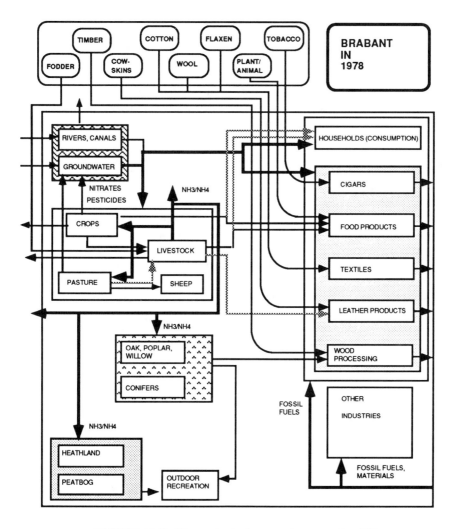

FIGURE 18.12 The ecological-economic system in 1978

urbanized area. 20% of the land has been urbanized, up from 10% in 1950. Most of the newly urbanized land was converted from agricultural. The decline of natural systems has ceased. The eastern part of the province has become a feedlot-agriculture region with many pigs, cattle and poultry. Fodder is imported and feed-corn is grown extensively, since it is tolerant of the excessive amount of manure that is dumped on the land.

The traditional industries have almost completely turned to imports, but together still only hold 31% of the market. New industries have come, but the economy is already changing towards a postindustrial economy. The renewable natural resources still in demand are clean groundwater, clean air, vital commercial forest, diverse natural communities, and land for agriculture, urban development, and as buffer for the natural communities. What is left of the resource systems, such as groundwater and forests is extensively used by the community, agriculture and industry.

Relatively new to the region is the pollution of air, water, soil and groundwater which in the 1980s becomes a widespread phenomenon. Feedlot farming practices produce large emissions of ammonium. Nearby forests and fens have suffered various forms of ammonium related damage. Most industrial activity in the region is relatively clean, but the air quality is affected quite intensively by the industrial emissions (SO_2, NO_x) from other regions. Total potential acid deposition in the eastern part of Brabant reaches 9000 H+-equivalents per hectare per year.

The high loads of nitrogen cause changes in the heathland by giving pioneer species a competitive advantage. Manure is spread over both pasture and cropland, leading to relatively high levels of soil nitrate, much of which leaches to the groundwater aquifer. Nitrate concentrations in several wells have recently exceeded World Health Organization standards. Considering the timelag involved, the concentrations are certain to increase over the next 10 to 25 years, even if infiltration is stopped now.

Besides the quality problem, the region is also facing a groundwater quantity problem. Most farmland is drained each spring, lowering the water table to allow machines to work the land and cattle to enter pastures early. During summers, shortfalls in soil moisture are circumvented by irrigation sprinklers pumping water from shallow groundwater reserves when recharge of the aquifer is limited because of the spring drainage. The last few wetlands in the region are threatened by dropping water tables.

SUMMARY AND CONCLUSIONS

The indicators document a sustained growth of the volume of the region's economy, at least since the beginning of the 20th century. The single important *nonrenewable* resource, peat, was mined in small quantities until the second half of the 19th century. Until then, it could be considered a renewable resource in view of the extent of the still-living peat moss fields. In a period of a few decades, this fuel source was almost completely mined. To a great extent the expansion of the industry in the region was due to the availability of this fuel source. Much of the peat, however, was exported out of the region.

The changing role of the *renewable resource* base in the region's economic development can be seen in the change in composition of the two dominant economic sectors in 1900: agriculture and industry. The population growth and economic expansion involved a major change in land use, resulting in the nearly complete disappearance of wild lands and natural forests.

Industry composition changed as resource importation expanded and as dairy and meat *export* decreased. New industries such as electronics were almost completely import-export based, i.e., they imported almost all resources and exported most products. The regional system provided labor, space and environmental services.

With increasing population and economic expansion, *environmental regulation* was slowly introduced, including zoning, pollution regulation and nature conservation. The pace of regulation, however, did not keep pace with economic development, allowing the degradation of air and water quality and forest vitality, and allowing groundwater depletion.

Succession of system components is seen in the shifts in land use, in the relative importance and composition of the economic sectors, and in the composition of the livestock herds.

The major environmental risks involved in economic development in North Brabant turned out to be: 1) nitrate and pesticide pollution of groundwater aquifers, 2) decrease of reserves of groundwater, and 3) loss of forest and wild lands as productive and recreational subsystems. An easy solution is not available in any of these cases. Another environmental risk, air pollution, was caused by farming practices and compounded by a reticence to change those practices. The problem was intensified by significant cross-boundary pollution that the region was powerless to control.

As to economic risks, the region's multinational electronics industry has had its ups and downs, due to competition. The agricultural output is presently being curtailed both because of limitations in European Community subsidies and expanding environmental regulations. In the longterm, it is expected that some of the agricultural land will be taken out of production. The growth of the government portion of the miscellaneous sector has basically stopped and the brief surge of the computer industry is almost over as well. At this moment, it is not quite clear what the next successional stage in the economic development will be.

The North Brabant region has clearly sustained its economic development during the period considered, expanding both the volume and the diversity of its economy, and increasing per capita welfare. To achieve this it has used and converted its resource systems and shifted to energy-intensive import-export based economic activities.

This positive score on the development side, has brought along some negative marks as well, which may be viewed as the cost of development. Renewable resource systems have degraded and environmental quality has dropped considerably by all standards. These changes have only just started to show economic and social impacts, and several of the economic sectors in the region are facing hard times.

The development perspective is, again, quite uncertain. In the past two centuries, the socioeconomic system has responded by changing its resource base, and replacing old sectors by new sectors, while moving towards an import-export based economy. It will be interesting to see what the next successional shift will be.

The Sustainability of Regional Development

Historical Strategies of Regional Sustainable Development
The case history of the North Brabant region has shown us that economic development of an open system may be sustained by a combination of strategies:
1. shifting from local to external resources
2. using renewable fuel resources to the extent that they become depleted (peat) and then shift to alternative external fuel sources
3. provide space and (ground)water for "resource import-product export" industrial activities
4. shift from one use of a renewable resource to another, with or without energy subsidies (e.g., heathland -> forest for poles -> forest for pulp and outdoor recreation)
5. shift from extensive-grazing to intensive-feedlot livestock farming

It appears that by moving from a predominantly autarkic(semiclosed) economy to an open economy, an increasing number of people can be supported while the average material welfare grows considerably. But, at the same time, regions become dependent on national and international economic trends, and the by-products of economic production can not be exported at the same rate as the main products. It appears that import-export based regional economies tend to become industrial production and waste disposal sites.

The listed strategies have one by one eliminated the development constraints. Ecosystems and soils with low economic productivity rates are converted and fertilized. Uncontrolled grazing productivity is replaced by controlled feedlot productivity. Wet seasons are countered with extra drainage, dry seasons with irrigation and sprinkling.

Environmentalists may point out that development processes destroy most of the natural ecosystems and pollute what is left. This may be so, and recent studies indicate that an increasing number of people are concerned with these costs of sustained economic development. At the same time, however, it seems that the human population is already adapting by shifting to alternatives for rest and recreation in nature, such as indoor-holiday centers and vacations abroad.

Future Development Options
If a region intends to develop sustainably, it must deal with the evident risks. Ideas for sustainable strategies of development might be taken from the past. The strategies listed above include several applicable ideas. For example, the structure of the agricultural sector has changed several times completely in a period of 150 years. Other structural changes, feared by most farmers, do not necessarily stop economic development. The conservative political stance of the present generation of farmers, which they consider protects their present and short term future income, is clearly not a basis for an adaptive strategy. A change towards an environmentally more acceptable agriculture may not be a step back, but turn out to be a deed of economic progress.

The groundwater issue must be addressed as soon as possible. With appropriate technology, nitrate can be removed from pumped water. In the long run, the source of the problem must be addressed. Infiltration of manure must be stopped. Attempts to convert the manure into dry fertilizer, which can easily be sold and transported, are increasingly successful. The use of groundwater for all purposes is reconsidered presently as well. A parallel supply of lower quality water for noncritical domestic and industrial purposes is being proposed.

International agreements on pollution reduction also are progressing, providing a small light at the end of the forestry tunnel. It may however be too late for productive commercial forestry. Another shift in resource use is, however, not without historic precedent. Introduction, natural invasion, settlement or evolution of species which tolerate acid soils may lead to forests which satisfy other demands, such as outdoor recreation and nature conservation.

A strategy with much promise for sustaining regional development appears to be a rational multiple use of renewable resources. In densely populated regions, or regions with low resources endowments, resources are often used simultaneously for multiple purposes. Many renewable resource systems can perform various functions, both through the production process—for example, timber from forests for paper—and through direct satisfaction of human needs as in outdoor recreation. In such situations, compatible combinations of resource use may contribute to overall sustainable development planning.

REFERENCES

Braat, L. C. 1991. Systems Ecology and Sustainable Development. In C. A. S. Hall, ed., *Maximum Power* . In press.

Braat, L. C. and W. F. J. van Lierop, eds. 1987. *Economic-Ecological Modeling*. Amsterdam: North Holland.

Brown, B. J., M. E. Hanson, D. M. Liverman and R. W. Merideth. 1987. Global Sustainability: Towards a Definition. *Environmental Management* 11:713-719.

Brown, L. R., W. U. Chandler, C. Flavin, J. Jacobson, C. Pollock, S. Postel, L. Starke, E. C. Wolf. 1987. *State of the World 1987*. New York: Norton.

Clark, W. C. and R. E. Munn 1986. *Sustainable Development of the Biosphere*. Cambridge: Cambridge University Press.

Costanza, R. and H. E. Daly. 1987. Towards an Ecological Economics. *Ecological Modelling* 38(1/2):1-8.

Daly, H. E. 1973. *Steady State Economics: The Economics to Biophysical Equilibrium and Moral Growth*. San Francisco: Freeman.

Forrester, J. W. 1971. *World Dynamics*. Cambridge, Mass.: Wright-Allen.

Goldsmith, E. 1972. *Blueprint for Survival*. Boston: Houghton-Miflin.

Goodland, R. J. A. and G. Ledec. 1987. Neo-Classical Economics and Principles of Sustainable Development. *Ecological Modelling* 38(1/2): 19-46.

Ikeda. S. 1987. Economic-Ecological Models for Regional Systems Policy. In L. C. Braat and W. F. J. van Lierop 1987:195-202.

Isard, W. 1972. *Ecologic-Economic Analysis for Regional Development*. New York: Free Press.

Meadows , D. H., D. L. Meadows, J. Randers and W. W. Behrens. 1972. *Limits to Growth*. New York: Universe Books.

Odum, H. T. 1983. *Systems Ecology: An Introduction*. New York: Wiley.

Odum, H. T. 1987. Models for National, International, and Global Systems Policy. In L. C. Braat and W. F. J. van Lierop 1987:203-256.

Opschoor, J. B. 1987. *Sustainability and Change*. Amsterdam: Free Univ. Press. (In Dutch)

Opschoor, J. B. 1989. *Indicators of Sustainability in The Netherlands*. IES-working report. Amsterdam: Free University Press. (In Dutch)

Pezzey, J. 1989. Economic Analysis of Sustainable Growth and Sustainable Development. Environment Department working paper no. 15. Washington, D.C.: The World Bank.

RIVM. 1989. Concern for Tomorrow: A National Environmental Survey. Bilthoven: RIVM.

Steetskamp, I. 1989. Economic Activities and Renewable Resources in the Region "Central and East Brabant" since 1850. M.Sc. Thesis. Institute for Environmental Studies, Free University Amsterdam. (In Dutch)

Turner, R. K. 1988. *Sustainable Environmental Management*. London: Belhaven.

World Commission on Environment and Development. 1987. *Our Common Future*. Oxford: Oxford University Press.

World Conservation Strategy. 1980. *World Conservation Strategy*. Gland, Switzerland.

Zucchetto, J. 1975. Energy-Economic Theory and Mathematical Models for Combining the Systems of Man and Nature, Case Study: The Urban Region of Miami, Florida. *Ecological Modelling* 1: 241-268.

Zucchetto, J. and A. M. Jansson. 1985. Resources and Society: A Systems Ecology Study of the Island of Gotland, Sweden. *Ecological Studies* 56. New York: Springer-Verlag.

19

NATURAL RESOURCE SCARCITY AND ECONOMIC GROWTH REVISITED: ECONOMIC AND BIOPHYSICAL PERSPECTIVES

Cutler J. Cleveland
Department of Geography
and
Center for Energy and Environmental Studies
Boston University
675 Commonwealth Avenue
Boston, MA 02215 USA

ABSTRACT

The neoclassical model of production assumes that capital and labor are primary inputs to production. Consistent with this assumption, the neoclassical model of natural resource scarcity assumes that real resource prices or capital-labor extraction costs are the appropriate empirical indicators of scarcity. In their seminal work, Barnett and Morse found that capital-labor costs per unit of extractive output declined throughout most of this century, a trend they attributed to "self-generating" technological change. A biophysical model of the economic process assumes that capital and labor are intermediate inputs produced ultimately from the only primary factor of production: low entropy energy and matter. A biophysical model of scarcity posits that direct and indirect energy costs of resource extraction increase with depletion because lower quality deposits require more energy to locate, upgrade, and otherwise transform into useful raw materials. I repeated Barnett and Morse's analysis from a biophysical perspective using energy costs to measure changes in the quality of extractive output in U.S. agriculture, forestry, fishing, and mining industries. In most cases, energy costs per unit of output increases with depletion. Results show that labor and capital costs declined because large quantities of surplus fossil fuel substituted for and increased the productivity of labor and capital. Substantial increases in energy costs were found in agriculture, fisheries and the mining of metals and fossil fuels. Energy costs per unit output in forest products declined. I discuss possible trends for future costs in light of the result that the energy cost of fossil fuels is also rising.

INTRODUCTION

Do the limits to our economic aspirations reside in humans and their institutions or in the natural environment? The adequacy of natural resources to meet human needs is an issue deeply rooted in Western intellectual tradition. Some argue that biophysical constraints, manifest as rising costs of natural resource extraction and degradation of key environmental services, will slow and eventually halt economic growth (Meadows et al. 1972) Others argue that human ingenuity, stimulated by isolated and short-run scarcities and channeled through technological change, will overcome any long-run economic constraints imposed by nature (Barnett and Morse 1963; Simon 1981). The energy and environmental events of the past two decades rekindled this debate and elevated it to new levels. The U.S. dependence on imported oil reared its ugly head again following the Iraqi "annexation" of Kuwait, and produced calls for drilling in the Arctic National Wildlife Refuge and a reversal of the recent presidential moratorium on offshore oil drilling. The *Exxon Valdez* oil spill, the continued decline in U.S. oil production, the debate over the national security implications of rising oil imports, the degree to which our National Forests should be used for timber production versus wilderness, aesthetic, or other purposes, the continued degradation of our best agricultural land via soil erosion, and whether "cold" fusion will make all our resource concerns irrelevant are additional examples of ongoing resource issues which produce little consensus.

The relation between resource scarcity and economic growth has most often been analyzed in the context of developed nations with industrial economies dependent on the depletion of nonrenewable stocks of fuels and minerals and on ever increasing control of renewable energy and material flows. Increasingly, however, the issue of scarcity and growth is central to the economic and environmental planning of less developed nations. The lesson of powering economic expansion by controlling the energy and material flows in nature has not been lost on the developing nations. China is now the world's leading coal producer. Throughout the developing world, agricultural "development" means using more fossil fuel derived fertilizers and pesticides to boost food output. Tropical rainforests, centers of the planet's biodiversity, are being harvested at alarming rates. Interwoven with these resource issues are complex social, political, institutional, and strategic considerations. A critical biophysical issue nevertheless remains: how is economic performance affected by changes in the quality and quantity of a society's resource base? Indeed, much of the debate about "sustainable development" focuses on integrating economic growth with the need to insure the long term provision of basic natural resources and environmental services (Repetto 1985; World Commission on Environment and Development 1987; Turner 1988).

These and other events prompted an ongoing dialogue between social scientists (primarily economists), who generally argue for the efficacy of human institutions in mitigating scarcity, and biophysical analysts who generally argue that basic physical and ecological laws constrain our economic choices in ways that are not accurately reflected in existing economic models (Cleveland et al. 1984; Hall et al. 1986; Daly and Cobb 1990; Ayres and Nair 1984). This dialogue has produced some important modifications to the models of both sides. Some economists acknowledge the discipline's historic lack of a sophisticated treatment of natural resources and have modified their models to account, for example, for the economic effects of environmental degradation (e.g., Pearce 1988). Likewise, some biophysical scientists have learned that the market mechanism can stimulate potent antidotes to resource scarcity and that human tastes and preferences are too adaptable to tie our future to a single resource. The interchange of ideas and analytical tools has led to the genesis of "ecological economics," a formal discipline dedicated to the interdisciplinary analysis of the nature-society relationship (Costanza 1989).

Despite the interdisciplinary advances, the gulf between economists and biophysical analysts remains large. The root cause of this void is different a priori assumptions about the driving forces behind the economic process. Such assumptions cannot be proven "right" or "wrong." Rather, they reflect different views of the relationship between human society and its natural environment, and therefore the different weights given to human institutions or biophysical constraints to economic growth. Those differences are in turn reflected in different theoretical models: the empirical models of resource scarcity derived from the theoretical models, and the interpretation of the historical record generated by the empirical models.

If a true synthesis of economic and biophysical perspectives of scarcity and growth is possible, it must begin with an explicit statement and understanding of their underlying assumptions. These assumptions are rarely discussed in public debates. Instead, much of the resource scarcity debate has degenerated into finger pointing and criticisms of various aspects of empirical models, such as data limitations and statistical techniques. Each group generates models, results, and conclusions based on the historical record that are heralded by adherents and ridiculed by opponents, regardless of the merits of the modeling process itself. If the results confirm one's a priori expectations about the scarcity or abundance of natural resources, then proponents argue that the analysis is accurate and unbiased, the methodology is rigorous and scholarly, and the analyst's integrity and qualifications are beyond reproach. If the results contradict one's a priori expectations, then opponents argue that the data used to make the estimate are biased and inappropriate, the methodology is fraught with bias, and the analyst obviously has an "axe to grind." Unfortunately, much of this attitude continues, producing a debate which has done little to further our understanding the critical underlying issue: what is the nature of resource scarcity and how do changes in scarcity affect our material well-being?

The purpose of this analysis is to develop a biophysical model of natural resource scarcity and apply it in an empirical analysis of scarcity trends in the U.S. mining, forestry, fishing, and agriculture sectors. The theoretical model and empirical results are compared to their counterparts in neoclassical economics. My purpose is not to replace or repudiate the standard economic perspective of resource scarcity. Rather, I argue that the economic model of scarcity is based on a set of plausible but arbitrary assumptions about the driving forces of the economic process which result in logical but arbitrary specification of empirical tests and interpretation of the historical record. The neoclassical model embodies important corrections of earlier perspectives that put too little weight on the ability of the market to stimulate substitution, recycling, technological change, and other antidotes to scarcity. Likewise, the biophysical model shows that the neoclassical model of resource scarcity ignores an essential feature of the extractive sectors: massive quantities of natural resources are poured into the resource-harvesting process, substituting in an unmeasured way for reductions in capital and labor use.

THE NEOCLASSICAL MODEL OF NATURAL RESOURCE SCARCITY

The conventional wisdom about the relationship between natural resources and economic growth during the nineteenth century earned economics the title as the "dismal science." The classical economists Malthus and Ricardo posited models in which the rate of population growth eventually surpassed the rate of growth of output from agricultural land, the primary natural resource input at that time. Ricardo and Malthus argued that the primary constraint to economic expansion resided in nature itself. In Malthus' model the constraint was a fixed supply of arable land, and in Ricardo's, it was declining land quality as production expanded. According to Ricardo and Malthus, society's path was inexorably fixed by those resource constraints. Natural resources became increasingly scarce with time; that scarcity opposed and ultimately halted economic growth.

The neoclassical model of natural resources rejects the assumptions underlying the classical school's dim view of society's economic future. The twin pillars of the neoclassical economic model of resource scarcity are Hotelling's (1931) theory of optimal depletion and Barnett and Morse's (1963) empirical analysis of resource scarcity in the United States. According to the neoclassical model, the relationship between natural resources and economic growth was not dictated by the iron law of diminishing returns as Malthus and Ricardo had envisioned. The relationship is more malleable, the exact trend in scarcity being determined by the relative strengths of cost-increasing resource depletion and cost-decreasing technological change. From this perspective, the iron law of diminishing returns becomes a testable economic hypothesis about the relationship between those two driving forces. Barnett and Morse tested for increasing scarcity in the United States from 1870 to 1957 by analyzing labor and capital costs and real price trends in the agriculture, mining, forestry and fishing sectors. They found no evidence for increasing scarcity, except for forest resources. Indeed, costs declined for most resources over that period despite massive physical depletion of the highest grade resources in most sectors.

According to the neoclassical model, the means of escape from increasing scarcity lies in the market mechanism. The reasoning goes as follows. The price of a resource growing scarce will eventually rise due to increases in its extraction cost and/or rental payment to resource owners. Price increases stimulate a host of resource-augmenting mechanisms: increased exploration for new deposits, recycling, substitution of alternative resources, increased efficiency of converting resources into goods and services, and, most importantly, technical innovation in resource exploration, extraction, processing, and transformation into goods and services. In the neoclassical model, long-run resource scarcity impinging on economic growth is a near impossibility because rising scarcity is assumed to automatically sow the very seeds for its amelioration. Numerous historical examples are cited which substantiate the observation that many resource shortages in the past eventually proved to be interesting but hardly monumental economic events (Rosenberg 1982). The ultimate driving force is sociotechnical change which Barnett and Morse describe as "self-generating," "automatic" and "self-reproductive" in modern economies.

The powerful arguments and analyses of Hotelling and Barnett and Morse have stood the test of time within economic circles. As Norgaard (1989) observed, there have been no major theoretical advances in the half-century since Hotelling, with improvements largely taking the form of interesting but secondary caveats to Hotelling's original assumptions. Empirical tests of the Hotelling model are scant, and those available give mixed results about the validity of the model's assumptions (Eagan 1987). Empirical analysis of resource trends since Barnett and Morse has focused on applying more sophisticated econometric tests to updated cost and price time series (Slade 1982; Hall and Hall 1984). Most important, the assumption that technological change will overcome any constraints imposed by nature is still firmly entrenched in neoclassical economic theory. (Houthhakker 1983; Baumol 1986; Bower 1987). Prominent economists (Baumol 1986) cling tenaciously to the optimistic conclusions of Barnett and Morse, seemingly unaware of alternative explanations of the same data trends (Petersen and Maxwell 1979), or that more recent econometric studies suggest that the tide may have already begun to turn (Slade 1982; Hall and Hall 1984).

A BIOPHYSICAL CRITIQUE OF THE NEOCLASSICAL MODEL

The identification and interpretation of differences between the neoclassical and biophysical models of natural resource scarcity begins with a comparison of their underlying assumptions about the role of natural resources in the economic process. In this section, I discuss differences in two critical areas: the distinction between primary factors and intermediate inputs to production, and the importance of energy surplus.

Primary Factors Versus Intermediate Inputs

The neoclassical model of scarcity begins with a plausible but arbitrary assumption: labor, capital, and land (and sometimes energy) are primary, independent factors of production. The empirical tests derived from that model and the interpretation of their results are consistent with the standard view of the economic process in which inputs to production (land, labor, and capital) and outputs (goods and services) cycle endlessly between households and firms. Interpretation of the results of those empirical tests are consistent with the assumption that human ingenuity and institutions, particularly "self-generating" technological change, are the primary means by which humans have avoided the potentially deleterious economic impacts of rising scarcity.

A biophysical perspective begins with a different but equally plausible set of assumptions grounded in the first principles of ecology and thermodynamics. The first is a distinction between "primary factors" of production and "intermediate inputs." A primary factor of production is one which is necessary for production but which cannot be physically produced inside the economic system. It is independent of internal influences. Intermediate inputs are also necessary for production but they are produced or recycled by some combination of primary factors and other intermediate inputs. Based on this definition, low entropy energy-matter is the only primary factor of production because, as the first law of thermodynamics instructs us, that is the only factor of production that cannot be produced within the economic system. The services of labor and capital and certain properties of land (i.e., soil fertility) are produced from low entropy energy-matter. Technological change is not a self-generating phenomena, but instead is an internal, interdependent, and intermediate input requiring energy directly and indirectly for its production. Knowledge is not free (it's not even cheap) and the manifestation of human knowledge in technology requires energy. Viewing knowledge as the ultimate scarce resource is nothing more than a Maxwell's demon argument. A biophysical perspective of the economic process does not ignore inputs other than energy as some critics have erroneously charged (Allesio 1981; Huettner 1981). It merely considers capital, labor, and technology to be interdependent, internal, and intermediate, not independent, external, and primary (Costanza 1980; Cleveland et al. 1984; Hall et al. 1986).

The role of "land" as an input to production is frequently ignored in neoclassical models, relegated to explaining the minor part of production not explained by capital and labor inputs (Daly and Cobb 1990). More importantly, even when land's role is considered, it does not stem from the "original and indestructible powers of the soil" that Ricardo wrote about or the other energy, natural resources, and environmental services that only land can provide. Consistent with the anthropocentric bias of the neoclassical model, land is viewed as property, not a physical input. The study of land as an input to production in neoclassical economics has evolved into the study of property relations. Reduced to property, the abstract notion of land is no different than capital and labor which are the respective property of capitalists and landlords. As Daly and Cobb (1990) observed, when land became solely a property relation in the neoclassical model, distinct from other property relations in only minor ways, then it is merely one commodity among others. The "forces of nature" the classical economists were originally concerned with have disappeared from view in the neoclassical model.

The Importance of Energy Surplus

A biophysical perspective of scarcity highlights a glaring omission of the neoclassical model: the neoclassical model ignores the massive quantities of energy used in the resource harvesting process itself. Barnett and Morse's capital-labor index explicitly excludes energy and all other "produced" inputs because they are assumed to a be intermediate inputs. The resource transformation process is a work process in which natural resources are discovered, extracted, refined, and otherwise transformed into useful raw materials and eventually into goods and services. Each stage of that transformation process requires the use of energy to increase the organizational state, and hence economic usefulness, of natural resources. It follows, therefore, that natural resource quality can be defined in physical terms: the quantity of energy used to upgrade a unit of resource to a socially useful state. An inexorable relationship exists between energy costs and resource quality: lower quality resources have a lower degree of organization (higher entropy) and require more energy to upgrade to a given level of organization.

The remarkable decline in labor costs of extracted resources documented by Barnett and Morse was made possible by a simple physical substitution. Large quantities of surplus fossil fuel energy were substituted for labor in the resource transformation process. The increase in the quantity and quality of energy used per laborer increased labor productivity. Indeed, the principal driving force behind the transition to industrial society was the switch from relatively low-quality, animate sources of energy such as biomass, human labor, and draft animals to high-quality, inanimate energy sources such as fossil and nuclear fuels. The economic advantage gained by this substitution is rooted in the simple thermodynamics of energy conversion. Humans convert the chemical energy in food to mechanical work with about 20% efficiency. A laborer consuming 3000 kilocalories per day can thus perform about 600 kilocalories of useful mechanical work per day. An internal combustion engine burning gasoline at 10% efficiency can perform an order of magnitude greater amount of useful work. A laborer subsidized by an energy convertor burning fossil fuels can obviously mine more coal, cut more timber, catch more fish, and harvest more fish per day compared to earlier unsubsidized efforts of labor. Thus, one does not have to appeal to the nebulous and empirically unsupported concept of "self-generating" technological change to describe the decline in labor costs of natural resources. The answer lies in the development of technologies which empower the efforts of labor with increasing quantities of fossil fuels.

The unprecedented ability of fossil fuels to subsidize labor and the extraction of other resources derives from the large net energy surplus those fuels deliver to the extractive sector and society at large. Like any work process, the extraction of oil and coal requires energy directly in the form of fuels and electricity and indirectly in the fuel embodied in labor, capital and other materials. The quantity of energy available to non-energy sectors of the economy after the energy costs of energy extraction and processing have been paid is net or surplus energy. The absolute size of the surplus energy and the efficiency with which it is used sets broad but distinct limits on the quantity of work done in the economy and thereby on the quantity of goods and services produced (Cottrell 1955; Cleveland et al. 1984; Hall et al. 1986).

Fossil fuels revolutionized the economic process because the size of the surplus they deliver dwarfs the surplus of earlier animate energy sources. When large fossil fuel surpluses became available, societies with access to those fuels experienced unprecedented rates of economic growth and diversification. In the extractive sectors, large surpluses of cheap, high-quality fossil fuels quickly replaced the efforts of human labor in the mines, forests, and agricultural fields. Subsidized with diesel fuel to power tractors and pump groundwater for irrigation, refined petroleum to produce pesticides, herbicides, and insecticides, and energy embodied in tractors, combines and genetic research, modern farmers can feed substantially more people than pre-industrial farmers. Labor displaced from agriculture and other extractive tasks was free to move to other sectors, promoting social and economic diversification and growth. A positive feedback was established whereby surplus energy was invested in machines, tools, and other energy convertors, some of which were used to capture an even larger energy surplus, which was then used to produce more machines, and so on (Christensen 1989).

The period covered by Barnett and Morse's empirical analysis of real resource prices and labor-capital costs coincides with the rapid transition of the economy driven by low-quality wood to one powered by high-quality fossil fuels. In 1870, the beginning of their analysis, wood accounted for about 20% of total U.S. fuel use. By 1910, coal had almost completely replaced wood. Barnett and Morse's analysis ends in 1957, by which time oil and natural gas had largely replaced coal as the dominant fuel source. The period spanned by Barnett and Morse's analysis was therefore dominated by two complete cycles of substituting high-quality fuels for low-quality fuels. It is not surprising, therefore, that the price and capital-labor cost of other resources declined. Larger energy surpluses of ever higher quality fuels reduced the labor-capital cost of extracting all other resources.

RESOURCE SCARCITY FROM A BIOPHYSICAL PERSPECTIVE

The biophysical model of natural resource scarcity derives from two cornerstones of the biophysical model of the production process. First, high-quality resource deposits require less work to locate, upgrade, and refine than their lower-quality counterparts. In each stage of the resource transformation process, some high-quality economically useful energy is degraded into low-quality, economically useless energy. That energy conversion process achieves useful transformations of natural resources. Crude oil is discovered, extracted and refined into gasoline. Copper is discovered, extracted, concentrated and smelted into pure copper metal. Timber is felled, sawed, and ultimately configured into lumber. Green plants concentrate highly dispersed molecules of carbon, hydrogen, nitrogen and other elements into carbohydrates and proteins. In thermodynamic terms, all such transformations involve a major change in the state of the system, which means a change in the internal energy of the system. For a given material and given amount of increase in order (or change in state of system), we can calculate the minimum amount of energy required to effect that change for each and every process. That energy can be supplied by a human laborer, a draft animal or an internal combustion engine, but the laws of thermodynamics tell us that there is absolutely no substitute for that minimum energy use.

Lower-quality resources are more random and dispersed and therefore require a larger change in state in order to upgrade them to a socially useful state. It follows, therefore, that lower-quality resources require more energy per unit to upgrade them to that socially useful state. Copper molecules that are more dispersed in their sulfide and silicate ores require more energy to transform them to 99.9% pure copper metal. Crude oil in reservoirs that lie at greater depths require more energy to lift to the surface. Carbohydrate production by plants requires more energy when grown in soil with more dispersed nitrogen, phosphorus and water molecules. Technological change cannot alter the thermodynamic minimum energy cost increases imposed by lower quality resources. Of course, minimum energy requirements can only be reached under theoretically ideal (i.e., non-real world) conditions. In theory, therefore, technological change can reduce actual energy costs toward the theoretical minimum even as a resource is depleted, although improvements become increasingly difficult to win.

The second important aspect of natural resource scarcity from a biophysical perspective is the physical basis of technological change in industrial economies. The technology which supplies the energy required to upgrade a resource can do so in many different combinations of energy input. Historically the most important sources of energy in the extractive sector have been the mechanical energy of humans, draft animals, and inanimate energy converters powered by fossil fuels and electricity. During the past century, fossil fuels have become dominant, replacing almost entirely the efforts of human and draft animals. Thus, it is essential to account for the direct and indirect energy used to deliver resources to society.

Empirical Measures Of Scarcity From A Biophysical Perspective

For obvious reasons, humans use natural resources in order of decreasing quality, given existing technological, legal, social and geographic constraints. Sometimes those constants are binding to the degree that the best-to-worst development pattern is not followed (e.g., Cleveland and Kaufmann 1990; Norgaard 1990). Over the long run, however, the pattern holds for most resources. The relation between energy costs and resource quality is formalized by the biophysical resource conversion function, a modification of the original concept of Barnett and Morse (1963). Let R represent "standard resources" that is, resources with homogeneous physical and locational attributes, measured in physical units (tons, acres, barrels). Let R_u be resources with heterogeneous, lower quality physical and locational attributes. The biophysical resource conversion function shows the amount of direct and indirect energy required to upgrade a unit of lower quality R_u to the equivalent of a unit of standard R. For example, a barrel of oil in a reservoir at 2000 feet can be transformed to the equivalent of a barrel at 1000 feet by investing the energy required to lift that barrel an additional 1000 feet. Land of lower fertility can be transformed to the equivalent of "standard" land by using energy in the form of fertilizer, irrigation water, tractors, etc.

The biophysical resource conversion function is given by

$$R = f_1(R_u, E_d + I_m + I_l + I_k) \qquad (1)$$

where E_d and I are the direct and indirect fuel and electricity used to upgrade the lower quality R_u into the standard quality R. Both E_d and I are measured in physical units. The subscripts m, l, and k refer to the energy embodied in materials, labor, and capital, respectively. Letting $E_i = I_m + I_l + I_k$, equation (1) reduces to

$$R = f_1 (R_u, E_d + E_i) \qquad (2)$$

The resource conversion function illustrates the increase in $E_d + E_i$ required to upgrade lower quality units of R as a function of cumulative depletion of R.

The biophysical resource conversion function includes the direct and indirect fuel use in resource extraction and transformation. Fuel use is measured in thermal equivalents (e.g., BTUs). Direct fuel use is the fossil fuel and primary electricity used directly in resource exploration, extraction, and transformation. Indirect fuel use is the energy used elsewhere in the economy to produce and maintain the capital, labor, and other materials used in the transformation process.

Measuring Output and Energy Costs

The next section applies the biophysical model of resource quality in a time series analysis of energy costs in the extractive sectors in the U.S. economy. Such a test requires an index which relates the quantity of resource extracted to the direct and indirect energy used in extraction. The following definitions are used:

Q = total output of the respective industry, measured in physical units wherever possible. When dissimilar units of output are aggregated, physical units are weighted by the relative values or unit prices in a base period.

E_d = the quantity of direct fuel use in the respective extractive industry.

E_i = the quantity of indirect fuel use in the respective extractive industry.

The quantity Q/E_d+E_i is the direct plus indirect fuel used to produce a unit of a resource. The methods used to calculate these quantities is reported in Cleveland (1990). In brief, output of individual minerals is in physical units (tons of copper, barrels of oil) as reported by the Bureau of Mines and the Census of Mineral Industries. Output from mining sectors (e.g., metal mining) is the production index published by the Federal Reserve Board. Output from the forest products sector is also the Federal Reserve Board production index. Output in the case study of fisheries is measured in physical units (tons of fish or BTU content of catch). Output in agriculture is measured in three ways: The Department of Agriculture's crop and livestock production index, Gross Domestic Product originating in the farm sector, and total calorific content of food grown on farms. These measures are discussed in more detail below.

Direct fuel use in mining and forest products is from the Census of Mineral Industries and Census of Manufacturers, respectively. Direct fuel use in the case study of fisheries is from Mitchell and Cleveland (1990). Direct fuel use in agriculture is from Cleveland (1990) who estimates fuel use from Department of Agriculture data on prices and expenditures for fuel and electricity. Indirect fuel use is calculated by multiplying the dollar expenditures on capital, labor, and materials by the appropriate energy intensity factor (Hannon et al. 1985). Energy intensities reflect the direct and indirect energy used to produce a dollar's worth of good or service. The energy cost of capital is calculated by multiplying the dollar value of capital consumption in the industry times the energy intensity of producing that capital. The energy cost of labor is calculated by multiplying the wages paid to labor in the industry times the average energy intensity of labor in the entire economy. The energy intensity of labor is estimated by the quantity of fuel and electricity purchased by households divided by total personal income (Cleveland 1990).

THE ENERGY COST OF EXTRACTIVE OUTPUT

Do the energy costs of natural resources increase over time in the United States due to changes in the relative strength of depletion and technical innovation? I test this hypothesis by calculating time series of Q/E_d+E_i in the extractive sectors of the economy: mining, agriculture, forest products, and fisheries. The time period covered in each sector is determined by the period for which data on energy costs are available.

Mining

The trend in Q/E_d+E_i in the U.S. mining sector from 1919-1982 is shown in figures 19.1-19.3. The metal mining and fossil fuel sectors clearly show a rising and then falling trend in Q/E_d+E_i . In the metal mining sector as a whole, output per BTU increases from the early part of the century through the mid-1950s and then decreases by a factor of two by the 1980s (figure 19.1). Bauxite, silver and, iron ore show marked declines in output per BTU. Copper ore shows a similar but less dramatic decline between 1963 and 1982. The only exception to the general trend of diminishing returns is the lead and zinc industry, where energy costs decline steadily throughout the study period.

The nonmetal mining sector as a whole does not show diminishing returns (figure 19.2). Output per BTU increases between 1919 and the 1950s, and remains relatively constant through the 1980s. Individual minerals, however, exhibit disparate trends. Clay minerals show a distinct decline in Q/E_d+E_i , while Q/E_d+E_i in the crushed and broken stone industry increases significantly. Most nonmetals do not show rising energy costs.

Output per energy in the fossil fuel sector is measured by energy return on investment (EROI) (Cleveland et al. 1984; Hall et al. 1986). EROI is the ratio of fossil fuel energy produced to direct and indirect energy costs of that production. The EROI for total petroleum (oil, gas, and gas liquids) and coal production increases in the early parts of the century and then declines in the 1960s and 1970s (figure 19.3). The EROI for petroleum production (crude oil, natural gas, and natural gas liquids), peaks in the early 1970s corresponding to the peaks in oil and gas production, and then declines by a factor of two by the early 1980s. The EROI for coal production (bituminous and anthracite) shows a similar decline which begins a decade earlier.

Agriculture

The relationship between energy use and food production has a rich intellectual history (e.g., Pimentel and Pimentel 1979). However, time series analysis of energy use in U.S. agriculture has been hampered by a lack of historical data on energy use by farms. This

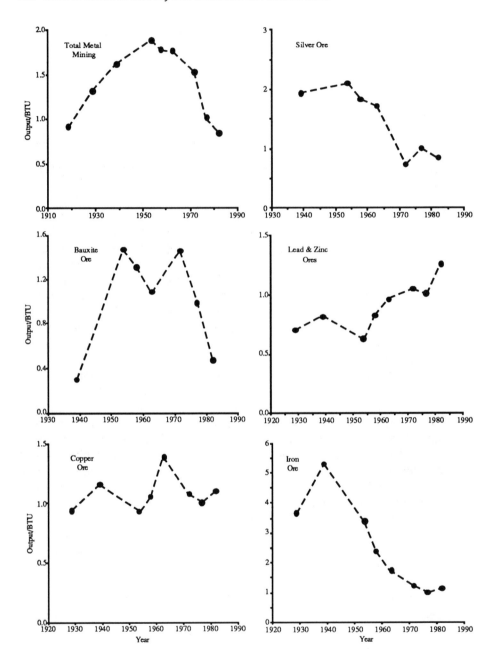

FIGURE 19.1 Output per energy input (Q/E_d+E_i) for the metal mining sector. Output for total metal mining is measured by the Federal Reserve Board production index. Output for individual metals is measured by tons of metal produced.

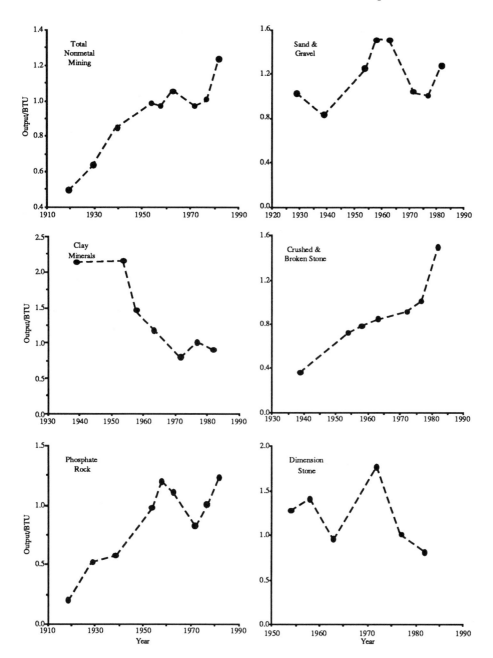

FIGURE 19.2 Output per energy input (Q/E_d+E_i) for the nonmetal mining sector. Output for total nonmetal mining is measured by the Federal Reserve Board production index. Output for individual nonmetals is measured by tons of mineral produced.

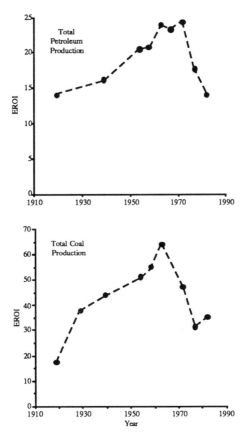

FIGURE 19.3 Output per energy input (Q/E_d+E_i) for the fossil fuel mining sector. Output for petroleum is measured by thermal equivalent of crude oil, natural gas and natural gas liquids produced. Output for coal is measured by thermal equivalent of bituminous and anthracite coal produced.

analysis presents new on-farm energy use data which is used to measure farm output and energy use in the United States between 1910 and 1988.

Measuring the output of the agricultural sector in "physical" terms is not straightforward because of the disparate qualities of food produced. Figure 19.4 shows three different measures of farm output: gross domestic product originating on farms, the USDA farm output index, and the total calories produced in principal crops. The output index has the desirable quality of including the physical quantity of crops and livestock produced. However, the physical quantities of output are weighted by their respective prices, and the price of agricultural commodities includes the cost of intermediate inputs (fuel, fertilizer, water, etc.) purchased by farmers. Thus, the output index can increase solely due to an increase in the price of inputs without physical output actually increasing. Gross farm product (GFP) avoids this problem because GFP is the difference between the value of farm output and the value of purchased intermediate inputs. In theory, at least, GFP should be a good proxy for the physical output of farms because it represents the value added in farming. The problem with this approach is that it ignores the critical role energy plays in expanding agricultural output because energy is relegated to the secondary role of a "produced input" in neoclassical income accounting. A third alternative is to measure output as the calorie equivalent of food produced. This method represents a primary attribute of agricultural output, namely the chemical energy embodied in food. However, measuring output in calories glosses over important quality differences between foods (e.g., protein content), and begs the question of how to account for the energy content of livestock raised on the energy content of grains.

Direct fuel use in agriculture is derived from USDA data on farmer's expenditures for fuel and electricity, the prices farmer's pay for those fuels, and individual fuel's shares of total direct fuel use. Cleveland (1990) developed an algorithm which calculates estimates of physical quantities of refined fuels and electricity used on farms based on the available expenditure, price and fuel share data. This data is available between 1910 and 1988 and is used to produce a much longer and consistent time series than is currently available (e.g., Gever et al. 1986). Indirect energy use in agriculture includes the energy cost of capital and labor used on farms, and also the energy embodied in manufactured inputs such as pesticides, herbicides, fertilizers, etc.

The trend in Q/E_d+E_i using the three output measures is shown in figure 19.5. All three indexes show a similar overall pattern: Q/E_d+E_i declines sharply between 1910 and 1973 and then increases between 1974 and 1988.

Forestry

The trend in Q/E_d+E_i in the forest products sector between 1950 and 1986 is shown in figure 19.6. Output is measured by the the Federal Reserve Board's production index for the logging and sawmill industries (SIC sectors 241 and 242). The trend in Q/E_d+E_i shows no overall trend. Output per energy use in the forest products sector generally declines between 1950 and 1973 and then increases 40% between 1974 and 1986.

Fisheries

Fisheries have garnered the least attention in empirical analysis of scarcity, due in part to the lack of historical data relative to other extractive sectors. National data on energy use in fisheries is not collected. However, Mitchell and Cleveland (1990) gathered detailed information on energy use and output in the New Bedford, Massachusetts fisheries between 1968 and 1988. Harvesting fish from the rich George's Bank region, the New Bedford fleet is the nation's largest in terms of value of catch and sixth largest in terms of poundage caught. Thus, output and energy use trends in the New Bedford fleet are probably representative of general national trends. This discussion is drawn primarily for the analysis of Mitchell and Cleveland (1990).

FIGURE 19.4 Alternative measures of output in the U.S. agricultural sector, 1910-1988. A) Thermal equivalent of the top 15 crops produced; B) Gross Farm Product (GFP) in billions of 1982 dollars; C) the Farm Output Index, 1977 = 100.

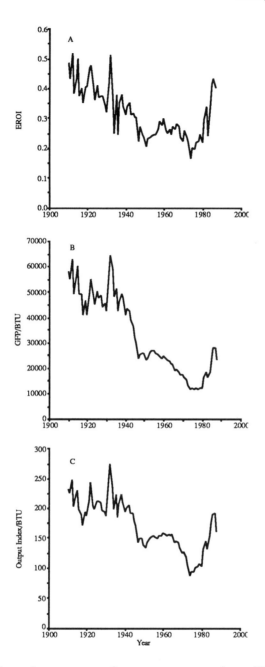

FIGURE 19.5 Alternative measures of output per energy input (Q/E_d+E_i) in the U.S. agricultural sector, 1910-1988. A) EROI is the ratio of the thermal equivalent of the top 15 crops produced to total energy use; B) Gross Farm Product per BTU of total energy use (1982 \$/BTU); C) the Farm Output Index per BTU of total energy use (1977 = 100).

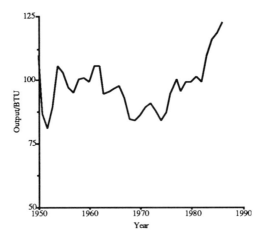

FIGURE 19.6 Output per energy input (Q/E_d+E_i) in the forest products industry (1977 = 100)

FIGURE 19.7 Output per energy input (Q/E_d+E_i) in the New Bedford fisheries.

Output in the New Bedford fisheries is measured by the energy content of edible fish protein (fish and shellfish) caught by the New Bedford fleet. Edible fish protein is assumed to be 50% of the round weight of fish. Energy use includes only E_d, the direct fuel use to power the fishing vessels, because information on indirect energy use is not available. Omission of indirect energy costs should not seriously bias the results because 80% to 90% of energy inputs to fisheries is the use of fuel to power vessels. The trend in Q/E_d+E_i between 1968 and 1988 suggests sharply increasing scarcity (figure 19.7). The industry now uses about 35 BTU of fuel to harvest 1 BTU of edible fish protein.

DISCUSSION OF RESULTS

Physical, economic, environmental and political factors act in concert to determine the quantity of resource extraction and its energy cost. The relative strength of these factors change over the life cycle of a resource. The net result of these force determines the energy cost of obtaining a unit of resource. A thorough explanation of the trends in the energy cost of natural resources presented in this analysis requires very detailed, resource-specific studies. I limit the discussion here to general observations and a few specific examples of well-studied individual mineral resources.

Mining

Resource depletion is clearly the major driving force in the metal mining industry. Declining ore grades and increased stripping ratios are driving up energy costs of metal mining and concentration. For example, the average grade of copper ore mined at the turn of the century was 2-3%, but by the 1980s had declined to about 0.5% (Hall et al. 1986). Resource depletion also increases average mine depth in new surface mines which in turn increases the amount of waste rock mined per ton of ore—known as the stripping ratio. The average stripping ratio of open pit copper mines increased from 1.9 to 3.1 between 1915 and the 1950s (Dale 1984) The stripping ratio for the metal mining sector increased from 1.23 to 1.86 between 1960 and the late 1970s (Gelb 1984). The stripping ratio for the nonmetal mining sector increased from 0.15 to 0.25 between 1960 and 1982, but improvements in recovery techniques apparently have more than offset that quality decline.

Depletion in the coal mining industry is a major force behind the decline in Q/E_d+E_i in the coal industry. The anthracite mining industry has extracted 40% of the original anthracite resource, or more than 60% of the reserve base, representing the most complete depletion history of any major U.S. mineral resource (Dale 1984). In the century spanning 1870 to the early 1980s, average anthracite seam thickness declined from 13.2 feet to 3 to 4 feet, and average mine depth more than doubled (Dale 1984). Seam thickness in underground bituminous does not show a commensurate decline because the resource base isn't as depleted, and because the shift to surface mining replaces to some degree thin seam underground mining. The rapid shift to surface mining after World War II produced a sharp decline in the quality of new surface mines. In major coal producing states such as Illinois, Kentucky and Ohio, the average overburden-to-seam-thickness ratio in surface mines more than doubled between 1946 and 1977 (Dale 1984). Reinforcing these cost-increasing effects is a 14% reduction in the average heat value of bituminous and lignite coal production between 1955 and 1982 (Hall et al. 1986).

The declining in Q/E_d+E_i in oil and gas production indicates clearly that depletion effects outweigh technical innovation in that industry. The effects of depletion on energy costs are evident in a variety of forms. Most important is the fact that the average size of new oil and gas fields have declined precipitously for almost half a century (Nehring 1981), resulting in a substantial increase in the quantity of well drilled (energy used) to add a barrel of oil to proved reserves (Hubbert 1967; Hall and Cleveland 1981; Cleveland

and Kaufmann 1990). Average well depth of new exploratory wells increased from 3200 feet in the 1930s to almost 6000 feet in the 1980s. Production from existing fields grows more energy-intensive as declining natural reservoir drive mechanisms are
replaced by human-directed energies. The number of producing oil wells using artificial lift methods increased 9% between 1950 and the 1980s (Cleveland 1988). Secondary recovery methods such as waterflooding and pressure maintenance through water and gas reinjection are increasingly used in older fields where natural drive energies are depleted. Energy intensive enhanced oil recovery (EOR) techniques now account for 10% of lower 48 U.S. oil production. For example, the "huff-n-puff" EOR method, used extensively in some California oil fields, burns one barrel of crude oil to generate steam which is used to squeeze three additional barrels from the ground. Neglecting all indirect energy costs, that is an EROI of 3 for that EOR method.

Agriculture

The E_d+E_i cost of producing food increased sharply between 1910 and the early 1970s (figure 19.5), suggesting that agricultural output grew "scarcer" in that period. That conclusion seems to fly in the face of the fact that the real price of food to consumers generally declined in this century, and the perception that technical innovation has made U.S. farmers the most productive in the world. Indeed, overproduction, rather than scarcity, has characterized U.S. agriculture in recent history. The trend in Q/E_d+E_i, however, is representative of trend characteristic of the entire extractive sector: technical innovation and improvements in labor productivity have been associated with vast increases in direct and indirect fossil fuel use.

Total energy use in U.S. agriculture increases more that 500% between 1910 and 1973 and then declines sharply through 1988 (figure 19.8). However, as recently as 1910, farm animals and human labor performed more than 90% of total useful work done on farms (Cleveland 1990). The discovery of cheap, high EROI oil and gas in the 1920s and 1930s stimulated the transition after from a coal, wood and animate energy sources to a petroleum based economy. Advances in petroleum refining and diesel engine technology during and after World War II, coupled with petroleum-intensive developments in biotechnology, accelerated the use of petroleum in agriculture. Animate energy sources performed less than 2% of work on farms by the 1960s and the energy costs of labor now represent only 1% of total on-farm energy use.

Indirect fossil fuel use dominates direct use in U.S. agriculture, although that relationship has undergone significant shifts. Direct fossil fuel and electricity use increases from 5% to 32% of total energy use between 1910 and World War II, reflecting the shift away from animate energy converters to inanimate converters powered by gasoline and diesel fuels. Between World War II and the first oil price shock, direct fuel use dropped to 17% of total energy use, reflecting the postwar shift to indirect energy use in the form of manufactured inputs (pesticides, fertilizers, etc.) and feed grain production dedicated to livestock production. There is a relative shift back towards direct fuel use between 1974 and 1988 due to a sharp drop in chemical use after the energy price shocks.

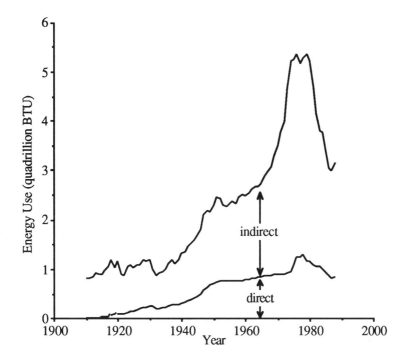

FIGURE 19.8 Total energy use in the U.S. agricultural sector, 1910-1988, in 10^{15} BTU.

All measures of Q/E_d+E_i decline precipitously between 1940 and 1950. That period is characterized by rapid increases in energy use which outpace advances in output. The relative shift from crops to energy intensive livestock production reinforced the decline in Q/E_d+E_i . The ratio of the USDA crop production index to the livestock production index declined 30% between 1910 and World War II. Output per BTU stabilizes between 1950 and 1970. Energy use slows in this period, while output continues to increase, in part because the shift from crops to livestock ceases and begins to swing in the opposite direction. Government conservation and stabilization programs may also contribute to the relatively stable Q/E_d+E_i in the 1950s and 1960s. The number of acres farmers were paid to set aside as a fraction of total harvestable cropland increases from 0 in 1954 to 20% by 1970. If, at the margin, farms withdraw from production the lowest quality land, then the average quality of harvested acres increased during that period. *Ceteris paribus*, such a trend is a positive influence on output per BTU.

The energy price shocks clearly had a major impact on the way U.S. farmers use energy on farms. After 60 years of decline, Q/E_d+E_i between 1974 and 1988 (figure 19.5). Total energy use declines 42% between 1974 and 1988 after increasing more than 100% between 1960 and 1974. Both direct and indirect energy use decline absolutely, but there is a pronounced relative shift away from indirect energy use. The reduction in indirect energy use is due principally to the severe curtailment in the use of energy-intensive agricultural chemicals, particularly fertilizers and pesticides.

The reduction in energy use in the wake of the energy price shocks does not cause a diminution in output. In fact, output increases between 1974 and 1988, with several bumper crops in the early 1980s. Output fell sharply in 1983 and 1988 due to hot, dry growing seasons and to stabilization programs. One interpretation of the data is that farmers have clearly improved the efficiency with which they use energy to grow crops and produce livestock through reduced tillage, more controlled and timely applications of fertilizer, pesticides and irrigation water, heat recovery systems, and other energy conservation measures (Stout and Nehring 1988). However, that conclusion must be tempered with the fact that government payments to farmers in price stabilization programs increased dramatically in the 1980s. One result was large fluctuations in the number of acres harvested.

An alternate way to measure the relationship between energy use and output in agriculture is to scale both inputs and outputs by the number of acres harvested. This relationship shows that prior to the price shocks the industry was experiencing diminishing output per acre as a function of energy use per acre (figure 19.9). Output per acre shows little if any increase at the high levels of energy use in the late 1960s. Between 1974 and the early 1980s, energy use per acre and output per acre declined at about the same rate as they had increased between 1910 and 1974. This suggests that although the energy price shocks reduced energy use, the underlying relationship between output per acre and energy use per acre did not change after the first price shock. A possible change in the relationship does appear after the 1980-1981 energy price shock. The outliers in figure 19.9 are the mid and late 1980s, when output per acre increased despite a decline in energy use per acre.

One interpretation of the apparent break in the energy-output link is that the second energy price shock reinforced the 1973-74 increases and accelerated the energy efficiency improvements that had already been set in motion. This interpretation is clouded, however, by climatic and political vagaries in the 1980s. Both 1983 and 1988 were extremely hot and dry years which reduced output. Bumper crops in the early 1980s depressed farm income and led to large scale government stabilization programs which significantly affected both energy use and output. Government payments to farmers increased from about $1 billion in 1980 to more than $18 billion in 1987 (expressed in constant 1982 U.S. dollars).

The number of acres set aside in conservation and stabilization programs increased from 0 in 1980 to 78 million in 1983, equivalent to 25% of total harvestable acreage. In 1983, the Payments-In-Kind (PIK) program alone set aside millions of acres of grain crops, contributing to the sharp output reduction in that year. Then, the apparent improvement in efficiency may be due to reduced "effort" by farmers, i.e., fewer acres planted. If lower quality land is removed first under stabilization programs, average yield per acre will rise as land is removed from production, other factors held constant. These relationships require more detailed analysis before the full magnitude of the energy price shocks can be assessed.

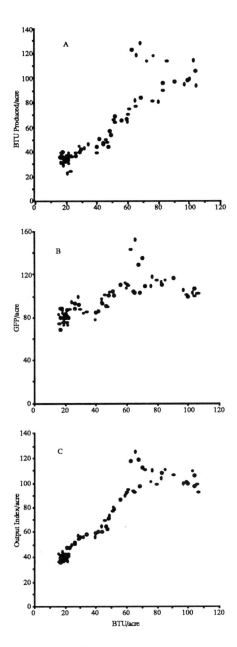

FIGURE 19.9 Alternative measures of the relation between output per harvested acre and total energy use per acre in the U.S. agricultural sector, 1910-1988. In all three graphs, the horizontal axis is the ratio of total energy use to total acres harvested. A) thermal equivalent of the top 15 crops produced to total harvested acres; B) Gross Farm Product per harvested acre; C) the Farm Output Index per harvested acre. All values are indexed to 1977 = 100.

Forest Products

The pattern of E_d+E_i in the forest products industry is similar to that in agriculture in several aspects. Intensification of energy use characterized the 1950 to 1973 period. In logging (SIC sector 2411), both direct and indirect fuel use replaced labor through the mechanization of timber harvesting methods. The use of chainsaws, skidders, tractors, hydraulic shears and other fossil fuel driven machines grew rapidly in the decades following World War II. That mechanization boosted total output and labor productivity, but diminished Q/E_d+E_i (figure 19.6).

The energy prices increases in the 1970s had a clear impact on the forest products industry. In the early 1970s, a large number of marginal, high cost sawmills closed due in part to the economic squeeze of the energy price increases. This restructuring increased the overall energy efficiency of the industry. In sawmills, technical improvements in wood loading and handling systems, improved saw blades, and computer-aid systems such as Best Open Face cutting techniques reduce the quantity of waste produced per unit of timber processed. Much of the waste that is produced is now recovered and incorporated in products such as particle board. Utilization of most of the entire physical volume of timber produced increases the dollar value of fuel product per unit of timber input. That increase in utilization rates is manifest as an increase in the FRB output index which is weighted by the value of output. More efficient logging techniques and higher utilization rates are also occurring in the logging sector. Tree limbs and bark left in the field as waste fifteen years ago now have economic value and represent an increase in the "output" of the logging sector. These and other improvements stimulated by higher cost energy increased Q/E_d+E_i in the forest products industry.

Interpretations of the trends in forestry must be qualified with the fact that the Census of Manufacturer's energy use data include only purchased fuels and electricity. Wood waste fuel use is excluded from Census data. Wood waste fuel use by the forest products industry is estimated from Department of Energy surveys of industrial wood energy use and a special 1985 energy consumption survey by the Census of Manufacturers. The degree of uncertainty associated with my estimates of wood waste energy is unknown and could be large, although general trends are probably accurate. Wood waste is the largest component of total energy use by the forest products sector, and has become increasingly important since 1970. Prior to 1970, a lot of wood waste from sawmills was burned on site. The passage of the Clean Air Act in 1970 severely restricted such practices, creating a strong incentive to use waste products as fuel. The energy price increases reinforced the shift towards wood waste fuel as firms looked for every possible means of reducing purchases of fuel and electricity.

Fisheries

In the New Bedford fisheries, Mitchell and Cleveland (1990) find that fuel use by the New Bedford fleet increased due to the increase in the total number of vessels in the fleet, the increase in average horsepower per vessel, and the increase in time required to travel to and from the point of harvesting. Increases in horsepower per vessel and travel time are due to the synergistic impacts of technological advances in the fishing industry and rising scarcity due to overfishing of the George's Bank fisheries. Advances in gear technology have gone hand-in-hand with larger, more powerful vessels. More horsepower is needed to drag larger scallop dredges across the sea floor and to tow larger dragging nets to catch fish. Larger vessels also lose less time to foul weather. Larger, more sophisticated gear systems require broader vessels with better balance which increases fuel consumption. As a result of these changes the average horsepower per vessel increased from 252 in to 624 between 1963 and 1988.

Depletion of fish stocks increases fuel consumption by driving up travel time. Vessels have to spend increasing amounts of time searching for scallop beds or schools of fish large enough to justify fishing effort. In the 1960s, a vessel could go directly to a scallop or fishing ground on George's Bank (about a 12-hour trip), fish for 8 days, and return to port. Vessels would therefore be away for 9 to 10 days. The average trip today is 12 to 13 days, with most of the extra time (and fuel) invested in steaming to and from the fishing areas. Some scallopers travel as far as the Carolinas, a 36-hour trip.

On the output side, total catch dropped more that 100% between 1965 and 1975 due largely to a reduction in fishing effort, particularly during the 1960s and early 1970s when heavy fishing pressure from foreign fleets reduced U.S. catch. Foreign fishing pressure was substantially reduced by the 1976 Fishery Conservation and Management Act (FCMA). Exclusion of foreign vessels and generous financial incentives available under FCMA stimulated entry into the New Bedford fleet. The number of vessels in the fleet nearly doubled between 1975 and 1985, and total catch increased more than 60%.

The net result of increasing number of vessels, increasing horsepower per vessel, and increasing scarcity of many commercial fish species is a sharp decline in Q/E_d+E_i (figure 19.7). Total catch in 1988 was 30% less than it was in 1968, but 300% more direct fuel was used to catch it. Unlike forestry and agriculture, the other renewable resource extractive sectors, technical change following the energy price shocks has not reduced E_d+E_i costs in fisheries. New engines are more fuel efficient per horsepower, but the large increase in total horsepower used and distance travelled has overwhelmed those improvements.

COMPARISON WITH ECONOMIC MEASURES OF SCARCITY

How do the trends in Q/E_d+E_i compare with the traditional unit cost measure used in economics? Barnett and Morse (1963) show that capital-labor costs fell in the metal and nonmetal mining sector between 1870 and 1957. Johnson et al. (1980) find that trend continues through 1970. In fact, Johnson et al. find that the rate of decline in capital-labor cost *increases* between 1958 and 1970. Hall and Hall (1984) find that the cost of ferroalloys show no significant trend between 1960 and 1980 while the cost of nonferrous metals increased in the 1960s and then decreased in the 1970s. My results generally support the findings for nonmetals but contradict the unit cost results for metals. Metals show a distinct increase in scarcity beginning between 1939 and 1954 (figure 19.1). My results are generally consistent with Slade (1982) who fit a U-shape time trend model to real resource prices and found that most metals are on the upward sloping portion of the long run price path. My results are also consistent with Hall and Hall's conclusion that coal and petroleum grew scarce in the 1970s.

The biophysical and economic models show opposite trends between the turn of the century and the 1970s. Capital-labor costs and real prices of agricultural output decline precipitously through 1980 (Barnett and Morse 1963; Johnson et al. 1980; Hall and Hall 1984). Conversely, E_d+E_i costs increase steadily through the first energy price increases in the 1970s, after which they decline.

Comparison of the biophysical and economic results in forestry and fishing shows mixed results. Labor costs in forestry increase between 1870 and 1957 but then declined between 1958 and 1970 (Johnson et al. 1980). Conversely, the biophysical model shows that E_d+E_i costs generally increased between 1950 and 1973 (figure 19.6). The real price of forest products increase in the 1970s (Hall and Hall 1984), while E_d+E_i decline. Labor costs in fisheries decline between 1870 and 1957, and then increase between 1958 and 1970 (Johnson et al. 1980). Energy costs rise sharply in the new Bedford Fisheries between 1960 and 1988.

CONCLUSIONS

Economic models based on traditional unit cost measures omit the massive substitution of fossil fuels for human labor that has underscored the nation's development of natural resources. One does not have to appeal to the nebulous and ill-defined concept of "self-generating" technical change to explain the decline in labor costs in the extractive sectors of the economy. The results of this analysis show clearly that technical change in the extractive sectors has historically taken the form of using more energy directly and indirectly to subsidize the efforts of labor. As a result, labor costs declined while energy costs increased sharply. These results are consistent with the empirical relationship between labor productivity and the energy subsidy of labor in non-extractive sectors of the economy (Cleveland, et al. 1984; Hall et al. 1986).

The energy price increases reversed the historic low price of fossil fuels relative to capital and labor. In some sectors, the rise in energy prices stimulated behavioral and technological changes which have reduced energy costs and not diminished output. Agriculture and forest products may be examples of this trend. The relationship between energy use and output in agriculture requires a more rigorous assessment before conclusions can be drawn. Agriculture and forestry develop renewable resources which are not necessarily characterized by a continuous decline in resource quality as a function of cumulative production. In other sectors, however, the effects of resource depletion dominate the effects of technical change to the degree that energy costs continue to increase. The fossil fuels and most metals are characterized by an overall march towards lower quality deposits as a function of cumulative depletion at the intensive and extensive margins of development. That continuous decline in quality carries with it an energy penalty which has become increasingly difficult to mitigate. Energy costs in metal and fossil fuel extraction continue to increase despite substantial improvements in energy efficiency in those industries (Dahlstrom 1986).

The increase in energy cost of fossil fuels has the most serious economic implications because they are used in the extraction and processing of all other resources. The effects of declining copper ore grade, for example, have been offset to a degree by using lots of cheap fossil fuels in copper mining and processing. The cost of many fish products has not increased despite severe overfishing due to the development of fishing techniques which use more oil per unit catch. The economy is no longer afforded this luxury because the energy costs of energy are now rising. As the quality of non-fuel resources decline, a positive feedback is established in which depletion accelerates the use of fossil fuels (Ayres and Nair 1984). This use accelerates the depletion of fossil fuel, which increases their own energy costs. Many factors can and do mitigate rising energy costs. Changing consumer preferences, the development of alternative fuels of equivalent or greater quality, or increases in the efficiency in the end use of energy and other resources offer the potential to offset rising energy costs. However, to the degree that technological change, fuel substitution, changing preferences, and other factors cannot or do not mitigate the higher energy costs associated with lower quality resources, rising energy costs have significant economic implications.

REFERENCES

Allesio, F. J. 1981. Energy Analysis and the Energy Theory of Value. *Energy Journal* 2:61.

Ayres, R. and I. Nair. 1984. Thermodynamics and Economics. *Physics Today* 35:62-71.

Barnett, H. and C. Morse. 1963. *Scarcity and Growth: The Economics of Natural Resources Availability*. Baltimore: Johns Hopkins University Press.

Baumol, W. J. 1986. On the Possibility of Continuing Expansion of Finite Resources. *KYKLOS* 39: 167.

Berndt, E. R. 1978. Aggregate Energy, Efficiency, and Productivity Measurement. *Annual Review of Energy* 3:225.

Bower, L. G. 1987. Present and Future Oil Usage. *Forum for Applied Research and Public Policy* 2:19.

Bureau of the Census. 1982 (and various earlier years). *Census of the Mineral Industries*. Washington, D.C.: U.S. Government Printing Office.

Chapman, P. F. and F. Roberts. 1983. *Metal Resources and Energy*. London: Butterworths.

Christensen, P. P. 1989. Historical Roots for Ecological Economics: Biophysical Versus Allocative Approaches. *Ecological Economics* 1:17.

Cleveland, C. J. 1988. Physical and Economic Models of Natural Resource Scarcity: Theory and Application to Petroleum Development and Production in the Lower 48 United States, 1955-1985. Ph.D. dissertation. University of Illinois, Urbana.

Cleveland, C. J. 1990. Energy Use in the Extractive Sectors of the U.S. Economy, 1910-1988. Manuscript, Center for Energy and Environmental Studies, Boston University, Boston.

Cleveland, C. J. 1991. Physical and Economic Aspects of Resource Scarcity: The Cost of Oil Supply in the Lower 48 U.S., 1936-1987. *Resources and Energy*.

Cleveland, C. J., R. Costanza, C. A. S. Hall and R. Kaufmann. 1984. Energy and the U.S. Economy: A Biophysical Perspective. *Science* 225: 890-897.

Cleveland, C. J. and R. K. Kaufmann. 1990. Forecasting Ultimate Oil Recovery and Its Rate of Production: Incorporating Economic Forces Into the Models of M. King Hubbert. *The Energy Journal 12:1*.

Cook, E. F. 1979. Limits to the Exploitation of Nonrenewable Resources. *Science* 191: 677.

Costanza, R. 1989. What Is Ecological Economics? *Ecological Economics* 1:1.

Costanza, R. 1980. Embodied Energy and Economic Valuation. *Science* 210:1219.

Cottrell, W. F. 1955. *Energy and Society.* New York: McGraw-Hill.

Dahlstrom, D.A. 1986. Impact of Changing Energy Economics and Mineral Processing. *Mining Engineering* (January) 45.

Dale, L. L. 1984. The Pace of Mineral Depletion in the United States. *Land Economics* 60:255.

Daly, H. E. and J. B. Cobb, Jr. 1989. *For The Common Good: Redirecting the Economy Toward Community, the Environment, and a Sustainable Future.* Boston: Beacon.

Eagan, V. 1987. The Optimal Depletion of the Theory of Exhaustible Resources. *Journal of Post Keynsian Economics* 9: 565.

Fisher, A. 1981. *Resource and Environmental Economics.* New York: Cambridge University Press.

Gelb, B. 1984. A Look at Energy Use in Mining: It Deserves It. *Proceedings of the 1984 Annual Meeting of the International Association of Energy Economists* (IAEE). Washington, D.C.: IAEE.

Georgescu-Roegen, N. 1979. Response to Daly and Stiglitz. In, V. K. Smith, ed., *Scarcity and Growth Reconsidered.* Baltimore: Johns Hopkins University Press.

Gever, J., R. Kaufmann, D. Skole, and C. Vorosmarty. 1986. *Beyond Oil.* Cambridge, Mass.: Ballinger.

Hall, C. A. S.and C. J. Cleveland. 1981. Petroleum Drilling and Production in the United States: Yield Per Effort and Net Energy Analysis. *Science* 21:576.

Hall, C. A. S., C. J. Cleveland and R. Kaufman. 1986. *Energy and Resource Quality: The Ecology of the Economic Process.* New York: Wiley-Interscience.

Hall, D. and J. Hall. 1984. Concepts and Measures of Natural Resource Scarcity with a Summary of Recent Trends. *Journal of Environmental Economics and Management* 11:363.

Hannon, B., S. Casler, and T. Blazeck. 1985. *Energy Intensities for the U.S. Economy— 1977.* Energy Research Group, Urbana, Ill. University of Illinois.

Harris, D. P. and B. J. Skinner, 1982. The Assessment of Long Term Supplies of Minerals. In V. K. Smith and J. V. Krutilla, eds., *Explorations in Natural Resource Economics.* Baltimore: Johns Hopkins University Press.

Hotelling, H. 1931. The Economics of Exhaustible Resources. *Journal of Political Economy* 39:137.

Houthakker, H. S. 1983. Whatever Happened to the Energy Crisis? *The Energy Journal* 4:1.

Hubbert, M. K. 1962. Energy Resources. A Report to the Committee on Natural Resources. Pub. No. 1000-D, National Academy of Sciences. Washington, D.C.: U.S. Government Printing Office.

Hubbert, M. K. 1967. Degree of Advancement of Petroleum Exploration in the United States. *American Association of Petroleum Geologists Bulletin* 51:2207.

Huettner, D. 1981. Energy, Entropy, and Economic Analysis: Some New Directions. *Energy Journal* 2:123.

Johnson, M. H., J. T. Bell, and J. T. Bennett. 1980. Natural Resource Scarcity: Empirical Evidence and Public Policy. *Journal of Environmental Economics and Management* 7:256.

Jorgensen, D. W. 1984. The Role of Energy in Productivity Growth. *The Energy Journal* 5:11.

Kaufmann, R. 1990. Substitution Between Capital and Energy in the U.S Forest Products Sector. Paper presented at a conference sponsored by the U.S. Forest Service, North Central Experiment Station, St. Paul, Minn.

Meadows, D. H., D. L. Meadows, J. Randers, and W. W. Behrens. 1972. *The Limits to Growth.* New York: Universe Books.

Menard, H. W. and G. Sharman. 1975. Scientific Uses of Random Drilling Models. *Science* 190:337.

Mitchell, C. and C. J., Cleveland. 1990. A Biophysical Model of Resource Scarcity for the New Bedford, Massachusetts Fisheries. Manuscript, Center for Energy and Environmental Studies, Boston University, Boston.

Nehring, R. 1981. *The Discovery of Significant Oil and Gas Fields in The U.S.* Santa Monica, Calif.: The Rand Corporation.

Norgaard, R. 1989. Economic Indicators of Natural Resource Scarcity. Manuscript, Department of Agricultural and Resource Economics. University of California at Berkeley.

Norgaard, R. 1990. Economic Indicators of Resource Scarcity: A Critical Essay. *Journal of Environmental Economics and Management* 19:19.

Pearce, D. 1988. The Sustainable Use of Natural Resources in Developing Countries. In R. K. Turner. ed., *Sustainable Environmental Management*, Boulder, Colo.: Westview.

Petersen, U. H. and R. S. Maxwell. 1979. Historical Mineral and Production Price Trends. *Mining Engineering* (January) 25.

Pimentel, D. and M. Pimentel, 1979. *Food, Energy, and Society.* London: Sage.

Repetto, R., Ed., 1985. *The Global Possible.* New Haven, Conn.: Yale University Press.

Rosenberg, N, 1982. *Inside the Black Box: Technology and Economics.* New York: Cambridge University Press.

Simon, J. 1981. *The Ultimate Resource.* Princeton: Princeton University Press.

Slade, M. E. 1982. Trends in Natural Resource Commodities Prices: An Analysis of the Time Domain. *Journal of Environmental Economics and Management* 9:122.

Solow, R. M. 1974. Intergenerational Equity and Exhaustible Resources. *Review of Economic Studies,* Symposium on the Economics of Exhaustible Resources. Special Issue: 29-45.

Steinhart, J. S. and C. E. Steinhart. 1974. Energy Use in the U.S. Food System. *Science* 184:307.

Stout, B. A. and R. F. Nehring 1988. Agricultural Energy: 1988 and the Future, *Phi Kappa Phi Journal* 68:32.

Turner, R. K., ed. 1988. *Sustainable Environmental Management.* Boulder, Colo.: Westview Press.

World Commission on Environment and Development. 1987. *Our Common Future.* New York: Oxford University Press.

Part III

Institutional Changes and Case Studies

20

ECONOMIC BIASES AGAINST SUSTAINABLE DEVELOPMENT

Colin W. Clark
Institute of Applied Mathematics
University of British Columbia
Vancouver V6T 1Y4, Canada

ABSTRACT

Three powerful sources of social and economic bias acting against sustainable development are discussed: common ownership of resource stocks, future discounting, and the effects of uncertainty. These mutually synergistic biases are omnipresent in contemporary societies. Although they have been widely recognized and discussed in the resource and environmental literature, the biases remain widely misunderstood or deliberately ignored by political leaders and decision makers. Unless the biases become recognized and are taken into account in the design of regulations and institutions, the hope of achieving sustainable development will surely be frustrated.

THE PROBLEM OF SUSTAINABILITY

The terms exponential population growth, resource overexploitation and environmental pollution have by now reached the status of cliches. The present rate of growth of the world's population is about 1.7% per annum. If this continues for another century, the population will reach 27.5 billion. Can such a figure be taken seriously? How well will the inhabitants of the late 21st century be living? Most nations hope to increase the standard of living of their citizens, and indeed rates of per capita resource use are increasing, albeit slowly in most developing countries. Can the earth support a population of this magnitude, living in splendor and abundance?

There is ample evidence that humanity is already beginning to have severe impacts on global resource and environmental systems. With a few notable exceptions, the impacts are not life-threatening, yet. But what does the next century portend? As far as that goes, what do the next two or three decades portend? Many scientists, especially biologists, are becoming increasingly concerned about the long-term sustainability of present, or soon-to-be-achieved levels of technological development. Global warming, deforestation, atmospheric ozone depletion, groundwater contamination and many other byproducts of contemporary society are already posing severe problems which are rapidly becoming worse. These problems have largely come to the forefront in the latter part of the present century, suggesting that we may be beginning to experience the inevitable conflict between population, technology and the finiteness of our planet. As in the metaphor of the exponentially growing population of lily pads in a small pond, we may now be beginning to approach the carrying capacity of the globe.

It is interesting to contrast these ideas with the treatment of resource scarcity given in neoclassical economics. Bluntly stated, limitations on technological development are either implicitly or explicitly rejected by the neoclassical theory (see, e.g., Underwood and King 1989). This ideology explains, for example, why most economists greeted the "limits to growth" models of the 1970s with such disdain. The underlying assumption of the economists' models (in contrast with the limits to growth models) was that increasing resource scarcity would always generate price signals which would engender compensating economic and technological developments, such as resource substitution, recycling, exploration, and increased efficiency of resource utilization. Empirical support for the models came from studies that showed little if any increase in real prices of basic resources. This was interpreted to mean that resources were not actually becoming scarce. It was also assumed that future scarcity of critical energy resources would be overcome as a result of technological developments such as nuclear fusion. Such ideas were, and still are abetted by an active public relations campaign on the part of the power industry.

In cases where resource depletion had occurred, as was undeniably so in many marine fisheries, the failure was attributed to the lack of private ownership of these common-property resources (Gordon 1954). Recommendations for the privatization of resource ownership were prevalent. While there are good reasons to expect that resource stocks will be better cared for under private ownership than under common-property, the belief that privatization will automatically resolve problems of overexploitation cannot be supported on either theoretical or empirical grounds (Clark 1973, 1990). Besides, privatization is simply not feasible for many vital resources; examples include the atmosphere, the oceans, groundwater, migratory birds, fish and animals. Indeed, in an increasingly crowded world, isolation of resource ownership, *including the internalization of externalities*, is probably becoming more rather than less difficult. The political realities are that exploiters of large resource stocks have every incentive to impose major external costs on the public at large, and these externalized costs can add up to nonsustainability.

The growth of population and technology has provided the background for the possibility of overexploitation of the earth's environmental resources. On the other hand, technology also has the potential for increasing the benefits that we obtain from the environment. Resources can be used more efficiently; pollution can be reduced and controlled; depleted fish stocks or forests can be rehabilitated by proper management. But these events require careful planning. The "frontier" mentality that has characterized man's approach to resources in the past is no longer adequate. The transition from the old attitudes to a new philosophy of resource husbandry will require major changes in institutions and methods of resource and environmental management.

BIASES AGAINST SUSTAINABILITY

Contemporary economic systems incorporate a number of powerful biases which operate against sustainable development. Preindustrial societies employed various institutions to ensure their sustained existence, but these are either no longer in existence, or are inadequate in modern society. An understanding of the biases seems essential for the design of novel institutions capable of achieving sustainability. Unfortunately, disregard of the sociology of these biases, whether by oversight or design, has characterized the operation of many of our existing resource-management institutions.

Three fundamental classes of anti-sustainability bias will be discussed: common ownership of resources, future discounting and effects of uncertainty. Ways in which these biases interact will also be described.

The Tragedy of the Commons

The exploiters of a common resource stock have little incentive for the conservation of that resource. Garrett Hardin, an American biologist, has called this the "tragedy of the commons" (Hardin 1968). His example was that of a common grazing ground: Each cattle owner will tend to add to his herd as long as doing so increases his income. But when all herders do the same, the inevitable result is degradation of the land and ultimate impoverishment of all herders. Economist H. S. Gordon (1954) had adduced the same argument for fisheries: As long as fish in the sea can be caught profitably, fishermen will wish to do so, and this may lead to severe overfishing. The more valuable the fish, and the more technologically efficient the fishing vessels, the more severe will be the overfishing. An equilibrium will be reached only when the fish stock has been so reduced that revenues from fishing barely cover operating costs. Gordon referred to this unfortunate situation as "bionomic equilibrium." Gordon's theory may have seemed somewhat academic in the 1950s; but by the end of the 1970s, the majority of the world's oceanic fisheries had become overexploited, and coastal states had proclaimed 200-mile exclusive fishing zones in an attempt to win control over their marine resources. Still, in many regions bionomic equilibrium remains the source of desperate poverty for many fishermen.

The common-property principle applies with equal force to environmental resources. The owner of a cottage on a large lake may be motivated to discharge his wastes into the lake untreated, since his own discharges will have a negligible effect on water quality. But when a large number of cottage owners behave in this way, the lake may become badly polluted. This example also suggests how such common-property problems might be resolved. Namely, the cottage owners must agree to control their discharges by installing individual or communal treatment systems. Obvious as this solution may seem, many difficulties may arise in putting it into practice. First, polluters must realize that the situation has become intolerable and that they are all to blame. They must understand that while their activities may have been individually rational, they have now become collectively undesirable. It must be recognized that voluntary actions are unlikely to succeed and that some form of *enforceable* regulations must be imposed. Before determining which system to adopt, it may be worthwhile spending some time and effort to determine the costs and benefits of the various available technologies for reducing pollution.

The above example illustrates, in microcosm, many of the basic principles and difficulties of sustainable development:
- Individually rational decisions may result in socially undesirable outcomes.
- These outcomes become more serious as the number of users increases.
- Preventing the undesirable outcomes requires joint action (usually undertaken by government), consisting of regulation and enforcement.
- The regulations inevitably impose costs on individuals.
- If properly designed, however, the overall benefits of regulation will exceed the overall costs.
- The design of the optimal system of reducing the undesirable outcomes is seldom evident, and may require considerable research.
- The incentive for individuals to "cheat" on the regulations is always present.

Resolving the tragedy of the commons is seldom simple. It is the very nature of the tragedy that it results from perfectly rational behavior on the part of individuals. However, there also exist more pernicious forms. To return to the example of the lake, suppose now that some company decides to build a chemical plant at one end of the lake. To maximize profits, the company will wish to discharge toxic wastes from the plant into the lake. The costs of this pollution will probably be borne entirely by the cottage owners; unless the plant requires clean water, the company will have little incentive to control its pollution. If cottage owners attempt to bring political pressure to bear to have the plant cleaned up, the company can always argue about the benefits of jobs.

Hardin (1985) calls this form of the tragedy of the commons the "Double C-Double P Game"— Communized Costs, Privatized Profits. Examples are legion, ranging from the usual cases of air, land and water pollution, to drug rackets, and even to terrorism. At a more mundane level, developers of suburban housing are able to attract buyers by offering relatively low prices, but hidden costs may exist in terms of increased traffic congestion on highways, ultimately requiring new construction. The costs of the new construction (not to mention traffic congestion), while resulting directly from the expansion of cheap housing in the suburbs, are borne by all homeowners. Profits accrue to the developers. Costs are commonized over the community. As a consequence, the amount of suburban development is higher than it should be. The CC-PP game also extends into the political arena. Developers lobby municipal councils but the public is disorganized and unable to lobby effectively. Local interest groups may occasionally be effective in stopping a development or diverting it to some other community, but the overwhelming social force is always for additional development.

Gigantic profits can accrue to the perpetrators of CC-PP games, so their prevalence is hardly surprising. Nor is it surprising to find that large political donations often originate with the CC-PP organizers. All this is perfectly legal, of course, but equally huge profits lurk beyond the bounds of legality. A beautiful racket recently exposed involved the mixing of PCB wastes with regular gasoline and trucking it into Canada where it was sold to unsuspecting motorists. The Mafia was said to be implicated in this racket and the mob appears to be engaged in many other similar waste disposal "businesses."

The tragedy of the commons constitutes perhaps the most powerful bias against sustainable development. As population and technology expand, the implications of our inability to solve the problem spread from local to global scales. Government institutions capable of dealing with common-property problems may exist at local and national levels, but they are often weak and subject to political influence. At the international level, no institutions having any powers of enforcement exist at all and we must rely on the cooperation and goodwill of each individual nation to deal with these problems. It seems certain that more future international agreements on air and water pollution will be reached. An interesting model is provided by the 1984 Montreal Protocol whereby leading industrial nations agreed to reduce CFC emissions (the cause of ozone depletion). In this example, the costs of inaction are potentially immense, while the costs of eliminating CFC production seem within reasonable bounds.

Uncertainty

Compounding the difficulties of the tragedy of the commons are two additional and related sources of bias against sustainable development. The first is uncertainty: At present, we simply do not know how serious the future effects of such phenomena as atmospheric warming, ozone depletion, acidic precipitation and tropical deforestation are likely to be. In the face of this vast uncertainty, convincing people to accept major changes in their lifestyles may be difficult indeed. Yet failure to initiate action soon enough may condemn us to irreversible economic disasters.

For example, if the worst of the suggested scenarios for global warming turn out to be realistic, we should clearly be taking drastic steps to reduce the emission of "greenhouse" gases immediately. This would require major reductions in the use of fossil fuels and other similarly drastic changes. Are such draconian measures justified given the degree of uncertainty surrounding the issue?

Several types of response to this question are possible:

1. Play safe: The worst scenarios would be intolerable (flooded cities, displaced populations, crashes in food production) and immediate policy changes are essential.
2. Wait: The danger is not imminent; future research will clarify the situation and indicate necessary remedies.

3.   Laissez faire: The market economy will react to future changes; development in technology and science will occur in response to future needs.
4.   Armageddon: The future is out of our hands; rely on God.

   We have no obvious way of determining which response is the best. It is ironic that the very technological developments that have made modern civilization possible have also led to unacceptable levels of uncertainty regarding our future.

Future Discounting

A further important bias against sustainability exists. As an illustration, imagine an isolated coastal community that relies on fishing as its sole source of income. Fish stocks have become depleted and current catches are much lower than they were traditionally. It is not a case of outsiders taking the fish, but rather that the local fishermen have been progressively overfishing. The local fishermen fully understand this and realize that the stock could be rehabilitated provided that they reduced their catches in the interim. Would they in fact wish to do so, assuming that the common-property problem could be resolved (for example, by allowing a fixed quota to each fisherman)? Or to take a more extreme scenario, suppose that the fishermen realize that if they continue their present fishing the stock will probably be completely wiped out. Would they want to reduce catches in order to save the fishery?

   It is clear, I hope, that the answer may not be an unqualified "yes" to either question. If the fishermen are extremely poor, any further reduction in catches may prove immediately disastrous. The need for immediate survival outweighs any consideration of the future. There is some poverty level below which sustainability becomes an unaffordable luxury. Vast numbers of impoverished people throughout the world face this dilemma every day of their lives.

   It is less obvious, perhaps, that the same argument may apply even if the fishermen are well off. By taking large catches now (especially if prices are good), a fisherman may increase his wealth, opening up opportunities for travel, for buying a new boat, or even for moving away from the fishing community altogether. These motivations are not unnatural or evil in any way, although one might argue that contemporary society encourages such spendthrift behavior.

   The preference for short-term over long-term gains can be understood in terms of the phenomenon of *discounting*: An individual who has a personal annual discount rate of i will be indifferent between a lump-sum immediate payment of $\$N$ and a payment of $\$N/(1+i)$ in one year's time. For example, at $i = 10\%$ per annum, $1000 next year will be considered equivalent to $909.09 today. Similarly, $1000 in 10 years' time is equivalent to $385.54 today, and $1000 in 100 years' time is equivalent to 7 cents today.

   Discounting is a normal procedure in financial analysis, although appropriate discount rates may be hard to determine in practice. Everyone discounts the future to some extent; generally speaking, the poor are forced to discount at higher rates than the well off.

   It may not be obvious that discounting has an effect on sustainable development and resource conservation, but in fact the effect is often extremely important. In the abstract, sustainability implies a concern for the future, and this concern is necessarily reduced by discounting. I have argued at length elsewhere (Clark 1973; Clark and Lamberson 1982) that the overexploitation of the world's whale stocks, some to the verge of extinction, could in some cases at least be interpreted not as an instance of the tragedy of the commons, but as a perfectly deliberate corporate strategy resulting from discounting. The profits from whaling were immense. Large species such as the blue whale cannot sustain an annual harvest of more than 2% or 3% of the breeding stock. With normal rates of return on investment in the neighborhood of 10%, the "optimal" strategy for the whalers was probably to simply wipe out the whales and invest the proceeds elsewhere. This seems to have been exactly what happened as members of the International Whaling Commission consistently set quotas far in excess of sustainable yields. Eventually, under pressure from conservation organizations, the IWC acted to prevent further depletion of the whale stocks, but this was long after the worst depredations had been completed. And, even today, Japanese whaling companies press for reopening of quotas on all species, even the blue whale, which is probably very close to extinction. (Japanese companies are also deeply implicated in the current crises involving elephants, rhinoceroses, sea turtles and tropical rainforests.)

The history of exploitation of the great whales is in many ways typical of contemporary resource and environmental issues; it combines aspects of common-property, uncertainty and discounting. Any degree of uncertainty (which was clearly a major factor in whale population dynamics) results in increased future discounting, usually known as "risk discounting." Resource and environmental problems are typically dominated by uncertainties of many kinds. If a given biological resource is available now, but uncertain in the future (whether because of unknown biological effects, or because it may be harvested by someone else), the self-interest of exploiters will favor immediate harvesting.

Thus, we find in many instances, no economic forces whatever acting in favor of sustainable development of the biosphere. Individually rational human incentives mitigate against sustainability. Immediate needs, jobs, economic growth and prosperity are the essential objectives of every nation's economic policy. Economic growth is assessed by the rate of growth of the GNP, with no attempt to account for the stripping of resource or environmental assets.

As noted above, the rich typically use lower discount rates than the poor. The rich man saves excess income for the future while the poor man is forced to spend everything in order to survive. The same phenomenon applies to nations. Tropical deforestation feeds the pulp mills of Japanese, European and American firms enhancing their wealth, but leads to ever increasing poverty and desperation in the Amazon, the Philippines and Thailand. The worldwide demand for protein-rich feedstuffs caused Peru to exploit its anchoveta population to the point of virtual extinction, thereby losing its second most important source of foreign income. In the Philippines, "dynamite fishing" has become standard practice among coastal fishermen, many of whom have been displaced from their forest homes by deforestation. (The dynamite, though illegal, is supplied by army personnel.) Dynamite destroys the reefs and the fishermen are forced to travel farther every year to eke out the meager catches on which their survival depends. "Cyanide fishing" is

ther common practice in the Philippines. The poisoned reef fishes float to the surface
d survivors are sold for export as aquarium specimens. The cyanide also kills the reefs.

The three biases against sustainable development are extremely powerful. They operate
everywhere at all times. Being based completely on individual rationality, they hinder
well-meaning attempts to control them. Leading decision makers barely understand them.
Yet, understanding and overcoming these biases may well be essential to the survival of
civilization as we know it.

STRATEGIES FOR ACHIEVING SUSTAINABLE DEVELOPMENT

There are three obvious prerequisites for coming to grips with the problems of sustainable development. The existence and severity of the problem must be recognized, the societal forces responsible for nonsustainability must be clearly understood, and effective
strategies for reversing the trend must be devised and applied. The most important requirement may be understanding: understanding that the earth is finite and depletable; understanding that everyday economic incentives are in many ways aligned against sustainability; understanding that poverty is incompatible with sustainable development; and
most important, understanding that we may not have much time to solve these problems.

Understanding is unlikely to come easily. It involves major reversals in the philosophical outlook that has dominated the development of contemporary civilization. Until
recently, resource and environmental limitations have simply not penetrated the consciousness of most bankers, economists, engineers, politicians or the public at large. The
emphasis has been entirely on further development; the adjective "sustainable" is a recent
innovation. Indeed, the acceptability of the phrase "sustainable development" to politicians may stem from the habit of equating "development" with "growth" On a finite
planet, sustainable growth is an obvious impossibility.

Outlooks, however, do change. Wide media coverage of such topics as tropical deforestation, the "ozone hole" and atmospheric and water pollution have intensified public
recognition of the limits to unregulated development. Extreme weather conditions in the
late 1980s have helped spread the belief that global climate changes may result from
human activities. Scientific evidence on global warming, however, remains inconclusive.

A wide gap exists between feelings that "something ought to be done" and the design
and implementation of effective economically productive countermeasures. Resource and
environmental systems are immeasurably complex, involving physical, biological,
social, economic and political components. Uncertainty and unpredictability are the rule
not the exception. Failure to account for the interaction of all components may lead to
inappropriate, expensive and unsuccessful attempts to achieve sustainable development.

Marine Fisheries

The history of oceanic fisheries management, particularly following the establishment of
200-mile zones, shows the effects of failing to come to grips with the realities of

common-property resource systems (Clark 1985). Massive increases in fishing capacity were caused primarily by governments' unwillingness to tackle the commons problem, encouraged by overoptimistic forecasts and often abetted by subsidization programs (always with the stated aim of increasing employment). The increases have resulted in declining fish stocks, massive regulatory bureaucracies, and perpetual headaches for management authorities. In Canada, the current outcome is large-scale layoffs in the fishing industry, even though ten years ago unbounded optimism prevailed—with politicians promising new jobs and prosperity, scientists forecasting greatly increased catches, and fishermen rushing to build new, larger, faster boats. The warnings of a few fisheries economists, who had seen it all before, were completely ignored. Meanwhile, the inevitable dynamics of the commons continued to operate, largely overlooked and widely misunderstood. (But perhaps not so widely misunderstood by the fishermen, who it is said had learned how to fish for government subsidies.)

What went wrong? All three of the biases discussed earlier play extremely important roles in fisheries. Most management programs simply made no attempt to respond to the tragedy of the commons. They were based primarily on the concept of limiting total catches to the estimated sustainable yield for each population, or in the case of depleted stocks, to temporary reductions in catch levels to allow the stocks to recover. Some of these programs proved effective in protecting fish stocks. Paradoxically, the more successful the rehabilitation, the more the number and size of fishing vessels seemed to grow. Occasional cyclically high prices for fish attracted further expansion of fishing capacity, thereby placing additional pressure on the stocks. When prices subsequently returned to normal, or when the fish population underwent the inevitable natural decline, the overexpanded fishing industry exerted pressure on the management authorities to maintain high catch quotas for reasons of "economic need." Where this was impossible, government subsidies were rationalized, allowing the managers to escape the responsibility of their mismanagement.

Resource economists have long maintained that taxation, not subsidization, of fisheries is the only equitable approach to rational management. Charging fishermen for the right to appropriate a valuable publicly owned resource overcomes the incentive for overfishing, and is analogous to charging loggers for the right to cut public forests and to charging oil companies for drilling rights.

A few countries, notably Iceland and New Zealand, have attempted to revise the traditional management systems in fisheries. Programs based on individual transferable quotas (ITQs) are now in operation in these countries. Such programs constitute a move towards privatization of rights to use the fishery resource. In principle, quotas could be sold to fishermen (a system of taxation), but few governments have had the courage to do so. Theoretically, at least, ITQs have the potential for reducing wasteful overcapacity, as well as for controlling catches to the sustainable level. Nonallocated quota systems (the traditional approach in fisheries), on the other hand, actually encourage overcapacity. They also force fishermen into a frantic "scramble" to obtain their share of the catch. For example, in the Pacific halibut fishery the fishing season is now completed in about three days of fishing per year, the annual quota being rapidly scraped up by a huge fleet.

Individual quota systems involve significant management difficulties, however. For example, an early attempt at ITQ management, in Canada's Bay of Fundy herring fishery, was a notable failure, doubtlessly because of lack of proper monitoring and control of the catches. For the quota system to succeed it must be rigorously enforced, since cheating on quotas would be highly profitable; the tragedy of the commons is still in operation here! The ideal system would be for the fishermen themselves to be responsible for enforcement. After all, any quota cheaters would be robbing their fellows.

The propensity for fish stocks to fluctuate unpredictably leads to difficulties in any management program. Catch quotas need to be adjusted annually to match the available stocks and to retain a viable breeding population. ITQs would therefore need perpetual adjustment, often with little advance warning. Equally important, fish stocks must be regularly assessed, often a difficult and expensive proposition. Fish in the ocean are not easily counted, and stock levels and appropriate quotas are often highly uncertain. Fisheries biologists typically argue that conservative quotas should be set, in order to ensure that stocks are not overfished. But the fishing industry seems inevitably to demand quotas consistent with the most optimistic stock estimates. Immediate access to the resource is seen to be more important than the possibility of future declines in resource abundance.

Fisheries thus provide a microcosm of the three biases against sustainability. Common-property, uncertainty and future discounting all playing major roles. How these biases can be effectively counteracted is still very much a topic of discussion among managers and fishermen.

Resource Management Principles

The example of the commercial fishery provides several insights into the general problem of achieving sustainable development. Sustainability will not likely be achieved without some form of regulation. But regulation which is not based on an understanding of the economics of common-property resources is virtually certain to fail to achieve economically efficient sustainable development. Some form of, or proxy for, private ownership of the resource seems essential, although community ownership might succeed in some circumstances. Ownership rights, however, must be enforceable; outsiders have to be excluded, and owners of shared resources must be prevented from exceeding their individual use quotas. In cases where significant externalities (effects external to the existing market) exist, direct government regulation and control are required. What are the options for regulation of a common-property resource?

Taxes or Quotas?

Allocated quotas provide one method for counteracting the tragedy of the commons. Quotas can be applied both to "goods" (i.e., resource quotas) and "bads" (pollution permits). But should the quotas be awarded free to deserving applicants or should the government charge a fee for the quotas? In fact, why bother with a complicated quota system at all? Why not just set a fee or tax for the use of the resource, and allow anyone willing to pay the fee to use the resource? For example, the government could auction rights to the resource, as is already common in offshore oil fields. The same principle could readily be applied to fisheries, forests, groundwater supplies and even pollution permits. The equivalence (in terms of economic incentives) of transferable quotas and taxes is a basic principle of the economics of regulation. However, the principle is fully valid only under specific assumptions, including lack of uncertainty. From the point of view of government revenues, taxes have the advantage of retaining the resource rents for the public sector, which is the original owner of the resource.

Clearly, there are practical limits to the use of resource or pollution taxes. For example, it would obviously be impossible to charge automobile owners a fee proportional to their emissions of undesirable gases if this required actual measurement of these emissions. Much simpler to require is the installation and maintenance of standard converters. On the international scale, taxes or fees are not feasible for the simple reason that no international authority with taxing power exists.

Individual quotas and community resource ownership incorporate the advantages of decentralized decision making (Berkes 1989). They encourage pride of ownership and can engender the economically efficient use and conservation of resources. Many important problems of sustainable development, however, demand centralized government control; this is the case whenever significant externalities exist, as in all cases involving pollution. Taxes, fees and various charges then become important instruments of regulation. Resource users will try to minimize the charges against them, whether legally or illegally. Attempts to limit pollution can make illegal pollution highly profitable. But society is not willing to legalize cocaine use simply because dope smuggling is a multibillion dollar business. Why should we be willing to accept life-threatening environmental insults because pollution laws might encourage the Mafia to take up pollution rackets?

Natural Capital

Any resource or environmental asset is a form of "natural capital," whose value to society is by definition equal to the amortized value of future benefits that can be derived from use of the asset. The owner of an asset would, in his own best interest, be fully aware of its value, and be motivated to manage the resource to maximize its asset value.

The uncontrolled use of a common-property resource, on the other hand, implies an implicit valuation of *zero*. This is simply another way of expressing the economics of common-property resources but it has implications that may be far from obvious. For example, Daly and Cobb (1989, pp. 81-83) have attempted to adjust GNP figures for the United States, 1945-1980, to account for depletions of natural capital. The surprising result is that *net* economic growth in the United States has been virtually flat over the past 25 years. If this is correct, the conclusion is devastating: All apparent economic growth, so beloved of politicians and economists, has been a delusion resulting from simply running down our inventory of natural capital. The same bookkeeping applied to many other countries would probably indicate negative economic growth, particularly for tropical countries in which deforestation has been prominent.

The likelihood that political and financial leaders remain completely ignorant of these elementary facts is a frightening commentary on contemporary development economics. At least the phrase *sustainable* development, which seems to have impinged on the consciousness of many world leaders, carries the implication that normal economic biases may be aligned against sustainability so that deliberate policy changes are required if the world's vital supply of natural capital is to remain intact. Why has natural capital not been taken into consideration in national accounting systems? What is our natural capital worth, and how is it that we haven't attempted to answer this question?

In a general sense, natural capital is worth the survival of mankind, but we have not made much of an attempt to quantify the value of forests, wetlands, agricultural soils, unpolluted waters, and so on. Until recently, natural capital was so abundant that we had little to worry about. We have therefore been unprepared for the transition to global scarcity. Much of natural capital is common-property, implying the lack of any incentive for evaluation. Also, many types of natural capital do not enter into market transactions, so that no direct value can be measured. A real concern for sustainable development, it seems to me, would demand a major effort at inventorying the nation's supply of natural capital. I would put this endeavor on par with the project of mapping the human genome.

Environmental impact assessments could be required henceforth to identify all potential losses to natural capital. Development projects would not be approved unless it could be demonstrated that they would actually enhance the national economy, counting losses as well as gains. Our current development path is transparently nonsustainable; reversing the trend will require monumental efforts and major changes in the way we live and think. But our very survival is at stake.

Acknowledgments

Garrett Hardin kindly suggested numerous improvements to the original manuscript. My research is supported in part by NSERC grant 83990.

REFERENCES

Berkes, F., ed. 1989. *Common-Property Resources*. London: Belhaven Press.
Clark, C. W. 1973. The Economics of Overexploitation. *Science* 181:630-634.
Clark, C. W. 1985. *Bioeconomic Modelling and Fisheries Management*. New York: Wiley-Interscience.
Clark, C. W. 1990. *Mathematical Bioeconomics: The Optimal Management of Renewable Resources*. 2d ed. New York: Wiley-Interscience.
Clark, C. W. and Lamberson, R. H. 1982. An Economic History and Analysis of Pelagic Whaling. *Marine Policy* 6:103-120.
Daly, H. E. and J. B. Cobb, Jr. 1989. *For The Common Good: Redirecting the Economy Toward Community, the Environment, and a Sustainable Future*. Boston: Beacon.
Gordon, H. S. 1954. The Economic Theory of a Common-Property Resource: The Fishery. *Journal of Political Economy*, 62:124-142.
Hardin, G. 1968. The Tragedy of the Commons. *Science* 162:1243-1248.
Hardin, G. 1985. *Filters Against Folly*. New York: Viking-Penguin.
Underwood, D. A. and P. G. King. 1989. On the Ideological Foundations of Environmental Policy. *Ecological Economics* 1:315-334.

21

ASSURING SUSTAINABILITY OF ECOLOGICAL ECONOMIC SYSTEMS

Robert Costanza
Coastal and Environmental Policy Program
Center for Environmental and Estuarine Studies
University of Maryland
Box 38, Solomons, MD 20688 USA

ABSTRACT

Assuring sustainability of ecological economic systems depends on our ability to make local and short-term goals and incentives (like local economic growth and private interests) consistent with global and long-term goals (like sustainability and global welfare).

Traditional sustainable cultures have used systems of taboos, religious mores, etc., (arrived at largely through trial and error) to bring long-term goals and constraints into the local, short-term decision making process. Our global environmental crisis is such that we don't have the time or flexibility to use trial and error or to comprehensively instill the appropriate taboos and mores. Institutions and policies that can use our current, uncertain, scientific understanding of the possible future consequences of current activities to adjust the local, short-term decision-making process quickly and effectively must be developed. This requires:

- Establishing a *hierarchy* of goals for local, national and global ecological economic planning and management. Sustainability should be the primary long-term goal, replacing the current focus on GNP growth. Local economic growth in this hierarchy is a valid goal to the extent that it is consistent with sustainability. Ecological economists can help to develop and popularize this hierarchy of goals, and the underlying world view. The goals can be operationalized by having them accepted as part of the political debate, and implemented in the decision making structure of institutions that affect the global economy and ecology (such as the World Bank).
- Developing better *global ecological economic modeling* capabilities to allow us to see the range of possible outcomes of our current activities. Ecological economists can play a major role in this.
- Adjusting current prices to reflect long-run, global costs, *including uncertainty*. To paraphrase the popular slogan, we should: model globally, adjust local incentives ac-

cordingly. In addition to traditional education, regulation and user fee approaches, a flexible assurance bonding system can specifically address uncertainty.

• Developing policies that lead to no further decline in the stock of *natural capital*. These policies will encourage the technological innovation that optimists are counting on, while conserving resources in case the optimists are wrong.

A HIERARCHY OF GOALS AND INCENTIVES

No complex system can be managed effectively without clear goals, and appropriate mechanisms for achieving them. In managing the earth, we are faced with a nested hierarchy of goals that span a wide range of time and space scales. In any rational system of management, global ecological and economic health and sustainability (see chapter 1 for definitions) should be "higher" goals than local, short-term national economic growth or private interests. Economic growth can only be supported as a policy goal in this context to the extent that it is consistent with long-term global sustainability.

Unfortunately, most of our current institutions and incentive structures deal only with relatively short-term local goals and incentives. This would not be a problem if the local and short-term goals and incentives simply added up to appropriate behavior in the global long-run, as many assume they do. Or, in other words, if they were *consistent* with global and long-term goals. Unfortunately, this often is not the case. Individuals (or firms, or countries) pursuing their own private self-interests in the absence of mechanisms to account for community and global interests frequently run afoul of these larger goals and can often drive themselves to their own demise.

Goal and incentive inconsistencies have been characterized in many ways, beginning with Hardin's (1968) classic paper on the tragedy of the commons and continuing through more recent work on "social traps" (Platt 1973; Cross and Guyer 1980; Teger 1980; Costanza 1987; Costanza and Schrum 1988; Costanza and Perrings 1990). Social traps occur when local, individual incentives that guide behavior are inconsistent with overall system goals. Examples are cigarette and drug addiction, overuse of pesticides, economic boom and bust cycles, and a host of others. Research on social traps can show how people behave in traplike situations and how to best avoid and escape from social traps (Edney and Harper 1978; Teger 1980; Brockner and Rubin 1985; Costanza and Schrum 1988) The idea emerging from this research is that, where social traps exist, the system is not inherently sustainable, and steps must be taken to harmonize goals and incentives over time and space scales. Explicit steps must be taken to make global and long-term goals incumbent on and consistent with local and short-term goals and incentives.

This is in contrast to natural systems, which are forced to adopt a long-term perspective by the constraints of genetic evolution. In natural systems, "survival" generally equates to sustainability of the species as part of a larger ecosystem, and natural selection tends to find sustainable systems in the long run. Humans have broken the bonds of genetic evolution by the expanded use of learned behavior that our large brains allow and extending our physical capabilities with tools. The price we pay for this rapid adaptation is a partial isolation from long-term constraints and a susceptibility to social traps.

Another general result of social trap research is that the relative effectiveness of alternative corrective steps is not easy to predict from the "rational" models of human behavior prevalent in conventional economic thinking. The experimental results indicate a need for more realistic models of human behavior under uncertainty which acknowledge the complexity of real world decisions and our limited information processing capabilities.

The Interface Between Ecological and Economic Systems

Ecological systems play a fundamental role in supporting life on earth at all hierarchical scales. They form the life-support system without which economic activity would be impossible. They are essential in global material cycles like the carbon and water cycles. They provide raw materials, food, water, recreational opportunities, and microclimate control for the entire human population. In the long run, a healthy economy can only exist in symbiosis with a healthy ecology. The two are so interdependent that isolating them for academic purposes has led to distortions and poor management.

Ecological systems are also our best current models of sustainable systems. Better understanding of ecological systems and how they function and maintain themselves can yield insights into designing and managing sustainable economic systems. For example, there is no "pollution" in climax ecosystems[1]—all waste and by-products are recycled and used somewhere in the system or harmlessly dissipated. This implies that a characteristic of sustainable economic systems should be a similar "closing the cycle" by finding economic uses and recycling "pollution," rather than simply storing it, exporting it, diluting it, or changing its state, and allowing it to disrupt existing or future ecosystems.

In the realm of behavior and the study of decision making, we are finding more and more that human behavior is part of a continuum of animal behavior, and that experimental studies of human and other animal behavior can shed much light on human behavior. The subfields of experimental economics and evolutionary economics are based on this idea and have begun to bear some fruit.

Understanding the linkages between ecological and economic systems and treating them as a whole, integrated system is therefore critical to sustainability (Costanza et al., this volume, for a definition and discussion of sustainability).

ECOLOGICAL ECONOMIC MODELING AND ANALYSIS

Valuation of Ecosystem Services

Valuation of ecosystem goods and services in units comparable with economic goods and services is essential for communicating and implementing our understanding of eco-

---

[1] A climax ecosystem is a mature, stable system that does not have a tendency to succeed into any other ecosystem. If the natural ecosystem in an area is removed (without damaging the soil structure too much) the area will generally progress through a succession of ecosystems (i.e., grass, shrubs, pine forest) until it again reaches the climax system.

logical economic linkages. Some argue that attempts to value ecosystems are not appropriate or necessary, and should be avoided. How, after all, they would ask, can one put a value on human life, environmental aesthetics, and a host of other "intangibles"? In the minds of some, even to think of the problem in these terms is distasteful. But distasteful as they may appear to some, I contend that these valuations are unavoidable, and, like the equally distasteful "death and taxes," to deny their existence can only cause sorrow and confusion in the long run. Without "value" measured in units that can be compared with other things, humans too often regard ecological goods and services as "free." This has produced unsustainable policies at every decision-making level.

We must make choices, and whenever we do, we at least implicitly rank or value the alternatives. Decisions about how to build highways *imply* a particular valuation of human life (whether we are willing to acknowledge that valuation or not) because spending more money could save lives. To achieve sustainability, we must bring these often difficult choices and valuations to the forefront, and confront their complexities and uncertainties directly rather than denying their existence.

The conventional economic view *defines* value as the expression of human preferences, with these preferences taken as given and with no attempt to analyze their origins or patterns of change. Using this idea, one can extend existing markets using questionnaires and observations of behavior to derive the willingness of current individuals to pay for environmental services. This approach yields useful information, but it assumes (among other things) that the individuals being questioned are well informed about the environmental resource or service under study, and that the current generation of humans are the only agents we need survey. These assumptions are not always valid or appropriate, especially for the more global environmental services and with an eye toward sustainability.

An alternative view (Costanza 1980; Cleveland et al. 1984) postulates a biophysical basis for value, that may only be perceived by humans and incorporated in their preference structures in the long run. It represents a *theory* of how preferences come to be formed and how they change. This view suggests that things are costly to produce to the extent that they are *organized* relative to their environment, and that this relative cost or organization forms the basis (in an evolutionary, long-term, thermodynamic sense) for perceived economic value.

The relationship of this biophysical cost to economic value is controversial (Huettner 1982; Costanza 1982). It is not a direct *mechanistic* connection but a more probabilistic and *evolutionary* one, and this make it less obvious and more difficult to present and test than many theories. But the theory *is* amenable to empirical testing (unlike the *definition* of value as utility), and to the extent that it has been tested it appears to be surprisingly accurate (Costanza 1980; Costanza and Herendeen 1984; Cleveland et al. 1984).

This ecological value theory implies biophysical limits to value based on the laws of thermodynamics and information theory. It also allows a system of cost accounting that is applicable with equal facility to both economic and ecological systems.

More detail on the willingness-to-pay (WTP) and biophysical approaches to valuation of ecosystem services is given below. The bottom line, however, is that we need a pluralistic approach to valuation that can utilize to full advantage the imperfect information gained from several different approaches to valuation, combined with ecological economic modeling and accounting, to form an integrated picture.

Ecological goods and amenities are valued by individuals for a variety of reasons. *Utilitarian (or use) value* refers to the value of using an ecosystem's products and amenities to derive both current and future benefits. These benefits include commercial outputs such as timber, outdoor activities and experiences, wildlife and aesthetics (for examples of raw material evaluation, see Bartlett 1984). Individuals may also be willing to pay now for the certainty of having a resource available in the future. Such an option price includes an amount equivalent to the expected use value plus a premium, similar to a risk premium, which a person would pay over and above the expected use value. This premium is referred to as *option value*, and is due either to uncertainty surrounding the individual's preferences or to uncertainty regarding the price or availability of the resource. This premium may be positive, negative, or zero (as in the case of preference uncertainty) but it will always be positive in the case of supply availability for a risk-averse person. (See Greenley et al. 1981; Bishop 1982; and Brookshire et al. 1983 for the theory and empirical studies of option value). The passage of time will likely reduce the uncertainty surrounding resource usefulness. When resource use is irreversible, individuals would be willing to sacrifice current irreversible use until uncertainty about its cost has been reduced. They would be willing to pay for increased information. This payment is termed *quasi-option value* (Arrow and Fisher 1974; Conrad 1980). It is not attributable to risk aversion, like option value, but is due to the value of information. This value arises in the case of resource use decisions that create irreversible damages, such as species extinction or large scale deforestation. A final, pure non-use value is what a person may be willing to pay simply to know that a resource exists even when there is no intention of use. This *existence value* has nothing to do with preserving options for future use or paying to delay use until more information is available (see Randall and Stoll 1980; and Brookshire et al. 1983).

In practice, the measurement of these value concepts has remained difficult and largely limited to the valuation of environmental commodities and amenities which produce direct benefits to humans. An approach to the measurement of indirect benefits is Norton's (1986) concept of *contributory value*, which assigns value to environmental resources not due to their *direct* value to humans, but according to their *indirect* role in maintaining and accentuating the ecosystem processes which support these direct benefits. These include the maintenance of atmospheric and aquatic quality, the amelioration and control of climate, flood control, the maintenance of a genetic library, and the supportive role of food webs and nutrient cycling. Contributory value recognizes both the long time horizons involved in many ecosystem processes and the synergism which can result when two or more species interact to create benefits which the individual species are not capable of creating independently.

Though empirically elusive, contributory value does provide a useful framework for conceptualizing how natural ecosystems might be evaluated. However, as Randall (1986) contends, human preferences are focussed more on life forms than on life processes. This bias is further distorted by the fact that humans, in general, will assign higher preferences to species with commercial value, to wild relatives of domesticated species and to those which are most familiar or easy to empathize with, such as large mammals. Lovejoy (1986) refers to this bias as vertebrate chauvinism, while others point to interspecies inequity (Costanza and Daly 1987). If it is accepted that each species, no matter how uninteresting or lacking in direct usefulness, has a role in natural ecosystems (which do provide many direct benefits to humans), it is possible to shift the focus of valuation from the imperfect perceptions of individuals to the contributory value of ecosystems, as expressed through their ecological relationships. One might argue that this contributory value is an estimate of the value individuals would place on environmental services if they were fully informed about the functioning of the environment in their behalf.

Assessing the contributory value of ecosystems involves the ability to understand and model the ecosystem's role in an integrated ecological economic system and its response to perturbations. The models must be at a level of detail and resolution that allows the assessment of impacts (marginal products) on economically important ecosystem commodities and amenities. Several types of ecological and economic modeling can be used for this purpose, which we define under the general heading of "ecological economic" models. They range along a continuum from multiple regression models (Farber and Costanza 1987) to static, linear input-output models (Hannon 1973, 1979; Isard 1972; Costanza and Neill 1984; Costanza and Hannon 1987) to more sophisticated nonlinear, dynamic spatial simulation models (Costanza et al. 1986, 1990). Braat and van Lierop (1985) provide a summary of ecological economic models currently in use.

The point that must be stressed is that the economic value of ecosystems is connected to their physical, chemical, and biological role in the overall system, *whether the current public fully recognizes that role or not*. The public is most likely far from being fully informed about the ecosystem's true contribution to their own well being, and they may therefore be unable to directly value the ecosystem's services (Costanza 1984). However, scientists may be able to derive estimates of the values that a fully informed public would produce by analyzing the structure and function of ecosystems and their indirect contribution to long-term human well being.

Ecological Economic System Accounting and Network Analysis

Ecology is often defined as the study of the relationships between organisms and their environment. The quantitative analysis of interconnections between species and their abiotic environment has therefore been a central issue. The mathematical analysis of interconnections is also important in several other fields. Practical quantitative analysis of interconnections in complex systems began with the economist Wassily Leontief (1941) using what has come to be called Input-Output (I-O) Analysis. More recently, these concepts, sometimes called Flow Analysis, have been applied to the study of interconnections in ecosystems (Hannon 1973, 1976, 1979, 1985a,b,c; Costanza and Neill 1984). Related ideas were developed from a different perspective in ecology, under the heading of Compartmental Analysis (Barber et al. 1979; Finn 1976; Funderlic and Heath 1971; Hett and O'Neill 1971). Isard was the first to attempt combined ecological economic system I-O analysis (Isard 1972) and combined ecological economic mass balance models have been proposed by several other authors. (Daly 1968; Cumberland 1987). I refer to all variations of the analysis of ecological and/or economic networks as *network analysis*.

Network analysis holds the promise of allowing an integrated quantitative treatment of combined ecological economic systems. One promising route is the use of "ascendancy" (Ulanowicz 1980, 1986) and related measures (Wulff et al. 1989) to measure the degree of organization in ecological, economic, or any other networks. Measures like ascendency go several steps beyond the traditional diversity indices used in ecology. They estimate not only how many different species there are in a system but, more importantly, how those species are organized. This kind of measure may provide the basis for a quantitative and general index of system health applicable to both ecological and economic systems.

Another promising avenue in network analysis is its use in "pricing" commodities in ecological or economic systems. The "mixed units" problem arises in any field that tries to analyze interdependence in complex systems that have many different types and qualities of interacting commodities. Ecology and economics are two such fields. Network analysis in ecology has avoided this problem in the past by *arbitrarily* choosing one commodity flowing through the system as an index of interdependence (carbon, enthalpy, nitrogen, etc.). This choice ignores the interdependencies between commodities and assumes that the chosen commodity is a valid "tracer" for relative value or importance in the system. This unrealistic assumption severely limits the comprehensiveness of an analysis whose major objective is to deal comprehensively with whole systems.

There are evolving methods for dealing with the mixed units problem based on analogies to the calculation of prices in economic input-output models. Starting with a more realistic *commodity by process* description of ecosystem networks that allows for joint products one can use *energy intensities* to ultimately convert the multiple commodity description into a pair of matrices that can serve as the input for standard (single commodity) network analysis. The new single commodity description incorporates commodity and process interdependencies in a manner analogous to the way economic value incorporates production interdependencies in economic systems (Costanza and Hannon 1989). This analysis would allow valuation of components of combined ecological and economic systems as a compliment to subjective evaluations.

Multi-Scale Ecological Economic Modeling

The high degree of interdependence between ecosystems and economic systems at local, regional, and global scales is becoming more obvious with each passing day. At the individual country level, the interdependence between ecosystem health and economic performance is beginning to be recognized, but has yet to be quantified to the point where it can begin to influence policy and planning in a rational way.

An integrated, multiscale, transdisciplinary suite of models can now be developed and usefully applied to these problems, and a host of other related problems. Rational planning and management for sustainability requires a pluralistic approach (Norgaard 1989) and an ability to integrate and synthesize the many different perspectives that can be taken. There is probably not one *right* approach or paradigm, because, like the blind men and the elephant, the subject is too big and complex to touch it all with one limited set of perceptual tools. Rather, we need to extend our view to cover the pluralism of modeling approaches that may shed light on the problem, and also develop the ability to use *all* of the available light to view and understand the system.

We need to develop an integrated, multi-scale, transdisciplinary, and pluralistic (IMTP), approach to quantitative ecological economic modeling. While this approach has frequently been suggested (see Norgaard 1989), it is difficult to implement within narrow traditional academic disciplines using traditional funding mechanisms. An IMTP approach would allow relationships between scales and modeling approaches to be directly investigated, and would result in a deeper understanding of the systems under study. We need to develop new ways of *scaling,* or using information at one scale to build models at other scales. This scaling is essential to building global ecological economic models, since no direct experiments or comparative analysis is possible at the global scale.

The objectives of an IMTP approach include modeling: a) human impacts on ecosystems, b) economic dependence on natural ecosystem services and capital and c) integrated interdependence between ecological and economic components of the system.

Ecosystems are being threatened by a host of human activities. Protecting and preserving these ecosystems requires the ability to predict the direct and indirect, temporal and spatial effects of proposed human activities, the ability to separate these effects from natural changes, and the ability to appropriately modify the short-term incentive structures that guide local decision making to better reflect these impacts (Costanza 1987). Predicting ecosystem impacts requires sophisticated computer simulation models that represent a synthesis of the best available understanding of the way these complex systems function (Costanza et al. 1990).

The more general objectives of human impact modeling are to predict ecosystem response as a result of various site specific management alternatives and natural changes. Development of this capability is essential for regional ecosystem management and also for modeling regional and global ecosystem response to regional and global climate change, sea level rise resulting from atmospheric CO_2 enrichment, acid precipitation, toxic waste dumping, and a host of other potential impacts.

Several recent developments make this kind of modeling feasible, including the ready accessibility of extensive spatial and temporal data bases from remote sensing, historical aerial photography, and other sources, and advances in computer power and convenience that make it possible to build and run predictive models at the necessary levels of spatial and temporal resolution (Costanza et al. 1990).

ADJUSTING INCENTIVES TO INCLUDE ECOLOGICAL COSTS AND UNCERTAINTY

Effectively communicating information about long-term impacts to the inherently short-sighted decision making process has always been a problem. Areas that need to be explored include: a) alternatives to our current command and control based environmental management systems that promise to do a much better job; b) modification of government agencies and other institutions to better account for and respond to environmental impacts; c) the appropriate role for economic incentives and disincentives; systems to account for and manage transnational environmental impacts; d) the best methods for incorporating the enormous uncertainty about environmental impacts into the decision making process; and e) the sociological, cultural, and political criteria for acceptance or rejection of policy instruments (Cumberland 1990).

One example of an innovative instrument currently being researched is a flexible environmental assurance bonding system designed to incorporate environmental criteria and uncertainty into the market system and to induce positive environmental technological innovation (Perrings 1989; Costanza and Perrings 1990). In addition to direct charges for known environmental damages, an assurance bond equal to the current best estimate of the largest potential future environmental damages would be levied and kept in an interest-bearing escrow account. The bond (plus a portion of the interest) would be returned if—and only if—the firm could prove that the suspected damages had not occurred or would not occur. If damages did occur, the bond would be used to rehabilitate or repair the environment, and to compensate injured parties. By requiring the users of environmental resources to post a bond adequate to cover potential future environmental damages (with the possibility for refunds), the burden of proof is shifted from the public to the resource user and a strong economic incentive is provided to research the true costs of environmentally innovative activities and to develop innovative, cost-effective pollution control technologies. This is an extension of the "polluter pays principle" to "the polluter pays for uncertainty as well."

A promising approach to developing and testing proposed new instruments, like the assurance bonding system outlined above, uses specialized computer games which test the responses of human players in order to address particular behavioral questions (Costanza and Schrum 1988).

MAINTAINING NATURAL CAPITAL TO ASSURE SUSTAINABILITY

Natural capital produces a significant portion of the real goods and services of the ecological economic system, so failure to adequately account for it leads to major misperceptions about economic health. This misperception is important at all levels of analysis, from the appraisal of individual projects to the health of the ecological economic system as a whole. It is particularly important at the level of national income accounting, however, because of the importance of these measures to national planning and sustainability.

Interest in improving national income and welfare measures to account for depletion of natural capital and other mismeasures of welfare has grown recently (see Ahmad et al. 1989). Daly and Cobb (1989) have produced an index of sustainable economic welfare (ISEW) that attempts to adjust GNP to account for depletions of natural capital, pollution effects, and income distribution effects. Figure 21.1 shows two versions of their index compared to GNP over the 1950 to 1986 interval. What is strikingly clear from figure 21.1 is that while GNP has been rising over this interval, ISEW has remained relatively unchanged since about 1970. When depletions of natural capital, pollution costs, and income distribution effects are accounted for, the economy is seen to be not improving at all. If we continue to ignore natural capital, we may well push the economy down while we think we are building it up. We are consuming our capital and endangering our ability to sustain income.

A simple policy proposal can accomplish much toward achieving sustainable development (Costanza and Daly, *in prep*). The necessary steps are:

1. Strive to hold throughput (consumption of Total Natural Capital (TNC)) constant at present levels (or lower truly sustainable levels) by charging for TNC consumption, especially energy, very heavily.

2. Put a substantial percentage of the charges into an assurance bond to cover the uncertainty about natural capital values, and seek to raise most public revenue from the remainder.

3. Compensate by reducing the income tax, especially on the lower end of the income distribution, perhaps even financing a negative income tax at the very low end.

Technological optimists who believe that energy efficiency can increase by a factor of ten should welcome this policy which raises resource prices considerably and would provide powerful incentives for just those technological advances in which they have so much faith. Pessimists who lack that technological faith will nevertheless be happy to see the throughput limited since that is their main imperative in order to conserve resources for the future. The pessimists are protected against their worst fears; the optimists are encouraged to pursue their fondest dreams. If the pessimists are proven wrong and the enormous increase in efficiency actually happens, then they will be even happier (unless they are total misanthropists). They will have gotten what they wanted, but at a cost below what they expected or were willing to pay. The optimists, for their part, can hardly object to a policy that not only allows but also provides strong incentives for the technical progress on which their optimism is based. If they are proven wrong, at least they should be glad that the rate of environmental destruction has been slowed.

Agreement on this policy seems politically possible, and does not hinge upon the precise valuation of natural capital. The valuation issue remains relevant because the policy recommendation is based on the perception that we are at or beyond the optimal relative size of man-made vs. natural capital. The system includes an assurance bond to incorporate our admittedly large uncertainty about the precise value of natural capital, and also provides incentives to learn more.

To be effective, this policy of no net loss of natural capital must be implemented immediately, before the natural capital is irreversibly destroyed. Instead of being mesmerized into inaction by uncertainty, we should fold uncertainty into the system in the form of a refundable assurance bond. By implementing this policy, we can make the global and long-term goal of sustainability consistent with local goals and incentives and assure a sustainable ecological economic system.

REFERENCES

Arrow, K. J. and A. C. Fisher. 1974. Environmental Preservation, Uncertainty, and Irreversibility. *Quarterly Journal of Economics* 55:313-19.

Barber, M., B. Patten, and J. Finn. 1979. Review and Evaluation of I-O Flow Analysis for Ecological Applications. In J. Matis, B. Patten, and G. White, eds., *Compartmental Analysis of Ecosystem Models. Statistical Ecology*. vol. 10. Bertonsville, Md.: International Cooperative Publishing House.

Bartlett, E. T. 1984. Estimating Benefits of Range for Wildland Management and Planning. In G.L. Peterson and A. Randall, eds., *Valuation of Wildland Benefits*. Boulder, Colo.: Westview.

Bishop, R. 1982. Option Value: An Exposition and Extension. *Land Economics* 58:1-15.

Braat, L. C. and W. F. J. van Lierop. 1985. *A Survey of Economic-Ecological Models*. International Institute for Applied Systems Analysis, A-2361 Laxenburg, Austria.

Brookshire, D. S., L. S. Eubanks, and A. Randall 1983. Estimating Option Prices and Existence Values for Wildlife Resources. *Land Economics* 59:1-15.

Brown, B. J., M. E. Hanson, D. M. Liverman, and R. W. Merideth, Jr. 1987. Global Sustainability: Towards a Definition. *Environmental Management* 11:713-719.

Brockner, J. and J. Z. Rubin. 1985. *Entrapment in Escalating Conflicts: A Social Psychological Analysis*. New York: Springer-Verlag.

Clark, Colin W, 1973. The Economics of Overexploitation. *Science* 181:630-634.

Cleveland, C. J., R. Costanza, C. A. S. Hall, and R. Kaufmann. 1984. Energy and the United States Economy: A Biophysical Perspective. *Science* 255:890-897.

Conrad, Jon M 1980. QuasiOption Value and the Expected Value of Information. *Quarterly Journal of Economics* 94:813-20.

Costanza, R 1980. Embodied Energy and Economic Valuation. *Science* 210:1219-1224.

Costanza, R 1982. Economic Values and Embodied Energy: Reply to D. A. Huettner. *Science* 216:1141-1143

Costanza, R. 1984. Natural Resource Valuation and Management: Toward an Ecological Economics. In A. M. Jansson, ed., *Integration of Economy and Ecology: An Outlook for the Eighties*. pp. 7-18. Stockholm: University of Stockholm Press.

Costanza, R. 1987. Social Traps and Environmental Policy. *BioScience* 37:407-412

Costanza, R. 1989. What is Ecological Economics? *Ecological Economics* 1:1-7

Costanza, R. and H. E. Daly, 1987. Toward an Ecological Economics. *Ecological Modelling* 38:1-7

Costanza, R. and H. E. Daly. In preparation. Natural Capital and Sustainable Development.

Costanza, R and S. C. Farber 1985a. Theories and Methods of Valuation of Natural Systems: A Comparison of Willingness-to-Pay and Energy Analysis Based Approaches. *Man, Environment, Space, and Time*. 4:1-38.

Costanza, R and S. C. Farber. 1985b. The Economic Value of Coastal Wetlands in Louisiana. Final Report to the Louisiana Department of Natural Resources. Center for Wetland Resources, Louisiana State University, Baton Rouge.

Costanza, R. and B. M. Hannon. 1989. Dealing with the "Mixed Units" Problem in Ecosystem Network Analysis. In: F. Wulff, J. G. Field, and K. H. Mann, eds., *Network Analysis of Marine Ecosystems: Methods and Applications. Coastal and Estuarine Studies Series. pp. 90-115*. Heidelberg: Springer-Verlag.

Costanza, R. and R. A. Herendeen. 1984. Embodied Energy and Economic Value in the United States Economy: 1963, 1967, and 1972. *Resources and Energy*. 6;129-163.

Costanza, R., and C. Neill, 1984. Energy Intensities, Interdependence, and Value in Ecological Systems: A Linear Programming Approach. *Journal of Theoretical Biology* 106:41-57

Costanza, R. and C. H. Perrings. 1990. A Flexible Assurance Bonding System for Improved Environmental Management. *Ecological Economics*. (In press)

Costanza, R. and W. Shrum. 1988. The Effects of Taxation on Moderating the Conflict Escalation Process: An Experiment Using the Dollar Auction Game. *Social Science Quarterly* 69:416-432.

Costanza, R, C. Neill, S. G. Leibowitz, J. R. Fruci, L. M. Bahr, and J. W. Day. 1983. Ecological Models of the Mississippi Deltaic Plain Region: Data Collection and

Presentation. U. S. Fish and Wildlife Service, Division of Biological Services, Washington, D.C. FWS/OBS-82/68.

Costanza, R., F. H. Sklar, and J. W. Day, Jr, 1986. Modeling Spatial and Temporal Succession in the Atchafalaya/Terrebonne Marsh/Estuarine Complex in South Louisiana. In D. A. Wolfe, ed., *Estuarine Variability.* pp. 387-404. New York: Academic Press.

Costanza, R., F. H. Sklar, and M. L. White. 1990. Modeling Coastal Landscape Dynamics. *BioScience* 40:91-107

Cross, J. G., and M. J. Guyer. 1980. *Social Traps.* Ann Arbor: University of Michigan Press

Cumberland, J. H. 1987. Need Economic Development be Hazardous to the Health of the Chesapeake Bay? *Marine Resource Economics* 4:81-93.

Cumberland, J. H. 1990. Public Choice and the Improvement of Policy Instruments for Environmental Management. *Ecological Economics* 2:149-162.

Daly, H. 1968. On Economics as a Life Science. *Journal of Political Economy* 76:392-406.

Daly, H. E. and J. B. Cobb, Jr. 1989. *For The Common Good: Redirecting the Economy Toward Community, the Environment, and a Sustainable Future.* Boston: Beacon.

Edney, J. J., and C. Harper. 1978. The Effects of Information in a Resource Management Problem: A Social Trap Analog. *Human Ecology* 6:387-395.

Farber, S. and R. Costanza. 1987. The Economic Value of Wetlands Systems. *Journal of Environmental Management.* 24:41-51.

Finn, J., 1976. The Cycling Index. *Journal of Theoretical Biology* 56:363-73

Funderlic, R and Heath, M., 1971. Linear Compartmental Analysis of Ecosystems. Oak Ridge Natl Lab, ORNL-IBP-71-4.

Greenley, D. A., R. G. Walsh, and R. A. Young. 1981. Option Value: Empirical Evidence from a Case Study of Recreation and Water Quality. *Quarterly Journal of Economics.* 95:657-73.

Hett, J and R. O'Neill. 1971. Systems Analysis of the Aleut Ecosystem. US-IBP, Deciduous Forest Biome Memo Report, 71-16, September.

Hannon, B. 1973. The Structure of Ecosystems. *Journal of Theoretical Biology* 41:535-46.

Hannon, B. 1976. Marginal Product Pricing in the Ecosystem. *Journal of Theoretical Biology* 56:256-267.

Hannon, B. 1979. Total Energy Costs in Ecosystems. *Journal of Theoretical Biology* 80:271-293.

Hannon, B., 1985a. Ecosystem Flow Analysis. *Canadian Journal of Fisheries and Aquatic Sciences,* 213. in R. Ulanowicz and T. Platt, eds., *Ecological Theory for Biological Oceanography.* pp. 97-118.

Hannon, B. 1985b. Conditioning the Ecosystem. *Mathematical Biology* 75:23-42.

Hannon, B. 1985c. Linear Dynamic Ecosystems. *Journal of Theoretical Biology.*116:89-98.

Hardin, G. 1968. The Tragedy of the Commons. *Science* 162:1243-1248.

Huettner, D. A. 1982. Economic Values and Embodied Energy. *Science* 216:1141-1143

Isard, W. 1972. *Ecologic-Economic Analysis for Regional Development.* New York: The Free Press.

Leontieff, W. 1941. *The Structure of the American Economy, 1919–1939.* Oxford: Oxford University Press.

Lovejoy, T. E. 1986. The Species Leave the Ark. In B. G. Norton, ed., *The Preservation of Species.* Princeton: Princeton University Press.

Liverman, D. M., M. E. Hanson, B. J. Brown and R. W. Merideth, Jr. 1987. Global Sustainability: Toward Measurement. *Environmental Management* 12:133-143.

Norton, B. G. 1986. On the Inherent Danger of Undervaluing Species. In B. G. Norton, ed., *The Preservation of Species.* Princeton: Princeton University Press.

Norgaard, R. B. 1989. The Case for Methodological Pluralism. *Ecological Economics* 1:37-57.

Pearce, D. 1987. Foundations of an Ecological Economics. *Ecological Modelling* 38:9-18.

Perrings, C. 1987. Economy and Environment: A Theoretical Essay on the Interdependence of Economic and Environmental Systems. Cambridge: Cambridge University Press.

Perrings, C. 1989. Environmental Bonds and the Incentive to Research in Activities Involving Uncertain Future Effects. *Ecological Economics* 1:95-110.

Pezzey, J. 1989. Economic Analysis of Sustainable Growth and Sustainable Development. Environment Department working paper no. 15. Washington, D.C.: The World Bank

Platt, J. 1973. Social Traps. *American Psychologist* 28:642-651

Randall, A. and J. Stoll. 1980. Consumer's Surplus in Commodity Space. *American Economic Review* 70:449-55.

Randall, A. 1986. Human Preferences, Economics, and the Preservation of Species. In B. G. Norton, ed., *The Preservation of Species*. Princeton: Princeton University Press.

Teger, A. I. 1980. *Too Much Invested to Quit*. New York: Pergamon.

Ulanowicz, R. E. 1980. An Hypothesis on the Development of Natural Communities. *Journal of Theoretical Biology* 85:223-245.

Ulanowicz, R. E., 1986. *Growth and Development: Ecosystems Phenomenology*. New York: Springer-Verlag.

World Commission on Environment and Development. 1987. *Our Common Future*. New York: Oxford University Press.

Wulff, F., J. G. Field and K. H. Mann. 1989. *Network Analysis of Marine Ecosystems: Methods and Applications. Coastal and Estuarine Studies Series*. Heidelberg: Springer-Verlag.

22

LOCAL AND GLOBAL INCENTIVES FOR SUSTAINABILITY: FAILURES IN ECONOMIC SYSTEMS

Stephen Farber
Department of Economics
Louisiana State University
Baton Rouge, LA 70803 USA

ABSTRACT

The failure of economic and political institutions to achieve sustainability through the destruction of ecosystems is attributable to short time horizons, failures in property rights, concentration of economic and political power, immeasurability, and institutional and scientific uncertainty. This paper illustrates how these factors have interacted in both developed and developing countries to destroy ecosystems. The major emphasis is on the agricultural and forestry policies in these countries, illustrating how population, foreign exchange, and internal political stability pressures have created policies that lead to ecosystem destruction. It points out the role that developed countries have played in creating incentives for ecosystem destruction in developing countries. It suggests the use of quantity-based regulatory instruments to protect the sustainability of basic needs-satisfying services of ecosystems, such as global climate. It considers various policy instruments to enhance sustainability incentives. It outlines a global carbon emissions allowance system that may increase time horizons of developing countries and reward them for changing their agricultural and forest policies. It proposes an alternative to traditional discounting procedures for project evaluation when ecosystem sustainability is at risk. It also proposes a means to increase incentives for private donations of lands to public management.

INTRODUCTION

The interaction between ecologic and economic systems has become increasingly important as we realize the impact of economic decisions on the sustainability and quality of the ecosystem. These impacts are apparent in developed and developing countries alike. In both cases, the historic and current treatment of ecosystem services as free goods is beginning to impose large economic costs. We can find evidence of local ecosystem damages such as soil erosion, salinization, deforestation, species loss, and toxic overloads in both developed and developing countries. We are beginning to find evidence of global ecosystem damages such as loss of the protective ozone layer, buildup of greenhouse gases, and levels of air pollutants that threaten health and environment.

The threat of ecosystem damage in developed countries is less immediate than the threat in developing countries. In developed countries, population growth rates are not as high. Greater knowledge and wealth allow more opportunities for substitutes, albeit costly, for lost ecosystem services, such as fertilizers for lost soil fertility. Greater wealth and, perhaps, political stability in developed countries is also likely to allow longer time horizons in public policy. On the other hand, poverty levels in developing countries dictate a myopic perspective, placing sustainability at risk.

Examples wherein economic institutions and public policy provide incentives for ecosystem damage are not difficult to find. These incentives directly impact sustainability of ecosystem service flows. The purpose of this paper is to outline the basic causes of failures in incentive systems to sustain ecosystems, and to suggest incentive systems which are directed toward sustainable ecosystems, both locally and globally.

FAILURES IN ECOSYSTEM SUSTAINABILITY INCENTIVES

Economic and political institutions have failed to provide proper incentives for sustaining ecosystems, because of five major, possibly interacting, causes:
  1. Short time horizons
  2. Failures in property rights
  3. Concentration of economic and political power
  4. Immeasurability
  5. Institutional and scientific uncertainty
For example, the classic problem of overgrazing the commons is attributable to property rights failure. When access to the commons is not limited through some form of property rights, each farmer knows that the grass not grazed today will be gone tomorrow. Failures in property rights, in this case, lead to short time horizons. Concentration of wealth in the hands of developed countries allows the exploitation and destruction of ecosystems in poor countries through, for example, timber harvests and mineral extraction. Concentration of wealth within developing countries skews public policy toward benefiting the wealthy and politically powerful, often at the expense of natural systems upon which the poor rely. The rich become richer, and the poor become poorer, forcing even shorter time horizons on the populations and political systems of developing countries. Ecosystem effects of policy may be more multidimensional than economic effects, which have the common denominator of money. The need for simplicity in policy making may favor the economic over the ecosystem effects simply because of measurability. Institutions may be (purposely?) ignorant of direct and indirect effects of policies on sustainability, or ignorant of actions within their jurisdiction on other areas. Scientific uncertainty about biophysical and geological relations will also impede sustainability

policies. Scientific uncertainty may be exploited to develop false claims in order to thwart sustainability policies, such as acid rain programs in the United States.

Any attempt at sustainable development, locally or globally, will require sustaining the flow of ecosystem services and correcting the above failures. Before considering corrective measures, it would be useful to highlight several particularly severe cases of ecosystem destruction attributable to one or more of the above failures.

Service flows from the global ecosystem, such as temperature, rainfall, and photosynthesis, are at risk of being unsustainable due to the inherent "commons" nature of the global ecosystem as a waste sink. Rights to waste disposal capacity do not have to be bought by users. Countries have incentives to independently reduce this ecosystem destruction when the scale of feedback effects from their activities on their own economies becomes significant. When a country such as the United States accounts for 20% of global carbon emissions from fossil fuels (Brown et al. 1990), and is likely to bear large costs under some global warming scenarios, one could anticipate unilateral policy proposals such as those introduced to the U.S. Congress in 1989. Countries have little incentive to unilaterally reduce damage to the ecosystem when doing so will not reduce the damaging behavior of others; or, worse, if policies such as reduced fossil fuel use would significantly change trade balances. The formation of multilateral consortia with mutually agreed upon goals and allocations of rights to use the ecosystem is a solution to this commons problem. The Montreal Protocol, the Helsinki Declaration on CFCs, Scandinavian agreements on CO_2 emissions, and European agreements on acid rain precursors are examples of the successful implementation of this strategy (Brown et al. 1990). These consortia agreements are easier to reach when the benefits and costs are "fairly" distributed, or when costs are low.

Concentration of economic power, both between and within countries has the potential for destruction of local ecosystems. Concentration alters the terms of contracts and policies in favor of preferences, benefits, and costs of the more powerful party. Whether local sustainability is enhanced depends on the goals of the powerful party. In certain instances, sustainability may be in the best interests of the powerful party, even when it may be at odds with local interests. However, there may be cases where contracts and policies are guided by the powerful party toward less sustainability than is locally desired, or that would be attained under more equal bargaining. This may occur in several instances. When an exploiting party has substitute ecosystems available, it can exploit one and then move to the next. Examples would include Japanese lumber firms harvesting in one country and moving to the next; or mining firms exploiting one area and moving to another. In these cases, the benefits of sustainability to the exploiter are low and the costs are high. Exploiters may have shorter time horizons than local interests. This may be due to the exploiter facing higher interest rates, the need to show quick and measurable returns, and concern for the potential loss of their economic power For example, aid to developing countries may need to show quick and visible returns in order for the legislatures of the granting country to continue support; or exploiting firms may be highly leveraged, like the Japanese lumber firms; or the economic power of the exploiter may be contingent upon temporary and unstable political conditions. Finally, the exploiters may seek to increase their bargaining power in order to further enhance their goals. Military intervention and the destruction of life support systems is an extreme example of this. More moderate examples would include policies directed toward shifting peasants from rural to urban areas in order to provide a cheap labor supply. Rapid exploitation of agricultural and forested areas would be consistent with the urbanization goal. This type of exploitation may occur in countries where wealthy industrialists are major forces in determining public policy.

Traditional methods of selecting projects that enhance wants are not valid when project costs include diminished sustainability of a basic ecological support system. The traditional use of discounting in evaluating projects is attributable to the argument that funds for one project could be used for another. One uses the rate of return available from an alternative set of projects to judge the project in question. Using rates of return on alternative projects implies perfect substitutability in benefits between projects. This may be reasonable for developed countries where basic needs are satisfied and benefits from projects are satisfying higher level wants. These wants are more easily tradeable for one another, such as better roads, more bridges, and greener lawns. It is not appropriate for some projects in developing countries where many costs of projects do not merely satisfy wants, and are not easily substitutable for one another. For example, one cannot measure the cost of constructing a dam by using the rates of return on traditional alternative feasible projects when the dam displaces a critical life support ecosystem. The developed nations impose these discounting criteria on project selection in developing countries. The result is economic development projects that undermine sustainability of ecosystems. Such projects include short-lifetime dams which displace critical ecosystems forever, irrigation projects which result in irreversible salinization of farmlands, and high-yielding, monoculture crops supported by intensive fertilizer and pesticide use which result in eventual health and environmental risks (Reid 1989).

Agricultural and forestry policies in developing countries provide good examples of short time horizons, property rights failures, and wealth concentration leading to ecosystem destruction. Timber is seen as a good short run source of foreign exchange, especially in countries where traditional commodity export prices have fallen. For example, Indonesia's expansion of its hardwood plywood industry was motivated by lower petroleum earnings. The result is deforestation at the rate of nearly one thousand hectares per year (Repetto 1990, pp. 40-41). Three government induced institutional factors combine to hasten timber harvesting in developing countries. First, concessions, or development rights, are frequently given to politically connected or wealthy individuals. This has been the case in countries such as the Ivory Coast, the Philippines, and Indonesia (Repetto 1990). In countries where political change can void property rights of concessions, political instability would motivate any individuals receiving these concessions to rapidly develop or sell them. Second, severance royalties and fees are considerably below their revenue maximizing rates. For example, potential revenues to Indonesia could have been three times higher than actual revenues during 1979-1982. During this period taxes and royalties were collected on only three-fourths of this country's harvested acres. The full cost of harvesting is further subsidized by income tax write-offs. Third, concession agreements are often for periods shorter than those necessary for sustainable yield management (Repetto 1990, p. 39). This creates a "tragedy of the commons," since what you do not harvest today will belong to someone else tomorrow. Concentration of economic and political power may be the root causes of these non-sustainable policies.

Developed countries contribute to short time horizons for timber policy in developing countries. For example, Japan accounts for nearly one-third of the tropical timber trade. The high leverage financing of the Japanese firms requires quick cash flow to pay financing charges (Repetto 1990, p. 40). This is another example of high discount rates in developed countries being imposed on the management of developing countries' assets. Another example arises from European and Japanese protection against importation of processed wood. Although it may be misguided policy, many developing countries are attempting to develop downstream processing of timber in order to create jobs and add value. In order to be cost-competitive in protected markets, these countries must substitute cheap timber inputs for expensive sawmill capital. The result is harvest and sawmill operations that require very high timber input per unit of processed output. In the Ivory Coast, for example, the processing industry requires 30% more timber than efficient mills (Repetto 1990, p. 39).

Agricultural policy in both developed and developing countries has affected ecological sustainability. For example, in the United States, price support systems which include allowances for keeping lands fallow create incentives to farm remaining lands more intensively with heavy pesticide and fertilizer use. Ground and surface water contamination and air pollution result. More intensive tillage enhances erosion potential. Intensive irrigation causes aquifer drawdown and salinization. In the United States, pumpage rates on one-fifth of the nation's irrigated land exceed recharge rates, and salinization has irreversibly damaged over one-fourth of irrigated cropland (Brown et al. 1990). This is an example of institutional ignorance, and a concentration of political power that skews policy toward farm interests. Although the ostensible purpose of this policy is to assure a long-term food supply, longer term ecosystem effects may negate that policy. Again, the time horizon is too short to incorporate ecosystem effects.

In developing countries, the particularly short-sighted agricultural policies are driven by a combination of survival needs due to population pressures, shortages of foreign exchange, and potential political revolution. These factors result in short time horizons in public policy, an exploitation of the commons mentality, and attempts by ruling political parties to pacify small numbers of individuals who control the wealth of a country.

In order to reduce urban population pressures, which are attributable to high population growth coupled with weak urban economies, countries have provided incentives for opening up and developing their frontiers. Countries such as the Philippines and Brazil, where an incredible 8 million hectares per year is deforested (Repetto 1990, p. 40), have land tenure policies whereby individuals obtain title or rights of occupancy only when they have cleared the land. In Thailand, lack of ownership rights has led to extensive rather than intensive land use. Only 20% of the total area of private land in Thailand has title deeds. The result is that farmers, since they have no collateral, are unable to obtain loans for land improvements or fertilizer purchases. Instead of continuing to farm their plots more intensively, they simply migrate and clear new land. For example, forested area in Thailand decreased from 53% in 1961 to 29% in 1986. Where soil quality was poorest, forests declined even more. For example, forested area fell from 42% to 14% during 1961-86 in Northeast Thailand (Phantumvanit and Sathirathai 1988). These poorly defined property rights systems directly result in forest ecosystem destruction.

A good example of poorly defined property rights in developed countries occurs in the wetlands of Louisiana, US. Mineral rights revert to the state when wetlands become open water if there has been no mineral development on the property. The cheapest methods to avoid the loss of mineral revenues have been of two types: hurried development of oil and gas in areas likely to revert to open water, causing extensive wetlands damage and hastening erosion and salt water intrusion; and construction of ring levies around the property to maintain it as private property, interfering with normal estuarine processes.

In countries such as Brazil, tax and subsidy policies directed toward increasing cattle ranching allow cattle barons to buy up individuals' cleared plots. These displaced individuals can simply move on to clear other forested areas. The visible result is the conversion of nearly 9 million hectares of forest to cattle ranches, many of which are now abandoned (Reid 1989, p. 32). The short time horizon of public policy is illustrated by the fact that, while net present values of land in the Amazon are greater under agricultural uses over 10-year time spans, over longer time spans the net present value of Amazonian land is greater under low impact extractive uses, such as rubber tapping, nut gathering, hunting, and fishing (Reid 1989, p. 33). The net global value of maintaining Amazonian forests would be even greater when including the global costs associated with the carbon releases resulting from deforestation. In fact, agricultural and forestry policy in Brazil have made it the fourth largest discharger of carbon in the world (Brown et al. 1990).

Competition in markets may shorten time horizons. There is evidence that more competitive firms face higher interest rates and undertake fewer long-term projects (Farber 1981). Similarly, competitive forces in agricultural markets may induce farmers to take short-term perspectives for financial survival. Farmers must maintain yields and cash flow to satisfy bankers and to make a sufficient return on their land investments in both developed and developing countries. This has led to the adoption of high-yield crops, monoculture farming, and reduction in genetic diversity. The result is increased variability in crop yields under varying climate and economic conditions. For example, the loss of 15% of the 1970 maize crop in the United States to leaf blight has been cited as an example of the risks associated with genetically narrow, high-yield cropping (Crosson and Rosenberg 1986, p. 125).

Fortunately, countries such as the United States with a "diversified portfolio" of agricultural products can tolerate this variability. Developing countries may not be so fortunate. There is a particular need to maintain genetic diversity in developing countries, but it is in these countries where the short time horizons, competitive pressures, government policies directed toward obtaining foreign exchange, and ignorance are most severe and are likely to lead to the narrowest portfolio of gene pools through adoption of a narrow range of monoculture crops. High-yield crops are especially vulnerable to variations in climate, water, and soil quality (Barbier 1989, p. 14). If development projects in these countries lead to greater variations in climate and to the development of poorer quality soils, as might be the case where extensive forests are converted to croplands, the use of high-yield crops may create considerable risk to agriculture. Since intensive fertilizer use will be essential to the success of high-yield crops on poor soils, the economic success of agriculture will become highly dependent on fertilizer prices, which depend on the price of oil and gas and are outside the control of individual countries. The result may be serious foreign exchange problems.

POLICY INSTRUMENTS

Policy instruments must be presented in the context of sustainability goals. There is considerable ambiguity associated with the operational meaning of sustainability: sustainability of what and for whom. It is unlikely that there will ever be any global agreement on anything but sustainability of a level of welfare sufficient to satisfy very basic needs of survival. Hopefully, this would include sustainability of basic life-support ecosystems, biodiversity sufficient for robust ecosystems and future information needs, and sustainability of renewable resource systems at levels of regeneration sufficient to provide for substitution options to future generations. In any case, sustainability will likely come to mean the perpetual availability of some basic needs-satisfying ecosystem services. In this context, we are concerned with finding policy instruments designed to guide the economic system in a manner consistent with basic sustainable ecosystem goals. These are instruments that would guard against the purposeful or inadvertent permanent damage done by the economic system to life-support systems, biodiversity, and the renewable resource base.

Policy instruments differ primarily on the basis of whether they are price-based or quantity-based. Any instrument which sets a direct price on behavior (whether it is legal or illegal behavior) is price-based. Obvious examples are taxes on emissions, fees on hazardous waste, severance taxes on harvested trees, or fines for illegal behavior. Quantity-based instruments set direct restrictions on volumes of inputs, emissions, harvest, or technologies. They may be enforced with price instruments, such as fines.

When information is perfect and enforcement costless, the two instruments are identical. However, it can be shown that when there is considerable uncertainty regarding the control cost of meeting regulations, such as the cost of pollution control, and the marginal benefits of control increase dramatically as the level of control diminishes, quantity-based instruments are superior to price-based instruments (Weitzman 1974). Other than this preference for quantity-based instruments to satisfy the goal of basic needs sustainability, the particular instrument is likely to depend upon the nature of the institutional failure. For example, short of remedying the immediate causes of short time horizons, quantity-based use rules, such as fishing quotas, can force longer horizons on ecosystem users. Quantity or price restrictions on resource use can each correct property rights failures by creating implied rights. When ignorance is the cause of damage to an ecosystem, private actors can be assessed refundable fees or be required to post variable bonds in advance of potentially damaging behavior, thus inducing them to discover the true risks. Restricting access to resources, a quantity instrument, until more information is known is another way to reduce ignorance. In addition, price-based regulatory systems can be layered on top of the quantity-based needs sustainability regulations. Limiting the number of emissions permits and allowing trading of excess permits is an example.

In some cases, institutional changes are needed in addition to specific regulatory instruments. For example, traditional environmental regulation has been on a media-by-media basis in the United States (air, water, land). Regulatory actions taken in one media often had adverse consequences on others. The problem was poorly defined jurisdictions, or regulatory property rights. Efforts are now being taken to broaden jurisdictional concern. Louisiana is currently attempting to require the agency which grants tax exemptions to incorporate environmental factors in exemption decisions.

Instruments may differ in their enforcement costs. For example, it is cheaper to monitor technologies in place than discharges. Price-based instruments, such as a CO_2 tax, may be impossible to enforce globally. Ease of enforcement, as well as the ability to capture rents under regulation (Buchanan and Tullock 1975), may explain why U.S. regulation has been primarily directed toward technology and quantity-based instruments. Philosophical reasons, such as price-based systems being viewed as providing "rights to pollute," also influence the choice of quantity-based instruments.

An efficient instrument for CO_2 control at the global level may be quantity-based carbon emissions allowances. Countrywide carbon emissions are measurable with some degree of accuracy, given estimates of fossil fuel use and satellite imaging of forest cover and burning. Trading of allowances would have some favorable international wealth distribution effects. The global distribution of allowances would implicitly be a form of income distribution. Income from excess allowances would be a source of foreign exchange for developing countries. This may reduce pressures within these countries to engage in short-sighted agricultural policies. As total allowable global emissions are reduced over time, the value of trees for earning emissions credits would increase. This would place a higher value on forests than currently, and may place additional pressures on countries to deal with population, land tenure, and frontier development on a sustainable basis. This new source of wealth may stimulate longer time horizons in developing countries.

Improving the definition of property rights can enhance the likelihood of sustainability in many instances. This is most noticeable in countries where land tenure rights are short-lived or where rights are obtained only after ecosystem destruction, such as land clearing, has occurred. Land reform, whereby households are given too small a plot of land on which to practice integrated pest management, crop rotation, and erosion control, is also an example of inefficient property rights. We see in this example that improperly set quantities create property rights problems.

Shortsighted development projects in developing countries, induced by traditional discount policies of international lenders could be partially avoided by requirements that such projects meet two tests. First, the traditional test would consist of positive net discounted benefits, where costs include full ecosystem costs. The discount rate that should be used would not be the rate of interest in developed countries, or even the rates of return on private investments in the country where the project is to take place. Instead, the discount rate should be the rate of time preference in the project country, under the argument that these funds could have been used to satisfy basic consumption needs. This rate of time preference could be assumed to equal the real rate of interest in the project country. The second test would be that the project have a greater internal rate of return than ecosystem enhancement investments of similar scale. Due to the "public good" nature of many ecosystems benefits, there is no reason to expect that social rates of return on ecosystem investments would equal any observable market interest rate.

An important alternative to the discounted benefit-cost criteria for project evaluation would be to consider the magnitude of least cost methods of maintaining a safe minimum standard for preservation of ecosystem functions (Bishop 1978). This is a potentially powerful evaluation method when poor measurability, long term effects, and uncertainty are critical factors in evaluating the costs of ecosystem destruction. For example, an annual cost of $3 million to protect the habitat of the California condor is rather small (Randall 1986, p. 98), although it may appear to be a large cost in a developing country.

Significant social spillovers may not be capturable privately within large ecosystems, and traditional private discounting in investment and harvest decisions may result in destruction of significant portions of these ecosystems. Examples of the latter include wetlands and forest systems. In these cases, public ownership and management of the ecosystems could be superior to private ownership. Restrictions on private use of ecosystems, such as takings, zoning, or restrictions on activities to protect endangered species, are quantity-based methods of preserving ecosystems. Such methods may impose large costs on those from whom rights are taken, while benefits accrue to the public at large. There have been instances where such procedures have created incentives for landowners to destroy and not report endangered species on their property (Carlton 1986, p. 261). There may be more effective methods for preserving important ecosystems and endangered species.

Debt-for-ecosystem swaps are proving to be successful in developing countries. A problem in developed countries, however, is that traditional voluntary procedures for conversion may result in too little conversion. For example, in the United States, private land donations are increasingly being made to environmental preservation groups, such as the Nature Conservancy. A major difficulty with these transfers has been the land assessment process for tax purposes. Land must be assessed by one of three methods: comparable market value, replacement value, or income value. Such methods may leave ecosystems substantially undervalued since they may have significant social values beyond any of these valuation methods. For example, in the United States, Louisiana, wetlands are currently selling for $200-300 per acre, primarily for hunting rights, yet some studies show reasonable estimates of their social value to exceed $2500 per acre (Costanza et al., 1989 p. 354). If donations of ecosystems could be made at their estimated social value, this would increase the rate of conversion of socially significant ecosystems from private to public ownership, as well as increase research directed toward the difficult process of social valuation of ecosystems.

In the case of endangered species, landowners could be compensated for the costs they incur from having to restrict land usage in order to protect the species habitat.

Alternatively, property which protects such species could have tax exemptions. These compensation or subsidy schemes would not only be equitable, but would provide private individuals with greater incentives to protect and identify endangered species, compared to purely restricted use methods.

SUMMARY

This paper has illustrated how short time horizons, failures in property rights, concentration of economic and political power, failures of measurability, and ignorance combine to reduce the likelihood of ecosystem sustainability. The failure of ecosystem sustainability, in turn, increases economic costs of satisfying needs and wants. Failures of development policies in developing countries to consider long-term ecosystem impacts will be particularly costly, since these ecosystems are the primary source of wealth to these countries. Sustainability of crucial ecosystems may be critical for sustainability of the entire economic and political structure of the developing countries. These short-sighted policies are partially the fault of developed countries, who use the resources, impose the financial return requirements on investments, and set the restrictions on development project selection.

Some policy instruments may be easily adapted to solving problems of ecosystem destruction in developing countries. Creating a global market for the right to carbon emissions, through carbon allowances or carbon dioxide emissions reduction credits, would increase the value of preserving forest ecosystems, provide needed foreign exchange to developing countries, and institutionalize a wealth transfer system. Increasing the costs of using forested areas as population relief valves, may induce countries to undertake more responsible population control policies. The initial allocation of such rights could assure a net income transfer from developed to developing countries. The value of these rights would increase with time as the total global emissions limits were reduced.

Development project selection criteria can be more responsive to long-term ecosystem costs. This can be accomplished by using lower discount rates, undertaking more extensive quantification of ecosystem benefits, and considering the opportunity costs of Safe Minimum Standards of ecosystem preservation.

The incentives to adequately manage or to convert privately-owned critical ecosystems, or portions of ecosystems, to public ownership can be increased if private values of these ecosystems more closely approximated their full social value. By using full social values as the basis for tax purposes when land is donated to conservation groups or to the public, the incentives to donate would be increased. Such valuation procedures would have the added effect of increasing research and development in ecosystem benefits. Alternatively, private owners could be compensated for lost income if use restrictions are imposed on critical ecosystems.

REFERENCES

Barbier, E. B. 1989. Sustaining Agriculture on Marginal Land. *Environment* 31:12-39.

Bishop, R. 1978. Endangered Species and Uncertainty: The Economics of a Safe Minimum Standard. *American Journal of Agricultural Economics* 60:10-18.

Brown, L. R., A. Durning, C. Flavin, H. French, J. Jacobson, M. Lowe, S. Postel, M. Renner, L. Starke, and J. Young. 1990. *The State of the World, 1990*. New York: Norton.

Buchanan, J. and G. Tullock. 1975. Polluters' Profits and Political Response: Direct Controls versus Taxes. *American Economic Review* 65:139-147.

Carlton, R.L. 1986. Property Rights and Incentives. In B. G. Norton, ed., *The Preservation of Species, pp. 255-267*. Princeton: Princeton University Press.

Crosson, P. R., and N. J. Rosenberg. 1988. Strategies for Agriculture. *Scientific American* 261:128-136.

Costanza, R., S. C. Farber, and J. Maxwell. 1989. Valuation and Management of Wetland Ecosystems. *Ecological Economics* 1:335-361.

Farber, Stephen. 1981. Buyer Market Structure and R&D Effort: A Simultaneous Equations Model. *Review of Economics and Statistics* 63(3):336-345.

Phantumvanit, D., and K. S. Sathirathai. 1988. Thailand: Degradation and Development in a Resource-Rich Land. *Environment* 30:11-32.

Randall, A. 1986. Human Preferences, Economics, and the Preservation of Species. In B. G. Norton, ed., *The Preservation of Species*, pp. 255-267, Princeton: Princeton University Press.

Reid, W. V. C. 1989. Sustainable Development: Lessons from Success. *Environment* 31:7-35.

Repetto, R. 1990. Deforestation in the Tropics. *Scientific American* 262(4):36-45.

Weitzman, M. 1974. Prices vs. Quantities. *Review of Economic Studies* 41:477-491.

23

INTERGENERATIONAL TRANSFERS AND ECOLOGICAL SUSTAINABILITY

John H. Cumberland

Professor Emeritus
University of Maryland
Department of Economics
and
Center for Environmental and Estuarine Studies
Chesapeake Biological Lab
Solomons, MD 20688-0038 USA

ABSTRACT

Neither market economies nor centrally directed economies will necessarily maintain stocks of nonrenewable resource and environmental-ecological systems at levels adequate to assure intergenerational sustainability. The prevention of long-run decline in welfare may require making intergenerational transfers. Planning effective means for making transfers to future generations must be centered not only upon monetary and fiscal measures, but more importantly upon real resources relevant to ecological sustainability. Among the sustainability-relevant resources which might be transferred are capital equipment, knowledge, genetic stocks and reserves of nonrenewable resources. Transfers of these kinds of assets will probably be necessary for sustainability, but they may not be sufficient to assure intergenerational equity. Another type of sustainability-relevant transfer which should be explored is that of large-scale complex functioning ecologies such as estuaries, rain forests, game preserves, and river basins. The protection of these ecologies and their transfer to future generations would involve costs and benefits whose intragenerational as well as intergenerational distribution both merit detailed evaluation. A public choice approach is suggested in order to develop transfer policies which are scientifically valid, economically efficient and distributionally equitable between regions, interest groups, and generations.

INTRODUCTION

While the process of defining sustainable growth and sustainable development continues, one of the earliest definitions, "Sustainable development is development that meets the needs of the present without compromising the ability of future generations to meet their own needs" (World Commission on Environment and Development 1987) is a useful working concept. Valuable refinements and extensions have been added by Pezzey (1989), Pearce, Markandya, and Barbier (1989), and others. One of the precursors to this work was Tietenberg (1988, p. 492), who pointed out that in considering long time spans, conventional discounting so undervalues the future that a concept of sustainability as an alternative to discounted present value should be used in making provisions for future generations. While the conventional neoclassical analysis of welfare for future generations, based upon projecting recent trends, assumes increasing material wealth over time, the growing literature in ecological economics (Costanza 1989) has drawn attention to the fact that the external effects of economic growth in depleting ecological capital may require that current generations make transfers to future generations in order to achieve intergenerational equity. Pearce et al. (1989, p. 3) point out that:

1. Compensation for the future is best achieved by ensuring that current generations leave the succeeding generations with at least as much capital wealth as the current generation inherited;
2. Compensation for the future should be focused not only on man-made capital wealth, but should pay special attention to environmental wealth. That is, future generations must not inherit less environmental capital than the current generation inherited.

The purpose of this paper is to examine the case for transferring environmental wealth between generations, to examine the appropriate form for intergenerational transfers should these be undertaken, and to examine policy instruments for achieving the transfers.

ARGUMENTS FOR AND AGAINST INTERGENERATIONAL TRANSFERS

The case for intergenerational transfers is not strongly supported by neoclassical economics. The comfortable assumption is made that long-term historical trends of increasing per capita real income can reasonably be extended far into the future. The reasons for this include: growing accumulation of capital equipment, increased knowledge, improved technology, and, more recently, the increased adoption of market economies. Troublesome doubts have been introduced from time to time from the field of natural resource economics over concern for exhaustion of nonrenewable resources. However, such concern over depletion has usually been dismissed in the neoclassical model as an irrelevant geological concept which should be replaced by the more operational economic concept of market driven optimal rates of utilization, rendering physical depletion irrelevant. Occasionally, it was recognized that increasing consumption of particular nonrenewable resources could eventually result in such high costs of extraction that the resource was no longer economic, but such concerns were generally treated as inconsequential on the assumption that substitution would assure the availability of alternative resources.

Other arguments against the need for compensation and intergenerational transfers have been advanced. One of the most thoughtful critiques of concern over sustainability is that based upon the positive and accelerating accumulation of knowledge, and the assumption that this accumulation of knowledge represents a possible escape from the forces of entropy (Mancur Olson). Another counter position reported by Pezzey (1989) is simply that if efficient policies are introduced for the management of natural and environmental resources, no additional policies for compensation and intergenerational transfer need be introduced.

The internally consistent theoretical structure of economic analysis has provided a powerful critique of some of the more naive doomsday alarms about running out of natural resources and has diverted the mainstream of economic analysis away from concern for the future availability of conventional natural resources. However, a few cracks in the classical foundation of economic theory have been introduced by Kneese, Ayres and d'Arge (1970) and Georgescu-Roegen (1971), who were among the first to confront economic equilibrium theory with the realities of mass balance, entropy and irreversibility. More serious challenges to economic assumptions appeared with the emergence of ecological economics and its emphasis on the sensitive and intricate web of living relationships needed for the earth's support system (Daly 1977). If conventional economic development continues to destroy irreversibly some of the strands of this web, intergenerational sustainability cannot be assured. To the extent that there are nonsubstitutable components of the earth's life support system, such as fresh water and oxygen, the neoclassical market model is an incomplete and misleading blueprint.

Even economic purists have to accept, from the logic of their own models, the potential seriousness of market failures in the form of externalities, public goods, and open access common property resources. It is not necessary to accept the more philosophical, deep ecology attacks upon the neoclassical system, such as that of Sagoff (1988), to recognize the magnitude of welfare losses from market failures when they appear in the form of widespread global pollution externalities, and widespread abuse of the global commons. It should also be noted that in addition to the positive intergenerational transfers now occurring in the form of growing knowledge, technology, and capital stock, negative intergenerational transfers are also being made to the future in the form of climate change, radioactive waste generation, species extinction, toxic waste storage, contamination of aquifers, and other environmental damage.

Thus, while no conclusive case has yet been made for intergenerational transfers, the growing concerns over the empirical evidence of global environmental damage including the greenhouse effect, ozone depletion, deforestation, acidification, all point to the wisdom of designing some form of environmental insurance measures against irreversible global damage. Arrow and Fisher (1974) drew attention to the positive expected value of delaying irreversible development because of the high probability of having better information in the future on alternative technologies and other options. Leon Taylor (1990) suggests that the demand for environmental protection will grow over time as affluence increases because the degree of relative risk aversion appears to rise with wealth.

The case for intergenerational transfers and compensation is strengthened by the growing empirical evidence of human modification of the global habitat. As long as the present rate of population increase continues at approximately 2% annually, the global population will double every 35 years, or about once per generation, implying a global population of forty billion souls in the next century. However, even a century is not a sufficient time span over which to plan for ecological sustainability. As soon as a finite time span is contemplated, even one as long as a century, rational planning allows for capital consumption, especially towards the latter part of the period. The planning period needed to protect ecological resources for truly sustainable sustainability verges upon perpetuity. Therefore, in addition to thinking globally and acting locally, sustainability requires that we think in terms of perpetuity and act intergenerationally.

If exponentially growing human populations occupy the most fertile and productive areas of the earth, then most of the globe's most sensitive ecosystems will be threatened. Furthermore, if the trend toward the adoption of market economies results in the expected increases in production, consumption and development, very significant pressures will be put upon renewable and nonrenewable resources alike. Additionally, mass balance and entropy forces will generate extraordinary waste loadings to be discharged back into the environment, some of them highly toxic and environmentally detrimental.

Straight-line extrapolation of present trends would be both excessively pessimistic and unrealistic. The growing environmental pressures from population growth, resource depletion, and waste disposal can be expected to generate price changes and other feedback signals which should, to an unknowable extent, reduce population growth, conserve resources, generate substitution, encourage recycling and improve waste disposal methods. Capital stock should increase, technology should advance, and above all, human knowledge should expand. However, despite all of these feedback mechanisms and corrective adjustments, crowding, resource pressures and waste disposal problems will inevitably increase, as will pressures on sensitive ecological systems. Therefore, in the light of these fundamental demographic and environmental trends, current production and consumption patterns are not sustainable, if variables such as space per capita, amenities, and natural endowments per capita are defined as significant dimensions of the quality of life. Even with global crowding, it's possible that real per capita incomes could expand, at least in terms of priced market goods and services (Kahn, Brown and Martel 1976). However, environmental and ecological resources per capita must, by definition, decline over the next one hundred years. The question is, by what means is it possible for future generations to be compensated for these negative contributions to their future heritage, should society wish to make such bequests, and what policy instruments can be devised for this purpose?

ALTERNATIVE OPTIONS FOR INTERGENERATIONAL TRANSFERS

Pearce et al. (1989, p. xiv), have indicated that the achievement of intergenerational equity will require that earlier generations leave to succeeding generations all forms of wealth, including,"—a stock of knowledge and understanding, a stock of technology, a stock of man-made capital *and* a stock of environmental assets—no less than that inherited by the current generation."

This list of the forms of wealth that might be bequeathed to later generations provides an agenda with which to begin an analysis of intergenerational compensation. The first item, a stock of knowledge and understanding, is not only highly resistant to entropy, but under ideal circumstances, can actually result in negative entropy. Furthermore, knowledge and understanding, if properly applied, can be a major factor in generating the corrective behavior which may offset the environmental and ecological damage which would otherwise occur. However, the historical record suggests that human beings individually and in large numbers are not always capable of applying knowledge rationally in ways that assure and improve their well-being. If population pressures, war, and environmental damage continue, the acquisition of appropriate knowledge, and the ability to apply it may not occur at an adequate rate. The problems of irrationality, irreversibility, and the market failures involving externalities, and public goods (nonexclusiveness and nonrivalry) may grow at rates greater than the acquisition and application of knowledge needed to offset them. Generation of knowledge is a necessary condition for sustainability, but without wisdom and effective application, is hardly sufficient.

A growing stock of improved technology can also provide a valuable bequest to future generations. However, not all technology is environmentally benevolent. Examples of technology without which the sum of human welfare might have been greater are the automobile, the bulldozer, all terrain vehicles, and weapons of mass destruction. Although technology assessment is a subjective and inexact science, I suggest one possible criterion for evaluation: The most damaging technologies are those which require the largest scale of operation and which most increase the rate of entropy, while the technologies which on net balance add the most to human welfare are those that combine the minimum amount of material and energy input with the maximum amount of human intelligence and creativity. Early examples of welfare enhancing technologies are the compass, the sextant, and the chronometer, which opened vast new opportunities to humanity, using minute quantities of energy, plus extraordinary inputs of human intelligence and innovation. More recent examples of low entropy technology which can reduce energy requirements and waste disposal problems are electronic miniaturization, the computer, solar energy, and, potentially, genetic engineering. To the extent that these technologies expand the role of human intelligence and reduce entropy, the transfer of technology to future generations can result in great increases of human welfare. However, since technology, like knowledge, is inherently neutral, and is capable of being used for environmentally constructive or destructive purposes, technology transfer is not automatically a beneficial inheritance to offer future generations.

Stocks of manmade capital also provide opportunities for bequests to the future. Unfortunately, not all of this capital is consistent with ecological sustainability. Clearly, the automobile and all of its associated support systems have greatly depleted the earth's stock of resources and ecological assets. Readers can probably add a personal list of manmade capital assets and technologies without which the world would be better off. But it should also be recognized that much of the capital that civilization has been acquiring during recent millennia will constitute a benevolent heritage to the future, including communication systems, information transfer systems, housing stocks, plant and equipment, urban systems, health care facilities, educational institutions, public utilities, and mass transportation systems. A particularly interesting case is the investment in plants and equipment which future generations may use to meet their material needs. The key to the future value and the long run intergenerational benefits of this capital is its adaptability. Detroit-style automobiles will be highly irrelevant to human needs in the year 2090, unless the equipment in these plants can be economically adapted to producing the very different consumer needs of that future period, such as small solar powered vehicles. Capital equipment, such as machine tools, are flexible in adapting to changing current needs for capital equipment. However, they may become less useful in a post-industrial, ecologically sensitive age. Since all capital equipment is subject to depreciation, its inheritance value is limited by the extent to which amortization processes can continually adapt it for meeting contemporary needs. The capital equipment needed in the future will depend much less on machine tools and much more on flexible research and development facilities, such as those which pioneered the silicon chip.

To the limited extent that we can anticipate the capital needs of the next century, it appears likely that the really scarce resources will be environmental capital in the form of natural resource systems, such as watersheds, aquifers, waste removal systems, wetlands and basic life support systems. Manmade capital may have some role to play in restoring free flowing rivers, water supplies, forests, lakes, and wetlands, but once such natural phenomena have been consumed, depleted, or destroyed, it is difficult to conceive of how capital can be used to restore them. One example is the costly effort to return the Everglades to their original state. Bulldozers and draglines may be able to fill in some of the irrigation ditches and canals which are now regarded as ecological disasters, but it is not clear that the use of manmade capital can ever reclaim all that has been lost in the Everglades. The extinction of species like the Florida panther is a form of ecological irreversibility and investment in high density settlements is a form of economic irreversibility, given the high cost and low probability of restoration.

The most important item on the list by Pearce et al. of inheritable wealth, is the stock of environmental assets. Bequeathing to the future environmental assets which, to the extent possible, are pristine and undamaged by industrial society, appears to offer the most unequivocal opportunity to compensate future generations for the ecological damage that industrial society is spreading over the globe. Given the limitations of knowledge and of the ability to apply it, the shortcomings of technology, and the limitations of what can be accomplished with manmade capital, environmental assets assume major significance as bequests.

In designing a plan for intergenerational transfers appropriate for achieving sustainability, it is essential to include all four of the forms of wealth described by Pearce, et al. Furthermore, these should be, for maximum effectiveness, bequeathed in appropriate proportions according to a Cobb-Douglas production function or other appropriate function in quantities and combinations designed for efficient productivity in achieving sustainability. Although the functional form of future production functions for human welfare can not be known with any certainty, it seems reasonable to assume that the supply of environmental assets will be both inelastic, and a source of diminishing returns. Human capital and manmade capital may well turn out to have high elasticities of substitution for environmental capital on a national basis, but this is most unlikely for the total global economy. Unfortunately, given the inevitable uncertainty about the future and about necessary and sufficient conditions for sustainability, no precise guidelines can be set down for designing an optimal environmental inheritance to leave for future generations. A more realistic strategy would be to provide for flexibility and adaptability by bequeathing large scale, functioning, diverse, resilient resource systems.

The central point of this paper is that the stock of environmental resources to be passed along to future generations should emphasize such resources as large-scale living ecologies containing species diversity, complex interrelationships between species, and, above all, the capability of supporting evolutionary processes over sufficiently long time frames that species can evolve and adapt to both manmade and natural changes in climate and other environmental conditions. Just as this author's candidates for intergenerationally sustainable ecosystems transfers would be estuaries such as the Chesapeake Bay, others will have their own preferences and priorities. Obvious candidates include rain forests, wetlands, lakes, river basins, grasslands, polar regions and coral reefs. However, the ultimate selection of the highest priorities for protection of sustainable ecologies should be made by transdisciplinary teams including not only ecologists, but other representatives of life sciences, earth sciences, physical sciences, and social sciences preferably with insights also from the arts and humanities.

After the identification of the scientific principles and priorities for selecting sustainable ecologies for intergenerational transfer, the challenge will remain of designing the most effective policy measures for acquiring and protecting these ecologies.

CRITERIA FOR ACQUIRING AND PROTECTING SUSTAINABLE ECOLOGIES

A major challenge is that of gaining acceptance for large-scale current sacrifices which will produce uncertain benefits in an uncertain future. Another complicating factor is the need for international consensus on goals and for global cooperation in implementation. The fact that serious intragenerational inequalities exist in the distribution of current income and wealth will make it difficult to achieve consensus on the need for intergenerational transfers and will complicate the problem of apportioning sacrifices. A related problem is that, in an uncertain future, the continuity of a commitment to pass on ecological resources cannot be guaranteed for future generations that are not parties to the agreement. Therefore, intermediate generations may be tempted to consume all or part of an inheritance which was intended for the more distant future. There is the danger of a prisoners' dilemma in which uncertainty about the action of intermediate generations could reduce the welfare of more distant future generations. However, as successful experience is gained in protecting intergenerational transfers, uncertainties could be reduced and welfare gains increased.

Well-known public goods problems could pose additional difficulties in making intergenerational transfers, to the extent that future benefits were shared by all regardless of which group made the sacrifice to provide them. In the case of global public goods like the atmosphere and oceans, those groups making current sacrifices to protect the resources could not reap the entire benefits. This free-rider problem could reduce incentives to sacrifice unless measures could be designed to spread the burden widely.

Therefore, in choosing policy instruments for acquiring and protecting sustainable ecologies, new courses must be charted utilizing what limited insights are available from the fields of public choice and policy science. It is unlikely that acceptable policies and instruments can come from any one discipline such as economics, with its primary focus on efficiency, or ecology with its limited institutional content, or from any other monocentric discipline. Therefore, it seems self-evident that policy instruments for intergenerational transfers must be drawn from a transdisciplinary approach.

Given the fact that making bequests requires sacrifices and therefore involves scarcity problems, economic efficiency concepts can be helpful in achieving the maximum amount of resource protection for a given amount of resources available, or assist in achieving specified resource endowments at minimum total cost. The field of economics can also offer some limited insights into problems of distribution and equity. An especially important concept is that of Pareto improvement, which suggests that policies are most likely to gain acceptance if they can be designed so that there are no losers, or alternatively so that the gains from the policy are great enough to compensate the losers.

The criteria for ecological bequeaths must be based upon good science, and should emphasize protecting species diversity and reducing entropy. Finally, in order to gain acceptance, policies for making intergenerational transfers must be realistically based upon acceptance by the major interest groups involved. In summary, policy instruments for intergenerational transfers should be based upon:

1. scientific validity
2. distributional equity
3. economic efficiency
4. transdisciplinary collaboration, and
5. interest group acceptability

POLICY INSTRUMENTS

Society has already begun the process of making intergenerational environmental transfers in the form of wilderness areas, wildlife sanctuaries, parts of the polar regions, and similar set-asides. These programs have been initiated not only by local, state, national and international governmental organizations, but also by nongovernmental organizations (NGOs), such as the Nature Conservancy. Significantly, many families and individuals have demonstrated the value they place upon intergenerational environmental transfers through their willingness to bear the opportunity cost of holding land and resources in their natural state. These public and private initiatives in protecting living ecologies offer guideposts for the much greater future efforts which will be necessary for achieving sustainable global environments.

In cases where governments already own very large tracts, such as in the western United States, the task of acquisition and set-aside can be relatively easily accomplished. Setting aside tracts currently held by governments has the advantage of not requiring additional expenditure, but it must be recognized that there is an opportunity cost equal to the value of the highest alternative use to which the asset could be put. The least-cost way of protecting valuable ecologies is through simple appropriation, but this approach fails the equity test. In cases where high priority ecologies are in private hands, a wide array of policy instruments for acquisition is available. The most straightforward method is through purchase, which has the equity advantage of fully compensating current owners, but has the budgetary disadvantage of being very costly. The funds available for acquiring ecologies can be stretched through the purchase of easements strong enough to protect the desired ecological features but sufficiently permissive to grant current owners lifetime estates or limited use in return for long run protection.

In the cases where funds are raised by the government for acquisition, the cost to current generations is made explicit through the taxing and budgeting process, and in democratic societies can be achieved only through consensus. Transfers of funds from the general public to the current owners of the ecologies are made explicit under this procedure. An important economic consideration is what the taxpayers must give up in order to make the transfer possible, and what the recipients of the funds do with the proceeds. Thus, when government purchases of ecological assets occurs, redistribution occurs not only between generations, but within current generations.

The taxing and expenditure problem also draws attention to other fiscal aspects of transfers. Depending upon methods used for raising revenue, government purchase of environmental assets during periods of high economic activity could add to inflationary pressures, unless offsetting tax measures were introduced. Consequently, the timing of purchase is important. During periods of reduced economic activity, government purchase of ecological preserves could be used as counter-cyclical policy. The real cost and sacrifice resulting from government purchase of ecological preserves could be significantly reduced by closely integrating this policy with monetary and fiscal policy.

Affluent sovereign nations are finding it to be in their national interest to protect certain living ecologies, not only for the sake of future generations but because of ecological benefits provided to current generations. Less affluent nations face severe economic problems in financing such programs. Investing in international aid programs designed to protect unique ecosystems in developing nations could yield significant returns in future global welfare. One means of financing ecological programs would be to authorize an international organization to impose an international emissions charge on emissions of selected pollutants which have a global impact, such as CFCs and carbon dioxide. Basing charges on emissions would have the equity advantage of linking payment to sources of damage, and the efficiency advantage of providing economic incentives to reduce emissions which cause the damage. The acceptability of a schedule of international emission charges should be based upon research on damage functions and treatment costs functions by scientific panels of scientists with impeccable credentials (Cumberland 1972), such as Nobel laureates. Revenues could be used for acquiring ecological preserves and conducting research on global ecological sustainability.

One of the most important considerations in making intergenerational transfers over long time periods with high rates of uncertainty is to design mechanisms which lock the transfers tightly to real ecological resources. For example, one possibility for facilitating the acquisition of ecological assets would be through setting up a process by which Pigouvian charges on pollution or other environmental externalities could be converted into real terms through off-setting acquisition of natural assets. Examples are already emerging, such as the concept of no net loss of wetlands. Another example is the proposed legislation in Maryland requiring developers to replant trees for every one destroyed. Communities whose sewage adds to the nutrient overloading of estuaries could be required to set aside wetlands and marshes capable of assimilating the nutrients. A required cost for cutting rain forests could be setting aside and reforesting a comparable amount of previously cleared rain forest. Pearce et al. recommend that a significant portion of the economic rent from the depletion of exhaustible natural resources be "reinvested in resources which will compensate for the exhaustible resource when it is depleted" (1989, p. 145). In the real world, exact replacement of any ecological asset may be impossible, but the requirement of "closest equivalent ecological replacement" would have the advantage of drawing attention to the costs of irreversible losses. Severance taxes on minerals and other nonrenewable resources could be used to set aside resources capable of producing the nearest substitute. For example, a severance tax on fossil fuels should be set at the level needed to provide sufficient biomass to produce an equivalent amount of energy. If the burden of proof were put on the extractor of the fossil fuel to demonstrate the production of a comparable amount of fuel for energy from renewable sources, the proper incentives would have been established both to minimize the depletion of the nonrenewable resource and to find the least cost way of providing the substitute. Fines and penalties for spillage of oil and toxic substances could be levied in the form of requiring comparable areas and resources to be set aside and protected from exposure to damage. The central point is that in making intergenerational transfers for sustainability, trust funds and transfer mechanisms should be assessed in real terms so that activities which generate withdrawals from the ecological balance sheet would be required to bear the full cost of providing equivalent replacement in terms of functioning ecological assets.

CONCLUSIONS

Early New England sea captains, who made fortunes by harvesting whales nearly to extinction, attempted to ensure the perpetuation of their wealth for future generations by setting up spendthrift trusts to prevent unsustainable levels of consumption by intermediate generations. Some of the great New England fortunes have survived, but the task of providing a heritage of sustainable development for future generation cannot be solved by mere trust funds and financial arrangements.

The problem of finding the appropriate policy instruments for multigenerational transfers is complicated by problems of international pollution externalities, global public goods problems of nonrivalry, and nonexcludability in consumption and problems of equity between and within generations. Policy instruments to solve these problems must meet the criteria of:

1. scientific validity
2. distributional equity within and between generations
3. economic efficiency
4. international acceptability

In an age of exponential population growth, resource depletion, species extinction and global environmental damage, the most adaptable and essential heritage which can be set aside for the future is large scale functioning ecologies, such as rain forests, estuaries, wild life sanctuaries, river basins, grasslands, wetlands, polar regions, and coral reefs.

The important criterion for protecting ecologies is that they be large enough to provide contiguous habitats of air, land, and water needed by far ranging species, with sufficient diversity to support continuing evolutionary processes and ecological regeneration. The policy instruments needed for intergenerational transfers of living ecologies should include economic incentives both to reduce environmental damage and to insure that part of the proceeds which result from activities which do result in environmental damage are used to acquire and protect offsetting environmental assets. Examples include severance taxes on extraction of nonrenewable resources, international emission charges and pollution fines. Guidelines for making transfers to future generations should include levying on those current activities which generate ecological damage, fees calibrated in real terms for acquiring living ecologies sufficient to offset current damage.

Acknowledgments

The author is grateful for helpful comments by Henry Peskin and Leon Taylor on earlier drafts, but is solely responsible for remaining errors.

REFERENCES

Arrow, K. J. and A. C. Fisher. 1974. Environmental Preservation, Uncertainty, and Irreversibility. *Quarterly Journal of Economics* 55:313-19.
Ayres, Robert V., and Allen V. Kneese. 1969. Production, Consumption, and Externalities. *American Economic Review* 59:282-297.
Costanza, Robert. 1989. What is Ecological Economics? *Ecological Economics* 1:1-7.
Cumberland, John H. 1972. Establishment of International Environmental Standards—Some Economic and Related Aspects. *Problems in Transfrontier Pollution* pp. 213-229. Paris: OECD.
Daly, Herman E. 1977. *Steady State Economics.* San Francisco: W. H. Freeman.
Georgescu-Roegen, Nicholas. 1971. *The Entropy Law and the Economic Process.* Cambridge, Mass.: Harvard University Press.
Kahn, Herman, William Brown and Leon Martel. 1976. *The Next 200 Years: A Scenario for America and the World.* New York: William Morrow.

Kneese, Allen V., Robert V. Ayres, and Ralph C. d'Arge. 1970. *Economics and the Environment, a Materials Balance Approach*, Washington, D.C.: Resources for the Future.
Olson, Mancur. personal discussions.
Pearce, D. W., A. Markandya and E. B. Barbier. 1989. *Blueprint for a Green Economy*. London: Earthscan.
Pezzey, John. 1989. *Economic Analysis of Sustainable Growth and Sustainable Development*, Policy Planning and Research Staff, Economics Department, Working Paper no. 15, March. Washington, D.C.: The World Bank.
Sagoff, Mark. 1988. *The Economy of the Earth,* pp. x, 271. Cambridge/New York/Melbourne: Cambridge University Press.
Taylor, Leon. 1990. Relative Risk Aversion and the Demand for Environmental Protection, Unpublished manuscript.
Tietenberg, T. 1988. *Environmental and Natural Resource Economics*. 2d ed.. Glenview, Ill.: Scott Foresman.
World Commission on Environment and Development. 1987. *Our Common Future*. New York: Oxford University Press.

24

ECONOMIC STRATEGIES FOR MITIGATING THE IMPACTS OF CLIMATE CHANGE ON FUTURE GENERATIONS

Ralph C. d'Arge
and
Clive L. Spash
Department of Economics
Box 3985 University Station
University of Wyoming
Laramie, Wyoming 82071 USA

ABSTRACT

In previous papers, the authors have argued that there are likely to be adverse economic effects to future generations resulting from tropospheric warming (d'Arge, Schulze, and Brookshire 1982; Spash and d'Arge 1989). It has also been proposed that on intergenerational equity grounds, transfers of real resources are needed among generations to bring about compensation for losses in the future (Spash and d'Arge 1990). In this paper, these results are extended to the case where there are both gainers and losers in the future. In this case, the present can, at least, propose that part of the future loss be offset by part of the future gain.

If individuals in present society gain in knowing they have tried to compensate future generations, then the optimal strategy for the present is to start a climate compensation fund now, even though there are uncertainties as to the magnitude of losses or the distributions of gainers or losers. This is an application of the Andreoni "warm glow" hypothesis appropriate to certain classes of public goods (Andreoni 1989). It is argued that intergenerational climatic impacts are one type of "warm glow" public good.

The general conclusions of this paper is that, based on intertemporal efficiency considerations, there should be an intergenerational transfer of resources for compensation, unless the present generation severely restricts the buying of fossil fuels and introduces other practices that will eliminate anthropogenically induced climatic warming. If this is impossible, compensation is justified by most ethical and economic principles. Because of classical and new problems in valuing public goods, it is currently impossible to quantitatively estimate the amount of optimal compensation. Thus, a problem exists of estimating actual necessary compensation which should be based on: actual damages to future generations, reliable wealth or commodity transfer mechanisms, and the value to the current generation of the transfer.

INTRODUCTION

For the second time in the twentieth century, the first being the introduction of nuclear weapons and the possibility of a "nuclear winter," mankind is confronted by the possibility of nonsustainability due to climate change. That is, continued tropospheric warming due to fossil fuel burning is likely to place the global climate outside of the limits of recorded history. Whether human adaptation is great enough in modern societies to offset substantially warmer climates is, at least, debatable. Certainly, the goals of sustainability outlined by Kneese of "continued well-being of people in developed countries, improvement in the well-being of people in developing countries and protection and maintenance of a safe and attractive environment," become highly questionable (Kneese 1990).

This paper aims to explore three ideas with respect to global warming. The first is to examine the evidence on the hypothesis that global warming will have substantially different regional impacts, with some regions gaining and others losing significantly. The second idea to be evaluated is the extent and speed at which the climate warming will occur. According to recent calculations, past burning of fossil fuels has led to substantial warming potential for the near future, and present emissions of CO_2 are accelerating the potential for greater warming. Average global temperature is forecast to change by as much as 30%. The third idea to be examined is, if the other two are correct, the efficient and appropriate set of actions for the multiple set of global societies to take now.

REGIONAL IMPACTS OF GREENHOUSE WARMING

One of the main sources for predicting the impacts of global warming is climate simulation modeling using General Circulation Models (GCMs). GCMs typically solve 200,000 equations on each run and are extremely complex. Yet, the results from different GCMs often agree well with each other and with historical temperature data, over large scales (global, hemispherical, zonal). More than 100 independent studies have given estimates of average global warming within the 1.5°C to 4.5°C range for a scenario indicating a doubling of CO_2, with values near 30°C tending to be favored (MacDonald 1988, p. 437).

However, as results are examined on successively smaller scales, eventually focusing on subcontinental regions, significant differences arise between models. GCMs treat the world as uniform blocks (at their most detailed based on a 500 kilometer grid size) ignoring topographical and meteorological variations within these blocks. The use of averages over areas of greater regional extent than those of concern, can be misleading. Two models can have precisely the same average for a regional variable but still give significant spatial differences for that variable within the region. Thus, GCMs are not ready to be used for quantitative prediction even at the level of a multi-state region, let alone a particular state, county, or city (Grotch 1988, p. 253).

Therefore, the following is a qualitative picture of the regional effects of global warming based on broad latitudinal zones. This outline of regional impacts is partially derived from the conclusions of the workshops held in Villach, Austria (September 28-October

2, 1987), and Bellagio, Italy (November 9-14, 1987), under the auspices of the Beijer Institute, Stockholm (World Meteorological Organization 1988; see Jaeger 1989 for a summary). In addition, we identify specific areas as being particularly susceptible to the expected adverse effects of global warming.

The regional temperature and precipitation changes due to climate change via the greenhouse effect are summarized in table 24.1. The temperature changes, presented as multiples of the global and annual average temperature changes, are taken from computer modeling results for the Northern Hemisphere. A qualitative assessment of precipitation changes is given, but is supported by a number of studies. The general picture is one of far greater annual average changes in temperature towards the poles, and variable interregional changes.

In the Northern Hemisphere, above latitude 60° north, temperature change is predicted to be 2 to 2 1/2 times faster and greater than the global average. Withdrawal of summer pack ice could leave the Arctic ice-free around Spitzbergen and along the north Siberian coast. A positive feedback occurs as the duration of snow cover is reduced, affecting albedo (amount of reflected solar energy) and causing further warming. Increased precipitation in the high-latitude regions of the Northern Hemisphere leads to greater runoff into the Arctic Basin. The growth of Boreal forests will be stimulated and the rate of decay of organic material enhanced. Most effects in the region carry both benefits and costs, so that neither net benefits nor costs can be identified as potentially dominant. For example, while opportunities for increased use of the Northeast and Northwest passages should be beneficial, there are potentially serious effects on national and international security.

The mid-latitude zone, from 30° to 60°, is anticipated to warm more than the global average but less than the high latitudes. Winter warming is greater than summer, while summer rainfall decreases. The main impact is expected to be on unmanaged ecosystems, forests, and agriculture. The reproductive success of many tree species will be reduced and both tree and plant mortality will increase. The extent and timing of forest dieback depends on the estimated sensitivity of the climatic response and actions taken to control greenhouse gas emissions. Major effects on forests could begin between the year 2000 and 2050, or given serious global restrictions on gas emissions, be pushed back to 2100. A faster rate of temperature change worsens the expected forest damage.

Global food supplies, as a whole, should be maintained via adaptation based on agricultural research, except in the case of rapid warming. Negative effects on productivity in the lower-latitudes (around 30°), due to greater evapotranspiration, may be balanced by positive effects in the higher latitudes (around 60°), due to a longer growing season. However, irrigated agriculture in the semi-arid areas of the mid-latitudes will probably be adversely affected, unless substantial offsetting irrigation developments are implemented.

The semi-arid tropical regions, from 5° to 35° north and south, already suffer from unevenly distributed seasonal average precipitation with high spatial and interannual variability. This latitudinal belt as a whole is expected to suffer reduced precipitation. There has been a pronounced downward trend in precipitation since the early 1950s for this region; resulting in prolonged drought and aggravating the ongoing desertification process. Future climatic changes could worsen these current problems.

TABLE 24.1 Regional Scenario for Climate Change

Region	Temperature Change (as a multiple of global average)		Precipitation Change
	Summer	Winter	
High Latitudes (60°-90°)	0.5x - 0.7x	2.0x - 2.4x	Enhanced in Winter
Mid Latitudes (30°-60°)	0.8x - 1.0x	1.2x - 1.4x	Possibly reduced in summer
Low Latitudes (0°-30°)	0.9x - 0.7x	0.9x - 0.7x	Enhanced in places with heavy current rainfall

Source: World Meteorological Organization (1988), table 1

In humid tropical regions global warming leads to a 5% to 20% increase in rainfall. Water levels along coasts and rivers are increased by sea-level rise, greater frequency of tropical storm surges, and rising peak runoff. Large areas of flooding and salinization are likely to result. In addition, the spatial and temporal distribution of temperature and precipitation could change.

The outline of potential regional effects of global warming, given above, implies numerous positive and negative impacts. The evidence shows that climate change will have different impacts for different regions and countries. Global warming would alter precipitation patterns, evapotranspiration, and the length of growing seasons, affecting agriculture and forestry. It is also likely to increase the rate of sea-level rise threatening capital values in low lying areas. Thus, as has been previously recognized (see d'Arge et al. 1982; Kosobud and Daly 1984; Broadus et al. 1986; Barbier 1989; Glantz 1990) the benefits or costs of induced climatic change are likely to be unevenly distributed between nations. In particular, sea-level rise and changes in semi-arid regions imply greater costs for the developing nations of the world.

Sea-level rise is already of concern with half of the global population inhabiting coastal regions, while average sea-level increased at 0.01 meters per decade during the last century. An acceleration of this rate would occur as global warming melted land ice and caused thermal expansion of sea water. Estimates of sea-level rise vary but a probable scenario is 0.9 to 1.7 meters by 2100 (Thomas 1986; Titus 1986). Jaeger (1989) describes typical impacts as beach and coastal margin erosion, wetland loss, increased flood frequency and severity, and damage to coastal structures and water management systems.

Low lying developed countries already possess sea defenses which could be modified to account for the dangers of a 1 meter sea-level rise. Goemans (1986) estimated the cost of protecting against such a rise for the Netherlands, where extensive coastal protection already exists, at $4.4 billion. In the United States the cost of maintaining shores under threat on the East coast is in the range of $10 to $100 billion, for a 1 meter rise (Jaeger 1989, p. 101). Park et al. (1986) have predicted a loss of 40% to 73% of U.S. coastal wetlands by 2100. This might be reduced to 22% to 56% if new wetlands are allowed to form on what are currently inland areas.

In the developing countries serious impacts would be felt by a 1 meter sea-level rise or less. Flood disasters are particularly threatening for the delta regions of South Asia and Egypt. Egypt would have 12% to 15% of its arable land area flooded, affecting 7.7 million people, 16% of its population (Broadus et al. 1986). While in Bangladesh 8.5 million people would be affected, 9% of the population, and 11.5% of the land area would be flooded (Broadus et al. 1986). The 177,000 Maldivians are in a situation typical of many low lying island populations. They could be totally displaced by even a small sea-level rise due to the encroachment of storm surges. The Maldives are coral islands on average less than a meter and a half above sea-level ("Maldives" 1988, p. 36).

The ability of nations to adapt to sea-level rise is determined by their capital resources and technical knowledge. Developed nations are therefore better equipped to respond, at least initially, to predicted sea-level increases. Many less developed countries will find their current floods worsening, and resulting in direct impacts such as human suffering, permanent dislocation, and loss of life. As Jaeger (1989, p. 101) has stated, "In developed countries, lowland protection against sea-level rise will be costly. In developing countries without adequate technical and capital resources, it may be impossible."

Another area in which developing countries stand to suffer disproportionately is the semi-arid regions. The important arid zones of the world are found around latitude 30° north and south, where most of the large deserts are situated e.g., the Sahara (North Africa), the Syrian desert and Arabia (West Asia), Death Valley and surrounding areas (U.S.), and the Australian deserts (Wallen 1966). Only twelve nations are wholly arid while twenty-seven nations have some territory within the arid zone but the larger part of their land area outside of it (see table 24.2). The peripherally arid nations are those in this latter group for which aridity is significant only at the regional level.

Characteristically, the equilibrium of water, soil, geological, and vegetative processes in arid lands is a delicate one, so that a slight shift in one aspect may initiate a disastrous chain of events (White 1966, p. 15). The scarcity and variability of rainfall are dominant elements in the complex of physical factors affecting arid lands. The natural limits of arid lands are set by climatic conditions influencing the surplus and deficit of water for plant growth. The swing from crop success to utter failure is quick and frequent and the greatest risk is in semi-arid areas. Except for the United States and the desert oil exporting nations, most arid or semi-arid nations have little of the capital foundation and/or technical expertise required to develop strategies to protect against drought or extreme variations in climates.

High temperatures over long periods increase the energy available for evapotranspiration so that the potential water loss is greater than the amount available. Such high temperatures can intensify the chemical processes in a plant causing death. Extreme temperature may lead to crop failure and stress both animals and humans (Wallen 1966).

In the United States an increase in frequency, duration, and severity of droughts in the Great Plains would lead to a more rapid depletion of the Ogallala aquifer (partial source of groundwater to eight western states). The semi-arid western and midwestern states in the U.S. may be particularly susceptible to the effects of climate change on river systems (d'Arge 1975). Large decreases in surface runoff (40 to 76%) could accompany a $2^{\circ}C$

TABLE 24.2 The Arid Nations

Description	Number	% of Nation Arid/Semi Arid	Nations
Core	11	100	Bahrain, Djibouti, Egypt, Kuwait, Mawitania, Oman, Quater, United Arab Emirates, Saudi Arabia, Samalik, South Yemen.
Predominately Arid	23	75-99	Afghanistan, Algeria, Australia, Botswana, Cape Verde, Chad, Iran, Iraq, Israel, Jordan, Kenya, Libya, Mali, Morocco, Namibia, Niger, North Yemen, Pakistan, Senegal, Sudan, Syria, Turisia, Upper Volta.
Substantially Arid	5	50-74	Argentina, Ethiopia, Mongolia, South Africa, Turkey.
Semi Arid	9	25-49	Angola, Bolivia, Chile, China, India, Mexico, Tanzania, Togo, USA.
Peripherally Arid	18	< 25	Berin, Brazil, Canada, C. African Rep, Ecuador, Ghana, Lebanon, Lesotho, Madagascar, Mozambique, Nigeria, Paraguay, Peru, Sri Lanka, USSR, Venezuela, Sambia, Zimbabwe.

Source: Heathcote (1983) p. 9

temperature increase (Barbier 1989, p. 24). The effect of a CO_2 doubling on U.S. agriculture in the western states has been estimated by Adams et al. (1988) to induce crop losses of $6 billion to $33 billion (1982 dollars). However, almost half of the consumers' surplus losses from climate change in the United States fall on foreign consumers.

Adams et al. (1988, p. 348) note that climate change is not a food security issue for the United States. Even in the most extreme case their analyses indicate that the productive capacity of U.S. agriculture will be maintained at a level that avoids major disruptions to the supply of the modeled commodities. Consumers in the United States face slight to moderate price rises under most scenarios, but supplies are adequate to meet current and projected domestic demand. Exports, however, experience major reductions. Exported commodities in some scenarios declined by up to 70%, assuming constant export demand.

While arid regions of North America may suffer as atmospheric CO_2 doubles, present values the northern latitudes may simultaneously benefit from an increase in the length of the growing season and a shift northward and eastward in the agricultural belt. Agricultural adjustments could increase yields via the extension of winter wheat into Canada, a switch from hard to soft wheat in the Pacific Northwest due to increased precipitation, and an expansion of areas in fall-sown spring wheat in the southern latitudes due to higher winter temperatures (Rosenzweig 1985, p. 380). Kelejian and Varichek believe the United States to be slightly below optimal temperature for wheat and corn so that a small temperature increase (less than a double CO_2 climate) could increase U.S. yields by 5%. If the rest of the world's production fell, as they predict, the position of U.S. agricultural producers would improve through exports. However, more recent evidence shows other countries may benefit also.

Higher temperatures tend to favor yields of cereal crops in regions where temperature limits the growing season. For example, in the central European region of the USSR, wheat yields have been projected to increase a third under a double CO_2 scenario (Parry and Carter 1986, p. 275). Smit et al. (1988) have reviewed several recent studies and concluded that current evidence suggests a warmer climate could create a more favorable environment for wheat and grain corn in Canada, Northern Europe, and the USSR.

In the case of semi-arid regions of less developed countries, the impact of climate on economic activities is most severe in the agricultural sector of the desert and marginal areas, e.g., in Africa the Sudan and Sahel. Areas marginal to the desert can give a false sense of security to their inhabitants during successively wet years, but are highly vulnerable in periods of drought (Oguntoyinbo and Odingo 1979, p. 2). Continued wet-year land use and livestock practices during droughts contributes to desertification.

Moderate, or worse, desertification has occurred in 80% of the agricultural lands of the worlds arid regions. Irrigated lands suffer waterlogging and salinization, grazing lands lose plant cover, and rainfed croplands are desertified by soil erosion. Those areas currently undergoing severe desertification are agricultural economies with the livelihood of their people being 27% urban based, 51% cropping based, and 22% animal based (Dregne 1983, p. 20).

The population of the arid regions is approximately 700 million with 78 million living in areas where severe desertification has occurred. Among these people, 50 million have been estimated as already suffering a loss in ability to support themselves and are under pressure to migrate to overcrowded cities. The population of the moderately desertified portion of the arid lands is at least eight times that of the severely affected regions (Dregne 1983, p. 19-21).

Climatic changes due to the greenhouse effect will be imposed upon the already fragile ecosystems of these semi-arid regions. The prediction of reduced precipitation for the latitudinal zone from 5^0 to 35^0 north and south would, alone, aggravate the existing climatic problems of the arid regions. In addition, the temperature increase can be expected to add to crop and livestock stress and fatality. The size and subsistence level of population living in this region implies the possibility of a far greater cost than costs associated with expected shifts northward of agricultural production for North America and Europe.

The failure of crops in one region may be balanced by productivity increases in other regions, as mentioned earlier for the middle latitudes. Newman and Pickett (1974) have suggested that generally an equilibrium climate obeys a law of conservation, borne out by the history of global climatic variations. That is, whenever one large area gets too little precipitation, another gets too much. Thus, one strategy to respond to the problems of arid lands is to organize resource use to compensate for the inevitable shortages of production in one location by transfers of surpluses from other locations.

Global impacts of greenhouse warming on agriculture will be unevenly distributed with the North-South divide playing an important role in the ability to adapt. The major Northern Hemisphere food-exporting, and capital intensive, countries are in a better position than the low-income, food-importing developing countries. The wealthier northern nations while facing substantial changes in types of crops cultivated, water distribution and areas of available land, are more able to adapt. These countries "tend to have a surplus of land available for production as well as accumulated stocks of some produce, highly developed agricultural R and D infrastructure and techniques, efficient marketing credit and information systems and extensive water management and control systems." (Barbier 1989, p. 25). If adaptation raises costs and reduces supplies in the food-exporting North, the food-importing South is likely to suffer the most.

There are 65 low-income, food-deficit, countries with the greatest risk group being those who have failed to increase per capita food production in the past. Unfortunately, most of this latter group have extensive agricultural areas located in arid and semi-arid regions. As a result global climate change threatens to make a bad situation worse. Currently there are excess supplies of world food production but 730 million people in developing countries are denied enough energy from their diet to allow them an active working life (Barbier 1989, p. 26).

INTERTEMPORAL IMPACTS OF GLOBAL WARMING

Mean global temperature has in the past been much warmer than at present: during the Holocene climatic optimum (5000 to 6000 B.P.) $1^{O}C$ warmer; during the last interglacial warming (125,000 B.P.) $2^{O}C$ higher; during the Pilocence (3 to 4 million years B.P.) $3^{O}C$ to $4^{O}C$ higher (MacDonald 1988). However, during the last 10,000 years, from the warm Holocene to the Little Ice Age, the mean temperature of the northern hemisphere has varied by no more than about $2^{O}C$ (Gates 1983). In recent years, mean global temperature has been about $15^{O}C$. Thus, mean global temperature has varied by less than 14% over this time.

The greenhouse effect represents a potentially drastic temperature increase over a relatively short space of time. The earth's surface temperature has increased between 0.5^{O} and $0.7^{O}C$ (3-5%) since 1860 (Abrahamson 1989, p. 10). Hansen et al. (1986) estimate the warming at most mid-latitude northern hemisphere land areas will be $0.5^{O}C$ to $1.0^{O}C$ by 1990 to 2000, and $1^{O}C$ to $2^{O}C$ by 2010 to 2020. Toward the end of the next century, the planet could warm up by $4.5^{O}C$ (or a 30% increase) (Smagorinsky 1983). Within 100 years, the earth could be a lot warmer than it has been in several million years and temperatures will certainly surpass anything observed during recorded history. Such changes are well beyond known human capabilities of adaptation and raise fundamental questions regarding what constitute limits to sustainability.

The expected consequences of sea-level rise and agricultural and forestry damages suggest an overall negative impact due to a 2°C or more increase in average global temperature. Those who have contended that human-induced warming could be beneficial appear to be considering the period before temperature rises by this much. In 1938, Callendar argued that, for small increases in global temperature due to fossil fuel consumption there would be benefits of fuel use, the expansion of the northern margin of cultivation, and delayed return of the "deadly glaciers" (Callendar 1938, p. 236). A slight warming of 0.5°C has been estimated to have slight net benefits if air conditioning, agriculture, and water use are considered (about 16% of the global economy) (see d'Arge et al. 1975). More recently Idso (1983) has argued that escalated CO_2 concentrations would enhance crop productivity by increasing rates of photosynthesis and water use efficiency by decreasing rates of transpiration; thereby supplying more food for a growing population. Idso has concluded that increased levels of atmospheric CO_2 may actually be beneficial to our future well-being. During the initial period of warming this qualitative assessment may be accurate. The proportion of climate forcing due to CO_2 is greater than 50% compared to other greenhouse gases. Global warming is combined with a CO_2-fertilization effect which may benefit crops and trees. The extent of climate change will be initially within "normal" historical variability. Meanwhile, society benefits now from the unrestricted and low-cost use of fossil fuels and CFCs (ignoring the depletion problem).

However, as climate change becomes more extreme the benefits are likely to recede and the costs will dominate. The role of non-CO_2 greenhouse gases will become relatively more important over time. As a result, the beneficial CO_2-fertilization effect will diminish relative to damaging temperature increases. The more extreme temperature becomes, the greater the expected rise in sea-level and ecosystem stress. Yet the absolute temperature change is only part of the problem affecting the intertemporal distribution of costs and benefits. Just as important is the speed of climate change, which will determine how fast impacts escalate and the extent to which forests and unmanaged ecosystems are able to adapt.

There is reason to believe a faster pace of climate change will occur over time because emissions of the principle greenhouse gases are increasing at rates between 0.3% and 5.0% per year (Wuebbles et al. 1989). As a result, changes in the composition of the atmosphere are taking place within decades, and at an exponential rate (MacDonald 1988). At current emissions rates, the earth is being committed to an additional warming of at least 0.15°C and perhaps as much as 0.5°C per decade (Abrahamson 1989, p. 10).

In North America, the fate of numerous tree species will depend on whether they can shift north to cooler climates when their current range becomes uninhabitable. Each 1°C rise in temperature translates into a range shift of about 100 to 150 kilometers (d'Arge et al. 1975; Roberts 1989). The toll would be greater on forests near the southern limits of their current range. The rate of northward dispersal due to historical warming, shown by fossil records, for North American trees is about 10 to 45 kilometers a century. The record is held by Spruce which spread 200 kilometers a century. Thus, historical scales of forest expansion are much slower than scales of anticipated climate warming. Forest effects should be far more pronounced where climate warming is greater, e.g., Canada and Alaska.

The rate of sea-level rise also appears set to increase near the end of the next century. Titus (1989) has reviewed several studies which analyze sea-level rise; table 24.3 shows two forecasts for the next century. In the initial twenty-five years, the marginal rise is relatively small, regardless of the chosen scenario, compared to the last twenty-five years of the century. This implies a dramatic increase in flooding, salt water intrusion, erosion, and land loss as the century progresses.

The overall picture of climate change is one of initial benefits from slight global warming for most regions, but increasing likelihood of very large economic costs as the climate becomes warmer. Population migration will undoubtedly occur as land is lost to rising seas and storm surges, and as semi-arid land becomes unproductive. Wealthier nations with better endowments of fertile soils, less arid and semi-arid land further inland and/or above sea-level will be tempted to ignore less fortunate regions (Barbier 1989). The more extreme and rapid the temperature increases, the greater are the costs and the fewer are the benefits. Thus, not only will the damages of preceding generations' greenhouse gas releases be placed upon those in the distant future, but the cost of continuing to release those gases will escalate.

Within 50 years, we are likely to create an irreversible increase in mean global temperature of 1.5°C to 5°C. If no attempt is made to slow the rate of increase, further 1.5°C to 5°C temperature rise could occur in the following 40 years. Emissions prior to 1985 have already committed the earth to a warming of about 0.9°C to 2.4°C, of which about 0.5°C has been experienced. The warming yet to be experienced is the unrealized warming. This warming, 0.3°C to 1.9°C, is unavoidable (Ciborowski 1989, pp. 227-228). The main losers due to the greenhouse effect clearly appear to be the future generations of less developed countries with poor resource endowments.

TABLE 24.3 Total Estimated Sea-level Rise in Specific Years (in centimeters)

STUDY	Future Year				
	2000	2025	2050	2075	2100
Hoffman et al. (1986)					
Low	3.5	10	20	36	57
High	5.5	21	55	191	368
EPA (1983)					
Low	4.8	13	23	38	56
High	17.1	55	117	212	345

Source: Titus (1989), figure 12.2, p. 169

ECONOMIC MITIGATION STRATEGIES

A 20-30% increase in global temperatures, since it has never occurred during recorded history, is outside any practical known limits for sustaining the viability of the present world economy. Since the time frame of change is predicted to be much shorter than historical responses of plants and animals, there is little knowledge as to how ecological systems will adapt or even if they can. The incidence of positive and negative economic effects is unlikely to be uniform and may be highly pervasive in that generally, the poorer nations will become even more impoverished and the richer nations relatively richer.[1] Unfortunately, those nations which are contributing the bulk of current CO_2 emissions are also the nations which will benefit from its release in the short to medium term and stand to gain relatively in the long run. This is a kind of worst case situation for managing a global common property resource. The current polluters benefit in the future from their pollution and the future receptors have little or no current assets to induce polluters to cooperate and will have even less in the future. Thus, no one has the interest or capability to unilaterally attempt to negotiate a solution to this externality problem (see d'Arge and Kneese 1981). The dilemma is one example of the type of problem first posed by Samuelson on overlapping generations. (Samuelson 1958; see also Geanakoplos and Polmarchakis 1984).

Samuelson suggested the following kind of model structure. The young in each generation receive commodities or income which the old do not. Presumably, the young are productive (and own the assets necessary for producing) while the old are not productive. The old would benefit from income shared by the young and in turn these young would also be compensated by the next generation of young. Unfortunately, there is no mechanism in Samuelson's paradigm for each generation to know that it will be compensated when old, as it had compensated the previous generation when it was young. This is a form of intergenerational negative externality since each generation would be better off by transferring income to the old and in turn receiving such transfers.

The same type of transfer problem arises for the greenhouse problem. The rich nations activities now negatively affect the poor nations in subsequent generations. There is no inherent mechanism for the rich nations now to transfer income to the poor now or in the future, even though such compensation might make everyone better off, because, as in the Samuelson case, there is no guarantee that appropriate transfers will be made to the harmed parties in the future.

Without such transfer mechanisms, efficiency in an intertemporal sense is unlikely. The only case where intertemporal efficiency would occur is where each generation accepted as inviolate the distribution of wealth, and maximized its own welfare, regardless of impacts on other generations. The appropriate transfer mechanism appears to require

---

[1]The greenhouse effect is characterized here as leading to gains for the rich nations while the poor lose. In fact, as argued earlier there are likely to be serious impacts on all nations in the long run, but the rich nations will remain relatively better off. That is, as the cake shrinks the rich nations end up with a larger percentage. For simplicity in the following discussion and model we assume the rich gain absolutely, even in the long run.

Samuelson Paradigm

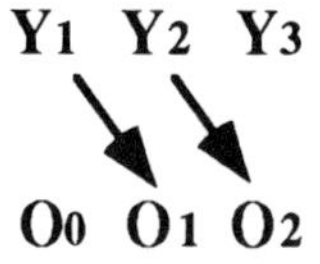

Y_1 Y_2 Y_3

O_0 O_1 O_2

FIGURE 24.1

CO_2 Paradigm

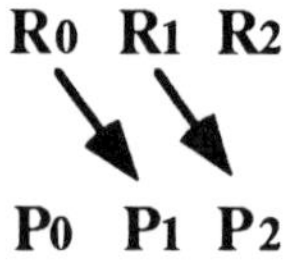

R_0 R_1 R_2

P_0 P_1 P_2

FIGURE 24.2

the existence of an autonomous regulatory agency over succeeding generations and, by definition, continuous governments. This raises the issue where, unlike Samuelson's paradigm that can be resolved within the context of an existing government, there is no such "natural" governmental unit for transfers across governments and time. (See d'Arge and Kneese 1981). That is, the CO_2 paradigm of inefficiency does not lead to a well defined prescription for its resolution. However, without some form of transfer, the rich nation's will become richer at the expense of poor nations in the future. This outcome is only consistent with a "super elitist" ethical view of the global commons where those with the most get the most and those with the least get even less. (See d'Arge, Schulze, and Brookshire 1982).

One of the key problems in the Samuelson overlapping generations model is to provide incentives to an early generation to voluntarily accept the risk and make a transfer regardless of whether a subsequent transfer is made to them, i.e., they are compensated for their transfer. In the CO_2 paradigm, there is no reward, except in terms of ethics, for the wealthier nations to take this first step. Perhaps, this is one of the reasons why we observe a continued call for more research and economic evaluation by the United States before taking decisive actions on the greenhouse problem. The only inducement for the wealthier nations, who will be made better off by climatic change, to negotiate might stem from some ethical belief that they should not harm others, without compensation, for their own gain, even though the other nations cannot stop them. Some economists have argued that negotiation might come about because of interdependent utility or value functions. That is, the wealthier nations citizens are made better off knowing the citizens of impoverished nations are not made worse off by their actions. However, this appears to be somewhat farfetched for impacts to unborn generation of other nations that will occur in 100 or so years. The presumption is that we are significantly affected by the negative effects we cause to other nations' citizens a long time in the future. With any reasonable individual discount rate for citizens today, the present value of even substantial future damages expressed in individual utility today must be extremely small. There is some preliminary evidence that individuals tend to value impacts on future generations, even their directs heirs, at practically nothing after a century (see Case 1986). If individual preferences provide no imperative for preventative actions and/or compensation now for distant future impacts, what can?

In recent years, it has been suggested by Andreoni (1988) and others that individuals when expressing values for public goods, are not deriving their "individualized" Lindahl prices for public goods but rather are expressing a value for concepts of "moral satisfaction." As Kahneman and Knetsch (1990) suggest, individuals are obtaining moral satisfaction as opposed to commodities which elevate their utility through direct or vicarious consumption. By such purchases, individuals make themselves feel good or have a "warm glow" by what they have donated to a good cause. There is substantial empirical evidence that the "warm glow" hypothesis is correct in that individuals are willing to pay a great deal to realize it (see d'Arge 1989). Individuals derive utility not from "consuming" a public good, but in effect, consuming an ideal. The actual quantities representing the amount of public good purchased or public bad removed is irrelevant. Rather, utility is derived from knowing the right thing is being done. Utility results from the purchase, not from consumption. The tradeoff between current wealthy individuals and future poor individuals is thereby not in terms of reduced wealth to the wealthy and reduced impacts to the future poor, but increased "warm glow" now for less impact on the future poor. Control costs borne by the present generation are compensated for through the purchase of moral satisfaction. Sustaining controls through intermediate generations, inclusive of contributions to future compensation, is more likely if these generations also receive "warm glow." In order to place these ideas in a more formal structure, a rather simple model is outlined. The model structure embodies, in elemental form, the concepts just outlined in a rather simplistic way.

THE MODEL

Let there be two time periods, 1 and 2, where 1 denotes the present and 2 the future. Assume there are two countries, R (Rich) and P (Poor), each with an endowment of resources which determines consumption in the two time periods. The wealth of each country's citizens can be represented by a single, quasi-concave utility function $u(\cdot)$. The normal argument is consumption in each country in each time period, e.g., C^{R1}. The first order conditions for an optimum can be simplified to

$$\frac{U_C^{R1}}{U_C^{R2}} = N^R(1 + X^R)$$

(1)

where N^R is the utility discount factor for country R from period 1 to period 2, and X^R is the productivity on investment earned from period 1 to 2 per unit of consumption foregone in period 1. Country P has an identical set of conditions as long as there is no externality between the two countries. This condition is commonplace, in that all that is established is: the marginal rate of substitution between present and future consumption equals the utility discounted productivity of foregone consumption.

Next introduce an external diseconomy, where C^{R1} negatively influences the levels of C^{P2}. If the two countries act independently then country R will continue to use first order condition (1) to maximize utility, ignoring the effect on country P. The first order conditions (1) under social optimality change because of the presence of the externality; the independent choices implied in (1) are no longer optimal. Joint maximization of utility across countries makes the optimal condition for country R:

$$\frac{U_C^{R1}}{U_C^{R2}} = N^R \left[(1 + X^R) + \frac{\theta f_C^{R1}}{\delta^{R2}} \right]$$

(2)

An additional term has been now added to the right hand side of the equation. This term includes δ^{R2} - a Lagrange multiplier for consumption in the future period, θ – a multiplier on the externality and F_c^{R1} - the reduction in C^{P2} resulting from a unit increase in C^{R1}. The addition of this term is a standard result in externality theory. The usual procedure is to levy a tax on C^{R1} equivalent to this negative effect and thereby bring about a change of first order conditions from (1) to (2).

The first order condition for the P country is:

$$\frac{U_C^{P1}}{U_C^{P2}} = N^R (1 + X^P) + \frac{\delta^{P2}}{\delta^{P2} \theta}$$

(3)

Note, a new term involving multipliers appears in the first order conditions for the P and R countries. For an optimum to be achieved, joint maximization is required over both countries, and first order conditions of both countires have been adjusted.

Now, let us add the further complication that not only does increased C^{R1} reduce C^{P2}, but also raises C^{R2}. Consuming fossil fuels at higher rates now not only reduces possible consumption of the poorer nations in the future, but raises income-consumption of the wealthier nations in the future. Thus, there is an intertemporal diseconomy coupled with an intertemporal positive economy. The first order conditions of country R acting independently become:

$$\frac{U_C^{R1}}{U_C^{R2}} = N^R (1 + X^R - g_c^{R1})$$

(4)

and the first order social optimum condition for country R is:

$$\frac{U_C^{R1}}{U_C^{R2}} = N^R \left(1 + X^R - g_c^{R1} + \frac{\bullet f_c^{R1}}{\delta^{R2}} \right)$$

(5)

where g_c^{R1} is the change in C^{R2} resulting from an increase in C^{R1}. Note that current investment through reduced current consumption now has a penalty attached to it reflecting the loss of wealth by not causing greater climatic change.

The important point in this simple model is that efficiency is impossible in the two county context unless 1) a Pigouvian tax is levied from outside, or 2) country R voluntarily agrees to impose the tax on itself, even though this would reduce its future wealth from climatic change. Neither of these possibilities seems very realistic since there is no international authority with the power to enforce the tax, and no apparent economic motive for the R country to voluntarily comply.

Now let us introduce the idea of Andreoni-Kahneman-Knetsch, that individuals desire to do the right thing and are willing to purchase "moral satisfaction." The utility function for the R country now includes an argument for not depressing consumption in the P country. Such an argument can be modeled in a number of ways, but the general idea is that citizens of R country will make contributions to P country to offset P country losses from R's activities, and citizens of R receive utility from the "act" of giving.

The effect of this is to create a new value in the optimum which partially or completely offsets the negative impacts on country P, and provides country R with a motivation not to indiscriminately harm country P. The first order conditions for country R acting independently become:

$$\frac{U_C^{R1}}{U_C^{R2}} = N^R (1 + X^R) - \frac{U_{CP2}^{R1} \cdot \frac{dC^{P2}}{dC^{R1}}}{\delta^{R2}} \tag{6}$$

which is structurally similar to (2). Since $\frac{dC^{P2}}{dC^{R2}}$ equals f_2^{R1}, we see that (6) and (2) are identical if U_{CP2}^{UR1} equals x in the joint maximization; this is unlikely except by chance.

The introduction of a moral motivation induces country R to independently take actions that near a social optimum. The 1990 Earth Day theme stressing individual actions and responsibilities is an example of potential desire to purchase moral satisfaction.

SUMMARY AND CONCLUSIONS

It has been tentatively concluded that two pervasive elements might emerge with regard to climate warming. The first is, on balance, that wealthier nations will benefit and poorer nations will be harmed. The second is that relatively small increases in global temperature will be beneficial globally, but because of the almost irreversible accumulation of CO_2, larger and harmful changes will result. Actions to offset or compensate for climate warming do not require people to accept a kind of worst ethical principle, where the rich become wealthier at the expense of the poor. We suggest such an outcome is unlikely because of the value citizens place on doing the right thing by purchasing moral satisfaction. How can this be accomplished? We suggest it can be partially accomplished through compensation to future generations through "moral satisfaction" donations now.

Further, individual actions on reducing use of fossil fuels, directly or indirectly, can contribute to reducing intertemporal inefficiency. Such individual actions are not enough, but at least they would be a start in the right direction.

REFERENCES

Abrahamson, D. E. 1989. Global Warming: The Issue, Impacts, Responses. In D.E. Abrahamson, ed., *The Challenge of Global Warming*, pp. 3-34. Washington, D.C.: Island Press.

Adams, R. M., B. A. McCarl, D. J. Dudek and J. D. Glyer. 1988, Implications of Global Climate Change for Western Agriculture. *Western Journal of Agricultural Economics* 13(2):348-356.

Andreoni, J. 1989. Giving with Impure Altruism: Application to Charity and Ricardian Equivalence. *Journal of Political Economy* 97(6).

Barbier, E.B. 1989. The Global Greenhouse Effect: Economic Impacts and Policy Considerations. *Natural Resources Forum* (February), pp. 20-32.

Broadus, J. M. J. D. Milliman, S. F. Edwards, D. G. Aubrey and F. Gable. 1986. Rising Sea-level and Damming of Rivers: Possible Effects in Egypt and Bangladesh. In J. G. Titus, ed., *Effects of Changes in Stratospheric Ozone and Global Climate*, vol. 4.

Callendar, G. S. 1938. The Artificial Production of Carbon Dioxide and its Influence on Temperature. *Quarterly Journal of the Royal Meteorological Society* 64:223-237.

Case, J. C. 1986. Contributions to the Economics of Time Preference. Ph.D. dissertation, University of Wyoming. Laramie, Wyo.

Ciborowski, P. 1989. Sources, Sinks, Trends, and Opportunities. In D. E. Abrahamson, ed., *The Challenge of Global Warming*, pp. 213-230. Washington D.C.: Island Press.

d'Arge, R. C., ed. 1975. *Economic and Social Measures of Biologic and Climatic Change. Climate Impact Assessment Program.* Prepared for the U.S. Department of Transportation, Washington, D.C.: U.S. Government Printing Office.

d'Arge, R. C. 1990. A Practical Guide To Economic Valuation of the Natural Environment. In *The Rocky Mountain Mineral Law Institute.* New York: Matthew Bender.

d'Arge, R. C., and A. V. Kneese. 1980. State Liability for International Environmental Degradation: An Economic Perspective. *Natural Resources Journal* 20(2):427-450.

d'Arge, R. C., W. D. Schulze and D. S. Brookshire. 1982. Carbon Dioxide and Intergenerational Choice, *American Economic Review* 72:251-256.

Dregne, H. E. 1983. *Desertification of Arid Lands.* Chur, Switzerland: Hardwood Academic Publishers.

Gates, D. M. 1983. An Overview. In E. R. Lemon, ed., *CO_2 and Plants: The Response of Plants to Rising Levels of Atmospheric Carbon Dioxide*, pp. 7-20. Boulder, Co: Westview.

Geanakoplos, J., and H. M. Polmarchakis 1984. Intertemporally Separable Overlapping Generation Economics, *Journal of Economic Theory* (34).

Glantz, M. H. 1991. Forthcoming. Assessing the Impacts of Climate: The Issue of Winners and Losers in a Global Climate Change Context. In J. Titus, ed., *Changing Climate and the Coast.* Washington, D.C.: U.S. Environmental Protection Agency.

Goemans, T. 1986. The Sea Also Rises. The Ongoing Dialogue of the Dutch with the Sea. In J. G. Titus, ed., *Effects of Changes in Stratospheric Ozone and Global Climate*, vol. 4.

Grotch, S. L. 1988. *Regional Intercomparision of General Circulation Model Predictions and Historical Climate Data.* Washington, D.C.: Department of Energy.

Hansen, J., A. Lacis, D. Rind, G. Russell, I. Furg, P. Ashcroft, S. Lebedeff, R. Ruedy, and P. Stone. 1986. The Greenhouse Effect: Projections of Global Climate Change. In J. G. Titus, ed., *Effects of Changes in Stratospheric Ozone and Global Climate*, vol. 1. Washington, D.C.: U.S. Environmental Protection Agency.

Heathcote, R.L. 1983. *The Arid Lands: Their Use and Abuse.* Harlow, England: Longman,.

Idso, S. B. 1983. Carbon Dioxide and Global Temperature: What the Data Show. *Journal of Environmental Quality* 12(2):159-163.

Jaeger, J. 1989. Developing Policies for Responding to Climate Change. In: D. E. Abrahamson, ed., *The Challenge of Global Warming*. Washington, D.C.: Island Press.

Kahneman, D. and J. Knetsch. 1991. In press. Valuing Public Goods: The Purchase of Moral Satisfaction. *Journal of Environmental Economics and Management*.

Kelejian, H. K. and B. V. Varichek. 1982. Pollution, Climate Change, and Consequent Economic Costs Concerning Agricultural Production. In J. H. Cumberland, J. R. Hibbs, and I. Hoch, eds., *Economics of Managing Chloroflurocarbons: Stratospheric Ozone and Climate Issue*, Baltimore: John Hopkins University Press.

Kneese, A. V. 1990. Confronting Future Environmental Challenges. *Resources* (Spring), vol. 99.

Kosobud, R. F., and T. A. Daly. 1984. Global Conflict or Cooperation over the CO_2 Climate Impact *Kyklos* 37:638-659.

MacDonald, G. J. 1988. Scientific Basis for the Greenhouse Effect. *Journal of Analysis and Management* , 7(3):425-444.

Maldives: A Sinking Feeling. 1988. *The Economist* (October 1):36.

Newman, J. E., and R. C. Picket. 1974. World Climate and Food Supply Variations, *Science* 186:877-881.

Oguirtoyirbo, J. S., and R. S. Odirgo. 1979. Climatic Variability and Land Use: An African Perspective. Presented at World Meteorological Organization, World Climate Conference, Geneva, February 12-23, 1979.

Park, R. A., T. V. Arentaro, and C. L. Cloonan. 1986. Predicting the Effects of Sea-level Rise on Coastal Wetlands. In: J. G. Titus, ed., vol. 4.

Parry, M. L. and T. R. Carter. 1986. Effects of Climatic Changes on Agriculture and Forestry: An Overview. In J. G. Titus, ed., *Effects of changes in Stratospheric Ozone and Global Climate*, vol. 1. Washington, D.C.: U.S. Environmental Protection Agency.

Roberts, L. 1989. How Fast Can Trees Migrate? *Science* 243 (February): 735-737.

Rosenzweig, C. 1985. Potential CO_2 Induced Climate Effects on North American Wheat Producing Regions. *Climate Change* 7:367-389.

Samuelson, P. A. 1958. An Exact Consumption Loan Model of Interest, With or Without the Social Contrivance of Money. *Journal of Political Economy* 66(6).

Smagorinsky, J. 1983. Effects of Carbon Dioxide. In: *Changing Climate* , National Research Council. Washington, D.C.: National Academy Press.

Smit, B., L. Ludlow, and M. Brklacich. 1988. Implications of a Global Warming for Agriculture: A Review and Appraisal, *Journal of Environmental Quality* 17(4) 519-527.

Spash, C. L., and R. C. d'Arge. 1989. The Greenhouse Effect and Intergenerational Transfers. *Energy Policy* v 17(2):April .

Spash, C. L., and R. C. d'Arge. 1991. Compensation of Future Generations for Adverse Climatic Changes. Submitted to *Georgetown International Environmental Law Review*.

Thomas, R. H. 1986. Future Sea-level Rise and its Early Detection by Satellite Remote Sensing. In J. G. Titus, ed., *Effects of Changes in Stratospheric Ozone and Global Climate*, vol. 4. Washington, D.C.: U.S. Environmental Protection Agency.

Titus, J. G., ed., 1986. *Effects of Changes in Stratospheric Ozone and Global Climate*. U.S. Environmental Protection Agency. Washington, D.C.: U.S. Environmental Protection Agency.

Titus, J. G. 1989. The Cause and Effects of Sea-level Rise. In D. E. Abrahamson, ed., *The Challenge of Global Warming*. pp. 161-195. Washington D.C.: Island Press.

Titus, J.G. In press. Changing Climate and the Coast. Washington, D.C.: U.S. Environmental Protection Agency.

Wallen, C. C. 1966. Arid Zone Metrology. In E. S. Hills, ed., *Arid Lands: A Geographical Appraisal*. pp. 53-76. London: Methuen.

White, G. F. 1966. The World's Arid Areas. In E. S. Hills, ed., *Arid Lands: A Geographical Appraisal*. pp. 19-30. London: Methuen.

World Meteorological Organization, World Climate Program. 1988. J. Jaeger, ed. *Developing Policies for Responding to Climatic Change*, A Summary of the Discussions and Recommendations of workshops held in Villach, Austria September 28-October 2, 1987 and Bellagio, Italy, November 9-13, 1987). World Meteorological Organization, WCIP-1.

Wuebbles, D. J., N. E. Grant, P. S. Connell, and J. E. Penners. 1989. The Role of Atmospheric Chemistry in Climate Change. *Journal of Air Pollution Control Association*, 39(1): 22-28.

25

THE ROLE FOR ECONOMIC INCENTIVES IN INTERNATIONAL ALLOCATION OF ABATEMENT EFFORT

Tomasz Zylicz
Poland's Ministry of Environment and Warsaw University
Ministry of Environment, ul. Wawelska 52/54
00-922 Warsaw, Poland
Fax: (48-22)25-4141 or (48-22)25-3355

ABSTRACT

There is a large body of literature on how to optimize the allocation of abatement effort within a state or a homogeneous region. It has been proved that under the standard assumptions the least cost solution can be arrived at by means of imposing Pigouvian taxes or trading permits. The efficiency proof rests, among other things, on the assumption that there exists an agency granted with the authority to tax or to initially allocate permits. Coasian bargaining may offer a solution in the absence of such an agency, but its applicability is limited. Thus, as long as a group of countries is lacking an appropriate supranational authority, the only viable emission reduction programs are those based on voluntary reductions from an historical reference level. The 30%-Club of European countries reducing their SO_2 emissions and the 50%-Club of the Baltic countries serve as examples here. A similar solution has been adopted in the case of limiting (freezing) NO_x emissions. On the other hand, the policy towards CFCs favors developing countries vis-a-vis the developed ones and grants the former with less stringent constraints on their future use of these chemicals. In any case, the implicit allocation of abatement effort does not have to comply with a supranational efficiency criterion. A question arises whether an incentive scheme could be applied in order to push these programs towards an optimum, i.e., beyond the point accessible by means of voluntary reductions. The author explores such possibilities discussing them in a broader context of institutional reforms in todays world.

SHORT COURSE IN THE POLITICAL ECONOMY OF POLLUTION

At first sight environmental pollution economics is nothing more than a classical externality problem. Emissions to the air or discharges to rivers, lakes, and seas mean an economy to the polluter and a diseconomy to its victim. The former easily disposes of a useless waste product, and saves on abatement costs. The latter suffers from the pollution and/or is forced to bear defensive outlays. Thus, there occurs an external transfer of costs from the polluting country to the polluted one(s). This distorts the cost-benefit analysis and results in a non-optimal allocation of global resources.

Arthur Pigou (1920) analyzed the externality problems thoroughly, and devised a solution now called a Pigouvian tax. The tax, which is paid by the polluting party, should be equal to the marginal external cost born by the pollutees. The polluter is thus informed about the full social (regional or global) costs of his operations, and the victims can be fairly reimbursed.

Until 1960, the Pigouvian tax was viewed as the only way to solve the misallocation caused by externalities. However, in his widely debated work, Ronald Coase (1960) pointed out that there is a sort of symmetry in the property rights to the environment. Unless it is explicitly stated otherwise, both parties—the polluter and the pollutee—have a priori equal rights to use their common environment. External costs do imply a misallocation of the resources, but this does not imply that the Pigouvian tax is necessarily the solution. If by an efficient solution of a pollution problem we mean to reduce[1] the emissions up to the point where marginal costs of this process meet its marginal benefits (in terms of reduced pollution losses), then an alternative to the tax is available. This alternative solution boils down to bribing the polluter to reduce his emissions. Pigouvian taxes assume the right to the environment is assigned to the pollutee. On the contrary, the bribing solution assumes the right is initially assigned to the polluter, but the pollutees may wish to purchase some of these rights. The key proposition—known as the Coase theorem—is that the efficient solution of the misallocation problem is independent of the initial property rights assignment. If the parties behave rationally, in both cases the abatement effort would be undertaken as long as its costs are justified in terms of benefits.[2]

Of course, mixed solutions are possible, as it is conceivable that property rights can be split between the parties. All in all, the only instance that the optimal solution may not be achieved is—according to Coase—when the polluter is not fully charged and the pollutees can not or do not wish to enter the negotiations.

The bribing solution frees the government from the burden of estimating and imposing the tax. Also, it motivates the pollutees to adjust their activities or habits so as to minimize unreimbursed losses. However, it makes the polluter less motivated to seek ecologically benign technological changes than by Pigouvian tax. Bribing (as the general principle) can be easily reduced ad absurdum when one considers new investments. Any

---

[1] It is assumed throughout this paper that reducing the emissions can be achieved either by decreasing the scale of activity, redesigning its technology or running an abatement equipment; in a broader sense "abating" is also used as a synonym for "reducing."

[2] Actually Frank Knight (1924, pp. 584-592), was the first to question the idea of Pigou. According to Knight, efficiency requires that the assignment of property rights be unambiguous; defining and protecting property rights (not levying Pigou taxes) is the true role of a state. However, it was not until the Coase 1960 article that the original Knight critique penetrated the economic literature.

firm planning to erect a new plant would be motivated to announce that its emissions would be as high as the balance of raw materials permits, so as to make the prospective victims willing to pay as much as possible. Having negotiated a compensation agreement the potential polluter then may totally abandon the project (to please the potential victims) and switch its funds to "campaign" elsewhere.[3] To a certain degree this kind of parasite behavior could be expected on the pollutee side too, but for all practical purposes it is definitely less relevant (Kamien et al. 1966). This, together with the contemporary notion of the right to a clean environment, resulted in the adoption of the Polluter Pays Principle (PPP) as a guiding rule for environmental policy in most countries.

However, as John Pezzey (1988) points out, two forms of PPP exist: the "standard" and the "extended" one. Under the standard PPP, it is assumed that the polluter bears the expenses of carrying out the abatement measures decided by a public authority. Whereas the extended PPP requires the polluter to compensate the pollutees (in addition to the cost of controls) for the damage resulting from residual pollution. Despite all the political rhetoric, in fact, only the first form is widely used in practice. It means that the right to the environment is actually split between the polluters and their victims.

Calculating a Pigouvian tax is an almost impossible task. Cumulative effects and synergism known in environmental sciences make it difficult to estimate the contribution of an individual polluter to a given loss (even provided the loss itself can be assessed properly). This is the reason why policy makers instead use the second best solution and impose a set of emission standards which polluters are obliged to meet. The administrative standards are effective but not efficient in allocating abatement efforts among potential polluters. Nevertheless, the reductions in atmospheric pollution which occurred in most developed market economies in the 1960s and 1970s were achieved in this way. Likewise lake and river clean-up—wherever it occurred—resulted from tough administrative measures. What some of economists and politicians objected to were the excessive costs of the abatement program induced by administratively allocated emission standards.

Conceived by Thomas Crocker (1966), a new solution was introduced in the late 1970s. It combines the certainty and effectiveness of an administrative standard with the efficiency of a market allocation. The solution—variously called emissions trading, marketable permit system, tradeable rights or the bubble policy—is based on emission standards set for regions rather than individual sources (plants or stacks). The term "bubble policy" derives from the fact that plants can be grouped—as if under a dome or "bubble"—for the purpose of aggregating pollution emissions and imposing only an overall control on total emissions leaving the bubble. The standard is thus equivalent to the sum of n issued permits, each valid for the emission of 1/n of it. The emission sources under the bubble compete for the access to the same stock of discharge permits. After an initial administrative allocation the firms may sell and buy them. Thus market forces start working and—under certain formal assumptions (Bohm and Russell 1985, pp.

---

[3] The procedure resembles that of Major Major's father—from Joseph Heller's *Catch-22* — who specialized in non-producing lucerne (alfalfa) in order to get a compensation, and having received the money, he eagerly invested it in buying more land to non-produce even more lucerne.

419-428)—should result in an optimal allocation of permits which go to those who value them most.

The marketable permits approach perfectly conforms with Herman Daly's (1984) principle of separating *scale* and *allocation* decisions in environmental policy. To regulate the environment means to decide to what extent a given resource should be used, and what portion of the resource should be allocated to any of its potential users. Only the second problem is suitable for market regulation. When it comes to the question of scale the market obviously fails as a guide, and the solution should be arrived at by means of a public debate and finally an administrative (ambient rather than source) standard.[4] This theoretical background and an enormous cost-reducing potential make marketable permits a preferred instrument of environmental policy.

Marketable permits outperform alternative solutions as far as technological progress is concerned since they establish the strongest incentives for firms to innovate in abatement activities (Milliman and Prince 1989)[5] However, profound problems arise in practical applications of the marketable permits principle. The costs of establishing a market for these may outweigh the (expected) savings. In particular, this may happen in the case where "thinness" of the regional market makes a competition problematic. In any case the actual efficiency of a marketable permits solution should be compared to that of alternative policy measures including standards and taxes.

This relatively simple framework for analyzing and solving problems of atmospheric emissions or water discharges collapses when it comes to transboundary pollution. There exist external costs, but there is no supranational body to impose Pigouvian taxes. There is a need for ambient quality standards, but the local concentrations depend on emissions which may originate outside the country. Finally, there is a possibility to globally optimize the abatement effort, but, again, this cannot be carried into effect for the lack of an appropriate supranational authority or mechanism to fairly distribute its costs and benefits. The economics and politics of the global commons require a much more complex approach.

THE "X-PERCENT CLUBS"

The European "acid rain" provides a good example of transboundary pollution which is now tackled without help of economic instruments. The exceptional perplexity of European airborne pollution stems from the fact that huge emissions accompanying economic activity concentrated on a relatively small area are matched with the conflicting sovereignties of no less than 30 nations. Each solves its environmental problems compromising certain political, social and economic goals while being exposed to certain

---

[4] See also David Pearce's (1987) concept of an "ecologically bounded economy." Pearce argues that a sustainable use of the environment cannot be endogenously derived from the market (nor the central planning) mechanism itself.

[5] Empirical evidence does not fully support this opinion, but on the other hand, as Liroff (1988, p. 259) puts it, marketable permits have "been the victim of unduly high expectations."

pressure from its sovereign neighbors. Obviously, a solution of the SO_2 problem in Europe requires a concerted action of all countries concerned. All of them lose due to the "acid rain" but some do less than others. The economic burdens of emission control differ greatly from country to country. One may then ask what are the chances of effective, concerted action on the part of all European countries.

Negotiated in Geneva in 1979, the Convention on Long Range Transboundary Air Pollution has been ratified by all members of the United Nations Economic Commission for Europe (ECE) except Romania and Yugoslavia. The Convention does not establish any liability rules for exported emissions. Its main significance consists in the development of the European system of monitoring (EMEP) and information exchange.

The 1985 Helsinki Protocol to the Convention was the first step towards the concerted action which was called for. The Protocol requires countries to reduce the total amount of SO_2 they emit by 30% of the level they emitted in 1980 by the year 1993. Regrettably, Greece, Poland, Portugal, Romania, Spain, Turkey, the United Kingdom, Yugoslavia (and the United States) have not signed the Protocol. At the same time, the environmentalists are urging for 75% reductions in SO_2 by 1995. This reduction would possibly save some of the coniferous forest in central Europe but it is highly improbable that the main polluters will accept such a reduction schedule.[6] The group of signatories of the Protocol has been referred to in the media as the "30-Percent Club."

How are states persuaded to accept commitments that are unprofitable in short-run, national economic terms? The problem is rather delicate, as hypothetical benefit and cost shares are highly diversified.[7] Moreover, any reference level for reductions is arbitrary so that bargaining is unavoidable. It is thus no surprise that the Helsinki solution was arrived at in 1985 after 6 years of negotiations, and still consensus among the polluters has not been universal.

A question arises to what extent the SO_2 reduction controversies are typical, and the adopted policy adequate for other cases of transboundary pollution. Three types or groups of air-polluting substances could be distinguished in this context:[8] (1) SO_2-like (i.e., toxic migrating) pollutants, (2) toxic nonmigrating pollutants, and (3) nontoxic migrating pollutants.

Group 1 consists of substances which behave like SO_2, i.e., directly cause damage due to their toxicity and may easily migrate over vast distances. Nitrogen oxides (NO_x) and suspended particulates belong to this group. Particulate matter emissions may or may not

---

[6] See Peter H. Sand (1987) for an outline of the history of the Convention traced back to the 1972 Stockholm Conference. Zylicz (1988) discusses specific problems that (former)centrally planned economies encounter in meeting the requirements of the Helsinki Protocol.

[7] In the case of SO_2 abatement cost differentiation results from the variety of fuels burnt as well as from system efficiency of various economies. Damages from emissions to a large degree can be explained in terms of natural sensitivity (to acidification) of various ecosystems (see Paces 1988).

[8] This is a rough classification for the purpose of our further argument only. A finer one would be required for a more detailed study of various policy instruments (see, e.g., Tietenberg 1988).

be toxic, depending on their chemical composition. Coarse particulates fall in the immediate neighborhood of the polluter. The fine (suspended) particles behave in the atmosphere more less like gases. Generally particulate matter emissions may belong to either of Groups 1 or 2 defined above.

Apart from coarse particulates—called also dust emissions—Group 2 consists of several gaseous pollutants. The non-migrating pollutants can be labeled as such either because they quickly fall down (as the dust does) or because they decay (or get neutralized in another way) before being transferred very far from the source. Carbon monoxide (CO) is an example of the latter category, as it transforms to nontoxic CO_2.

Group 3 consists of substances that are non-toxic yet seriously disturb the biosphere. Two such pollutants have been the focus of scientists' attention worldwide at least since the 1970s. These are carbon dioxide (CO_2) and chlorofluorocarbons (CFCs), a whole family of chemicals also known as "freons" after their trade name. There are two major sources of CO_2 releases into the atmosphere: biomass degradation, e.g., resulting from deforestation of the Earth, and fossil fuel combustion. CFCs entered the biosphere in the 1930s. A DuPont patented product, they have proved to be extremely useful in many applications. They are nontoxic, do not react with other materials and thus, up until the 1970s, were considered absolutely safe.

CO_2 together with several other "greenhouse gases" (methane, nitrous oxide, and CFCs) act as an energy trap resulting in increased Earth's temperatures. Raising the Earth's temperature would lead to a number of adverse effects. The most spectacular one would be melting some of the glaciers and ice caps around the poles, which would raise the sea level and thus submerge many estuaries and coastal areas. Ozone (O_3), although a pollutant when in the lower atmosphere, turns out to be beneficial in the upper atmospheric strata. The stratospheric ozone layer protects the biosphere against the excessive ultraviolet radiation. This layer has been gradually depleted through several human activities such as supersonic transportation, space rockets, and CFC emissions.

The third group of pollutants, poses a truly global challenge. Rising CO_2 concentrations affect the global climate, and depleting stratospheric ozone threatens the whole biosphere. It is rather unlikely that a natural compensation mechanism for the shrinking ozone layer will be found. Hence, control of CFC emissions is inevitable. Two kinds of CFC use can be distinguished: essential and unessential ones. The latter comprise those instances where—as in the case of cosmetics—the use of the dangerous chemical can be easily substituted with something else. A complete ban for such uses could and indeed should be introduced—coupled with ceilings per capita and freezing/reduction schedule for production aimed at the essential ones.[9]

A more complicated pattern emerges in the case of an anti-carbon dioxide policy. CO_2 cannot be avoided as long as fossil fuels are burnt. Some ecologists believe the additional

---

[9] From the purely economic point of view the distinction between "essential" and "unessential" uses can not be drawn since substitution—being only a matter of adequate price incentives—is always possible. However, given the fact that the world does not consist of perfect market economies, second best (administrative) solutions have been adopted in international agreements quite often.

increments in atmospheric CO_2 will be dissolved in the world ocean. However, this has not been observed and only appears to be true because of a time lag (Colinvaux 1978). (After all, the massive CO_2 releases began only a century ago). Rebuilding the planet's biomass—which has always played the role of a gigantic carbon store—seems a far more certain strategy for CO_2 removal from the atmosphere. Of course, planting trees, reforesting the Earth, is the most effective way to restore and even augment the biomass. Despite some optimistic opinions this is not to be expected as a natural outcome of the "greenhouse effect" (Brown et al. 1988, pp. 43-44). Rather, it must be the goal of concerted efforts of the whole global community. Let us also note, the more degraded the biosphere is, the harder this task is to accomplish.

To what extent can the international response to the challenge posed by SO_2 emissions be repeated in the case of each of the three groups of pollutants? What is happening now with respect to Group 1 is very much like a repetition of the Helsinki Protocol experience. A new Protocol on freezing NO_x emissions was negotiated within the Geneva Convention in 1988. Again, some countries (e.g., the United States) raise objections to the status quo as the baseline for the agreement. The status quo reference point "favors" nations that have done nothing so far and "penalizes" those which abated in the past. There are also problems reported by former Centrally Planned Economies (CPE) which may not be able to overcome some technical constraints by the mid-1990s. The technological obsolescence of the cars produced and used there, matched with far from saturated demand for passenger cars, make the United Nations ECE freezing requirements rather hard to meet. It can be expected that suspended particulates will emerge as the next item for the European agenda as soon as the nitrogen oxides question is settled. In this sense we can speculate that the model developed for SO_2 will be repeated in the case of any pollutant from the first group.

From the economic point of view the second group of pollutants is highly heterogeneous. It contains coarse particulates—the emissions which are easiest to abate—as well as some car exhaust gases which, on the contrary, are hard to control. Dust emissions have been virtually eliminated almost everywhere in developed market economies. This was made possible by two factors. Firstly, scrubbers are available (at a reasonable cost) which can remove as much as 99.99% of potential emissions. Secondly, stringent standards were imposed on industry in response to the citizens' demand for a "clear sky." Dust emissions in former CPEs are still vast despite the technical possibility and even economic affordability of investing in adequate abating equipment.[10]

As a rule, the nonmigrating pollutants are not subject to international controversies. There are, however, exceptions when a polluter is close to the country's boundary. Several East German energy plants (brown coal fired) erected close to the Polish border serve as rather rare examples of this sort of pollution.[11] Prevailing winds made these

---

[10] The paternalistic attitude of the state administration towards the prevailing state-owned sector of the economy has been the main reason for the lack of an effective control in these countries.

[11] It is ironic that two of these plants lacking an adequate abating equipment are called Volksfreundschaft and Friedengrenze.

emissions a clearly visible external cost for a number of Polish municipalities without prospects for a solution until recently.

Nations have no experience in dealing globally with pollutants. The recent moderate progress in the ECE negotiations over "acid rain" rests upon several factors. First of all, the consequences of "acid rain" in many European countries have reached an alarming level. Moreover, cause-effect relationships between acidification of the environment and sulfur depositions have been empirically confirmed. The continental network (matrix) of sulfur emissions and migrations has been computed and successfully tested against the monitoring data. Finally, the class of the parties concerned consists of "just" thirty nations that are economically better off (on the average) than the rest of the world.

None of these factors is valid in the case of carbon dioxide. Its consequences are still disputable and its control cannot be confined to a continent. Its net emissions are hard to allocate to specific countries because biomass production/destruction balances are not trivial to calculate. As for CFCs, the 1985 Vienna Convention for the Protection of the Ozone Layer did not succeed in establishing any control strategies. Some strategies were adopted in its 1987 Montreal Protocol, although they were almost immediately criticized as environmentally insufficient (Brown et al. 1988, p. 8).

Possible financial mechanisms for phasing out CFCs were recently discussed within the framework of the Montreal Protocol. Less developed countries are offered more time to adjust their economies to CFC-free technologies. Various incentive schemes were considered to speed up the process while leaving some room for flexibility (i.e., greater efficiency). The schemes were based on either charges and subsidies or marketable permits. In either case an implicit allocation of property rights to the atmosphere should have been agreed upon. Again, the historical status quo proved to be the only viable assignment of these rights. The Protocol—as amended in 1990—allows the signatories to trade their CFC quotas. It also provides additional incentives by compensating the costs of phasing out CFCs in the less developed countries. Both provisions have improved the efficiency of the global CFC abatement.[12]

Similar arrangements are contemplated for CO_2 and other greenhouse gases. Here what is peculiar is that the net emissions—at least those of carbon dioxide—can be controlled in two ways: through curbing fossil fuels use and through augmenting biomass. Emission charging in this case would immediately translate into subsidizing nature conservation and reforestation. To the extent that the less developed countries may have greater natural potential for storing carbon, this option could be attractive for their economies. On the other hand, only the developed countries' demand for energy has stabilized. It is thus a matter of detailed empirical studies to determine whether a given economy would financially benefit from a charging scheme. The same argument applies to marketable permits. Even if initially grandfathered, the permits should then be subject to a "shrinking" schedule in order to achieve a net emission decrease. Thus the question of net payments is open in both cases, and it will not be easy to work out an effective yet widely acceptable solution. The relative success of the CFC compensation scheme was

---

12 As Bohm (1990) points out, the way developing countries are being compensated for their *incremental costs* of meeting the Protocol requirements is still far from efficient.

fairly easy to achieve when it turned out that the costs involved were much lower than assessed previously (Markandya 1990). On the contrary, the costs of controlling greenhouse gases will be much higher.

Is then the global atmosphere doomed to the "Tragedy of the Commons"? The latter expression was coined by Garret Hardin (1968) to characterize the fate historically shared by many natural resources with unconstrained access: they were exploited up to the point of the total and irreversible destruction. There are, however, some encouraging historical precedents showing that the transition from a res nullius (no property rights) to a res communis (common property resource) is possible (Ciriacy-Wantrup and Bishop 1975). The recent European decisions targeted at the "acid rain" issue have made a step in the right direction. However, the economic and political analysis of these developments suggests that the establishment of a comprehensive system of property rights to control the access to the atmosphere will be a long and difficult process.

The preceding review focused on the atmosphere, but obviously there are many important cases of waterborne transboundary pollution as well. Here too one can talk about the three groups of pollutants and discuss possible ways of their international control. Sometimes there are well defined upstream and downstream parties, but sometimes the use of an aquifer resembles that of a common property open access resource, such as polluting the Baltic Sea. The sea receives considerable amounts of both toxic and nontoxic pollutants. The former have been responsible for contaminating the trophic chains which eventually affects human health. The latter contribute to eutrophication of the sea which—in the short run—can be coupled with the observed growing yields of Baltic fisheries; in the long run, however, it may lead to total ecological disruption, the first symptoms of which have been visible at least since the 1970s.

As in the case of the European "acid rain," considerable externalities exist, and the costs and benefits of cleaning up the sea are diversified. Instead of resolving the problem by means of Pigouvian taxation, the Baltic countries have been seeking a political consensus. An initial agreement was reached within the framework of the Gdansk Convention in 1973. This was followed by the Helsinki Convention a year later. The first meeting dealt with the living resources of the sea. A more comprehensive protection program was adopted in the second meeting. However, it was not earlier than 1988 that "Helcom," the executive body of the Helsinki Convention, established a 50-Percent Club. For a state to be a member, it must reduce discharges into the sea by 50%. Again, uniform proportional reductions proved to be the only politically viable allocation of the abatement effort despite the universal awareness of their economic inefficiency (see e.g., Hoel (1990) or Opschoor et al. (1990)).

THE VICTIM PAYS PRINCIPLE

Until recently, there were no clear indications of the Coase theorem working at a supranational level. Some Central European developments might have looked like the theorem's manifestations, but under closer scrutiny they were not. West Berlin investment in abating pollution in the G.D.R. was never quite considered as an external transfer from the West German policy perspective. On the other hand, Sweden's early initiatives aimed at modernizing some of the Polish energy plants assumed that Poland—the bribed polluter—would bear the expenses of the necessary equipment; the role of the bribing pollutee confined to taking care of the transaction costs. The precedent-setting debt-for-nature swap[13] to save a part of Bolivia's rain forest came closer to the Coasian idea. However, the very limited scope of the swap does not allow one to attach too much weight to the case. So far, managing the global commons has proved to be neither successful nor efficient.

Reluctance to subsidize the polluter is based on several points. First, any Coasian solution can be challenged on moral grounds. Moreover, its long-run efficiency is questionable as indicated in the first section of this paper. Additional doubts arise in the case of an economically inefficient polluter. Then not only the equity, but also the long- and short-run efficiency of a payment, becomes a matter of concern. No one wants to spend money if the effect is too expensive or uncertain. This explains the history of Sweden's involvement in subsidizing environmental protection in Poland.

As long as Poland remained a CPE, Swedish initiatives were confined to promoting end-of-pipe technologies almost exclusively. The donors correctly assumed that such investments could lead to overall emission reductions although not in a least-cost way. More efficient options would require better industrial practices as well as energy and material conservation measures. However, it has been known that in CPEs, the demand for energy (and other inputs) is subject to physical rather than financial constraints. Thus no technological improvements can result in lower aggregate energy consumption, since every ton of fuel which is physically available will be burnt anyway.[14] In other words, energy conservation in one plant would only result in higher energy availability elsewhere. In a CPE which—by definition—is a shortage economy, this might not necessarily result in lower overall demand. Even though less efficient, the results of end-of-pipe technologies are more environmentally effective.

The attitude towards various kinds of Swedish involvement in environmental protection in Poland changed radically after serious market reform had been initiated in 1990. Not only did the absolute level of aid increase, but also—for the first time—the donors started to believe that more cost-effective conservation measures may work out. As a result, completely new forms of financial assistance are now considered.[15]

Coasian arrangements have to be relied on in the case of greenhouse gases, as no effective solution would be possible without a wide participation of less developed countries. On the contrary, since there are only a few CFC manufacturers, the protection of the ozone layer could be worked out by partners from the developed economies. However, it would be a good way of channeling international development aid to assist in the adop-

---

[13] The 1987 swap subtracted just $650,000 from Bolivia's $2.8 billion foreign debt. [The purchase was made by Conservation International—a U.S. based environmental organization. (Shabecoff, 1987)].
[14] To avoid this, a Swedish partner once suggested a barter transaction: coal-saving technology for the coal saved (payment in kind).
[15] What is crucial is that the aid is not being confined to end-of-pipe technologies.

tion of environmentally "correct" substitute products and/or technologies. In fact, the solution now emerging within the framework of the Montreal Protocol is more of a development assistance project (to encourage compliance) than an incentive scheme. Its goals are based on percent reductions and include virtually no incentives to optimize the "abatement."

EXPERIMENTING WITH MARKET SOLUTIONS

"Harnessing market forces" has become the most distinctive feature of the American environmental policy over the last decade (Stavins 1989), and one can expect a similar trend in many other OECD countries (Hahn 1989). The newly prepared II World Conservation Strategy will explicitly endorse controlled market solutions to global problems of environmental abuse (Prescott-Allen 1989). The apparent revitalization of market mechanisms in the West has been coupled with courageous promarket reforms in the East. Several former CPEs are now in the process of restoring market institutions that were deliberately destroyed two generations ago. All these developments call for a very careful scrutiny of the role markets actually play or could play in managing natural resources.

The following discussion will focus on marketable permits rather than charges and subsidies. This is justified by the fact that any constraint in applications identified for a permit system refers to the equivalent charging/subsidizing scheme as well, but not necessarily vice versa. The theoretical equivalence[16] (and long-term optimality) of both schemes was convincingly outlined by Pezzey (1989). They are not equivalent, however, if environmental safety is taken into account. From this point of view, the physical outcome of charging schemes in the absence of overall quantity ceilings is less predictable than in the case of marketable permits because of financial uncertainties.[17]

The ever-growing acceptance of markets in natural-resource management is symptomatic for a more general process of substituting administrative standards with more flexible instruments. Praising and prescribing the latter, economists (both in the East and West) rightly point to inefficiency inherent in any administrative solutions. In doing so, however, they often neglect some critical assumptions that are necessary in order to validate their recommendations. Briefly, to arrive at many typical results of policy relevance, it is necessary to assume that the decision makers maximize their concave profit functions within budget constraints, and the market is in a competitive equilibrium with perfect information. There are many reasons to question realism of this even in a developed market economy of a single state. Additional doubts arise in the case of a group of states—especially when some of the partners are lacking well developed market institu-

---

[16] The equivalence proof relies on the notion of a "benchmark" or a base level of emissions; only the difference between actual emissions and the "benchmark" is subject to a uniform linear charge/subsidy. It is worth noting here that most economists unnecessarily assumed the zero base level.

[17] See Hoel (1990) for a thorough discussion of marketable permits versus taxes to internationally reduce CO_2 emissions.

tions.[18] Can we really expect that even a carefully designed permit market will move us towards an optimal allocation of abatement?

The most common confusion stems from the fact that different time horizons may—and in fact do—produce different optima. Taking into account long term strategic considerations, a firm or a state may simply prefer to forgo some opportunities offered by price incentives. The reluctance to engage in trading marketable permits under the US Clean Air Act provides a good example of this behavior. Even though trades occur under the bubbles, they are incidental rather than routine, and much less frequent than originally predicted. More importantly, as banking has continued to be the least vigorous of the trading activities (Hahn 1989, p. 100), nobody can claim that a regular market in environmental assets developed.

There are many reasons why the firms do not trade as much as they should from the static efficiency point of view. Apart from imperfect information, an important reason is sacrificing a quick buck now for a (supposedly) safer and more stable future. Any firm that offers an emission reduction credit for sale reveals an upper bound of its marginal cost of abatement. This information can then be used by the issuing authority to economically justify a lower sum of permits in the next period.

Thus, from the dynamic efficiency point of view, what a firm or a state earns from selling a marketable permit may not outweigh what it loses (or risks to lose) in the long run from the transfer or postponement of its property right. As a result, the optimizing behavior predicted on the base of a typical partial equilibrium model does not correspond with actual levels of trading. The less operational but more realistic long term profit maximization criterion helps to explain why permit markets have not been able to realize all the benefits that are potentially available. The reluctance to trade may well be observed on an international scale too.

Economists commonly assume that the profit or social welfare functions they maximize are concave (or, at least, quasi-concave). To the extent that profits negatively depend on costs, this typically requires convexity (or quasi-convexity) of the cost functions. Convexity in economic analysis is a technical assumption which guarantees that any market equilibrium produces the global optimum. Without convexity, the market fails as its agents may get stacked in an equilibrium position far from the optimum; they have no incentives to revise their decisions and yet the system as a whole has not exhausted all possible gains from its operation, i.e., it has not reached Pareto-optimality.

The convexity assumption is quite plausible in many instances: indeed, tightening controls usually implies more than proportional abatement effort. Economies of scale contradict this argument, but as long as they do not violate some weaker requirements (e.g., quasi-convexity), proofs of efficiency carry on and the market does not fail.

If individual decision makers act independently and no external economies are possible, convexity of their cost functions can be reasonably assumed. Troubles start as soon as cooperation becomes possible and results in substantial economies of scale. Joint indus-

---

[18] It should be noted that some typical general equilibrium assumptions are hardly met even in a Canada-US model (see, e.g., Merrifield, 1988) not to speak of models involving more heterogeneous groups of countries.

trial or municipal treatment plants serve as an example of this class. Here, some sort of administrative intervention would be necessary to reach the efficiency. Similar problems will arise in the case of an attempt to establish price incentives to control several environmental variables simultaneously. This is especially true in the case of synergistically interacting emissions (Zylicz 1989). However, even without synergism, but merely as a result of scale economies[19] from abating several pollutants simultaneously, the convexity assumption can be violated—thus calling again for an administrative instrument. It would be interesting to see whether this is just a "theoretical curio" or an empirically observed phenomenon.

Information proved to be another barrier to the successful operation of many permit markets in the United States. Potential buyers and sellers have a hard time identifying each other, and the related transaction costs are high (sometimes prohibitively). One can argue, however, that to a large extent this is just a matter of time. Having made certain transactions possible, perhaps one should wait a couple of years for adequate intermediary institutions to emerge naturally. It is a matter of adequate international cooperation to make sure that information will not emerge as a barrier to efficiency gains from trading permits.

The size of the market is still another potential barrier. Unlike the former, this is a physical rather than institutional constraint and it cannot be relaxed over time. In the case of the U.S. emissions trading policy, the size of a market is given by the size its bubble. This sharply constrains the market, yet it often leaves the number of agents which is large enough to approximate the perfect competition assumption. Typically, in the international context, the thinness of a market is not a problem. Pollutants from the second group (dust and certain gaseous pollutants) defined above may contradict this pattern. To some extent, the same applies to pollutants from the first group, as local ambient quality requirements do not allow for unconstrained emissions trading. Pollutants from the third group can be traded globally, and it is only here that market thinness will never threaten efficiency.

Countries with weak market institutions and strong state intervention encounter a problem studied under the heading of the "soft budget constraint." In Janos Kornai's (1986, p. 4) words: "The "softening" of the budget constraint appears when the strict relationship between expenditure and earnings has been relaxed, because excess expenditure over earnings will be paid by some other institution, typically by the State. A further condition of "softening" is that the decision maker expects such external financial assistance with high probability and this probability is firmly built into his behaviour."

The soft budget constraint is a way of abandoning or just weakening the market as a mechanism of allocation with economic efficiency as the sole criterion for decision mak-

---

[19] Many pollutants show a high degree of technological complementarity. Not only do they emerge from the same technological processes, but—to some extent—they can be abated with the same equipment at substantial cost savings. The flue gas desulfurization technique may serve as an example; it is aimed at reducing sulfur dioxide content of emissions, but simultaneously it is effective in removing nitrogen oxides which are produced during the combustion process.

ing. To some degree it exists in every market economy, but only in CPEs, it emerged as the routine way of sanctioning the outcomes of allocation processes which went out of control. As a result, any price incentives lose their edge, and decision makers do not have to pay much attention to them. This is what in fact has been happening in nonmarket economies. It should be emphasized that price instruments—especially charges and subsidies—require a healthy, competitive market environment to work. This is the necessary although, by far, not a sufficient condition for a successful application of today's sophisticated market methods.

SUMMARY AND CONCLUSIONS

Even in a local context, pollution cannot be confined to the problem of external costs. These costs are highly uncertain, and thus—as a rule—environmental policy is not confined to Pigouvian taxes. On the contrary, some theoretical findings as well as successful policy experiments have resulted in developing marketable permit systems, which neatly combine administrative and market instruments. This approach, however, has not yet been applied at an international scale.

When we start to analyze the atmosphere, the ocean, the climate or biological diversity as a global commons, the economics of pollution becomes a truly *political economy* of pollution. What would be efficient from an economic point of view turns out to be politically unfeasible; what happens to be a realistic solution is far from being ecologically satisfactory. This does not mean that cost-benefit analysis has been useless in explaining and predicting how nations perceive the pollution and react to it. Any policy solution is somehow rooted in economic considerations, but what we actually see is a complex compromise of the conflicting economic considerations of many sovereign states which is far from an ecological-economic optimum.

In this paper, particular attention has been paid to the "acid rain" controversies in Europe, and the way they are solved within the framework of the Convention on Long Range Transboundary Air Pollution. The Convention—together with its existing and prospective protocols—may be viewed as a model for some pollutants, while it is clearly impossible to copy it in many other instances of atmospheric pollution. Moreover, despite all the progress achieved or expected within X-Percent Clubs, a more economically efficient approach is called for.

It would be difficult to apply the Polluter Pays Principle internationally. The basic obstacle is the lack of an appropriate supranational authority to levy Pigouvian taxes (or initially allocate permits). The resistance to create such an authority stems from the fact that polluters would not fully benefit from their abatement effects. Even in a national context, there are objections to the strict adherence to PPP. Once applied, however, its political acceptability can be derived from hypothetical compensations to be paid by the gainers to the losers while making both parties better off.[20] As long as these payments

---

[20] The compensations do not have to take the form of direct payments. They can materialize as indirect benefits the polluter derives from saving his victims.

are hypothetical only, the notion of efficiency remains an abstract concept for those who may be potential partners.[21]

The lack of appropriate (global or regional) political institutions is thus the main obstacle to applying charges and marketable permits on a international scale. The separation of efficiency from equity can exist in academic textbooks only. No real progress can be achieved without practically integrating both points of view on welfare.

Meanwhile, bilateral Coasian bargains are being developed spontaneously. Their scope has been limited because of equity considerations but also because of the difficulty for the victim to make sure that the polluter uses the aid efficiently. One can assume, therefore, that the use of this solution will widen as soon as the polluting countries progress towards more efficient and manageable economies. This would help in achieving better international allocation of abatement before adequate multilateral political organizations make it possible to fully optimize the efforts.

REFERENCES

Bohm, P. 1990. Efficiency Aspects of Imperfect Treaties on Global Public Bads. Paper presented at the World Institute for Development Economics Research Conference. Helsinki.

Bohm, P., and C. S. Russell. 1985. Comparative Analysis of Alternative Policy Instruments. In A. V. Kneese, J. L. Sweeney, eds., *Handbook of Natural Resource and Energy Economics*, Amsterdam: North Holland, 1:395-460.

Brown, L. R., W. U. Chandler, A. Durning, C. Flavin, L. Heise, J. Jacobson, S. Postel, C. Pollock Shea, L. Starke, and E. C. Wolf. 1988. *State of the World 1988*. New York: Norton.

Ciriacy-Wantrup, S. V. and R. C. Bishop. 1975. "Common Property" as a Concept in Natural Resources Policy. *Natural Resources Journal* (October) 15:713-727.

Coase, R. H. 1960. The Problem of Social Cost. *The Journal of Law and Economics* (October) 3:1-44.

Colinvaux, P. 1978. *Why Big Fierce Animals Are Rare. An Ecologist's Perspective.* Princeton: Princeton University Press.

Crocker, T. D. 1966. Structuring of Atmospheric Pollution Control Systems. In H. Wolozin, ed., *The Economics of Air Pollution*. New York: Norton.

Daly, H. E. 1984. Alternative Strategies for Integrating Economics and Ecology. In A.-M. Jansson, ed., *Integration of Economy and Ecology. An Outlook for the Eighties.* pp. 19-29. Proceedings from the Wallenberg Symposia. Stockholm: University of Stockholm Press.

Hahn R. W. 1989. Economic Prescriptions for Environmental Problems: How the Patient Followed the Doctor's Orders. *Journal of Economics Perspectives* 3:95-114.

Hardin, G. 1968. The Tragedy of the Commons. *Science* 162:1243-1248.

Hoel, M. 1990. Efficient International Agreements for Reducing Emissions of CO_2. Memorandum from Department of Economics, no. 6. University of Oslo.

Kamien, M. J., N. L. Schwartz, F. T. Dolbear, Jr. 1966. Asymmetry between Bribes and Charges. *Water Resources Journal.* (First Quarter) 1966:147-157.

Knight, F. H. 1924. Some Fallacies in the Interpretation of Social Cost. *Quarterly Journal of Economics* (August(38:582-606.

Kornai, J. 1986. The Soft Budget Constraint. *Kyklos* 39:3-30.

---

21 That is why it is so much easier (although still difficult) to control interstate pollution within a federal entity (Ruhl, 1988). What can be achieved in a country like the U.S. is not necessarily possible to replicate in a group of nations.

Liroff, R. A. 1988. EPA's Bubble Policy: The Theory of Marketable Pollution Permits Confronts Reality. In R. C. Hula, ed., *Market-Based Public Policy*, pp. 242-261. London: Macmillan.

Markandya, A. 1990. The Costs to Developing Countries of Entering the Montreal Protocol: A Synthesis Report. Geneva: UNEP.

Merrifield, J. D. 1988. The Impact of Selected Abatement Strategies on Transnational Pollution, the Terms of Trade, and Factor Rewards: A General Equilibrium Approach. *Journal of Environmental Economics and Management* 15(3):259-284.

Milliman, S. R. and R. Prince. 1989. Firm Incentives to Promote Technological Change in Pollution Control. *Journal of Environmental Economics and Management* 16(4).

OECD. 1989. *Economic Instruments for Environmental Protection*. Paris: OECD.

Opschoor, J. B., O. J. Kuik and K. Blok. 1990. *Economic Aspects of Carbon Dioxide Reduction. A View from The Netherlands*. Leidschendam: Ministry of Housing, Physical Planning and Environment.

Paces, T. 1985. Acidification in Central Europe and Scandinavia: A Competition between Anthropogenic, Geochemical and Biochemical Processes. *Ambio* 6:354-356.

Pearce, D. W. 1987. Foundations of an Ecological Economics *Ecological Modelling* 38:9-18.

Pezzey, J. 1988. Market Mechanisms of Pollution Control: "Polluter Pays", Economic and Practical Aspects. In R. K.Turner, ed., *Sustainable Environmental Management. Principles and Practice*, pp. 190-242. London: Belhaven Press.

Pezzey, J., 1989. Efficient and Politically Acceptable Pollution: The Property Rights Approach. Paper presented at the Eastern Economic Association Fifteenth Annual Convention, Baltimore.

Pigou, A. C. 1920. *The Economics of Welfare*. London: Macmillan.

Prescott-Allen, R. 1989, Economic Sustainability. In *Managing for the Future*, ch. 3. II World Conservation Strategy. Gland: IUCN.

Ruhl, J. B. 1988. Interstate Pollution Control and Resource Development Planning: Outmoded Approaches or Outmoded Politics? *Natural Resources Journal* 28:293-314.

Sand, P. H. 1987. Air Pollution in Europe: International Policy Responses. *Environment* 29(10):16-20; 28-29.

Shabecoff, P. 1987. Bolivia to Protect Lands in Swap for Lower Debt. *The New York Times*, July 14.

Stavins, R. N. 1989. Harnessing Market Forces to Protect the Environment. *Environment* 31:5-7;28-32.

Tietenberg, T. 1988. *Environmental and Natural Resource Economics*. 2d ed. Glenview, Ill.: Scott, Foresman.

Zylicz, T. 1988. Will Poland Join the 30% Club? *European Environment Review* 2(1):2-5.

Zylicz, T. 1989. Marketable Permits for Synergistic Pollutants. Institute of Behavioral Science, University of Colorado, Boulder.

26

RETHINKING ECOLOGICAL AND ECONOMIC EDUCATION: A GESTALT SHIFT

Mary E. Clark
Center for Conflict Analysis
George Mason University
Fairfax, VA 22030-4444 USA

ABSTRACT

Two gestalts and their consequences are elaborated, contrasting neoclassical and ecological economics.

> Gestalt I: Linear Progress and Competitive Individualism
> Self-interest and greed, growth and progress
> Neoclassical theory and the problem of "values"
> Gestalt II: Community/Environment Relations
> Desired environmental and social goals
> Some living social models

The implications for education of gestalt II are:
- •Grounding in ecological principles of sustainable systems
- •Grounding in psychosocial factors creating sustainable societies, and in approaches to the self-generated change that most societies must undertake to become "sustainable"
- •Training in critiquing the limitations of present economic theory, thus bridging from present to future gestalts
- •Training in identifying the appropriate socioeconomic entities for establishing local, bioregional, and global sustainability

INTRODUCTION

In his illuminating little book, *The Culture of Technology*, British physicist/environ-mentalist/anthropologist/philosopher Arnold Pacey suggests that it is "time to change the topic of conversation" (Pacey 1983). It is time to see ourselves and our world in a new and different light. The topic of conversation in today's international arena has become "the global economy," and its "health" has become our central concern. This reified abstraction is the intellectualized icon that forms the focal point of our shared existence as a species; it has taken on a kind of sacred meaning.

Through the power of our electronic media to project the image of Western wealth to the rest of the world, we have managed to mesmerize nearly five billion people into thinking a) that the West has solved all the old economic and social problems, and is on to a great thing; and b) that they, too, can live like us if only they follow our prescriptions. We have created an almost global "cargo cult"—a belief that the wealth of the North is not only within the grasp of all, but once grasped it will also bring remission from all pain. Unfortunately, far too many people living in the North and in the South are failing to notice both the environmental destruction *and* the social disintegration that Western economic behavior causes.

I have not space here to provide detailed evidence for these two negative consequences of modern free-market, industrial systems. They are described in detail in my book, *Ariadne's Thread*, and by many others as well (Clark 1989, and references therein). I shall simply take these as "givens." This paper will analyze the attitudinal shifts necessary to move from conventional economic thinking to what, in my view, is needed for a sustainable global economy. This will entail considerable restructuring of the Western world view and of the assumptions on which it is based. For education, it will mean merging and restructuring current thinking in all the social sciences, especially economics.

TWO GESTALTS AND THEIR CONSEQUENCES

Each human being lives embedded in an environment and in a society, and is dependent on both. Among indigenous societies, cultural patterns are often interwoven with the natural patterns of the environment—from Hopi rain dances in the American southwest to the ritual festivities synchronized with the annual spawning of the prized palolo worms in the southwest Pacific. Although prehistory had its environmental faux pas, the preindustrial track record was far more environmentally sound than is that of modern industrial society. While this was in part owing to smaller population size and limited technical ability to wreak destruction, it was also due to a deep sensitivity toward the environment, coupled with effective communal management of the resource base. This contrasts sharply with the approach of modern industrial societies, which view the environment as infinite provider and bottomless sink, and society as an engine of production and consumption. Let us examine these two world views, or gestalts, for their ability to bring economic activity into line with both ecological and human psychic limits.

Gestalt I: Linear Progress and Competitive Individualism

"Man is an acquisitive animal whose wants cannot be satiated." So wrote economist Lester Thurow (1981, p. 120) about the only "man" he knows, modern Western man who has been carefully taught to be an avid consumer. His statement implies that, if we are to live within the limits of our planetary environment, we are doomed to eternal dissatisfaction. Yet there are cultures with no evidence of unlimited acquisitiveness. Indeed, if we were to examine all the assumptions underlying the Western world view that now dominates the so-called global economy, we would find most of them suspect. Here, I will tackle only a few.

Self-interest and Greed, Growth and Progress

Communal or organic societies, whether egalitarian or hierarchical, work well as long as individuals regularly receive their expected allocation of resources and the rules by which the system is justified are adhered to. But by 1500, the hierarchical societies of Europe were starting to disintegrate owing to a combination of population pressure, growing social oppression in the face of blatant opulence in the Church and the royal courts, and the ideas of equality resurfacing from classical times. Within three centuries, a thousand-year-old order had been overturned and a new world view, based on a new set of assumptions about human nature and social order, came into being. This new gestalt viewed societies as built consciously by contracts between self-centered, competing individuals; it saw property as belonging by right to particular individuals, and not to all the people nor to the state; it concluded that unfettered self-interest and competition will maximize a nation's wealth; and finally it defined utility and pleasure as the supreme goals of life. Accumulation of wealth and power were a sign of virtue; poverty, once pitiable, became a sign of sinfulness and sloth.

This newfound emphasis on the rights to private property and personal accumulation of material wealth, fortuitously catalyzed by technological innovation and particularly by the exploitation of fossil fuels, led to unprecedented economic expansion. But socially, the laissez-faire capitalist gestalt has two serious flaws. One is the polarization of wealth, which the unregulated system attempts to compensate through the "trickle-down" of constant growth, to be achieved by never-ending "progress" in technological efficiency, as well as through constant stimulation of throughput. This, of course, runs head-on into environmental limits, and has created the enormous backlog of unpaid bills to Mother Nature.

The other flaw is more subtle. When John Locke wrote three centuries ago that "The chief end of trade is Riches & Power which beget each other" (Locke 1962, p. 207), he little knew that modern society would convert it to "The chief end of *life* is riches and power which beget each other." Furthermore, to participate competitively in today's economic activity requires carefully honing oneself into an efficient cog in a giant, impersonal economic "engine." Living one's life means establishing a career ladder and climbing it; progress is measured by rank, power, and ability to consume. This pattern effectively destroys meaningful community, creating psychic angst throughout all levels of society (Clark 1989, ch. 11). By failing to meet basic psychological needs, the modern Western gestalt is chalking up enormous unpaid social bills. For a time, centralized welfare services alleviated some of the despair at the bottom, but welfare has greatly shrunk with the onset of fiscal "austerity." And although many psychiatrists earn a living by ministering to the unhappy wealthy, their efforts are largely fruitless; conspicuous consumption simply is not a substitute for meaningful existence. Like the fabled bluebird of happiness, the piece of the American pie, once attained, fails to satisfy. At the moment, however, neither environmental deterioration nor social deterioration is recognized as a cost that is directly attributable to the way we think.

Finally, there is another important (but often overlooked) consequence of the gestalt of competitive individualism and that is the role it has played in evolutionary and ecological thinking. As pointed out in *Ariadne's Thread* (p. 135-138), it is unfortunate for our understanding of ourselves and of our environment that evolutionary theory has given rise to two popular misconceptions. One is that evolution represents a continuum of "progress" from lower to higher, from lesser to greater, a grand ladder on which humans stand at the top. Although few biologists would argue for this view, it remains implicit in our overall Western world view, and leads to the West's utilitarian attitudes toward Nature, as well as its self-anointed belief in its own cultural "progress."

The other, even more pervasive misconception is that "competition" underlies all of Nature (including human nature), that it is the very basis of evolution. Now, not only does this belief permeate popular culture (reinforced by televised "nature" shows that emphasize predation and competition because these are filmable), it also subtly affects how ecologists themselves view Nature. In recent ecological journals, I found two to three times as many papers dealing with interorganismic competition as with reciprocal cooperation. Such a bias in science reinforces our cultural bias toward struggle and competition by making these seem "natural" and hence inevitable. We need to understand that evolutionary success is not a matter of outcompeting in a "win-lose" contest, but a matter of "fitting in" with the overall scheme of things, an idea well-summarized by evolutionist George Gaylord Simpson :

To generalize that natural selection is overall and even in a figurative sense the outcome of struggle is quite unjustified. . . . Struggle is sometimes involved, but it usually is not, and when it is, it may even work against rather than toward natural selection. Advantage in . . . reproduction is usually a peaceful process in which the concept of struggle is really irrelevant. It more often involves such things as better integration into the ecological situation . . . more efficient utilization of food, better care of the young, elimination of intra-group discords . . . that might hamper reproduction, [and] exploitation of environmental possibilities that are not the objects of competition or are less effectively exploited by others. (1949, p. 221-222)

Species that survive, do so because they fit in with their surroundings; failure to do so leads to extinction. In Nature, "conquering" one's environment is a meaningless concept.

Neoclassical Theory and the Problem of "Values"

Modern-day economics proclaims itself to be a science, capable of discovering laws that govern the way material wealth is produced, distributed and consumed. From these laws, it is supposedly possible to predict the behavior of economies, and thus to manage them. Because the discipline of economics has drawn arbitrary lines about its area of concern, it has formulated "laws" that either ignore or simplify the relations between economic activity and Nature on the one hand and the human psyche on the other. Since both of these are crumbling under present economic behavior, economists are searching for ways to incorporate Nature (and, to a much lesser extent, social psychology) into their economic calculus. The question is how do we "value" them.

Now, the central problem in economics has always been to find some way to measure the "value" of various economic activities, and so far no one seems to have managed to do this. Indeed, the definitions of "value" range from Adam Smith's attempted distinction between "moral" and "market" values, to Joan Robinson's assertion that "value" is always a metaphysical concept without use in economic theory: it is whatever we say it is. Once such polemics are out of the way, however, economists automatically revert to prices in the marketplace when measuring sectors of the economy, or doing input-output analysis, or accumulating national income statistics. Yet the multiple problems with prices as indicators of economic wealth or well-being are well known. Temporary gluts or scarcities can cause the prices of items with a constant intrinsic value—e.g., oil—to swing wildly. (This is why growth proponent Julian Simon feels justified in claiming that fossil fuels are not becoming scarce, although it is easy to show that oil-producing countries, when facing internal economic chaos, chose to deplete their oil reserves as fast as possible in an attempt to regain economic stability.) Prices also reflect only exchange value at the moment of purchase, not long-term value. An "expensive" car that lasts twenty years may in fact be cheaper than a sequence of three cars each lasting only seven years.

Another problem with prices as indicators of well-being is the equating of "junk" with "life-serving" goods and services. In the economists' calculus, sales of alcohol and tobacco are just as beneficial as those of milk and antibiotics. A single-value system—in this case, money—has the effect of making moral distinction impossible, yet failure to do so can create economic "costs" to the community as a whole down the road, such as alcoholism, lung cancer, unhealthy children and unnecessary deaths when wrong choices are consistently made. Closely allied to this is the problem of algebraic signs: the price of curing a social or environmental problem is *added to* our calculation of total annual income, rather than subtracted from it. "Everytime there is an automobile accident, the GNP goes up." Or, as one of our students recently said, "By viewing the downfalls of society as being career opportunities, students can hope to make a difference in the future." Healing our growing environmental and social ills—paying back our overdue accounts—will indeed offer rewarding work as we gradually restore natural and psychic wealth that often was not visible in the marketplace, but which we nevertheless profligately consumed.

The question is: How do we learn to value our social relations and our environment? More and more non-mainstream economists are recognizing that there are unpaid services that lie outside the present pricing system. Among the most useful insights are those of Hazel Henderson and Marilyn Waring (Henderson 1988; Waring 1988). Both note that the market economy—which includes the so-called "global economy"—comprises only a small portion of the goods and services upon which we depend. Aside from the services of the public sector (government, roads, education—which many economists wish to "privatize" as much as possible), there are the illegal, underground and barter economies, a huge volume of unpaid productive work done mostly by women and children, and finally the enormous services of Mother Nature herself (purifying water, recycling wastes, growing forests, creating soils, etc.). How do we learn to value these unpaid services and stop destroying them? How do we learn to stop overexploiting the environment and tearing our societies apart by destroying community in the name of efficiency?

The neoclassical "solution" has been—and continues to be—to seek ways of commoditizing everything, of assigning a price to all externalities. At first glance, this seems a useful solution. All that is needed is to assign the right values, and then apply a cost/benefit analysis. Yet as Goodland and Ledec (1986) and others (Goodland et al. 1989) have shown, this approach has some major—in my view, fatal—flaws. First, because the "price" of losing the Amazon forests or of widespread child abuse is not available in the absence of a market, one needs to invent a "shadow price." Unfortunately, these shadow prices tend to be estimated transactional prices, not the price of long-term loss of social "income" from the absent forest or the disabled child. They also fail to include the cost of restoring the lost object when its loss proves intolerable. (The cost of replacing irreversible losses is, of course, infinite.)

In an attempt to get a less arbitrary measure of "value," Costanza and his colleagues have suggested embodied energy (Costanza 1980; Cleveland et al. 1984) and more recently Howard Odum (1988) has begun to include not only the energy content but also the information content of various items, from a species going extinct, to an ecosystem, to a human being at various stages in life. While such approaches help to remind us just how valuable a thing we lose when a species goes extinct or when we destroy a whole people, their culture and their language, they cannot be applied to solving our economic overdrafts. First, as most systems analysts now agree, accurate estimates of all the information content in, say, a patch of rainforest or in an indigenous culture, are simply impossible.

Second, such an accounting system cannot possibly assign a numerical value to our *affective relations* with our surroundings or with each other. Such proposals—whatever quantitative valuing system one might choose to employ—ultimately lead to the prostitution of all meaning! While these kinds of value-signposts may be useful in bridging from our present gestalt to another, more viable one, they cannot excuse us from changing our entire view of economic activity. We can no longer claim "economic success" for a model that destroys both human communities and the environment. While free-market economies have successfully "outcompeted" centralized command systems (which suffered from many of the same internal diseases), they have yet to perceive their own narrow and ultimately self-destructive arguments.

Gestalt II: Dominance of Community/Environment Relations

The alternative gestalt is one in which economic activities and material consumption are not the central focus of society, but are seen rather as a link between Nature and human community. While physicists can blithely chase after a unified field theory of the universe without fear of seriously disturbing it, this is not true for economists. A single-value universe, whether expressed in constant 1990 U.S. dollars or in Odum's units of solar EMERGY, will not automatically produce viable societies or sustainable ecosystems. To accomplish this, we need to progress beyond the "scientific" ideal of quantification of values, to a deeper understanding of human nature, its needs, and its proper relations with Nature as a whole. For educational purposes, this requires a much broader set of disciplines and a pulling together of insights from every sector of human understanding. We begin by asking: What are desired environmental and social goals? This is then followed by actual examples of societies that have, at least partially, achieved these goals—and that thus serve as models for refashioning our own worldview.

Desired Environmental and Social Goals

The environmental goal can be defined in one word: sustainability. Yet alone, this does not get us very far. There is sustainable use of the *global* environment and sustainable use of *local* and *regional* resources. The first implies a "top down" management system, emanating from the present international power structures that are based on the Western economic gestalt. These centralized institutions may be necessary in the short run to begin responding to climate change and other global concerns. But in the long run (as I have tried to demonstrate in *Ariadne's Thread*), effective global management can only emerge from universally responsible management of local ecosystems, not from centralized management. The reasons are to be found in the available knowledge base: Who is capable of defining local sustainability? and in motivation: Who is concerned about maintaining local sustainability?

In their introduction to a remarkably useful book, *Common Property Resources*, Fikret Berkes and Taghi Farvar (1989) point out that local people, who have been in close touch over many generations with their environment, possess an enormous, not readily replaceable, fund of indigenous knowledge. Furthermore, in its usefulness this knowledge often far exceeds that currently available from scientific "experts." The current rush to exploit ecosystems that are still controlled by such traditional peoples threatens the survival of both the ecosystems per se and this priceless indigenous knowledge. While ecologists struggle to come up with adequate scientific measures of ecosystem sustainability, they are ignoring cultural wisdom that already exists. As Berkes and Farvar argue, the future challenge is how to integrate scientific and traditional knowledge in order to explore the possibilities for truly sustainable development.

In maintaining sustainable ecosystems, we also need to ask: Who is motivated to maintain them? To answer this question, one only has to observe the social and environmental wreckage left in the wake of multinational corporations as they hop about the globe, resembling the Mad Hatter in always getting a clean place at the table, leaving the dirty, used-up places for others. As the Mad Hatter explains, in Wonderland, time has stopped, so that "it's always tea-time, and we've no time to wash things between whiles." To which Alice replies, "Then you keep moving round, I suppose?" And the Hatter answers, "Exactly so, as the things get used up." "But when you come to the beginning again?" Alice asks, to which the reply is "Suppose we change the subject."

By discounting the future, the contemporary Western gestalt achieves exactly the same result, with the same kind of peremptory dismissal when questioned. Not so, traditional gestalts. A society that depends directly on its local environment for survival and comes to know it intimately, understands the need for community responsibility in its maintenance. This explains the emphasis in human ecological and development literature on local self-sufficiency as a means to escape from dependency on externally earned income for community survival. Global sustainability may well depend on the universal emergence of self-sufficient local economies managed by local communities. This, of course, does not preclude global trade, but places it in a completely different context than that of supposed (and highly suspect) "comparative advantage."

Turning next to desired social goals, we briefly touch on human needs theory.[1] When Abraham Maslow constructed a "hierarchy" of human needs starting with survival necessities and progressing through parental affection, personal identity within a group, and finally, to a kind of Sartreian self-actualization, he, like most other Western psychologists, took the *isolated* individual as his point of reference. John, Dick and Mary were each to be evaluated on their own, self-identified terms. Yet if we examine the sociology of non-human primates and pay attention to our own social context, we arrive at quite a different view of human needs. Our first needs are for bonding, for affection, for social acceptance, and these persist throughout life. We are hard-wired for them, even more than for food, drink and shelter. Why? Because it is through bonding that everything else we need in life is obtained, including not only biological necessities, but the psychological need for meaning. Without bonding we suffer not only psychologically, but physiologically (House et al. 1988). Brain and body are a *single* unit, designed for life imbedded in a secure and meaningful social context. Roger Hausheer (1980) sums it well:

[O]ne of the fundamental needs of men, as basic as those for food, shelter, procreation, security and communication, is to belong to identifiable communal groups, each possessing its own unique language, traditions, historical memories, style and outlook. Only if a man truly *belongs* to such a community, naturally and unselfconsciously, can he enter into the living stream and lead a full, creative, spontaneous life, at home in the world and at one with himself and his fellow men; enjoying a recognized status within such a natural unit or group, which itself must command full and unqualified recognition in the world at large; and thereby acquiring a vision of life, an image of himself and his condition in a community where concrete, immediate, spontaneous human relations may flower undistorted by neurotic self-questioning about one's true identity, and free from the crippling wounds inflicted by real or imaginary superiority of others.

If the survival of the planet now depends upon wise human behavior, and if such behavior cannot occur among people whose basic needs for community remain unfulfilled, then saving the planet and providing a future for *Homo sapiens* means, in the West, abandoning the premises of linear "progress," competitive individualism and insatiable acquisitiveness. These are a prescription for extinction; they cause society to destroy its resource base, not because of survival pressures, but because their logic forces such an outcome.

---

[1] These ideas have appeared in chs. 5 and 11 of Clark 1989, and in Clark 1990a and 1990b.

The alternative is a communally-based society, where individual competition gives way to shared goals. When the first social purpose is community sustainability, a quite different social ethic emerges. There is a strong sense of sharing—of work, of food, and of decision making. Furthermore, as occurred in the villages of Sri Lanka during the Sarvodaya movement, the Physical Quality of Life Index soars when communities feel in charge; birthrates drop and health improves. Nor is it considered "extravagant" to provide free education and to subsidize food prices (in the absence of universal access to jobs or land) as a form of community investment in the next generation (Vittachi 1985). Former Peace Corps volunteer, Terence Ratigan (1986), in his description of life in the villages of Sierra Leone, provides a picture of how a sense of social participation and an opportunity to share, create the sense of a meaningful life:

The high value placed on work may be seen in the common greeting, "Thank you for the work." In many different languages the greeting is used in passing whenever a person is seen working, going to work, or coming from work—in short, *constantly*. The important point is this: *there is no implication of ownership or personal benefit in the greeting*. Whoever expressed thanks for the work did not expect any direct benefit from the work being done. It seemed more a cultural recognition that work—any kind of work—contributed to the common wealth of the community. A farmer plowing a field, a blacksmith making a hoe, a woman carrying water, a weaver spinning thread; all deserved and received recognition for their contributions. . . .

The end result of invested labor was certainly important. But more than money, respect was the prized possession. And respect was earned through honest effort, and fulfillment of social responsibilities. A farmer whose rice harvest exceeded all expectations could not expect his bounty to go unnoticed. More visitors than usual could be expected around meal time. . . . A rich man . . . had to earn his respect. He was not envied merely for his wealth.

Some Living Social Models
Unfortunately, in the West, the "tragedy of the commons" is seen as an inevitable cause of resource degradation, curable only by assigning private ownership of resources to individuals—or to those pseudo-individuals called corporations. Besides the obvious fact that private ownership is no guarantee of sustainable husbanding of a resource, there are in fact numerous cultures where long-term, sustainable use of communally managed resources does occur. Indeed, as Berkes and Farvar (1989) observe in their introduction to *Common Property Resources*, "The truth is that traditional systems [of communal resource management] have been the main means by which societies have managed their natural resources over millennia on a sustainable basis. It is only as a result of this that we have any resources today to speak about." Furthermore, there is a deep interlocking between social coherence and sustainable resource use. When the social arrangements break down, the resource base diminishes, and the society itself may perish. The following are examples of successful interactions between biological and cultural landscapes, all taken from Birkes' book.

The West has long regarded coastal fisheries as an open-access resource, available to all nationals—with the result that many inshore fisheries have been seriously depleted.

Yet elsewhere, the idea of local coastal sea tenure has existed for centuries. In Japan, patterns of communal ownership of inshore fisheries evolved around individual villages, and today each of these widely varied systems of local traditional rights is preserved by national law. The whole pattern of resource use is highly complex, but the local cooperative, run and owned by member fishermen, functions to manage, regulate and plan for sustainable development of the resource (Ruddle 1989).

In Mexico, on the east coast of the Yucatan peninsula, a lucrative spiny lobster fishery has recently grown up. Mexico's long history of cooperative land tenure (*ejidos*) extends to tenure of coastal fisheries, and the government issues rights to fish only to cooperatives. Although many so-called cooperatives were established, only two have entered into communal management arrangements, monitoring their borders, allocating specific fishing areas to each individual, and restricting entrance of newcomers. Again, it is the group as as whole which makes rules and settles disputes. No one is free to use his "share" of the resource independently of the others (Miller 1989). Similarly, in the Maine lobster fishery, extralegal territories are communally established and maintained by so-called "harbor gangs" that allocate each fisherman's rights and defend the collective perimeter from interlopers. Such defended areas appear to be sustainable, although members will consider changing the agreed-upon rules if catches begin to decline (Acheson 1989).

In the context of broader resource bases, there are numerous examples of communal management. Among these are the traditional land-and-sea management systems of indigenous communities in Melanesia, where the kind of centralized, tributary societies found in Polynesia never evolved. In general, there is no "ownership," merely rights of use, which can be complex. Primary rights may be held by a few families, who in turn grant secondary rights to a large number of users. Fortunately, during colonial times, the British in Melanesia tended not to disturb such traditional systems and many still exist, although they are coming under increasing pressure from external "developers" on the one hand and from local population growth on the other (Baines 1989). Sustainable communal resource management was also once widespread on the Indian subcontinent, being controlled by village communities, "in which the different caste groups were linked to each other in a web of reciprocity." Unfortunately, British rule in India did disrupt communal organization and gave open-access rights to resources, resulting in their inevitable exhaustion. A few pockets of communal control persist, however, and offer useful models for reestablishing systems of sustainable use (Gadgil and Iyer 1989).

Thus, far from being destructive, communal resource use, when woven into the social fabric, is the best means for maintaining sustainable use. Indeed, there is a reciprocal relationship, since communal resource use strengthens social bonds. All this provides a strong argument for the usefulness of Gestalt II in thinking about future planetary management and human survival. The current view of neoclassical economics (Gestalt I), with its assumption that communal rights implies a free-for-all and that only private ownership of resources can preserve them is not only incorrect; when coupled with assumptions of a necessary competition among individuals to maximize "efficiency" in resource use, it becomes downright dangerous. The social controls existing under communal management are lacking, and so too is the psychic satisfaction that comes with membership in a sharing, meaningful society. Without the latter, is it any wonder that human material wants become "insatiable"? They can never be an adequate substitute for social bonding. Once again, quoting from Berkes and Farvar, "[C]ommunal-property systems are not evolutionary relics about to be replaced by 'newer and better' management regimes. They exist [in various places in the world today] because they provide certain advantages over other property-rights regimes."

IMPLICATIONS FOR EDUCATION

As must now be apparent, the thesis of this paper is that "economics"—especially neo-classical theory as developed in the Western gestalt—can no longer take center stage in our thinking, forcing our understanding of ourselves and our universe to conform to it. Instead, we need to reexamine our assumptions about human nature and human needs, about the meaning of "wealth" and "well-being," and about appropriate ways to achieve global sustainability (Daly and Cobb 1989; Robertson 1990). This will mean consider-able adjustment in the Western gestalt, and here education must play a significant role. Educators must recognize the need for a comprehensive, metadisciplinary approach to our entire educational endeavor. Appending ecological considerations onto neoclassical eco-nomic theory may be a temporary measure for slowing such crises as climate change, but it cannot produce a satisfactory long-term outcome.

The following are proposed curricular approaches to global education, of the sort needed by policy makers and the public at large.

Grounding in Ecological Principles of Sustainable Systems

In addition to a basic understanding of the principles of energy flow and dissipation and of material recycling, ecological insight should include an appreciation of the multiple ser-vices to the overall system provided by various "guilds" of organisms, such as nutrient and water retention, pollination, pest-controlling species, etc. A sufficient spectrum of examples needs to be provided from a variety of ecosystem types to give policy makers an awareness of the kinds of unexpected positive-feedback consequences that can occur through apparently innocuous interventions. (Remember that "positive-feedback" does not mean beneficial, but uncontrolled and destabilizing.) Although population fluctua-tions are still poorly understood, some consideration needs to be given to the principles of island biogeography and species survival.

A second area for understanding sustainability is through study of the knowledge base of indigenous societies that have successfully managed resource systems for millennia. As noted above, such groups carry a profound working knowledge of their natural envi-ronment, including awareness of interactions in the ecosystem that academic ecologists frequently fail to notice.

A third area is knowledge of the complex and reciprocal relations among soil, vegeta-tion and climate, so we work with local ecosystem patterns when trying to increase human carrying capacities. Major changes in vegetational types, for instance, are likely to create unstable systems that require excessive external inputs to maintain them. Such systems are not sustainable.

In this context, coming climate changes will cause major dislocations of vegetational types, which will result in significant changes in ecosystem productivities (increases in some localities, decreases in others). Although ecologists have long recognized that "equilibrium" is a poor term for describing ecosystems, that in fact they are constantly evolving, past rates of climate-induced change have been on the order of hundreds to thousands of years, not the decades foreseen for the coming greenhouse effect. This means that "sustainability" cannot be regarded as the maintenance of a fixed set of ecological parameters. Policy makers, academic ecologists, and indigenous peoples alike will be faced with a moving target, whose direction of movement may not be readily predictable.

Grounding in Psychosocial Factors Creating Sustainable Societies

Aside from a few indigenous cultures living sustainably, world societies are now living beyond their environmental means. To cure this, the causes of unsustainability must be appreciated, and these causes are quite different for the rich North and the poor South.

In the North, the Western gestalt of competitive individualism and constant economic growth has established institutions, such as private ownership of Nature and capital accumulation, that make change in direction difficult. Yet there is no doubt that these "successful" industrial societies with their massive per capita energy demands are unsustainable. Even with greatly increased energy efficiency, significant changes in lifestyles will be necessary. And, because virtually all the world's industrialized nations are, or shortly may be, under democratic control, it will be necessary to identify the sources of the economic expectations of their peoples and to explore means for bringing these into line with a sustainable level of economic throughput. Whether this is accomplished by appeals to self-interest, to a sense of moral commitment to the future, or to an aesthetic appreciation of Nature, it will entail a far higher level than now exists of popular understanding of how modern industrial economies interface with the natural world. The educational effort needed to bring about stable, self-generated, democratic change is enormous.

Causes of unsustainability in the South are of a different character. Although infected with the distant dream of a set of McDonald's golden arches in every village, most poor nations are immediately concerned with preventing further erosion of subsistence resources. Three interwoven factors are involved, all of which are likely to be ameliorated by restoring local community autonomy: population growth, unequal access to land and other subsistence resources, and continuing export of physical wealth owing to external indebtedness.[2] The current distress of the South results from a mixture of 1) past imperialism that actively suppressed development, often destroyed or disrupted indigenous societies and their former relations with their environments, and left behind a small group

---

[2] See Clark, *Ariadne's Thread*, Ch. 13, "Defusing the Global Powder Keg" for discussion of overall issues; George Foy and Herman Daly (1989) discuss these three problems in relation to three small Latin American countries.

of Westernized, often corrupt elites, programmed to a dependency relationship with the North; 2) attempts by the North to continue its economic exploitation of resources and cheap labor, often backed up by military interventions; and 3) misguided attempts to"develop" largely rural, agrarian societies through infusion of massive, capital-intensive technologies financed through indebtedness. The presumption of ensuing economic growth that would not only pay back the loans with interest but also trigger further economic expansion has proved disastrously wrong, bringing political unrest, environmental destruction, and economic bankruptcy.

To the extent that the North can be of any assistance in turning these things about, it can forgive outstanding loans, stop supporting oppressive, elitist governments, and *give* (not loan or sell) appropriate technical assistance to local communities to help them reconstruct and become once again self-reliant, cohesive and sustainable. The skills and insights needed to do these things will be distinctly different from those now taught in economics departments and graduate schools of business, and will include considerable grounding in cultural anthropology, co-evolutionary social theory, and human needs theory. I suspect that the world view changes needed in the South will be less, not more, difficult to get under way than those needed in the North.

Critiquing Present Economic Theory—Bridging from Present to Future Gestalts

[T]he ideas of economists and political philosophers, both when they are right and when they are wrong, are more powerful than is commonly understood. Indeed the world is ruled by little else. Practical men, who believe themselves to be quite exempt from any intellectual influences, are usually the slaves of some defunct economist.

These words, uttered a half century ago by John Maynard Keynes (1964), are even more apt today. Contemporary economic arguments are riddled with impossible assumptions that permit neat mathematical models which, in fact, model nothing in reality.[3] A short list of emergent beliefs includes the following: economies can only run on private greed; wealth is only produced by the commoditized private sector; competition results in efficiency; money and debt are forms of wealth; the future is worth less than the present; labor is just another commodity; and most disastrous of all, the costs of maintaining the planet and healing social wounds, both incurred as a result of current economic activities, are counted as "benefits." Inconvenient aspects of economic activity are simply ignored: the "free market" ignores how wealth is distributed; most people are not truly "free" but are dominated by the need to "make money" in a system they do not control; and social and environmental consequences are pushed off onto other disciplines. Economists who raise these issues are ejected from the fraternity of "respectable" professionals.

To bridge the gap from the disastrous present to a saner future, we must start by redefining "wealth," by clearly distinguishing between "use" and "consumption," and by

---

[3] See Leontief 1982. Also, see Clark 1989, ch. 12; Daly and Cobb 1989; Harrington 1974; Henderson 1988; Mark A. Lutz and Kenneth Lux 1988; Robertson 1990; Waring 1988.

noting the proper algebraic signs of "costs" and "benefits." Because important human values will necessarily be included, it is unlikely that future economic discussions will be cast in the form of linear mathematical models. How the actual institutions of monopoly capitalism are to be converted to a new set of humanly and environmentally sound activities is more difficult to foresee. There already exist models of change, however, in the plans for conversion of industries from military to civilian production, that offer clues as to how to begin (Gordon and McFadden 1984).

Identifying Appropriate Socioeconomic Entities for Establishing Sustainability

Theorizing from present to future is the easy part. Acting on that theory without causing disastrous social disruption will be the central challenge of the next quarter century. Without a widespread social awareness of the need to change, a relatively painless transition cannot be assured. Hence, a large part of education for the future must go beyond what is now labeled "economics," weaving social theory, economic theory, and ecological theory into a comprehensive new gestalt.

A growing group of thinkers from various disciplines is converging on the same overall vision: a globe with thousands of locally managed, self-reliant economies, based on ecologically meaningful boundaries and comprising culturally and historically integrated communities. The goal is to strengthen meaningful participation in a shared community while creating identification with the communally managed local resource base. Reestablishing and expanding traditional communal systems could help them regain community bonding while ensuring sustainable development. Even more important is discovering how to recreate a sense of community and identification with the local resource base in the world's large urban centers (Jacobs 1962; Mumford 1961). Through education (and perhaps other means) we can begin to "change the topic of conversation," shifting people's focus from individual to community, and from distant, centralized state power to a sense of local empowerment.

Trade in such a world is not precluded, but rather is made independent of survival, thus eliminating exploitation of both resources and people and creating, perhaps for the first time since written history began, a truly "level playing field" (Clark 1989, ch. 12-16). Nor does such decentralization mean an end to nation-states, nor to supranational organizations such as the United Nations. Such entities will still be necessary, not only to help adjudicate the inevitable disputes that arise among small units, but for overseeing the health of the biosphere as a whole, and for facilitating the exchange of ideas and information. Development, whether "upwards" or "downwards," must be carried out with the security of the planet foremost in mind. Decentralization *does* mean, however, a dispersal of the power now residing in nation-states and a softening of their boundaries, as the global village regains its human scale and develops its global awareness.[4]

---

[4] See Clark 1989, ch. 16; Johan Galtung 1990. Such thinking is promoted by the Green movement; see Fritjof Capra and Charlene Spretnak 1984.

REFERENCES

Acheson, James M. 1989. Where Have All the Exploiters Gone? Co-management of the Maine Lobster Industry. In F. Berkes, ed., *Common Property Resources*, pp. 199-217.

Baines, Graham B. K. 1989. Traditional Resource Management in the Melanesian South Pacific: A Development Dilemma. In F. Berkes, ed., *Common Property Resources*, pp. 273-95.

Berkes, Fikret, ed. 1989. *Common Property Resources*. London: Belhaven Press.

Berkes, Fikret and M. Taghi Farvar. 1989. Introduction and Overview. In F. Berkes, ed., *Common Property Resources*. pp. 1-17.

Capra, Fritjof and Charlene Spretnak. 1984. *Green Politics*. New York: E.P. Dutton.

Clark, Mary E. 1989. *Ariadne's Thread: The Search for New Modes of Thinking*. New York: St. Martin's Press.

Clark, Mary E. 1990a. Meaningful Social Bonding as a Universal Human Need. In John W. Burton, ed., *Conflict: Human Needs Theory*. New York: St. Martin's Press.

Clark, Mary E. 1990b. Each Culture Needs a Song, a Theme of Life from Which Its People Take Their Meaning. *Education* Spring.

Costanza, Robert. 1980. Embodied Energy and Economic Valuation. *Science* 210: 1219-24.

Cleveland, Cutler, Robert Costanza, Charles A. S. Hall and Robert Kaufman. 1984. Energy and the U.S. Economy: A Biophysical Perspective. *Science* 225: 890-7.

Daly, H. E. and J. B. Cobb, Jr. 1989. *For The Common Good: Redirecting the Economy Toward Community, the Environment, and a Sustainable Future*. Boston: Beacon.

Foy, George and Herman Daly. 1989. Allocation, Distribution and Scale as Determinants of Environmental Degradation: Case Studies of Haiti, El Salvador and Costa Rica. Environment Department Working Paper no. 19, September. Washington, D.C.: World Bank.

Galtung, Johan. 1990. Visioning a Peaceful World. In Mary E. Clark and Sandra A. Wawrytko, eds., *Rethinking the Curriculum*, pp. 195-214. Westbury, Conn.: Greenwood Press.

Gadgil, Madhav and Prema Iyer. 1989. On the Diversification of Common-Property Resource Use by Indian Society. In F. Berkes, ed., *Common Property Resources*, p. 240.

Goodland, Robert and George Ledec. 1986. *Neoclassical Economics and Principles of Sustainable Development*. Washington, D.C.: World Bank.

Goodland, Robert, George Ledec and Maryla Webb. 1989. Meeting Environmental Concerns Caused by Common-Property Mismanagement in Economic Development Projects. In F. Berkes, ed. *Common Property Resources*, pp. 148-164. London: Belhaven Press.

Gordon, Suzanne and David McFadden, eds., 1984. *Economic Conversions: Revitalizing America's Economy*. Cambridge, Mass.: Ballinger.

Harrington, Michael. 1974. *Twilight of Capitalism*. New York: Simon & Schuster.

Hausheer, Roger. 1980. Introduction in Isaiah Berlin, *Against the Current: Essays in the History of Ideas*, pp. xxxvi-xxxvii. New York: Viking Press.

Henderson, Hazel. 1988. *The Politics of the Solar Age*, 2d ed. Garden City, N.Y.: Anchor Press, Doubleday.

House, James S., Karl R. Landis, and Debra Umberson. 1988. Social Relationships and Health. *Science* July, 29:540-545.

Jacobs, Jane. 1962. *The Death and Life of Great American Cities*. Harmondsworth, England: Penguin Books.

Keynes, John Maynard. 1964. *The General Theory of Employment, Interest, and Money*, p. 383. New York: Harcourt, Brace & World.

Leontief, Wassily. 1982. Letter: Academic Economics. *Science* 217:194-197.

Locke, John. 1962. Bodleian Library, M. S. Locke, c. 30, F. 18. Quoted by C. B. MacPherson, *The Political Theory of Possessive Individualism*, p. 207. Oxford: Clarendon Press.

Lutz, Mark A. and Kenneth Lux. 1988. *Humanistic Economics, the New Challenge*. New York: Bootstrap Press..

Miller, David L. 1989. The Evolution of Mexico's Spiny Lobster Fishery. In F. Berkes, ed., *Common Property Resources*. pp. 185-198.

Mumford, Lewis. 1961. *The City in History*. New York: Harcourt, Brace & World.

Odum, Howard T. 1988. Self-Organization, Transformity, and Information. *Science* 242: 1132-1139.

Pacey, Arnold. 1983. *The Culture of Technology*. Cambridge, Mass.: M. I. T. Press.

Ratigan, Terence.1986. Former Peace Corps Volunteer, personal communication.

Robertson, James. 1990. *Future Wealth: A New Economics for the 21st Century*. London: Cassell.

Ruddle, Kenneth.1989. Solving the Common Property Dilemma: Village Fisheries Rights in Japanese Coastal Water. In F. Berkes, ed., *Common Property Resources*, pp. 168-184.

Simpson, George Gaylord. 1949. *The Meaning of Evolution*, pp. 221-222. New Haven, Conn.: Yale University Press.

Thurow, Lester C. 1981. *The Zero-Sum Society: Distribution and the Possibilities for Economic Change,* p. 120. Harmondsworth, England: Penguin.

Vittachi, Tarzie. 1985. Clues to Development in the Commonwealth Isles. *People* (IPPF Review of Population and Development) 12(4):4-6.

Waring, Marilyn J. 1988. *If Women Counted: A New Feminist Economics*. San Francisco: Harper & Row.

27

ECOLOGICAL ECONOMICS AND MULTIDISCIPLINARY EDUCATION

James J. Zucchetto*
National Academy of Sciences
2101 Constitution Avenue, N.W.
Washington, D.C. 20418 USA

ABSTRACT

This paper addresses the emerging concerns that exist regarding the environment, economic development, and education in the United States. It examines requirements for various levels of education, from the primary school system on up to high school, and to undergraduate and graduate level education. In this discussion, attention is focused on incorporating an ecological economic perspective into the curricula, on training teachers in this area, on the problems of doing interdisciplinary work and attaining tenure for junior faculty, and on attracting students into such cross-cutting endeavors. The paper is qualitative. It is still difficult to define what ecological economics might mean in the context of education and hence to recommend actions that might be taken to introduce such material into different educational levels.

INTRODUCTION

In this brief paper, I speculate on approaches that might be taken to develop the subject of ecological economics in curricula without much thought for the problems of implementation. As with many of the proposals for educational reform, the problem of dealing with entrenched bureaucracies and schools of education can easily prevent the implementation of proposed reforms. I can only derive from my personal experience as a student and a teacher what might constitute a valuable experience for a student undertaking studies in ecological economics.

---

* The views of the author expressed here do not represent the views of the National Academy of Science, the National Research Council or any of its constituent units.

The focus is primarily on problems of university undergraduate and graduate teaching; however, I also address some thoughts on what might be useful in U.S. primary and secondary schools, i.e., kindergarten through grade 12. In an attempt to address these issues, the following topics are considered: 1) what is ecological economics and what should it encompass; 2) what should be taught and which tools should be used in primary and secondary grades; 3) what should be taught at undergraduate and graduate level institutions; and finally, 4) the scope of research in this area since it is important to graduate and undergraduate programs.

WHAT IS ECOLOGICAL ECONOMICS?

In the aims and scope of the journal *Ecological Economics,* the study area is described as being "concerned with extending and integrating the study and management of 'nature's household' (ecology) and 'mankind's household' (economics)." This is a broad definition and could include a wide assortment of human activity These activities have implications for the biosphere and a wide range of biospheric phenomena that affect humans. To economists, this type of study might just mean environmental economics, i.e., evaluating impacts on the environment in terms of monetary measures and incorporating the environment in some fashion into the human calculus of values (internalizing the externalities). To some ecologists, it might mean incorporating and understanding human activity as extensions of the ecosystems of the world. Humans are considered another species in the system, albeit a species unique in many ways and very destructive to many other living species. These are all fair considerations, but in my own mind, I think of ecological economics as: *the careful and deliberate understanding of activity on the earth and its consequences, with the ultimate goal being that humans will somehow manage this activity for the long-term survival of itself and other life on earth.* I use activity in a very general sense, to include biological as well as physical processes associated with humans and the rest of the "natural" world. (I put natural in quotes to emphasize that many people separate humans from the natural world, although humans may be thought of as another manifestation of nature, i.e., there is no separation between man and "nature.") Ecological economics is truly multidisciplinary in character in that it requires an understanding of many fields of knowledge. It also incorporates the scientific method to validate claims where possible and seeks to develop paradigms and rules that supercede the individual disciplines.

Such a broad area of study runs the risk of "being everything and being nothing." It is especially difficult, although fascinating and interesting, to define for the purposes of education. In some sense, the student must be willing to be the ultimate generalist, willing to try to learn and incorporate any knowledge, whether associated with a given discipline or not, and discerning how this knowledge can be used to make the earth a habitable place not only for humans but also for all other life. The student even needs to consider how this natural world has been and might be manipulated and engineered by humans to create a "new" world including the possible production of new genetic information consciously introduced. But is such a broad program too much for most students? After all, most education and vocational training tend to go in the other direction, towards specialization. This tendency seems to increase as the quantity of information grows exponentially and society becomes increasingly complex; specialization and "wearing blinders" make it easier for individuals to deal with the complexity of current industrial society.

U.S. PRIMARY AND SECONDARY EDUCATION

In recent years, environmental issues have been introduced to many elementary school pupils. This introduction is often done with entertaining examples of pollution. Students can further explore issues by working on projects for annual science fairs. These exercises are seen as stimulating breaks in otherwise dull classroom exercises. But, of course, at this stage students need to learn the fundamentals associated with reading, writing, science, mathematics, and computer literacy. These are the tools with which they will be able to tackle all intellectual endeavors further on in their schooling. Hopefully, in the process they don't lose the ability to look at things afresh. Some initial ideas of ecological economics can be incorporated into ecology discussions that are a part of biology lessons, thereby explaining how some of these issues relate to everyday life. But children at this age are mainly interested in concrete matters and immediate needs and I suspect it would be difficult to maintain their attention to the more abstract ecological problems, especially since only a limited amount of time would be devoted to it each week. But every effort at this stage is worthwhile because the rest of the child's life will be based to varying degrees on the consumption of goods and information.

The difficulty of investigating the natural world increases as scale increases. At small scale, it is relatively simple and inexpensive to provide controlled experiments for students, for example, in physics, biology or chemistry. For economic systems and natural systems on a large scale, this becomes virtually impossible. Of course, for natural systems, microcosms (such as aquariums) can be established and allowed to self design or be externally manipulated to illustrate certain principles that are assumed to also operate at a larger scale. These are certainly of interest for students to configure and observe. Field trips are usually popular among students; they provide an escape from the dull routine of the classroom, provide new stimuli, establish camaraderie, and sometimes test the student's stamina and endurance. These are especially important for urban students who are not exposed to rural areas. The extent of these excursions into the actual natural and built environment are limited by the resources available to given school districts. Environmental consciousness can certainly be built up during the primary and secondary grades by visiting such sites as landfills, waste-to-energy plants, recycling plants, industrial and urban landscapes, power plants, and a select number of natural landscapes of interest. The issues should be cast in light of the costs and benefits, and also of how the individual student's life is connected to the consequences of resource use. In the process, an attempt should be made to present a simple systems view of the world. Such exercises in the elementary grades are probably conducted out of regular classes or with special trips or projects; at the higher grade levels, they are perhaps best conducted out of geography courses, other social studies courses, or biology courses. In an experimental school, or in one that gave students access to a number of elective courses, there could be a course "ecological economics" with the intention of addressing both ecological and economic subjects; the difficulty would be in finding an appropriately trained teacher.

Computers as Teaching Instruments

In addition to experiencing real examples and experimenting with microcosms, artificial simulation can complement, but not replace, real world experience. The use of computer simulation and graphics is a powerful approach. This is especially important for understanding the large systems of urban and natural landscapes. This approach can significantly help the student in formalizing notions of systems and how changes in one part affect others. Personal computer software and hardware exist that can simulate with superb graphics. These are especially attractive to students playing computer games. For example, a computer software package "Sim City" allows one to construct on the computer screen a city landscape. One can specify locations of residences, factories, airports, landfills, etc., and indexes of development and pollution are generated. It is fascinating to watch 10-year-olds absorbed in trying different patterns of development and seeing the effects through this simulation. It gives a real interactive understanding of the effects of development and the intra-city interactions. At this level of sophistication, the student is letting the computer program do all its calculations. The student develops an intuitive understanding of the implications of various kinds of urban development. Such a system, at a later stage of learning such as high school or college, or even graduate school, can be made more apparent to the student as the algorithms are revealed for how the various measures are calculated. The more sophisticated student could incorporate the environmental dispersion models that might be built in, the associated economic costs, and the functions developed to calculate any expressed "quality of life" indexes. A recent New York Times article (May 15, 1990) described a simulation game for a personal computer, "Balance of Nature," that incorporates problems of economics and environment. Unfortunately, many school districts probably do not have the computer capability to provide some of this experience to its students although some students will get access to it through their homes.

I don't think it is unreasonable to envision the use of modern communications and computer capabilities to bring activities as they occur ("real time") around the world into the classroom. Again, computer simulation packages coupled with, for example, video tapes could make an interesting simulation of investigating different ecosystems in the world. Such simulations could also give a good visual representation of the impact of human activity, as well as incorporating some notions of economics. Real time images of various areas in the world could be available to students studying in geography courses or studying ecology sections in biology courses. Is it not possible that we could coordinate international ecological "real time" lessons between different countries? This could enable students to obtain a global view of the world and make geography more interesting than only studying places and people out of books.

Presently curriculum is structured, in most school systems, so that studies are divided into various disciplines. Teachers have trouble finding time to deal with interdisciplinary matters such as how environmental issues relate to economics. The student is preparing for a future of tests and evaluations based on these disciplines and success means demonstrating some comprehension of them. The student starts learning a greater and greater number of facts about smaller and smaller subject areas. Is this of use to most students or only to the specialist? For example, one might ask does learning about smaller and smaller elements of organisms in biology foster the understanding of principles and the scientific method that can be applied to any subject?

High School Education

As the student enters the high school years, and if he or she is college bound, a certain number of years of science, mathematics, English, and social studies will be required. Some electives will be part of this education as well. The science and mathematics courses will be more formalized than the subject of economics at this point. Where does ecological economics fit in? If it isn't a course unto itself, then a better acquaintance of the interaction between human activity and nature can be accomplished by including it in various courses.

As mentioned above, ecological economics may fit with the least resistance into geography courses because those courses involve both the built environment and the ecosystems of the world. Economic geography, resource geography, etc., are already established fields, and one could integrate notions of ecological economics into such courses. Biology courses include ecology to one degree or another, although the share should be increased. In fact, ecology has matured to such an extent as a subject that it should probably be taught as a course in its own right. Students need to understand the science of the whole ecosystem of humans and nature as well as the details of genetics and molecular biology.

The more radical approach is to teach from the top down. That is, to frame an understanding and ask questions about the larger system and then have students seek answers to the broad questions from more detailed levels of information. But this may be too much to hope for. In physics and chemistry, why not give examples that accentuate environmental aspects? Learn physics by giving examples from the world of ecosystems. Chemistry courses could use numerous examples from the environment such as air, water, and soil chemistry.

The most radical approach to primary and secondary education would be to organize a whole school around concepts of ecological economics, an understanding of human activity and the environment, and the development of approaches to sustainability. A student would have to take the basic courses but at the same time constantly consider the bigger picture and think about ways to integrate humans and nature. This would be an alternative type of education; whether it could produce students that could compete in the marketplace is open to debate. Some international schools have worked along these lines for some years with promising results.

UNDERGRADUATE EDUCATION

As with the secondary school system, most universities will teach by dividing subjects into disciplines and most students will have to major in one of them. There are programs that might be called interdisciplinary or multidisciplinary (such as environmental studies, geography, planning, natural resources, or systems ecology programs) that would have a greater degree of flexibility with regard to course requirements than most traditional departments. However, this is not to say that a discipline-oriented approach is necessarily incorrect. It certainly helps to ground the student in some basic fundamentals, natural law, and a wealth of information. Too much flexibility in a program might result in a tendency for students to drift and not obtain the necessary tools and concepts to deal with the complex world of environment, technology, and modern economics.

In a traditional school, I would think that an ecological economics major should almost pursue two majors, one in ecology and one in economics. An ecology major alone would probably not have an appreciation of human economic systems and the decision making of humans and an economics major doesn't have the training in biology and ecology to understand natural systems. This student should also have a good grounding in physics, chemistry, and mathematics, the mathematics training including up to differential equations, probability theory, and statistics, as well as in good writing skills, some foreign language skills, and possibly some engineering courses as well.

Is this too much to ask of one student? I think it is possible but perhaps it entails a five-year program, which establishes the student with credentials in two disciplines that allows entry into graduate programs or the corporate world without too much problem. But it requires faculty that are willing to advise and be accepting of students that want to work outside of the faculty member's own expertise. Indeed, it would be preferable to have a faculty member who also bridges the two disciplines or is actively pursuing work on problems in this broad area of ecological economics. This population, however, is still very limited.

A different approach to an undergraduate education in ecological economics could be analogous to a "great books" approach to an education in the humanities in which a student learns through reading and analyzing the "great pearls" of literature. Perhaps a "great problems" education, with the guidance of a proctor or an oversight committee, could be structured in such a way that the student could address a limited set of issues that involve both economics and ecology. The challenge to an oversight committee would be to make sure that the student took appropriate courses for understanding scientific, mathematical and economic fundamentals. For example, a student studying acid rain could be guided through a learning process covering such topics as power plants, dispersion modeling, atmospheric chemistry, ecosystem interactions and modeling including soil chemistry and hydrology, and economic methods for calculating appropriate costs and benefits. Such an approach would probably be best in the latter part of the education, after the student has spent a couple of years learning the fundamentals in various subjects. The difficulty to the student might be the acceptability of the awarded degree if it is not in a traditional discipline.

As alluded to above, there is the problem of having the right kind of faculty for such endeavors. This might be somewhat less of a problem at small colleges where teaching is an important ingredient in making tenure decisions. At those universities where research counts more than teaching, faculty face the difficulty of being rewarded for doing "interdisciplinary" work, especially in a new field like ecological economics, and also obtaining grants to support such research. Grants would not be critical to support undergraduates but they would be necessary to foster faculty, activities, and graduate programs and students whose influence could filter down to undergraduates.

GRADUATE EDUCATION

In some ways, graduate education should more easily provide a chance to study ecological economics, but it may require taking more courses or spending more time in graduate programs. The time and course requirements will depend on how structured a given program is and course requirements. Studying for a master's degree in engineering over two years and preparing a thesis probably wouldn't give a student much time to study other subjects. The same would hold true in molecular biology or a number of other disciplines. However, if the student is willing, and the program allows, he or she could take a number of electives. This would allow for a broader education if the student is willing to spend extra time. It certainly becomes more difficult to make such arrangements, the further removed a student's home department is from the subject of ecological economics. Again, in an evolutionary sense, some traditional subjects could perhaps incorporate notions of the environment, e.g., engineering courses or molecular biology courses might incorporate analyses of the impact of the various technologies being studied in these courses. This approach would be difficult because it would depend on the decisions of individual departments and professors.

A better approach would be to design elective courses that cover the requisite interdisciplinary subjects. These types of courses will probably soon be very attractive because environmental protection and restoration are likely to be booming worldwide businesses. In addition, creating products with minimal environmental impact is also competitively sound. In programs more closely related to the subject at hand, such as biology, ecology, environmental engineering, economics, planning, and so forth, the student's freedom to choose courses will depend on the flexibility of the given program and department. The student must also be concerned that the courses taken represent an acceptable mix to a prospective employer, if the job market is of concern, or to a prospective graduate program, if a Ph.D. is the goal.

At the doctoral level, there is an even greater degree of freedom in the sense that there is more time and flexibility to take a variety of courses. However, the student will be embedded within a discipline and under the guidance of an advisor and eventually a dissertation committee. In departments like biology or economics, there will still be pressure from the program to take traditional courses and follow research that is built upon previous research. Generally, this pressure will be to do more specialized work in the respective disciplines. This pressure will be especially strong when a student is selecting a topic for a Ph.D. dissertation. I suspect it would be difficult for a Ph.D. student to have a wide choice of ecological economic subjects to choose from in most traditional academic departments; the exception would be if particular departments had professors working to some degree in related subjects or if there was active interest in environmental economics. The advantage to the student would be the value of the degree to further a career in the academic world or as credentials in private industry or government.

The flexibility to pursue an ecological economics dissertation would be greater in departments that incorporate a number of disciplines: programs or departments such as geography, planning, regional science, systems ecology, environmental science, resource economics, etc., in which there would be the possibility of designing a program of courses and a dissertation topic that might constitute what we would call ecological economics. The advantage of this approach would be that the student would obtain a degree in a currently recognized field, allowing entry into established departments if an academic career was desired, but at the same time the student would have flexibility to study the subjects that would lead to an ecological economics perspective.

The most radical approach providing an ecological economics education would be to develop a separate program or department in ecological economics. Such a department would have difficulty if it had to compete with a major economics department in a given university. Perhaps the department could be called something else like "Department of Environmental and Ecological Management." The management part would imply that economics is part of the program. Such a program would encompass both the master's and Ph.D. level (3-4 years of coursework and a dissertation) and would consist of courses along the following lines:

1. Courses in economics should include conventional macro and micro approaches, measurement and analysis of costs and benefits, price theory, environmental economics, consumer behavior, theory of the firm, and international economics.

2. Courses in environmental science and ecology should include systems ecology, population biology, environmental chemistry (soil, water, and air), some plant ecology courses, some specialized courses on either terrestrial or aquatic ecology depending on the student's interest, toxicology, genetics, molecular biology, etc.

3. Mathematics and engineering courses should include probability and statistics, systems modeling and engineering, research design, differential equations, numerical analysis, computer modeling, optimization, environmental engineering and control theory, etc.

4. Some additional social science courses related to perception and institutional aspects of environmental issues. Perhaps psychology, law and business courses would be appropriate, and foreign language skills would be important.

For a student who is directed with a goal of understanding the environment and its interaction with society, I think a study of such a set of courses would result in a solid background for addressing a number of research questions for the dissertation and for work after the student graduates. The only problem is finding support for the student in this area, which leads us to the next subject.

It is a truly educational experience if a student has a chance to work on a research project involving other researchers. This provides a good vehicle for learning how those from other disciplines think and what is required for interfacing approaches from different disciplines.

ECOLOGICAL ECONOMICS RESEARCH

The availability of research funding in such an area is critical to the production of professors and graduate students. This can also lead to the development of the field and the eventual training of teachers for primary and secondary grade levels as well. If such a program or department existed in a university, it would be difficult enough for a junior professor to be accepted by the broader university community and to have a chance of getting tenure. This situation is exacerbated if it is not easy to obtain grants in this area and support graduate students. Unfortunately, most government agencies that provide research money have responsibility for only one part of the broad arena that constitutes the proper study of ecological economics. For example, the U.S. Department of Energy concerns itself mostly with energy technologies and nuclear weapons and for the most part addresses the environment only in so far as it might evaluate emissions. There are some exceptions. For example, the Office of Policy, Planning and Analysis might address a variety of areas associated with a subject, for example, like alternative fuels. But I suspect the funding of different parts of the problem would go to different researchers or institutional groups.

The U.S. Environmental Protection Agency (EPA) is viewed as taking responsibility for environmental regulation and analysis. EPA is certainly examining costs and benefits of different approaches to improving the environment and this could be generally construed as related to ecological economics. This agency tends to divide activities into air, water and land, although there is a multimedia activity. An ecological economics approach to a study of society and environment would certainly consider important tradeoffs that would emerge given a thorough systems study of human activity and land, aquatic and atmospheric impacts.

A perusal of the U.S. National Science Foundation grants in ecology and economics, shows that they tend to be specialized with only a rare few that encompass both subjects. The grants tend toward specialists in the fields. There certainly is ecosystem-directed research but it doesn't normally incorporate the economic and management aspects.

Other agencies, private foundations, and state sources also support related activities. Given the bias towards specialization and discipline-oriented research that is part of our society, it is probably safe to say that there are few funding sources for this endeavor. The closest accepted disciplinary area is environmental economics; assessing the costs and benefits of various actions is of interest to a number of institutions. Whether many of these studies would constitute some fundamental activity that is appropriate to university research might be questionable since many such studies might be very applied in nature.

Many activities that are centered in interdisciplinary institutes at universities can be broadly construed to be associated with ecological economics. These institutes can compete for large grants that might entail the analysis of a complex problem involving participation by a number of researchers from different disciplines at a given university or from a number of universities. Such an arrangement is a good example, perhaps, of ecological economics management but it is probably not suitable for fully supporting an individual investigator or student in the field.

Environmental problems are becoming more complex as world population grows, technology development accelerates, economic activity increases, and international trade grows. The actions of some countries can now affect the global environment. The most telling recent example is the issue of global climate change. The addition of various gases to the earth's atmosphere, so-called greenhouse gases (such as carbon dioxide, methane, and chlorofluorocarbons) may lead to rapid warming of the earth over the next 50 years, resulting in potentially severe climate change and disruption of existing weather patterns. The science is poorly understood, as are the economic costs and benefits that might accrue. This issue has resulted in the creation of the International Geosphere-Biosphere Program (IGBP) with the aim of focusing the effort's of the world's scientists on the problem of global change. The task at hand requires that scientists work beyond their narrow disciplines. What is also required is the coordination of the activities of various agencies. For example, in the United States, mechanisms for interagency coordination have been set up under the Federal Coordinating Council for Science, Engineering, and Technology's Committee on Earth Sciences. The U.S. global change program entails expenses of about one billion dollars per year and involves the study of biogeochemical dynamics, ecological systems and dynamics, climatic and hydrologic systems, human interactions, earth system history, solid-earth processes, and solar influences. In some sense, it represents a program in ecological economics. As to whether the research money is structured in such a way as to train ecological economists as contrasted to specialists in existing fields, I cannot speculate on because of my ignorance of the detailed funding profile. But it is an exciting effort and can lead to better understanding of the activities on the earth.

SUMMARY

In summary, then, there are a number of approaches to this broad field of ecological economics. The approach which is most likely to be successful, at least in the short run, is an evolutionary one in which the subject is introduced within the structure of existing educational programs. At the primary and secondary school levels, this means the development of a consciousness about the environment and an understanding of how human actions need to be traded off against impacts on the natural world. An important component of this education is the use of computer simulation and systems concepts; these tools can be carried on to more advanced levels of education. At the secondary level, the chance exists to introduce the subject in some courses, like geography and ecology, to increase environmental systems examples in other courses such as biology, physics and chemistry, and possibly to develop specialized courses.

At the undergraduate level, a student needs to learn a lot about several different subjects. The student has the chance to get a broad background by taking perhaps a dual major, by obtaining a solid background in some disciplines, and in carefully structuring a set of studies. More latitude is possible if a student majors in a program that allows a breadth of areas to be addressed. However, the student still needs a firm grounding in fundamentals as outlined in the paper.

At the graduate level, a student may have more latitude in developing an ecological economics education because there is more time. However, it would probably be difficult in most standard disciplines because of the course demands of the given field and the need to have a dissertation acceptable to that field. In some case, there may be a traditional department with faculty members that can bridge fields, for example, an economist having ecological knowledge or vice-versa. The best chance for a student is to major in a department that has a good deal of flexibility, such as an environmental science department or a planning department, and carefully structure a program as outlined above. Of course, the student will later face the problem of acceptability in traditional economics departments or environmental or ecology departments.

A more radical approach has also been outlined. At the primary and secondary school levels, one could envision establishing private schools that orient students towards ecological economics. These students must develop all the basic skills but at the same time also devote these skills to the understanding and solution of ecological economics problems. At the undergraduate level, a "great problems" approach could be structured. At the graduate level, this approach could be continued or, at both levels, economic ecological departments set up with defined guidelines for course requirements. Such a department, however, might not be all that different from many of the interdisciplinary departments and programs that have emerged over the past 20 years. Do we need such an additional program?

As for research, it is critical to support professors in this area, train students, and eventually develop a set of concepts that can be used for schools of education to train the primary and secondary teachers. Some of this could be accomplished through training sessions for present day teachers. The research, of course, tends to be available from agencies that are structured along traditional lines. However, emerging global environmental problems, such as potential global climate change, may create demands for more people that are trained to take a broader perspective than that of people from the traditional disciplines. People will be needed who can apply this broad perspective to the understanding and solution of human development problems and to environmental repercussions of human actions.

REFERENCES

American Association for the Advancement of Science (AAAS). 1990. The Liberal Art of Science: Agenda for Action. Washington, D.C.: AAAS.

Clark, M. 1989. *Ariadne's Thread: The Search for New Modes of Thinking.* New York: St. Martin's Press.

Clark, M. and S. A. Wawrytko *Rethinking the Curriculum: Toward an Integrated, Interdisciplinary College Education.* Westport, Conn.: Greenwood Press.

Goodwin, N. Report on the 1983 College of the Atlantic Symposium: "The Ecological Paradigm in Education." Lowell, Mass.: Center for the Study of Global Development and Change.

Hufford, Terry L. 1990. Educational Receptivity before Science Literacy. *BioScience* 40(1):44-45.

Mayer, W. N. 1986. Biology Education in the United States During the Twentieth Century. *The Quarterly Review of Biology* (December) 61(4).

National Academy of Sciences. 1988. *Understanding Agriculture: New Directions For Education.* Washington, D.C.: National Academy Press.

Odum, E. P. 1988. *Ecology and Our Endangered Life-Support Systems.* Sunderland, Mass.: Sinauer Associates.

Rosen, Walter G., ed. 1989. *High-School Biology Today and Tomorrow.* Washington, D.C.: National Academy Press.

Thompson, J. 1987. *Environmental Systems. Teachers Guide.* International Baccalaureate School, Examination Office. United Kingdom: University of Bath.

28

ECOLOGICAL ENGINEERING: APPROACHES TO SUSTAINABILITY AND BIODIVERSITY IN THE U.S. AND CHINA

William J. Mitsch
School of Natural Resources,
The Ohio State University,
Columbus, OH 43210 USA

ABSTRACT

Traditional approaches to the control of environmental pollution involve large commitments of technology and long-term investments that make their continual use an economic, energetic, and possibly an environmental liability in a sustainable economy. Ecological engineering has been defined as the design of human society with its natural environment for the benefit of both. It is the prescriptive rather than descriptive discipline of ecology in that it utilizes ecological principles, the self-design or self-organizational capabilities of natural ecosystems, and the sustainability of solar based ecosystems rather than fossil fuel based technologies to achieve environmental quality. The concept of ecological engineering has co-evolved in both the United States (unsustainable, industrialized, moderately populated economy) and China (partially sustainable, agrarian, heavily populated economy). Examples of ecological engineering from each culture are presented and overall approaches are compared. Each culture has ecological principles guiding it but approachs built on different economic and social structures. The two approaches may possibly be merged into one ecological engineering paradigm for long-term sustainability of environmental quality.

INTRODUCTION

There is renewed interest in applying the concepts of ecology to solve large, seemingly intractable environmental problems such as the effects of global climate change and polluted oceans, lakes, rivers and streams. We are only now coming to the realization that we cannot solve all environmental problems through technology, particularly when we view a future economy which may depend on sustainable, not unattainable, technology. There are finite resources that prohibit us from attaining goals such as the complete elimination of pollution or complete control of floods on every stream and river. Furthermore, we have realized that solving one type of pollution or resource problem invariably leads to other pollution problems or loss of resources. Populations continue to rise and our remaining fossil fuels continue to be used at an accelerated rate. In short, the demands on our environment are becoming greater, and the use of technological options to solve those problems are becoming less likely.

Ecological Engineering: Definitions

Ecological engineering and ecotechnology may serve a significant role in a sustainable society. They have been recently defined as "the design of human society with its natural environment for the benefit of both" (Mitsch and Jørgensen 1989a) and as "the techniques of designing and operating the economy with nature" (Odum 1989a). Ecological engineering was first defined in the early 1960s by H. T. Odum as "those cases in which the energy supplied by man is small relative to the natural sources, but sufficient to produce large effects in the resulting patterns and processes" (Odum 1962) and as "environmental manipulation by man using small amounts of supplementary energy to control systems in which the main energy drives are still coming from natural sources" (Odum et al. 1963). Odum further elaborated on the breadth of ecological engineering by stating that "the management of nature is ecological engineering, an endeavor with singular aspects supplementary to those of traditional engineering. A partnership with nature is a better term" (Odum 1971). Ecotechnology, which is sometimes used as a term synonymous with ecological engineering, has been defined as "the use of technological means for ecosystem management, based on deep ecological understanding, to minimize the costs of measures and their harm to the environment" (Uhlmann 1983; Straskraba and Gnauck 1985).

Ecological Engineering as Design of Ecosystems

Odum (1962, 1971, 1983) has suggested that while we have several fields for the study of systems of humanity and its environment (e.g., landscape ecology, ecological economics, human ecology), study is not enough. We need to understand how to carry out sound design of ecosystems to deal with these environmental problems with a minimum commitment of scarce nonrenewable resources. Ecological engineering may offer the needed approach. It is the "prescriptive" discipline of ecology. Other fields of ecology are generally "descriptive" (Odum 1989a).

Ecological engineering is applied ecology in its best sense, designing natural systems according to all of the ecological principles we have learned over the past century. It is not only the study of nature (synecology and autecology), nor is it merely the usual concept of applied ecology, often limited to monitoring and assessing environmental impacts or managing natural resources. Both of these areas, namely basic and applied ecology, provide important fundamentals to ecological engineering but do not define it completely (figure 28.1). Nevertheless, ecological engineering has its roots close to ecology, just as

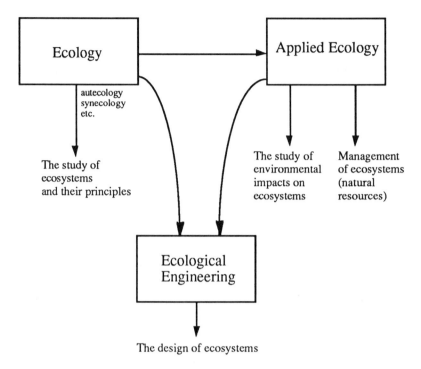

FIGURE 28.1   Relationships among basic ecology, applied ecology and ecological engineering.

chemical engineering is close to chemistry and and biochemical engineering is close to biochemistry. It should remain a branch of ecology.

Although not proposed as an additional field for traditional engineering, ecological engineering shares the concept of design with more traditional engineering (in that design of ecosystems is involved). Ecological engineering is based on the principles and the science of ecology, making use of them in the design of ecosystems. Thus we have used the terms ecological (using ecological principles) and engineering (for design by humans, combined with self-design by ecosystems). As stated in Mitsch and Jørgensen: "[Ecological engineering] is engineering in the sense that it involves the design of this natural environment using quantitative approaches and basing our approaches on basic science. It is technology with the primary tool being self-designing ecosystems. The components are all of the biological species of the world" (1989a).

Contrast with Other Technologies

Ecological engineering is not the same as environmental engineering, a respected field that has been well established in universities and the workplace since the early 1960s and

TABLE 28.1 Comparison of Ecotechnology and Biotechnology (from Mitsch and Jørgensen 1989a)

Characteristic	Ecotechnology	Biotechnology
Basic Unit	ecosystem	cell
Basic principles	ecology biology	genetics; cell
Control	forcing functions, organisms	genetic structure
Design	self-design with some human help	human design
Biotic diversity	protected	changed
Maintenance and development costs	reasonable	enormous
Energy basis	solar based	fossil fuel based

was called sanitary engineering before that. Environmental engineers are certainly involved in the application of scientific principles to solve pollution problems, but the concepts usually involve unit operations such as settling tanks, scrubbers, filters, and chemical precipitation. Ecological engineering and ecotechnology are also not to be confused with biotechnology, which involves genetic manipulation to produce new strains and organisms to carry out specific functions (table 28.1). Biotechnology has the cell as its basic unit, ecotechnology has the ecosystem. Ecosystems are manipulated by changing forcing functions or by introducing organisms to the system; cells are changed in biotechnology by manipulation of genetic structure. The most significant difference in ecotechnology and biotechnology is that the former is solar-based while the latter is dependent on our fossil fuel economy to survive.

CASE STUDIES

Because ecological engineering is an applied aspect of ecology, it is most useful to review case studies of projects that have historically been called ecological engineering to see if general principles and approaches emerge. One interesting aspect of ecological engineering is that definitions and subsequent "applications" of ecological engineering have coevolved in both the United States and China over the past few decades in relative isolation from one another. There are marked similarities in the use of the concept in both cultures, but some differences as well. Our recent research (Mitsch and Johnson

TABLE 28.2 Ecological Engineering Case Studies from the U.S. and China (discussed in this paper).

Ecological Engineering Project	Location	Purpose
United States		
Experimental Estuarine Ponds	Morehead City, N. Carolina	to investigate estuarine ponds subjected to waste-waters mixed with saltwater
Forested Wetlands for Recycling	Gainesville, Florida	to experimentally investigate forested cypress domes for wastewater recycling and conservation
Renovation of Coal Mine Drainage	Coshocton Co., Ohio	to study iron retention from coal mine drainage with *Typha* wetland
Restoration of Riverine Wetlands	Lake Co., Illinois	to restore Midwestern U.S. river floodplain and determine design procedures for restored wetlands
China		
Fish Production in Wetland Systems	Yixing County, Jiangsu Province	fisheries production synchro-nized to *Phragmites* wetland production and harvesting
Agro-ecological Engineering	Tai Chang Village Anhui Province	multiple-product farming with extensive recycling
Salt marsh restoration	China's east coast, esp, Wenling, Zhejiang Province	development of *Spartina* marshes on former barren coastline for shoreline protec-tion and food/fuel production
River Pollution Control	Suzhou, Jiangsu Province	use of waterhyacinths (*Eichhornia crassipes*) sys-tems for water pollution con-trol and production of fodder

1990) sought to review case studies of ecological engineering both in the United States (an unsustainable, industrialized, moderately populated economy) and in China (a par-tially sustainable, agrarian, heavily populated economy) and explore and compare the various definitions and principles of ecological engineering as they have developed in each setting. Discussed below and summarized in table 28.2 are several case studies in ecological engineering in the United States and China. Some of these case studies are given in more detail in Mitsch and Jørgensen (1989b).

Case Studies in the United States

Ecological engineering in the United States has stressed a partnership with nature and has been investigated primarily in experimental ecosystems rather than in full-scale applications. Some of the more significant experiments that have been conducted or are currently under way in ecological engineering relate to aquatic systems, particularly shallow ponds and wetlands. These ecological engineering experiments are summarized below.

Experimental Estuarine Ponds

H. T. Odum (1985, 1989b) reported on one of the first studies, done in the 1960s, described as an experiment in ecological engineering. Estuarine ponds were built in Morehead City, North Carolina to investigate ecological changes as the ponds received secondarily treated municipal wastewater mixed with salt water. Formally, the issue was "whether the self-organization process occurs readily there with new conditions from wastewater influence and how much time is required" (Odum 1989b). Three ponds received the sewage-salt water mixture and three ponds were controls, receiving only a tap water-salt water mixture.

The experiments demonstrated a rapid build-up of structure in the experimental wastewater ponds with heavy fringes of *Spartina* and blooms of the alga *Monodus* sp. in the fall and winter. Average annual net primary productivity was much higher in the wastewater pond (2.6-3.6 gO_2/m^2-day) than in the control ponds (1.1-1.2 gO_2/m^2-day). Self-organization was relatively rapid and the study concluded that while the estuarine wastewater ponds developed conditions that are often viewed as undesirable, they had well-organized structure and could be valuable in an ecological engineering sense such as for the design of aquaculture systems or the use of ponds as natural tertiary treatment systems. Odum (1989b) concludes that the design parameters such as inflows, loading rates, and other controls, along with the very important multiseeding of as many species as possible, were important considerations in the ecological engineering of these estuarine ponds and that these parameters and seeding set the boundary conditions for a relatively rapid process of ecosystem development.

Wetlands for Renovation of Domestic Wastewater

It is now common to see application of ecological engineering concepts with the use of wetlands to treat domestic wastewater (cf. Godfrey et al. 1985; Mitsch and Gosselink 1986; Hammer 1989). One of the first experiments to investigate this idea was conducted in the early 1970s in Florida by H. T. Odum, K. C. Ewel and colleagues (Odum et al. 1977; Ewel and Odum 1984). Whole ecosystem experiments to investigate the use of forested wetlands, particularly cypress domes dominated by *Taxodium distichum* var. *nutans*, for the treatment of high nutrient wastewater, were run for several years near Gainesville, Florida. Two cypress domes (0.5 to 1.0 ha in size) were subjected to treated wastewater from a trailer park at rates of about 2.5 cm/wk. A third dome received the equivalent amount of groundwater while a fourth served as a natural control. Measurements in shallow groundwater wells indicated that there was more than 90% removal of nutrients, organic matter and minerals by the wetlands receiving wastewater (Dierberg and Brezonik 1984). Furthermore, nitrogen and phosphorus concentrations in the foliage and branches of the trees increased as wastewater was added and decreased again after the treatment stopped (Straub 1984). This study, which also involved studies of wetland hydrology, soils, modeling, and wildlife, demonstrated that forested wetlands could be used in some cases to remove nutrients from wastewater with a minimum application of expensive technology. Table 28.3 summarizes the calculations done at that time on the savings that would result from using a cypress dome wetland as opposed to tertiary treatment. The ecological engineering alternative was a good match of fossil and solar energies, while the technological alternative expended an excessive amount of fossil fuel with no natural energy flows utilized.

TABLE 28.3 Economic comparison of ecological engineering approach with conventional means for controlling phosphorus (Mitsch, 1977).

	Ecotechnology with Cypress Wetland	Conventional Technology with Tertiary Treatment
Fossil Fuel Energy Cost, kcal/gal	3.28	25.3
Natural Energy Subsidy, kcal*/gal	3.30	~0

* fossil fuel equivalent

Wetlands for Renovation of Coal Mine Drainage

Wetlands have also been in ecological engineering applications of mine drainage pollution control from coal mines in eastern United States (Brooks et al. 1985; Hammer 1989). Mine drainage forms when pyrite (FeS_2), a compound commonly associated with coal deposits, becomes exposed to the oxidizing forces of air and water during the mining process. Sulfide oxidation results, producing dissolved iron, sulfate, large amounts of acid and ultimately an orange precipitate ferric hydroxide ($Fe(OH)_3$). Even after mining is stopped, mine drainage production may continue indefinitely. In one example, a 0.22 ha constructed wetland in eastern Ohio dominated by *Typha latifolia*, was evaluated for its ability to treat approximately 340 liter/min of coal mine drainage from an underground seep in eastern Ohio (Fennessy and Mitsch 1989a,b). Loading of mine drainage to the wetland ranged from 15 to 35 cm/day. Iron decreased by 50 to 60%, with slightly higher decreases during the growing season. Loading rates based on ecological engineering suggest that improved treatment of mine drainage is correlated with longer retention times and lower iron loading rates. A dynamic model which incorporated ecological and economic variables was developed from data from this site and others (Baker et al., in press). The model predicted that a wetland was an appropriate alternative as long as iron removal efficiencies of 85% or less were acceptable; higher efficiencies would require a larger wetland and hence greater cost than conventional mine drainage treatment. The advantage of the mine system is that it can self-design with little human intervention.

River Restoration

The restoration of entire rivers has been shown to be an elusive goal in many parts of the midwestern United States because of significant loads of sediments and other nonpoint pollutants. In an ecological sense, we have paid too much attention to the stream itself and not to the interactions of the river with its floodplain. One ecological engineering project on the Des Plaines River north of Chicago in Lake County, Illinois, involves the rescuing of a length of a river floodplain, recovering meanders back into the river channel, and establishing experimental wetland basins where the dynamics of sediment and nutrient control can be determined in experimental fashion. This project, begun in 1982 as the "Des Plaines River Wetland Demonstration Project" has as its goals "to demonstrate how wetlands can benefit society both environmentally and economically, and to establish design procedures, construction techniques, and management programs for restored wetlands" (Hey et al. 1989).

The project is being carried out at two scales. On the entire 182 ha site, woody and scrub vegetation has been reclaimed with native prairie species and oak savannas. Abandoned quarry lakes are connected to the river to give additional sediment trap efficiency and as backwater habitats for fish and shorebirds. Passive wetlands have been constructed with flap-gates to allow flooding by the river and trapping of water and sediments as the river recedes.

At the northern half of the site, experimental wetland basins have been constructed and instrumented for precise hydrologic control to investigate hydrologic design of wetlands subjected to high sediment loads (figure 28.2). Within this overall experimental design, there are several research questions that are being asked and can reasonably be expected to be answered by the research program just beginning. Among those question are:

- Will water quality improvement and sediment retention be lower in wetlands with high loading rates?
- Will the differences in major forcing functions (hydrologic flow-through) lead to different ecosystem development?
- How long will functions depend on initial conditions imposed by the wetland designers (e.g., plant stocks, organic soil added) and the physical conditions of original sediments?

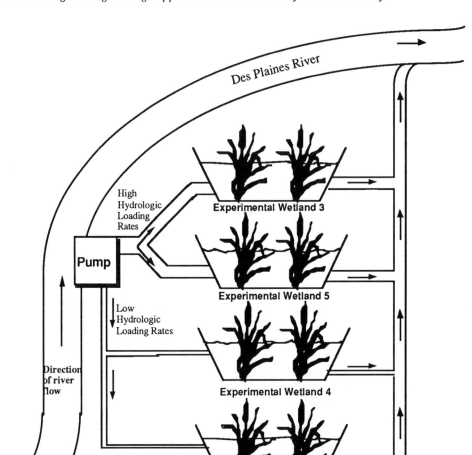

FIGURE 28.2   Des Plaines River Demonstration Project in Lake County, Illinois, showing experimental design for wetlands (Hey et al. 1989). Four of the wetlands have been constructed and water is currently being pumped through them in rates ranging from 8 to 46 cm/wk. Each of the wetlands will be maintained at similar depths (approximately 0.7 m avg. depth) so that depth will not be a variable in these initial experiments except in microhabitat studies. The different loading rates will result in different overall residence times for the wetlands.

Case Studies in China

Much of the approach to environmental management in China has remained an art, but in recent years there has been explicit use of the term ecological engineering to describe a formal "design with nature" philosophy. Ma (1988) suggests that ecological engineering was first proposed in China in 1978 and is now used throughout the whole country, with about 500 sites that practice agro-ecological engineering. At a recent Symposium on Acro-ecological Engineering in Beijing (Ma et al. 1988), Qi and Tian (1988) suggested that "the objective of ecological research [in China] is being transformed from systems analysis to system design and construction," stating that ecology has now a great knowledge base from observational and experimental ecology and is in a position to meet global environmental problems through ecosystem design, the main task of ecological engineering.

Several different applications of ecological engineering have developed in China. Ma (1988) has defined ecological engineering as: "a specially designed system of production process in which the principles of the species symbiosis and the cycling and regeneration of substances in an ecological system are applied with adopting the system engineering technology and introducing new technologies and excellent traditional production measures to make a multi-step use of substance." Agro-ecological engineering is defined as an "application of ecological engineering in agriculture" (Ma 1988). Ma and Yan (1989) further describe an ecological engineering system in the treatment of wastewater as "a series of techniques based on fundamental ecological principles established in a local region to adjust the structure and function in polluted bodies of water to maintain the ecobalance." They further elaborate that ecological engineering can have the highest economic benefits in wastewater treatment because it does not depend on great operation and construction costs and because it involves the regeneration and retrieval of abandoned resources (Ma and Yan 1989). Chung (1989) describes ecological engineering of coastlines as the design of a coastal human society with its environment, with its basis in general, estuarine, marine, and salt-marsh ecology developed over the past 80 years. Yan and Yao (1989) describe integrated fish culture management as ecological engineering because of its attention to waste utilization and recycling. Several case studies illustrate well the concepts of ecological engineering in the Chinese system (table 28.2) and are described in more detail below.

Fish Production in Wetland Systems

The Lake Go Reed Wetland and Fish Farm, located in Yixing County, China, integrates a *Phragmites* wetland with fisheries production. Water levels are manipulated in such a way as to maximize production of herbivorous grass carp (*Ctenopharyngodon idella*) and Wuchang fish (*Megalobrama amblycephala*) as well as to enhance the harvest of the reed grass (*Phragmites* sp.) which is used for fuel and other purposes. This application of ecological engineeringis accomplished primarily through water level manipulation that is synchronized with fish growing seasons and harvesting schedules (figure 28.3). A series of deep channels for fish overwintering extend through shallow wetland areas.

Surrounding the wetlands and channels are culture ponds for fish fry. When water is pumped from the lake it first passes through the surrounding fish ponds before entering the wetland-channel system. When wetlands are drained, the water is pumped directly to the lake. Pumped water is passed through screens to remove possible contamination of

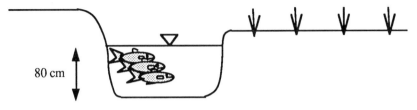

80 cm

Winter Season - Low Water Levels

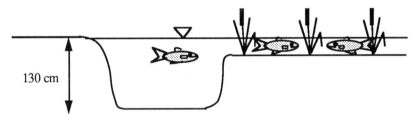

130 cm

Summer Season - High Water Levels

FIGURE 28.3 Seasonal patterns of water levels at Lake Go Fish-Wetland System in Yixing County, Jiangsu Province, China, showing winter and summer conditions. Water level manipulation maintains a dry wetland in January through March with about 80 cm of water in the channel. In April the water level in the channel is raised by pumping water from adjacent Lake Go, allowing some water to spill over into the wetland. The water level is maintained at about 130 cm (50 cm in the wetland in May) and water is continually pumped through October until there is up to 1.0 m of water in the wetland. From October to December, the water level is decreased until the wetland is dry again.

the system by lake fish. Yield of reeds is reported to be 75 mt fresh wt/ha-yr (with dry/wet = 30-50%). Fish yield is approximately 9 mt/ha with 60% as grass carp and Wuchang fish, 30% as silver carp and big head carp, and 10% as others including crucian and common carp. There are 5 fish channels for fish harvesting and culture. The system employs 104 people, most part-time. Some artificial chemicals are added to maintain the system. Artificial food is occasionally used for grass carp in the reeds and pesticides are sometimes employed in the open channels and ponds, but not in the wetlands.

Agro-ecological Engineering

Ma (1988) suggests that there are about 500 trial sites in ecological engineering or eco-agriculture in China in more than 20 provinces. Tai Chang Village, approximately 25 km west of Maanshan City on the south bank of the Yangtze River in Anhui Province, is one such demonstration site of work in acro-ecological engineering (figure 28.4). This description is based on the author's visit to the site in the spring 1989. The village where the experimental site is located is approximately 7 km^2 in size with 390 ha under cultivation. The population is about 3,408 with 894 families and 1,870 workers. There are 16 village-owned enterprises producing 11 million Chinese yuan a year. The fixed income is 1.4 million yuan a year with a per capita income of 1,030 yuan per year. Part of the village includes a demonstration site in acro-ecological engineering, including a pig farm, methane production and use, a fish pond, grape production and tree plantations of oranges, *Metasequoia*, mulberry and camphor (figure 28.4). Maintaining silk worms as they consume mulberry leaves and produce silk is another cottage industry in this acroecological engineering design. The emphasis is on recycling with pond detritus used as fertilizers for the plantations and pig waste fermented to produce methane for cooking.

Other examples of acroecological engineering are given by Wu et al. (1988) and Ma and Liu (1988) for villages in Hunan Province and Chai et al. (1988) for Jiangsu Province. For example, Wu et al. (1988) discuss the production of the "three materials" of fertilizer, forage, and fuel at Wu Tang village (pop. 1,505) through acro-ecological

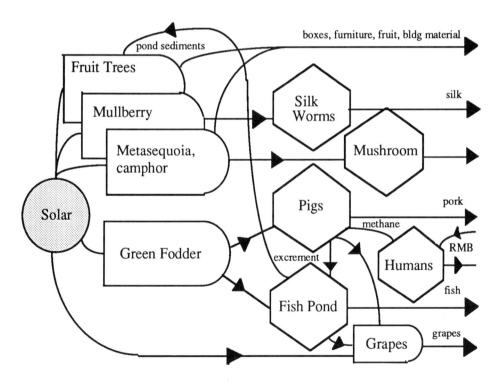

FIGURE 28.4 Acro-ecological engineering system in Tai Chang Village, Anhui Province, China (based on author's visit to village in May 1989).

engineering. While 90% of the energy of the system comes from solar-based organic production, only 10% comes from industrial energy (no energy quality ratios used). Furthermore, 88% of the total organic production is consumed in the village, while only 12% is exported. This is described as a "self-supporting system" that does not depend greatly on outside industrial energy.

Salt Marsh Restoration

Salt marshes, principally dominated by *Spartina anglica*, have been constructed along the east coast of China to stabilize the coastline, accelerate accretion of sediments for reclamation of tidal land for agricultural or industrial development, produce green manure, foodstuff and fuel, and partially control stream siltation and pollution (Chung 1982, 1985, 1989). Much of the salt marsh development along the coastline of China can be traced to a few individual plants imported from England and Denmark by Chung (Chung 1989). *Spartina anglica* C. E. Hubbard is an amphidiploid form of *S. townsendii* and is a hybrid between the English *S. maritima* and the North American *S. alterniflora* (Chung 1983). It was chosen partially because of its lower-elevation habitat, its rapid growth rate, and its wider seaward distribution. Propagations from seeds, sprigs, and rhizomes were all successful in initial experiments. Large *Spartina* plantations were developed by first transplanting plants to rice paddies, a planting technique well developed in China.

The reclaimed *Spartina* marshes along China's coastline have provided many benefits to humans (figure 28.5). The planting of *Spartina* in one location led to the accretion of 80 cm of sediments in 7 years at one location and 66-68 cm over control after 4 years at another site, leading to reclamation of the sea for crop production and coastal stabilization. The *Spartina* grass increased aeration and organic content and decreased the salt content of the soil in addition to dissipating wave energy and slowing currents. The newly created salt marshes were also used as a habitat for migratory birds, waterfowl, domestic fowl, nereids (worms), and crabs and as a pastureland for cattle and pigs. The grass is also harvested as a source of animal fodder, as an effective "green manure" for rice fields, as a source of fuel and as a source of marsh gas (methane) for cooking and illumination. More recently, *Spartina alterniflora* has been investigated by the same research group in China as a source of mineral supplements (esp. F, V, Cr, Mn, Fe, Co, Ni, Cu, Zn, Se, Sr, Mo, Sn, I) for beer and water (Chung, pers. comm.).

River Pollution Control

A riverine water hyacinth (*Eichhornia crassipes*) ecosystem was investigated in a series of ecological engineering experiments on a section of the Fumen River in an eastern suburb of Suzhou, Jiangsu Province (figure 28.6; Ma and Yan 1989). From May through December 1984, 2.7 ha of the river were planted in water hyacinths. The benefits of this system includes the clean-up of polluted river water, particularly for nutrients, organic matter, and heavy metals, and the production of green fodder for a number of consumers. The production of hyacinths, estimated to be about 9,000-10,000 g/m^2-yr, absorbs

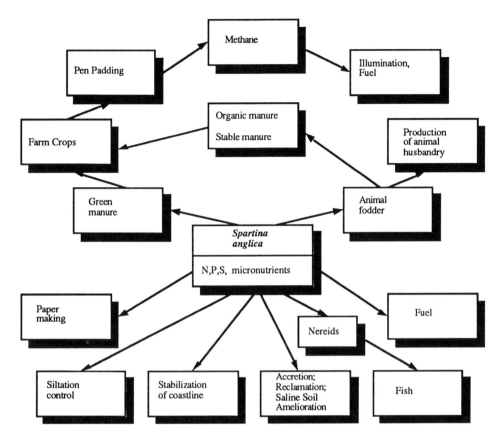

FIGURE 28.5 Benefits of reclaimed coastal wetlands in China with *Spartina anglica* restored marshes (from Chung 1989).

1,580 kg/ha nitrogen, 358 kg/ha phosphorus, and 198 kg/ha sulfur annually (Ma and Yan 1989). In the experimental hyacinth marsh, concentrations of COD (chemical oxygen demand), total nitrogen, total phosphorus, ammonia-nitrogen, ortho-phosphates, organic nitrogen, and organic phosphorus decreased during the growing season. Apparently microbial communities that develop on the root systems of the floating water hyacinths lead to a significant reduction of organic matter in the water. Approximately 2,500 metric tons (fresh weight) of water hyacinths were harvested from the site to be used as fodder for fish, particularly for grass carp (*Ctenopharyngodon idella*) and Wuchang fish (*Megalobrama amblycephala*) in culture ponds and for ducks, swine, and snails. Investigations of heavy metal uptake by fish and ducks that consumed the water hyacinths showed generally safe levels. The researchers recommend the following "ecotechniques" for increasing the efficiency of this ecological engineering system (Ma and Yan 1989):

1. Ensure sufficient planting area of water hyacinths for absorption of nutrients and for supplying matrices for sessile microorganisms among the plants' suspended roots;
2. Adjust the density of the water hyacinth population to maintain its rapid population growth rate and net production;
3. Harvest the water hyacinth on time (one-third to one-half of the standing crop is harvested every 7 to 10 days);

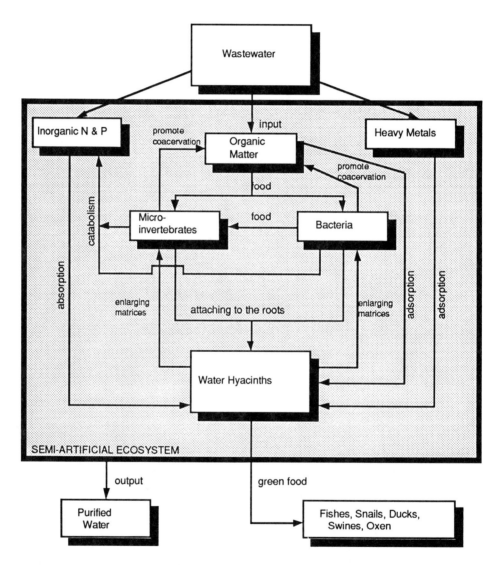

FIGURE 28.6   Model depicting processes and values of water hyacinth river restoration program (from Ma and Yan 1989).

4. Utilize the water hyacinth fully, as food for fish, ducks, snails, oxen, and pigs;
5. Add artificial matrices to supplement the root biomass to increase the number of sessile organisms capable of degrading organic pollutants; and
6. Inoculate bacteria to accelerate biodegradation of special organic pollutants such as phenol and petroleum oil and their products.

COMPARISON OF APPROACHES

There are some similarities and some fundamental differences in the practices of ecological engineering as developed in the United States and in China. Both systems use the principles of ecology as underpinnings to their approaches. But as principles of ecology are not uniformly agreed upon (e.g. Glesonian versus Clementian ecosystem development or energetic versus population theory, etc.) this statement does not assume commonality. Some of the important differences are summarized in table 28.4 and discussed below.

TABLE 28.4 Contrast between Ecological Engineering Concept in the U.S. and China

	U.S. and West	China
Basis	principles of ecology	
Design principles	self-design with some human work	heavy intervention by humans to maintain system
Emphasis on	changing forcing functions to produce outcome	changing ecosystem structure to produce outcome
People, #/ha	< 1	> 10
Subsidy type	fossil fuel economy supporting ecologists	human energy of resource managers
Subsidy size	should be small	sometimes surprisingly high, especially humans
Recycling	acceptable	absolute requirement
Commercial yield	not usual	always food production
Species diversity	often monospecific	"multi-layered" with many niches filled
Values considered	• aesthetics • natural resources conserved • non-market values, e.g. water pollution control	•commercial products
Economic values	willingness to pay, contributory, or replacement	utilitarian
Experience	30 yrs (since early 1960s)	3,000 years

Design Principles

Ecological engineering, as developed in the United States, is based on self-design or self-organization in which it "develops a pattern that maximizes performance, because it reinforces the strongest of alternative pathways that are provided by the variety of species and human initiatives" (Odum 1989a). The Chinese approach to ecological engineering does not as clearly rely on self-design. The fish-wetland system at Lake Go or the water hyacinth system in Suzhou involve significant manipulation of ecosystem structure to achieve desired ends. Human intervention into the process often overwhelms the self-designing capacity of the natural components. In the fish-wetland system, fish are grown in nurseries and added to the canals and wetlands to feed. In the water hyacinth system, the plant is added to the river each spring and harvested each fall; otherwise the winter is too cold for it to survive. This requires a significant amount of ecosystem structure manipulation rather than reliance on self design. Chinese systems might be considered *designer* ecosystems.

Subsidies and Human Intervention

There is a clear principle in Western ecological engineering that wasteful use of natural resources, including fossil fuels, should be avoided and that human management and investments of non-biological energies should be lightly applied. At one extreme of this concept is ecological engineering, as discussed by Kalin (1989), to deal with mining wastes as the creation of a "walk-away condition" where self-design and natural energies would be the only long-term investments. Chinese ecological engineering systems, while reliant on an economic system that is not as dependent on fossil fuels, are nevertheless much more management-oriented. This is partially because the low wages and provision of housing and other life-support to workers in the Chinese system allow many humans to be involved in the process of ecological engineering without the increased cost of operation that would be the case in Western systems. Throughout our visits to fish ponds, acroecological systems, and other ecological engineering systems in China in 1987 and 1989, workers were everywhere to be seen. This is a condition that could not be replicated in Western systems because of the high cost of labor.

Ecosystem Structure

Because constant manipulation and management of ecosystem structure are the methods of ecological engineering in China, the systems are often designed with multigradational and multilayered approaches. In other words, every possible niche (depths in a pond, places in a food-chain, etc.) is filled with different species and thus ecological diversity is often high. While high diversity systems are often used in Western ecological engineering, it is more the exception rather than the rule. When systems are developed for dealing with pollution, such as Odum's (1989b) coastal ponds or the *Typha* wetlands for mine drainage control (Fennessy and Mitsch 1989a, b), low diversity is often the case, even when multiple-seeding is attempted. Multilayered approaches are not as common in U.S. systems.

Recycling

The idea of recycling material so that every component of the ecosystem is used and reused is a cornerstone of Chinese ecological engineering. For example, wastes from pig farms are used to produce methane and pond sediments are used to fertilize plantations. In some cases, there is gradational use of resources by the diversity of organisms in the system, causing few waste products to result from its proper use. Ecological engineering in the West, while often citing the role of recycling of minerals and the complete utilization of resources as important, does not often specifically include recycling as purposeful design. It does, however, acknowledge the importance of self-organization of species in a way so as to maximize power through recycling, food web development, hierarchical relationships, and population control (Odum 1989b). Recycling is more implicit that explicit in Western ecological engineering.

Values and Economics

The desired output of Western ecological engineering is often an ecosystem that provides a non-commercial value to society such as water pollution control or habitat development. In the Western world, these values are often estimated from *willingness-to-pay*, *replacement value*, or *contributory value* approaches (cf. Costanza et al. 1989). In other words, a direct economic benefit does not accrue to individuals or organizations but to society as a whole. Nor are the benefits necessarily from products from the ecosystem. In contrast, Chinese ecological engineering usually has a *utilitarian value*, that is, "the value of using an ecosystem's products and amenities to drive both current and future benefits" (Costanza et al. 1989). This value is especially associated with producing food for the increasing population in China.

SUMMARY

There are similarities in the Chinese and Western views of ecological engineering. Both views are based on a realistic and ecologically-sound judgment of the needs of society and of the political and economic constraints in each system. Western society is now at a point where it cannot continue to use a fossil-fuel based economy to subsidize an unwise environmental protection program that commits us to structural maintenance and infrastructure for a very long time. Ecological engineering offers a possible solution to many cases where solar-based ecosystems can be developed to deal with otherwise difficult pollution problems. The Chinese, while concerned about their environment and possible contamination problems, are just as interested in systems that produce food and fiber in an economically sound way. Nevertheless, a certain empiricism in the Chinese methodology of designing ecosystems, developed in the first fish culture ponds in the Yin Dynasty about 3,100 year ago, remains to be interpreted for the improvement and continuing definition of ecological engineering. The two countries were developing the concepts and principles of ecological engineering with some communication but otherwise independently. While it is not possible to coordinate all terminology and concepts, at least an identification of common theories (presently under different names) and establishment of new theories in a common scientific language will facilitate the future application of ecological engineering projects. The Western scientific community could benefit from the scholarly review of Chinese ecological engineering theories, some now published only in Chinese, and from seeing how an overpopulated country has developed many ecological engineering approaches out of the necessity of conserving resources and maximizing their use of the landscape. The Western world could also benefit from this most detailed glimpse into an ecotechnological approach that could be useful if we are ever confronted with the necessity to develop a sustainable economy in periods of low energy. If one difference can be stated, Chinese use space (solar-based systems) for ecological engineering systems with time at less of a premium (labor is relatively inexpensive and information has a long residence time); Westerners maximize the use of time (labor costs are high and information is abundant and decays rapidly), with space at less of a premium. Our theories on ecological engineering will be truly general if they span this wide cultural difference.

Acknowledgments

This research was partially supported by the National Science Foundation, Grant INT-8802521 and by The Ohio State University, School of Natural Resources.

REFERENCES

Baker, K. A., M. S. Fennessy and W. J. Mitsch. 1991. Designing Wetlands for Controlling Coal Mine Drainage: An Ecologic-Economic Modelling Approach. *Ecological Economics*. In press.

Brooks, R. P., D. E. Samuel, and J. B. Hill, eds. 1985. *Wetlands and Water Management on Mined Lands*. Proceedings of a Conference, October 1985. Pennsylvania State University, University Park, Pa., 393 p.

Chai Tukui, Shi Wenchao, Lu Tienong, and Ye Maoxin. 1988. Benefit Analyses of Acroecological Engineering in Dongxu Village of Jiangsu Province. In Ma Shijun, Jiang Ailiang, Xu Rumei, and Li Dianmo, eds., *Proceedings of the International Symposium on Acroecological Engineering, August, 1988*. Beijing: Ecological Society of China.

Chung, Chung-Hsin. 1982. Low Marshes, China. In. R.R. Lewis, ed., *Creation and Restoration of Coastal Plant Communities*, pp. 131-145. Boca Raton, Fla.: CRC Press.

Chung, Chung-Hsin. 1983. Geographical Distribution of *Spartina anglica* Hubbard in China. *Bulletin of Marine Science*. 33:753-758.

Chung, Chung-Hsin. 1985. The Effects of Introduced *Spartina* Grass on Coastal Morphology in China. *Z. Geomorph. N. F. Suppl. Bd.* 57:169-174.

Chung, Chung-Hsin. 1989. Ecological Engineering of Coastlines With Salt Marsh Plantations. In. W. J. Mitsch and S. E. Jørgensen, eds., *Ecological Engineering: An Introduction to Ecotechnology*, pp. 255-289. New York: Wiley.

Costanza, R., S. C. Farber and J. Maxwell, 1989. Valuation and Management of Wetland Ecosystems. *Ecological Economics* 1:335-361.

Dierberg, F. E. and P. L. Brezonik. 1984. The Effect of Wastewater on the Surface Water and Groundwater Quality of Cypress Domes. In K. C. Ewel and H. T. Odum, eds., *Cypress Swamps*, pp. 83-101. Gainesville: University Presses of Florida.

Ewel, K. C. and H. T. Odum, eds. 1984. *Cypress Swamps*. Gainesville: University Presses of Florida.

Fennessy, M. S. and W. J. Mitsch. 1989a. Design and Use of Wetlands for Renovation of Drainage from Coal Mines. In W. J. Mitsch, and S. E. Jørgensen, eds., *Ecological Engineering: An Introduction to Ecotechnology*, pp. 231-253. New York: Wiley.

Fennessy, M. S. and W. J. Mitsch. 1989b. Treating Coal Mine Drainage with an Artificial Wetland. *Res. J. Water Poll. Control Fed.* 61:1691-1701.

Godfrey, P. J., E. R. Kaynor, S. Pelczarski, and J. Benforado, eds. 1985. Ecological Considerations in Wetlands Treatment of Municipal Wastewaters. New York: Van Nostrand Reinhold.

Hammer, D. A. ed. 1989. *Constructed Wetlands for Wastewater Treatment: Municipal, Industrial and Agricultural*. Chelsea, Mich.: Lewis.

Hey, D. L., M. A. Cardamone, J. H. Sather, and W. J. Mitsch. 1989. Restoration of Riverine Wetlands: The Des Plaines River Wetland Demonstration Project. In W. J. Mitsch and S. E. Jørgensen, eds., *Ecological Engineering: An Introduction to Ecotechnology*, pp. 159-183. New York: Wiley.

Kalin, M. 1989. Ecological Engineering and Biological Polishing: Methods to Economize Waste Management in Hard Rock Mining. In W. J. Mitsch and S. E. Jørgensen, eds., *Ecological Engineering: An Introduction to Ecotechnology*, pp. 443-461. New York: Wiley,

Ma Jinzhong and Liu Houpei. 1988. Analysis of the Functions of Paddy-Field Ecosystem Engineering in Southern Mountain and Hilly Areas of Hunan Province, China In Ma Shijun, Jiang Ailiang, Xu Rumei, and Li Dianmo, eds., *Proceedings of the International Symposium on Acroecological Engineering, August, 1988*. Beijing: Ecological Society of China.

Ma Shijun. 1985. Ecological Engineering: Application of Ecosystem Principles. *Environmental Conservation* 12:331-335.

Ma Shijun. 1988. Development of Acroecological Engineering in China. In Ma Shijun, Jiang Ailiang, Xu Rumei, and Li Dianmo, eds., *Proceedings of the International Symposium on Acroecological Engineering, August, 1988*, pp. 1-13. Beijing: Ecological Society of China, Beijing.

Ma Shijun, Jiang Ailiang, Xu Rumei, and Li Dianmo, eds. 1988. *Proceedings of the International Symposium on Acroecological Engineering, August, 1988*. Beijing: Ecological Society of China.

Ma Shijun and Yan Jingsong. 1989. Ecological Engineering for Treatment and Utilization of Wastewater. Pages 185-217 In. W. J. Mitsch, and S. E. Jørgensen, eds., *Ecological Engineering: An Introduction to Ecotechnology*. New York: Wiley.

Mitsch, W. J. 1977. Energy Conservation Through Interface Ecosystems. In R. Fazzolari and C. B. Smith, eds., *Proceedings: International Conference on Energy Use Managment*, pp. 875-881. Oxford: Pergamon.

Mitsch, W. J. and J. G. Gosselink. 1986. *Wetlands*. New York: Van Nostrand Reinhold.

Mitsch, W. J. and S. E. Jørgensen. 1989a. Introduction to Ecological Engineering. In W. J. Mitsch and S. E. Jørgensen, eds., *Ecological Engineering: An Introduction to Ecotechnology*, pp. 3-12. New York: Wiley.

Mitsch, W. J. and S. E. Jørgensen, eds. 1989b. *Ecological Engineering: An Introduction to Ecotechnology*. New York: Wiley.

Mitsch, W. J. and D. L. Johnson. 1990. *Ecological Engineering: Theory and Development in the U.S. and China*. Final Report to the National Science Foundation, Grant INT-8802521. Washington, D.C.

Odum, H. T. 1962. Man in the Ecosystem. Procedings of Lockwood Conference on the Suburban Forest and Ecology. *Bulletin of Connecticut Agricultural Station* 652:57-75.

Odum, H. T. 1971. *Environment, Power, and Society*. New York: Wiley.

Odum, H. T. 1983. *Systems Ecology: An Introduction*. New York: Wiley.

Odum, H. T. ed. 1985. *Self Organization of Ecosystems in Marine Ponds Receiving Treated Sewage.* North Carolina Sea Grant Office Publ. UNC-SG-85-04. Raleigh: North Carolina State University Press.

Odum, H. T. 1989a. Ecological Engineering and Self-Organization. In W. J. Mitsch and S. E. Jørgensen, eds., *Ecological Engineering: An Introduction to Ecotechnology,* pp. 79-101. New York: Wiley.

Odum, H. T. 1989b. Experimental Study of Self-Organization in Estuarine Ponds. In W. J. Mitsch, and S. E. Jørgensen, eds., *Ecological Engineering: An Introduction to Ecotechnology,* pp. 291-340. New York: Wiley.

Odum, H. T., W. L. Siler, R. J. Beyers, and N. Armstrong. 1963. Experiments with Engineering of Marine Ecosystems. Publication of the Institute of Marine Science of the University of Texas 9:374-403.

Odum, H. T., K. C. Ewel, W. J. Mitsch, and J. W. Ordway. 1977. Recycling Treated Sewage Through Cypress Wetlands. In F.M. D'Itri, ed., *Wastewater Renovation and Reuse.* pp. 35-67. New York: Marcel Dekker.

Qi Ye and Tian Han-qin. 1988. Some Views on Ecosystem Design. In Ma Shijun, Jiang Ailiang, Xu Rumei, and Li Dianmo, eds., *Proceedings of the International Symposium on Acro-ecological Engineering. August, 1988.* Beijing: Ecological Society of China.

Straskraba, M. and A. H. Gnauck. 1985. *Freshwater Ecosystems: Modelling and Simulation.* Amsterdam: Elsevier.

Straub, P. A. 1984. Effects of Wastewater and Inorganic Fertilizer on Growth Rates and Nutrient Concentrations in Dominant Tree Species in Cypress Domes. In K. C. Ewel and H. T. Odum, eds., *Cypress Swamps,* pp. 127-140. Gainesville: University Presses of Florida.

Uhlmann, D. 1983. Entwicklungstendenzen der Oktechnologie. Dresden: Wiss. Z. Tech. Univ. 32:109-116.

Wu Jin Fu, Yan Fu, Yang Shi Jie, Liang Hai Quan, and Yu Wu Jiao. 1988. A Primary Study on Coordinated Ecological Engineering of Fertilized and Forage and Fuel in Wu Tang Village of Changsha County. In. Ma Shijun, Jiang Ailiang, Xu Rumei, and Li Dianmo, eds., *Proceedings of the International Symposium on Agro-Ecological Engineering, August, 1988.* Beijing: Ecological Society of China.

Yan Jingsong and Yao Honglu. 1989. Integrated Fish Culture Management in China. In W. J. Mitsch and S. E. Jørgensen, eds., *Ecological Engineering: An Introduction to Ecotechnology,* pp. 375-408. New York: Wiley.

29

ON THE SIGNIFICANCE OF OPEN BOUNDARIES FOR AN ECOLOGICALLY SUSTAINABLE DEVELOPMENT OF HUMAN SOCIETIES

AnnMari Jansson
Department of Systems Ecology
Stockholm University
S -10691 Stockholm, Sweden

ABSTRACT

This paper addresses the question of how natural and artificial boundaries affect environmental quality and performance of systems of nature and humanity. The point is made that a sustainable design of a regionis one with free flows of energy, money and information between nations. When countries with different resources, cultures and specialities are able to interact through an open exchange, development constraints due to limiting resources in single countries can be released. Examples are drawn from the Baltic region, where recent weakening of political boundaries have led to increased movements of capital, people and ideas between the seven littoral states. This will open up new possibilities for international cooperation to solve problems of resource scarcities and environmental pollution in a mutually rewarding way. Increased compatibility of economic systems may facilitate self-organizing processes that increase the efficiency of resource use and pollution control. Only by developing feedbacks that enhance the life-supporting capacity of the regional ecosystems can industrial society be successfully maintained.

INTRODUCTION

The traditional forms of national sovereignty are increasingly challenged by the realities of ecological and economic interdependence. Because the impacts of human interventions in Nature spill across political boundaries, the traditional distinctions between matters of local, national and international significance are no longer valid. In regional systems, shared by many nations, sustainable development can only be secured through international cooperation. Up to now, analyses that would have suggested possibilities for finding common solutions have often been blocked by the lack of open communication or exchangeable currencies. The dismantling of the frontiers between socialist and capitalist economies in Europe enables a necessary shift to a broader view of the pressing environmental and resource problems, to which the respective systems were not providing answers.

This paper addresses the question of how a multinational region with a diversity of ecosystems and human cultures can be sustainably managed by considering the capacities and constraints of the environment for supporting economic activities and how sustainability is influenced by transboundary exchange. Intentionally, the author implies different meanings of the term "open boundaries." It refers to ecosystems and political systems as well as to openness in human attitudes. Examples are drawn from the Baltic Sea area where the economic interests of many nations overlap and compete. In this area, the merging of the planned and the free-market economic systems is on its way, parallel to the completion of the internal market of the European Community. These changes will influence the institutional structures of all European countries and eventually make them more open to the concepts of *ecological economics*, which "implies a broad, interdisciplinary and holistic view of the study and management of our world" (Costanza 1989).

ECOLOGICAL BOUNDARIES AND INTERNATIONAL RELATIONSHIPS

In Nature, closed systems with complete isolation are rare and temporary (Odum 1983). A state far from equilibrium is maintained through the open flow of energy and matter across the system boundaries (Prigogine 1982). At the landscape scale, boundaries often coincide with large scale physiographic features, such as mountain ranges or coastlines.

Political and legal boundaries seldom coincide with the natural ones. But this mismatch is probably not as culturally predestined as one might think. It mainly reflects the organizational pattern that has emerged after the industrial revolution, when mineral fuels became the dominating energy source for human societies and supported the growth of larger cities and populations. Human land use has homogenized the natural topography of the landscape and new artificial boundaries such as stone walls, pipe lines, drainage canals and roads have been established instead. In previous periods, patterns of human settlements conformed more closely to the biophysical structures and energy flows of the landscape because societies were directly dependent on their supporting indigenous ecosystems (Jansson 1991).

Boundaries between ecosystems are usually gradual although some sharp changes do occur at forest edges, river banks and at the contact zone between the land and the sea. These abrupt borderlines have always attracted humans because they provide a diversity of natural resources and environmental services. Many of the world's largest cities are on a coast at a river mouth. The river has been the central channel of activity along which most agricultural production has taken place. The density of settlement has decreased away from the river valleys and has been lowest at the divides. Early urbanization has created cities in which the spheres of influence have corresponded to the drainage areas (Rikkinen

1980). A glance at the map of the Baltic Sea region shows that this general organizational pattern is largely maintained (figure 29.1).

In the past 200 years, the use of coastal areas for settlements, industries, energy facilities and recreation has greatly increased, as has the upstream manipulation of river systems through dams or diversions for agricultural and municipal water supplies. The effects on inland and coastal waters of this exponential urban industrial and agricultural growth are severe all around the world, particularly on land-enclosed seas like the Baltic that have only a narrow opening to the ocean (Falkenmark and Mikulski 1988).

FIGURE 29.1. The Baltic drainage basin, measuring 1.63 million km^2. The heavy line marks the catchment area ("ecological boundary") and the thin lines show political boundaries.

Moreover, the effects of industrial expansion are not contained within the boundaries of a single drainage area or the economic zones of any one nation. They are coupled to the larger realms of the biosphere, to international trade and to the forces of military conflict. The densely populated, energy intensive, urbanized regions have large, unbounded life-supporting areas outside their political territory which provide food protein, water recycling, waste assimilation, air purification and so on (Odum 1989). Therefore, in order to establish which intensity of resource use is sustainable for a given geographical region, the internal economic-ecological interactions as well as the external influences have to be considered.

Of major ecological-economic concern in many geographical regions is the decline of forests due to transboundary air pollution, exploitation for tree biomass and conversion to agriculture. Because forests occupy such large areas, they often account for a dominant part of carbon fixing in a landscape, and are vitally important for controlling land runoff and soil erosion. Acidification of forest soils due to air pollution and sulfuric and nitric acid deposition have diminished the productivity of forests and may speed up (in a nonlinear fashion) the leaching of nutrients and of toxic metals from forest soils to aquatic systems. Thus, the environmental conditions of the Baltic drainage basin are very much related to the ability of societies to reduce the amount of airborne acids and to improve the health and vitality of forests. Because forests are such slowly evolving ecosystems, actions taken in the near future will determine the nature of forests several decades into the next century.

GENERAL DESCRIPTION OF THE BALTIC SEA DRAINAGE AREA

The Baltic Sea drainage basin, representing 15% of the area of Europe, forms an interlocked economic and ecological system in which the interests of some 70 million people in nine countries (Denmark, Estonia, Finland, Germany, Latvia, Lithuania, Poland, Sweden and the USSR) meet, overlap and conflict. The Baltic Sea provides the ultimate sink for the by-products of human activities in this land area, which is about four times larger than the sea itself. This means that the impacts on the marine environment of land use changes, pollution outlets and exploitation of natural resources are much stronger in the Baltic than in most other coastal seas. It was also in the sea itself that the environmental impacts of the rapid industrial growth after World War II were first observed (Fonselius 1972). Depletion of oxygen in bottom waters, loss of top-carnivores and extensive algal blooms were early warnings of pollution overload, which had already appeared by the late 1960s. These problems initiated large investments in pollution control of point sources in countries like Sweden, but the pollution load to the Baltic continued to increase in the 1970s and 1980s. This increase was mainly due to the increased deposition of airborne pollutants from a wide range of sources on land. About 25% of the deposition of nitrogen comes from countries outside the Baltic region (Iversen et al. 1987) while the sulfur deposition is dominated by emissions from East European heavy industries.

In the 1980s, many West European countries halved their SO_2 emissions, while most East European countries increased their emissions. This increase was brought about by a vast consumption of fossil energies extracted from coal mines in Poland, lignite deposits in East Germany and oil-shales in Estonia, combined with gas and electricity delivered at a low price from the Soviet Union.

The use of fossil energy in the Baltic drainage basin has increased about thirty times in this century. Between 1960 and 1980 the per capita use doubled from 3.3 to 6.5 tons of hard coal equivalents and in 1985 the average energy consumption was 284 tons per km^2 (Serafin and Zaleski,1988). In general much more energy is used per economic output in East European countries than in Western economies. Poland, for example, consumes three times as much energy per economic output than Sweden (table 29.1).

One of the most alarming consequences of extensive coal-burning is the acidification of forest soils. Forest decline attributed to air pollution has become a major concern in European societies in the 1980s. The East and Central areas are most severely affected. Potential harvest loss due to air pollution has been estimated at 16% of the total potential harvest in Europe. Germany, Poland and Sweden would suffer the most (Nilsson et al. 1990). International Institute for Applied Systems Analysis studies show that efforts to curtail air pollution are insufficient for arresting European forest soil acidification.

OPEN BOUNDARY TRANSFORMATIONS OF THE BALTIC SEA DRAINAGE AREA

The militarization and barriers to trade and scientific and cultural exchange around the Baltic Sea have led to regional disintegration (Serafin and Zalesky 1988). When comparing the economic well-being of the nine Baltic countries, as measured by traditional

TABLE 29.1.   Some Demographic, Economic and Energy Data for the Baltic Sea Drainage Basin (Westing 1989 and World Resources 1987)

Country	Baltic drainage basin (M ha)	Pop-ulation millions	Pop-ulation (no./km2)	GNP/ cap (k$/yr)	Energy/ cap (GJ/yr)	Energy, Baltic basin (PJ/yr)	Energy/ GNP 1984 (MJ/$)
Denmark	2.2	3.0	136.4	11.5	133.0	399.0	15.4
Finland	30.3	4.8	15.8	11.0	146.0	700.8	20.2
E. Germany	2.7	1.0	37.0	7.9	223.0	223.0	28.0
W. Germany	1.2	1.0	83.3	11.1	163.0	163.0	19.3
Poland	28.1	36.4	129.5	4.5	133.0	4841.2	30.0
Sweden	42.7	8.1	19.0	12.4	137.0	1109.7	14.4
USSR	56.0	17.0	30.4	6.8	176.0	2992.0	26.0
TOTAL	163.2	71.1	43.6		146.7	10428.7	

standards, they are found to vary substantially. The domestic product is generally much lower in E. European economies than in comparable market-type economies (table 29.1). In comparison with the rest of the world, military expenditures per capita have been a severe burden in the whole Baltic region due to political tensions along the line between the 2 major military blocs, which had separated East and West Europe (Westing 1989).

The strength of the economic relations among the Baltic countries and with the rest of the world, as demonstrated by their trade relations, show that only about 20% of total exports remain within the region. A very small proportion of the trade is flowing between Baltic states with different (capitalist or socialist) economies (Vesa 1989).

Before 1989, discussions about cooperation between the countries around the Baltic Sea were strictly confined to matters of scientific and environmental interest. Technological and economic cooperation were practically nonexistent. Agreements were made to reduce sulfur emissions and to protect the Baltic Sea from pollutants, but the socialist economies were unable to invest money in environmental technology to achieve common goals. Finally, the precarious financial and environmental crisis in Eastern Europe helped trigger the chain reaction that led to the disintegration of the communist bloc.

The perceived difference in standard of living and access to economic capital between the two systems was particularly strong for people living on either side of the Berlin Wall. This artificial borderline effectively stopped the flows of goods and money but could not prevent television news and other information from spilling over to the other side. A couple of months after the fall of the Berlin Wall, about 100,000 people left their homes on the East side and moved to the West, primarily for socioeconomic reasons.

The isolation of the socialist economies has been disastrous for the management of both marine and terrestrial resources, because most ecological knowledge about environmental degradation was considered as classified information. As the boundaries to the socialist countries have opened, the full extent of the ecological disaster in Eastern Europe has emerged. The worst environmental dangers that have been identified so far, are in the coal mining and industrial districts of East Germany and Poland and in the areas of oil-shale and phosphorite extraction in northeastern Estonia. In these areas, the pollution loadings to the atmosphere, soil and water have reached such high levels that they constitute direct threats to human health and therefore require immediate action.

The most radical solution would be to close down the heavily polluting industries and evacuate local populations from contaminated districts. The spontaneous emigration from East to West Germany can perhaps be seen as a first step in such a process, even though the driving motive has been economic rather than ecological. The social consequence of such migrations should be weighed against the costs of supplying appropriate cleaning devices to existing industrial plants, removing poisonous waste and increased health care.

FROM REPARATIVE TO PREVENTIVE MEASURES

The degradation of ecosystems causes severe economic long-term costs and losses, not only through reduced forestry, fisheries, and agricultural yields, but also through pollution and health problems. Both the free market and the centrally planned system share a common belief that technological solutions will always subdue nature and extricate humans and their economy from Nature's limitations. *This is contrary to ecological understanding, which states that the economy must adjust its volume and composition to the opportunities and constraints produced by Nature.* Only a fast spread and an acceptance of ecologically viable technologies can pave the way for new economically sustainable societies.

Crucial to sustainability is developing a safe and sustainable energy pathway, because energy transformation and utilization have so many different kinds of environmental effects. It is a political, economic and ecological failure when forests die due to air pollution and acidification of soils. Work has to be initiated to improve the quality of air, soil and water and to reestablish the ecological productivity of forest land and other natural ecosystems, so that they can function as effective filters for pollution, buffers in the hydrological cycle, and for recreation purposes.

The opening of the rigid political boundaries between socialist and capitalist societies in Europe affords an opportunity to shift emphasis from a basically reparative to a more anticipatory and preventive environmental strategy (Landberg and Larsson 1990). Although immediate economic investments, and technical measures are necessary to reduce the acute social and environmental problems in Eastern Europe, they are not sufficient. *A preventive strategy needs to consider the interdependence between environmental functions and economic development to reach ecological sustainability in the region.*

The development of environmentally benign societies requires a complete change in attitudes of technicians, entrepreneurs, politicians and the public towards the values and functions of natural ecosystems. It will no longer be considered feasible to replace Nature's services with technical solutions, e.g. in sewage and waste treatment. Instead, knowledgeable investments in a new kind of ecological technology will be required, which takes advantage of the self-organizing power of Nature. *The leading principle should be to increase the diversity of landscapes and species composition in order to promote the development of an ecological system that is more resilient to climatic fluctuations and human impact.*

An increased interest in ecological housing and management can be expected, which means that special areas near settlements will be maintained for composting household wastes and for ecologically treating sewage water. In particular, wetlands will be used to perform these functions. Drainage and other activities which increase freshwater runoff will be restricted. The wasteful use of artificial fertilizers in agriculture and the practices of intensive animal raising which pollute both surface waters and groundwaters will no longer be profitable as the supply of cheap fossil energy decreases. This means that the huge surpluses of food produced by West European agriculture will disappear. Bearing in mind the food shortages in East Europe and the continuing growth in world population, ideas of converting crop land to forests will probably no longer be justified.

ENERGY, ENVIRONMENT AND ECONOMY

The economic and environmental problems of the socialist countries of Europe are indeed shared by all nations in the Baltic region. International programs are needed to coordinate national goals in order to deal more effectively with the problems of air pollution, acidification of soils, forest decline, and toxic contamination of marine and fresh water systems. Energy, economy, environment and health are so closely interlinked that systems solutions must be found at the local, national and regional level. But, for more than 40 years, the prevailing belief of people living in this region has been that the political barrier and strong military forces would continue to separate the societies in East and West for many more years. This left both sides almost totally unprepared for the present situation in which the two economic systems are going to amalgamate. Undoubtedly, a market economy will gradually emerge, but there is no clear strategy how to facilitate this process without high social and environmental costs.

In general, the market economies seem to develop more efficient mechanisms for dealing with environmental problems than the control mechanisms in the centrally planned economies (Henry 1990). In both systems, however, economic incentives have been misused; for example, energy prices have been subsidized, thereby promoting wasteful use of resources. In most countries, public bodies like energy facilities have enjoyed legal political and economic advantages. Given the dominant role of governments as energy distributors and the substantial changes required in the existing energy mix in all Baltic countries, it is unrealistic to believe that a more energy efficient system will be achieved by market pressures alone. The importance of energy prices for national energy policies and economic competition indicates that in order to encourage energy saving measures a common price policy coupled to international trade agreements is needed for the Baltic region as well as for the whole of Europe (Bern 1990).

The market economy is distinguished from centrally planned economies by diversity, competition and flexibility. However, both types of economic systems have generally failed to solve their environmental problems, even though the consequences of that failure have turned out to be much more serious in the East.

Bearing in mind the wasteful consumption of resources and the production of toxic substances occurring in all modern industrial societies, it is difficult to understand the political declarations, spoken with implicit faith, that as soon as the mechanisms of a free market are let loose, all environmental problems will be solved automatically. Economic systems are no easily exchangeable modules. They are connected with the infrastructure and communication network of the society and are deeply rooted in social and cultural patterns. The transformation of such basic structures is bound to take a long time.

CONCLUSIONS

Countries within a region may have differing problems and differing interests, but the marine environment is one in which problems and interests are clearly shared. In many estuarine areas around the world international conventions on environmental issues have succeeded in bringing together antagonistic nations to cooperate for the mutual benefit of all (UNEP 1988). The Baltic Marine Environment Convention, signed in 1974, served in many ways as a model for other regional seas. Since the formulation of these international agreements, it has become obvious, however, that the water quality of coastal seas is affected not only by human activities along the coasts but throughout the drainage basin. *This means that the required environmental actions are much more complicated than previously thought and sometimes require intervention into the economic and political structure of the societies involved.*

Irrespective of existing political boundaries or barriers to trade, there are a multitude of ecological linkages both within and between different regions of the world. The ecological and economic effects of environmental disturbances associated with natural resource use and energy consumption crosses national frontiers. *One of the most crucial steps towards an ecologically sustainable development is therefore to reach an international agreement on the upper limit of the use of energy for stabilizing feedbacks to the natural resource base.* It is by clarifying the role and magnitudes of these relationships that *ecological economics* can contribute to the development of more sustainable human societies.

Acknowledgments

The final composition of this paper has benefited from discussions at the ISEE workshop at the Aspen Institute, and the many constructive comments made by Leon Braat, Mary Clark, Ramon Margalef, James Zucchetto, and Thomasz Zylicz, who kindly reviewed my manuscript. The analysis of the Baltic Sea region was possible through research grants from the Swedish Council for Planning and Coordination of Research and the Bank of Sweden Tercentenary Foundation.

REFERENCES

Bern,L. 1990. The Perspective of a Swedish Industrialist. In B. Aniansson and U. Svedin, eds., *Towards an Ecologically Sustainable Economy*, Report 90:6, pp. 119-121. Stockholm: Swedish Council for Planning and Coordination of Research.

Costanza, R. 1989. What is Ecological Economics? *Ecological Economics* 1:1-7.

Falkenmark, M. and Z. Mikulski. 1988. The Baltic Sea as an Example: Hydrological Conditions as a Determinant of the Ecosystem in a Semi-Enclosed Sea. *Nature and Resources* 24(1):14-25.

Fonselius, S. 1972. On Eutrophication and Pollution in the Baltic Sea. In M. Ruivo, ed., *Marine Pollution and the Sea Life*, pp. 23-28. London: Fishing News, Ltd.

Henry, C. 1990. Designing with Nature and Using Market Mechanisms. In B. Aniansson and U.Svedin, eds., Towards an Ecologically Sustainable Economy, Report 90:6, pp. 56-60. Stockholm: Swedish Council for Planning and Coordination of Research.

Iversen, T., J. Saltbones, H. Sandnes, A. Eliassen, and O. Hov. 1987. Airborne Transboundary Transport of Sulfur and Nitrogen over Europe: Model Description and Calculations. EMEP, MSC-W, Report 2/89., Oslo: The Norwegian Meteorological Institute.

Jansson, A. M. 1991. Ecological Consequences of Long-term Landscape Transformations in Relation to Energy use and Economic Development. In C Folke and T. Kåberger, eds., *Linking the Natural Environment and the Economy: Essays from the Eco-Eco Group*, pp. 97-110.

Landberg, H. and P. Larsson. 1990. Preface. In B. Aniansson and U. Svedin, eds., *Towards an Ecologically Sustainable Economy*, Report 90:6, pp. 3-4. Stockholm: Swedish Council for Planning and Coordination of Research.

Nilsson, S., O. Sallnäs and P. Duinker. 1990. Forest Decline in Europe: Forest Potentials and Policy Implications. Biosphere Dynamics Project, International Institute for Applied Systems Analysis. Laxenburg, Austria. Draft report.

Odum,E. P.,1989. *Ecology and Our Endangered Life-support Systems*. Sunderland, Mass.: Sinauer.

Odum, H. T., 1983. *Systems Ecology, An Introduction*. New York: Wiley.

Prigogine, I. 1982. Order Out of Chaos. In W. J. Mitsch, R. Bosserman, R. K. Ragade and J. A. Dillon, Jr., eds., *Energetics and Systems*. Ann Arbor, Mich: Ann Arbor Science Publishers.

Rikkinen, K. 1980. The Baltic's Urban Systems, *Ambio*, 9(3-4): pp. 138-144.

Serafin, R. and J. Zaleski, 1988. Baltic Europe, Great Lakes America and Ecosystem Redevelopment, *Ambio* 17(2): pp.99-104.

United Nations Environment Programme. 1988. Status of Regional Agreements Negotiated in the Framework of the Regional Seas Programme. Rev. 1. Nairobi, Kenya: United Nations Environment Programme.

Vesa, U. J. 1989. Political Security in the Baltic Region. In A. H. Westing, ed., *Comprehensive Security for the Baltic. An Environmental Approach*, pp. 35-45. Nairobi, Kenya: United Nations Environment Programme.

Westing, A. H. 1989. Environmental Approaches to Regional Security, In A. H.Westing, ed., *Comprehensive Security for the Baltic. An Environmental Approach*, pp. 1-14. Nairobi, Kenya: United Nations Environment Programme.

World Resources 1987. *A Report by the International Institute for Environment and Development and The World Resources Institute*. New York: Basic Books.

30

INTEGRATED AGRO-INDUSTRIAL ECOSYSTEMS: AN ASSESSMENT OF THE SUSTAINABILITY OF A COGENERATIVE APPROACH TO FOOD, ENERGY AND CHEMICALS PRODUCTION BY PHOTOSYNTHESIS

E. Tiezzi, N. Marchettini and S. Ulgiati
Department of Chemistry, University of Siena,
Pian dei Mantellini 44, 53100 Siena, Italy

ABSTRACT

The role of entropy and the limits of nature induce us to reconsider our conceptions of evolution, progress and the production of material things. The correct use of science does not lie in dominating nature, but rather in living in harmony with it. Traditional economists continue to promote unlimited growth and trust blindly in technology; but nature has cycles that follow other rules and other times scales. Traditional economists think of specialization as a positive value and a source of efficiency; but nature teaches us that specialization, carried too far, is a threat to the stability of living systems.

Economics and industry have demonstrated their incapacity to think in ways that transcend details and short-term thought. Thermodynamics and biology force us towards a state of minimum production of entropy and conservation of resources. Therefore, to maintain the energy flow at a low level, slowing down the entropic process, we must look toward a more decentralized, small-scale organization that uses renewable resources.

An integrated agro-industrial system that produces energy and chemicals from biomass fulfills this goal. Indeed, the organization of an integrated system requires the definition of the optimal "exchange area" between agriculture, industry, by-products recycling, and return of organic matter to the soil. The choice of the basic parameters of the project (time, scale, distance) can minimize the use of non-renewable energy sources and materials and maximize the use of solar energy. An integrated system, fulfilling production capacity and requirements of a given area, is less vulnerable to market fluctuations, inflation and macroeconomic factors.

Flow analysis of the exchanges taking place in the system makes it possible to evaluate the socioeconomic and environmental advantages over traditional systems of production when planning and developing legislation.

INTRODUCTION

In modern society, the production of goods and services is largely dependent on fossil energy input and is highly specialized in the production of marketable products. The production process of each sector have particular by-products which are not needed or recognized for energy or commodity value by the producing sector. An example of a valuable by-product is the low-enthalpy heat produced during the generation of electricity by thermal power stations. This "waste heat" is useless in the electricity production sector. However, outside the sector, for example in civil and agro-industrial uses, it can become a precious energy source when used for district heating. This use outside the sector conserves other fossil resources.

The task, therefore, is to identify links within and between sectors that will make the maximum use of the energy and information content of each by-product. Such an approach, similar to the cogeneration of electricity and heat, introduces a sort of "cogeneration of goods and services" into the economic system.

AGRICULTURE AND COGENERATION

The aim of agriculture has traditionally been to provide food for people. The introduction of feedstocks for animals produced meat for human consumption and also animal muscle power for ploughing and transport. This constituted a cogeneration of food and work sustained by solar energy flow. Additional materials were produced in the process: construction materials (wood), energy other than food (wood fuel), fibers (cotton, wool, etc.), chemicals (essences, drugs, etc), fertilizers (manure, etc.). In the past, the total productivity of agricultural systems was very low, in keeping with the level of technology before the advent of industrial agriculture. An increase in productivity was sought and obtained with massive fossil fuel input in all phases of production—from working and fertilizing the soil to processing and storing food. On the strength of cheap fossil fuels, agriculture specialized in producing food (especially high-yield crops) and no longer utilized the by-products. Thus, animal power was replaced by machines, wood for construction by concrete, cotton by synthetic fibers, manure by chemical fertilizers and so on. The transition from preindustrial cogeneration to specialization was short and fast. However, this system has resulted in the accumulation of enormous quantities of by-products of different origin: farm residues, food surplus, waste heat, nitrogenous fertilizer in groundwater, etc. Disposal of these products poses a serious new environmental problem.

A STABILITY PROBLEM

Modern monoculture requires operations with high energy costs and high environmental impacts. Its vulnerability to perturbations is very high. Related industries (e.g., fertilizer and fiber industries) also become unstable, because they depend so heavily on oil and are affected by its price on the international market. But most important, the system as a whole becomes unstable, partly due to the inefficient use of the environment as a source of primary materials and as a waste dump.

In the absence of intrinsic stability based on internal factors, modern agriculture introduces external stabilizers (herbicides, pesticides, fertilizers, irrigation) in order to make up for loss of ecological diversity. In other words, the agricultural product is obtained with massive external fossil input instead of by the maintenance of the complex and diversified structure of the system of production. The second law of thermodynamics warns us that any decrease in entropy in part of a system is paid for by a more than equivalent increase in other parts of the system. Monoculture, for example, is unstable because crops are more vulnerable to parasites due to the simplification of the ecosystem; techniques of monoculture render the environment unstable because the pesticides cause the selection of resistant insect species and the fertilizers pollute the water-bearing strata. Yet without this chemical input, the monoculture could not exist.

The result is an increase in instability, i.e., less elasticity of the system toward any sort of external perturbation.

THE NATURAL ECOSYSTEM IS AN INTEGRATED SYSTEM

There is nothing surprising about the fact that the stability of the natural ecosystem is based on its capacity to create a network of links, such that the residues of a living species become the substrate for another. This gives a triple result:
1. Each species finds its primary material and the energy necessary for its development.
2. The residues produced by a species do not accumulate.
3. A continuous cycle of primary elements (carbon, nitrogen, oxygen, etc.) and trace elements is created so that they do not become immobilized in stable structures.

In nature, therefore, only the network of connections between system components is stable, whereas the elements linked together are limited in time and space. This means that the natural ecosystem extends the concept of cogeneration (as of electricity and heat) to all elements in the system. In this way, the overall yield of solar energy conversion is maximized.

When we artificially force an ecosystem in order to maximize a single product, it is at the expense of the overall equilibrium of the connecting network; the classical example is the modern agricultural monoculture.

It is worth noting that the principles of thermodynamics do not prevent us from trying to increase the productivity of a system as a whole, flows of energy and materials permitting, by modifying the input flow of information. For example, by reintroducing certain crop rotations in multiyear cycles to replace the annually repeated monoculture, soil depletion is avoided while high productivity is maintained. The monoculture offers only a single product, particularly in demand at a given time and place. Crop rotation poses a problem for the producer because he must grow certain products which for market or organizational reasons are in less demand. Note that this problem is solved in the natural ecosystem with an increase in biological diversity. Man's resistance to this versatility is seen in the extreme specialization of modern farms and zootechnology.

Specialization aimed at maximizing a single product has been possible only by the input of fossil energy in addition to solar energy. The second principle of thermodynamics confirms that this leads to an increase in entropy outside the monoculture system higher than the entropy decrease in the agricultural product.

For any agricultural technique, it is well to remember that the main problem is not the maximization of productivity, but the maximization of stability, or at least a convenient compromise. Hence, it is not a matter of expecting an increase in productivity due to a miraculous change in natural processes but rather one of developing the farm ecosystem as a cultural organization which appreciates differences and "residues."

Obviously, any form of production will be limited by the density of solar energy flow, land quality, temperature, etc., and by inherent limits in the system. The limits must be identified and observed in order to avoid upsetting the stability of the system as a whole.

COMPARISON OF DIFFERENT LAND USAGE

We shall now examine different photosynthesis-based systems: a conventional mono-culture-based system and an "integrated system" based on the cogeneration characteristic of natural systems. Both can be compared to the productivity of a mature forest ecosystem, and to preindustrial subsistence farming that still exists in many parts of the world.

We shall examine the flows of energy and material across the boundary of four different thermodynamic systems defined as follows:
1. Farmland with non-industrialized agriculture (Ecosystem A);
2. Farmland with industrialized agriculture (monoculture, Ecosystem B);
3. Farmland with "integrated" agriculture (Ecosystem C);
4. A mature forest ecosystem (Ecosystem D).

For each system we shall consider the flows of energy and materials connected with the operations of production (e.g., energy incorporated in agricultural tools) and product processing (e.g., drying forage or producing ethanol).

Each ecosystem can be seen as a territory with energy inputs and outputs. Territories are differentiated by energy yields, primary materials, and degrees of ecological stability.

PRE-INDUSTRIAL AGRICULTURE

Subsistence or traditional agriculture (figure 30.1) existed in Western countries (North America and Europe) until about 50 years ago, and still exists in most developing countries. Apart from direct solar input, it relies on human and animal muscular energy.

Fields are fertilized by burying residues, spreading manure, and periodic rotation of leguminous species (soy, vetch, lupins). The products of crop rotations are used partly for human consumption and partly for animal feed.

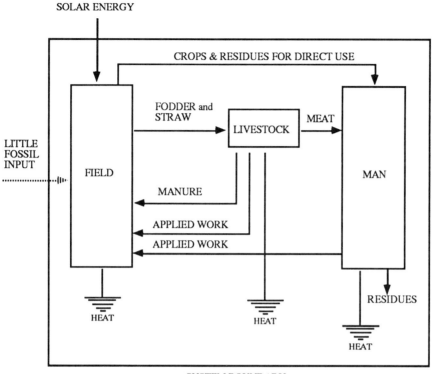

FIGURE 30.1 Traditional farming: input of solar energy and minimal fossil fuel feed a complex system based on natural recycling of all residues. Productivity is not high, but E.R. (output/input) may be very high due to the low input. The residues not recycled is negligible.

Table 30.1 gives typical corn production figures for several non-industrialized countries (Leach 1976). It has been estimated (Anderson 1979) that good subsistence farming without fossil input (Guatemala, table 30.1) could support the food requirements of a world population of 6.6×10^9.

INDUSTRIALIZED AGRICULTURE

Since 1945, the use of chemical fertilizers and pesticides derived from oil has spread throughout the United States and Europe. At the same time, the use of increasingly powerful tractors and machinery has reduced the necessity for human muscle power and has replaced animal power. The optimal use of these machines has meant vast expanses of monoculture.

TABLE 30.1 Energy Balance of Maize Production by Subsistence Farming
(Leach 1976, modified)

	input (10^5Kcal/Ha.y)	output (10^5Kcal/Ha.y)	E.R.[°]
Guatemala*	2.8	38.2	13.6
Guatemala†	9.8	38.2	3.9
Nigeria§	3.6	36.0	10.0
Messico†	6.9	33.9	4.9
Philippines+	6.7	33.9	5.1

* only human muscular energy
† with the use of animal muscle power and simple farm machinery
§ with the use of animal muscle power and moderate use of chemical fertilizers
+ with the use of animal muscle power, simple farm machinery and the moderate use of
   chemical fertilizers
° Energy Ratio (output/input)

Because of the low cost of oil, chemical fertilizers have supplanted crop rotation and fertilization with manure. The use of farm machinery has amplified the use of chemical fertilizers. Economic factors favor the use of machinery and chemical fertilizers, because the increase in monoculture farms has separated the large-scale external livestock farms from cropland making the produced manure uneconomical to transport, creating waste. In this case, scale has its price.

Figure 30.2 is a typical scheme of industrial farming (maize monoculture). Table 30.2 (Pimentel et al. 1973, modified) gives an energy balance of U.S. corn production in two different periods. The type of industrial agriculture practiced in the United States in 1970 can sustain an estimated world population of 30 billion (Anderson 1979). Apart from the problems associated with the stability of such a large population, farming of this kind would require the yearly consumption of 2% of current world resources of fossil fuels. At this rate, reserves would only last another 50 years, provided that the environmental consequences (erosion, greenhouse effect, etc.) did not further aggravate the situation.

MATURE FOREST ECOSYSTEM

The mature forest ecosystem evolves in such a way that incident solar energy is used with maximum efficiency. Ecological diversity, i.e., the number of different species the ecosystem contains, is high. A close network of internal energy exchanges allows each species to find its "own" solar energy source either directly or via the food chain. Energy passes from one trophic level to another, partly degenerating into heat and partly feeding the growth of higher organisms. As it moves up the food chain, energy accumulates in the form of organic material and genetic information.

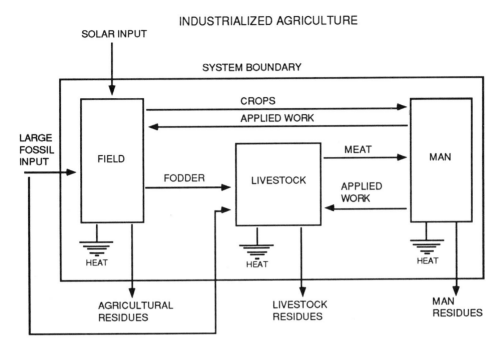

FIGURE 30.2 Industrialized agriculture: photosynthetic productivity is enhanced by the massive input of fossil energy. High specialization and the physical distance between the various sectors of production and use prevent the recycling of residues. Their accumulation in the environment constitutes a loss of precious resources and an increase in entropy.

TABLE 30.2 Energy Balance of Maize Production in the United States
(Pimentel 1973, modified)

	input (10^5Kcal/Ha.y)	output (10^5Kcal/Ha.y)	E.R.§
USA (1945)	22.8	76.9	3.4
USA (1970)	71.2	183.7	2.6

§ Energy Ratio (output/input)

To speak of the energy balance of this type of ecosystem and to compare it with agricultural ecosystems is rather strange, because solar energy is the only input into the forest ecosystem (figure 30.3) but it is by convention ignored in the farm energy balance. Even the output of the forest ecosystem is difficult to evaluate, and, at any rate, it is not exploited by man (unless the forest is cut) but rather by other organisms. If we consider output to be net primary productivity (organic material produced in a year by the photosynthesis of a hectare of forest), we see at once that it is impossible to calculate the

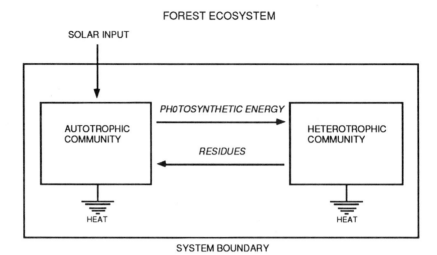

FIGURE 30.3 Forest ecosystem: the stability and productivity of this ecosystem are enhanced by the total recycling of all residues made possible by the very complex organization of trophic levels. The only input is solar energy. There is no accumulation of wastes.

Energy Ratio (E.R.), since input, by convention, is zero. The net primary productivity of the Hubbard Experimental Forest (White Mountain National Forest, New Hampshire, US) has been estimated at 468.0×10^5 Kcal/ha.year which is about 0.4% of the annual incident solar energy (Gosz et al. 1978). This is virgin forest of maple, beech and yellow birch. Table 30.3 gives some forest ecosystems figures (Woodwell et al. 1978, modified).

TABLE 30.3 Net Primary Productivity and Mean Total Biomass per Hectare of Forest Ecosystems (Woodwell et al. 1978)

	net primary productivity (10^5Kcal/Ha.y)	total biomass (10^7Kcal/Ha.y)
Tropical rain forest	876	179.37
Tropical seasonal forest	640	138.66
Temperate evergreen forest	515	140.29
Temperate deciduous forest	482	120.50
Boreal forest	318	79.87

THE INTEGRATED AGRO-INDUSTRIAL SYSTEM

Let us define the "integrated agro-industrial system" (figure 30.4) to be a group of farms and production techniques which uses a minimum of non-renewable materials and energy

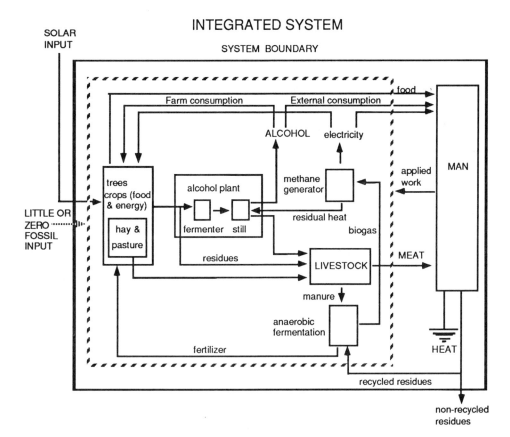

FIGURE 30.4 Integrated agro-industrial ecosystem: bioenergy (alcohol and biogas) production imparts increased complexity to the agro-zootechnical subsystem and maximizes yield by virtue of lower energy input and higher energy output. Bioenergy production is the core of the system: the introduction of new "trophic levels" renders the whole community stable. The fertilizing properties of the residues and rational organization of the system enable fossil input to be greatly reduced and perhaps even avoided. The result is high productivity and E.R.

and a maximum of solar energy in its different forms, to produce a given set of products (food, energy, fibers, fertilizers, chemicals and basic manufacturing materials). The system achieves this by operating at an appropriate scale for the fundamental parameters of the project (time, size, distance, etc.) and by recycling as much energy and material as possible. Such a system is based on a high information flow from outside (technology, experience, know-how, etc.); its organization, which is much more complex than that of the monoculture, is similar to the model of cogeneration of the natural ecosystem, and as such is aimed to maximize overall stability rather than a single product.

In the mature forest ecosystem, all residues return as fertilizer via the food chain of soil microorganisms. Similarly in the Integrated System (I.S.) farm, animal and processing residues return to the soil so that it is not depleted. In the forest ecosystem, there is no concept of "waste" since every organic residue constitutes the energy basis for another organism. Likewise in the I.S., energy extraction is maximized by exploiting all possible uses of organic residues (production of ethanol and biogas by fermentation, direct combustion, extraction of chemicals, etc.).

Just as the virgin forest ecosystem does not contribute to atmospheric CO_2, the I.S. annually recycles the CO_2 created. Hence it does not require extra fossil input (or requires much less) and it does not increase atmospheric CO_2 or enhance the greenhouse effect.

Of course, it is not easy to choose the right scale for the system so that it does not lean excessively toward one aspect of production (whether food, zootechnical or energy production). The system must also maintain soil fertility by an adequate return of minerals and organic material. It should not be imagined that the total production of the I.S. can exceed the physical limit of solar energy flux. Nevertheless, there is a good margin between the productivity of an integrated ecosystem which is capable of ensuring year-round field cover without depleting the soil and that of the monoculture which is capable of high yields, but only for a few months. After those few months, the field lies barren and does not capture any further solar energy.

It is not known whether the selection of new higher-yield species can contribute to this plan without adversely affecting stability. Efforts to organize an integrated network of energy exchange may provide a new theoretical basis for the inclusion even of these "new" crops.

In short, different levels of integration can be postulated:

- *Inside the individual farm*: Crop rotation maximizes the quantity of biomass produced without increasing fertilizer input or prejudicing soil fertility. The Washington University Center for the Biology of Natural Systems (CBNS 1980) proposes a practical system of rotating suitable forage crops to increase the carbon content of the biomass produced without decreasing the protein fraction. The extra carbon can be fermented to ethanol.
- *On a local scale*: Farms with different activities can coordinate the methods and means of harvesting and storage of residues, for the optimal use of organic fertilizer, feedstocks and energy (biogas, electricity) within the farms.
- *In processing plants*: Plants can work with farms in order to transform biomass of varying complexity into ethanol at different times of year, and simultaneously produce feedstocks, organic fertilizer and extra electrical energy for supply to local users.
- *In time*: The different productive sectors need to be coordinated so as to guarantee a constant flow of primary material through the system and a constant return of nutrients to the soil, without exaggerated peaks or gaps.

At the present time, all these expectations depend on a real desire to place agriculture at the center of a completely new picture of production. We are convinced that agriculture is capable of crossing the watershed of productive specialization (a small number of often surplus crops) and moving to integrated differentiation; a change made possible by the development of biotechnology.

AGRICULTURAL DATA FROM THE SIENA AREA (ITALY)

Table 30.4 shows the energy balance and C/N ratio of a corn monoculture for feedstocks in a farm with a high level of mechanization near Siena (Orsi and Sarfatti 1983). The figures are based on 3 years of monoculture. The output/input ratio of 2.18 is typical of corn production.

Table 30.5 gives an estimate on agricultural data for different land use (triennial crop rotation) (Giolitti et al. 1989). It is possible to obtain surplus biomass for ethanol production without a decrease in feedstock production. The proposed crop rotation yields an E.R. of 4.7 over three years and a total C/N ratio of 55.2 if by-products are included. The E.R. drops to 3.6 if by-products are not included.

TABLE 30.4 Energy Balance of a Maize Monoculture: Cerreto a Merse, Siena
(Orsi and Sarfatti 1983)

	input (10^5Kcal/Ha.y)	output (10^5Kcal/Ha.y)	E.R.*	N (Kg/Ha)	C (Kg/Ha)	C/N
Triennial	549	1,200	2.18	436	13,600	31.2
Yearly average	183	400	2.18	145.3	4,533	31.2

* without considering by-products

TABLE 30.5 Energy Balance for Triennial Crop Rotation; Proposed for an Irrigated Area near Siena (Giolitti et al. 1988)

Year	Crop	Product 10^2 Kg/Ha (d.w.)	Byproduct 10^2 Kg/Ha (d.w.)	Input 10^5 Kcal/Ha	Output[a] 10^5 Kcal/Ha	Output[b] 10^5 Kcal/Ha	N 10 Kg/Ha	C 10 Kg/Ha
1	fodder beet	102	44	136	503.7	721	0.42	58.6
2	barley +	38.25	38.25	71	165	315	0.82	30.6
	corn silage	134.8	---	125	290	290	0.60	53.9
3	greenpeas + sweet	33.0	66.0	91	113	313	2.29	39.6
	sorghum	177.8	---	73	708	708	0.46	71.1
Total (3 yr)		485.85	148.25	496	1,779.7	2,347	4.59	253.6
Year avg.		161.95	49.42	165.3	593.2	782.3	1.53	84.53
E.R.					3.6	4.7		

C/N = 55.2
a: without considering by-products
b: including by-products

These figures are obtained from mean values for crops typical of irrigated areas in southern Tuscany (Siena, Grosseto). When compared with the data of table 30.4, we find that there is increased photosynthesis per hectare with less fossil energy input (about 10% less over 3 years) and no loss of nitrogen content for animal feed.

However, this is not yet an I.S.—the data is given only to show photosynthetic potential. An accurate experimental evaluation, which aims to increase the integration of system components and increase the amount of solar energy fixed by photosynthesis, is currently in progress.

ECONOMY AND DISECONOMY

The four ecosystems described above may be examined in different ways; Table 30.6 clarifies some of them. Subsistence agriculture (Ecosystem A) has a very high output/input ratio due to the fact that fossil fuels are used sparingly or not at all. However, annual yields per hectare are very low and it is unthinkable to propose reintroducing subsistence agriculture, especially in countries where the ratio of fertile land to population is low.

Industrial agriculture (Ecosystem B) offers a high output per hectare, but at the expense of overall yield (output/input≈2.5), and yields tend to decrease over time. Besides the trend toward decreasing yields, industrial agriculture has several other serious disadvantages. Traditional methods of monocultural cultivation pollute groundwater and take a heavy toll on soil structure, increasing erosion. A yearly loss of more than 10 tonnes per hectare of fertile soil has been measured in the most industrialized areas, a serious concern when compared with a maximum soil formation rate of 2-3 tonnes/ha (Triolo 1988) under the best possible conditions. Since erosion leads to diminished productivity and water pollution, it is reasonable to attribute these diseconomies to the monoculture system.

Other hidden costs of industrial agriculture should be pointed out, including the problem of agricultural surplus and the production of feedstocks. Surplus is a chronic problem of monoculture. From 1980-1983, the production and conservation of the surplus alone cost the EEC the equivalent of 11 million tons of oil a year or 1% of the total internal energy requirements of the food and agriculture sector (Caporali and Rossini 1988).

The production of feedstocks costs the equivalent of 2.5 million tons of oil, 30% of the indirect energy requirements of the agriculture and forestry sector (CNEL 1982). This is because feedstock production has become an autonomous sector aimed at producing a single product.

Crop rotation and recycling of energy and nutrients within the I.S. (Ecosystem D) solves many of the problems of industrial monoculture. The I.S. has a stable or even increasing E.R. because other "trophic levels" are used to reduce waste and increase overall yield. Rotation of leguminous and pasture crops does not alter soil structure or deplete it of nitrogen, so erosion is prevented and no new limiting factors arise.

In the last 40 years, crop development has been directed toward selecting high-yield varieties for monoculture. If this trend were reversed and efforts made to develop varieties

TABLE 30.6 Energy Balance of Different Ecosystems (solar input not included)

	Input (10^5Kcal/Ha.y)	Output (10^5Kcal/Ha.y)	E.R.
Forest	0	400-800	---
Subsistence farming	2.8-3.8	87.9-88.2	4-14
Maize monoculture (US, 1945)	22.8	76.9	3.4
Maize monoculture (US, 1970)	71.2	183.7	2.6
Maize monoculture (Siena, 1983)	183	400	2.2
Crop rotation (Siena, 1988)	165.3	593.2+	3.6
Integrated System (our estimate for Siena area)	123.9*	750-800§	6.0-6.4

+ without considering residues

* based on a chemical fertilizer input which is half the total input. We estimate that chemical fertilizer input can be halved by recycling and crop rotation.

§ including the energy value of by-products not considered in the previous ecosystems (mostly hay), since the inclusion of bioenergy production enables the utilization of certain residues. Output does not include the fraction of residues returning to the field as fertilizer, which was already detracted from input.

which could be integrated into an agro-zoo-industrial system, the yield could probably increase further without causing environmental imbalances. The aim is therefore to increase the output/input ratio while maintaining a high productivity per hectare and reducing environmental impact. Table 6 gives an output estimated by our group for an I.S.; it is comparable to that of a forest ecosystem. The E.R. is also very high.

The production philosophy of the I.S. could eliminate surpluses or at least use any occasional surpluses for energy production. Moreover, I.S. is expected to produce feedstocks using less energy and at a much lower environmental cost.

RECYCLING RESIDUES AND MINIMIZING WASTES

Another fundamental difference between the four ecosystems under consideration is their different capacities to use the energy of residues and avoid accumulation of waste. Let us define "residue" as any by-product of a given process, and "waste" as a residue which cannot be used as a source of energy or primary material in the system in question.

Subsistence agriculture (Ecosystem A, figure 30.1) produces a very limited quantity of waste. This is the logical consequence of the ecosystem's minimal energy flow, and its ability to reuse most residue. The surrounding environment is able to accommodate the small quantities of waste.

The forest ecosystem (Ecosystem C, figure 30.3) is based on the same principles—every residue is primary material for another living species. Recycling is carried on with maximum efficiency and no wastes are produced.

On the other hand, the industrial monoculture (Ecosystem B, figure 30.2) is characterized by the concentration of production in places away from areas of consumer concentration. Moreover, only certain products are viewed as desirable, so that all residues inevitably become wastes. By raising livestock separately from cropland, the system creates higher energy costs because feed must be brought in and waste must be disposed of. The disposal of animal excrement becomes an inefficient and costly use of energy.

Hence, the production of residues which become non-recyclable because of factors of scale, integration, and failure to include "trophic levels," may be regarded as a fundamental parameter in the evaluation of an ecosystem.

Residues produced but not reused are destroyed through incineration or dumping; the environmental problems that this creates may be ignored but are still an unnecessary increase in entropy that is without a negative entropy counterpart. Any residue has an information content which in natural ecosystems is at least partly passed up to the next trophic level. The destruction of a uselessly produced or unrecycled residue involves a loss of information without any advantage for the ecosystem.

Maximum recycling is envisaged within the I.S. (Ecosystem D, figure 30.4). If the dimensions and spatio-temporal integration of the various processes of production and processing are well planned, the quantity of wastes is minimal and so is the use of fossil energy sources. This, therefore, constitutes a step toward greater productive maturity, similar to that seen in natural ecosystems.

BIOENERGY: A FUNDAMENTAL PHASE

The production of bioenergy is a fundamental phase in the transition from industrial agriculture to the I.S. The monoculture model does not recognize the energy or chemical value of farm residues and therefore turns them into polluting wastes. Within an I.S., residues find their place as substrates for the production of ethanol and biogas, contributing positively to the energy balance. After the production of ethanol, it is possible to recover high protein feedstocks, and after the production of biogas there are residues with high fertilizer value.

Thus, biomass processing plants, on an appropriate scale, are the core of the I.S.; they introduce new "trophic levels" and new feedback for the maximum exploitation of solar energy. Here, scale is key to reducing energy costs within the system.

CONCLUSION

Industrial agriculture has not responded adequately to the needs of world population and has created many problems of overall stability. The way of life based on concentrated production, and the agricultural processes produce large quantities of polluting wastes.

When these wastes are disposed of, precious resources are lost. A changeover to new models of production based on minimizing waste and increasing complexity-stability of the ecosystem as a whole, must therefore not be delayed.

We propose an agro-industrial integration that allows for the simultaneous production of food and energy while making maximum use of solar energy. This approach is ready to be implemented and offers at least a partial solution to waste and energy problems.

We are not saying that this proposal can meet the food and energy requirements of the world if the population continues to grow at the present rate. However, the model of the I.S. is, on the whole, an appropriate "technology-methodology" which can be adapted to individual territorial conditions. This will make it possible to avoid the complete destruction of precious environmental resources during the transition to an inevitable stabilization, or hopefully a decrease, in world population.

REFERENCES

Anderson, R. E. 1979. *Biological Paths to Self-reliance. A Guide to Biological Solar Energy Conversion*. New York: Litton Educational Publishing/Van Nostrand Reinhold.

Caporali, F. and P. Rossini 1988. *Una Valutazione del Costo Energetico delle Eccedenze Agricole della Comunità Europea*. Rome: Renagri.

CBNS (Center for the Biology of Natural Systems, Washington University, U.S.A.). 1980. The Technical Potential for Alcohol Fuels from Biomass. In *Farm and Forest Produced Alcohol. The Key to Liquid Fuel Independence*. Paper submitted to the Subcommittee on Energy of the Joint Economic Committee, Congress of the United States, August 22, 1980, Washington, D.C.: U.S. Government Printing Office.

CNEL (Consiglio Nazionale dell'Economia e del Lavoro Italia). 1982. *Modello Energometrico del Sistema Agro-silvo-alimentare*. Rome: CNEL Printing Office.

Giolitti, A., S. Ulgiati and G. Barbera. 1989. Agricultural Productivity, Biomass Energy and Land Use: A New Approach for Evaluation of Energy Balances. In G. Grassi, G. Cosse and G. dos Santos, eds., *Proceedings of 5th European Community Conference on "Biomass for Energy and Industry," Lisbon, 1989*. 1_1503-1506. London: Elsevier Applied Science.

Gosz, J. R., R. T. Holmes, G. E. Likens, and F. H. Borman. 1978. The Flow of Energy in a Forest Ecosystem. *Scientific American* 238(3):93-102.

Leach, G. 1976, *Energy and Food Production*. Guilford, Surrey: IPC Science and Technology.

Orsi, M. and P. Sarfatti. 1983. La Monosuccessione nella Maisicoltura: Risultati in Toscana. *L'Informatore Agrario*, 39(9):24759-24769.

Pimentel, D., L. E. Hurd, A. C. Bellotti, M. J. Foster, I. N. Oka, O. D. Sholes, and R. J. Whitman 1973. Food Production and Energy Crisis. *Science* 182:443-449.

Triolo, L. 1988. *Agricoltura, Energia, Ambiente*. Rome: Editori Riuniti.

Woodwell, G. M., R. H. Whittaker, W. A. Reiners, G. E. Likens, C. C. Delwiche, and D. B. Botkin. 1978. The Biota and the World Carbon Budget. *Science* 199:141-146.

31

GOVERNMENT POLICY AND ECOLOGICAL CONCERNS: SOME LESSONS FROM THE BRAZILIAN EXPERIENCE

Clóvis   Cavalcanti
Institute for Social Research
Joaquim Nabuco Foundation
Recife, PE 52071 Brazil

ABSTRACT

In this paper, I offer some empirical background that I hope will be used in discussions about how to attain the kind of genuine human progress which interests the population of not only Brazil but the entire planet. This paper discusses Brazil's formation and Amazonia's recent development to show the kinds of government policy that were conceived and implemented, and their corresponding ecological implications. No exhaustive list of the characteristics of the policies adopted is given, but actual situations and the relevant environmental issues are discussed. The paper's overall conclusion stresses the unsatisfactory way—to say the least— that ecological problems have been dealt with in Brazil since colonial times. Some explanations for that behavior are suggested.

INTRODUCTION

In a country like Brazil, which struggles to overcome the barriers of economic back-wardness and to build a modern society, the idea of linking ecological concerns to government policies forces one to consider a vast array of problems and situations. Actually, the Brazilian government's initiatives, over the course of almost 500 years, constitute a succession of actions which have always had huge environmental implications. The present picture contains examples of all the evils which afflict ecology all over the world. That picture, in the Brazilian case, is aggravated by the "pollution" of misery, and by the immense inequality in the country. The literature about environmental questions and the international press frequently offer dramatic information about the relationship between Brazil and its ecology. This has been especially true in the last few years, and has been brought on by the use of slash-and-burn techniques in the Amazon region to create pastures from forest. This was done in the framework of large agricultural and cattle-raising projects that were encouraged by government incentives.

It does not make any sense to refer here to all that is happening, or has happened in Brazil with respect to the ecological balance and its disruption as a consequence of government policies. It seems more interesting to study some aspects of the Brazilian panorama in an attempt to find an empirical background for discussions about what should be done to attain genuine human progress. Obviously the choice of the paradigmatic aspects—or cases—to illustrate the real Brazilian environmental situation can be made on the basis of various criteria.

I will refer here only to two basic "situations," that is, to two moments in Brazilian history which permit some observations and teach some lessons about the links between government policies and ecological concerns in Brazil. The two situations are a) the historical formation of Brazil; and b) the recent development of the Amazon region.

In table 31.1, I have considered the two reference situations and introduced two others. For each reference situation, I have briefly described the elements of government policy which characterize each situation and the corresponding environmental concerns. My goal is to give a model of reality, without giving an exhaustive list of the policy elements implemented or all their implications for the environment and society. My intention is also to offer a picture that includes the vision not only of the economist, but also of the student of ecology, of political science and of sociology. However, table 31.1 is only one of many possible representations of the Brazilian situation. In Brazil two-thirds of the working population make less than 120 dollars per month (cf. Jaguaribe et al. 1989, p. 17). In such a context, Lutzenberger's assertion (see Daly 1982, Table 1, p. 415) that "ecological concern and social justice are the two faces of a coin" makes sense. In effect, if one accepts the very reasonable premise that continuous growth cannot be ecologically viable, that is, that perpetual growth is something unsustainable, then to ignore or to postpone the task of redistributing income and wealth magnifies social injustice. At the root of environmental concerns in Brazil, one finds the dilemma of distribution problems that create poverty and misery.

The first situation described in table 31.1 refers to the historical formation of the country and underlines the question of slavery. The second situation refers to a vision which was created in the beginning of the 1970s, the so-called "Brasil Grande Potência" ("Great Brazil"). It focused on eliminating the "misery pollution," which was then considered more serious than environmental pollution. The third situation in the table, the notion of "development of the Amazon region" is based on the idea that, if the GDP rises, the income level of the population also rises. The situation classified as "population" in table 31.1 takes into account the social problems of the urbanization model that was pursued and the process of "slumization" (*favelização*) that resulted.

The ecological criterion which I try to adopt in interpreting Brazilian problems, is similar to that which the sociologist Gilberto Freyre proposed in 1937, when he wrote his important essay *Northeast: Aspects of the Influence of Sugar-Cane on the Life and*

TABLE 31.1 Brazilian Government Policies and Ecological Concerns in Four Situations

---

Reference Situations
1. Historical formation of Brazil (colonial period)
2. Great Brazil Project (starting in the early seventies)
3. Development of the Amazon region (from the early seventies on)
4. Population (after World War II)

Government policies regarding the reference situations:
1. Forest exploitation, monoculture, slavery, annexation of Indians' land, arrival of settlers without ties to the new land, hunting of Indians
2. Integration of the Amazon, industrial and infrastructure megaprojects, fiscal and credit incentives, attraction of polluting industries to increase the fiscal pie (GDP) for later redistribution (fight against poverty), alcohol-fuel program
3. Integration of the Amazon, construction of hydroelectric plants, creation of infrastructure, mining and agricultural big projects, settlement, Indian zoning, Great Carajás Project, blind forest exploitation
4. Approval of population increase (geopolitical reasons), urban development attracting rural populations

Ecological concerns regarding the proposed situations:
1. Destruction of nature, use of axe and fire to clear forest, disbandment of Indian groups (hunting of Indians), lack of reverence for the environment, greed
2. Rejection of the conclusions of the Stockholm conference (1972), creating pastureland from forests, limited use of conservationist practices, wasteful use of resources, abusive employment of agrotoxics, CFC production, neglect of energy crisis, use of excessively polluting autos
3. Destruction of ecosystems, killing of Indians, destruction of biodiversity, wanton use of fire, soil depletion and desertification, removal of native populations, water pollution, generalized violence, lack of observance of environment protection laws, disrespect for indigenous cultural values and empirical knowledge
4. Slumization (*favelização*), deterioration of urban living, insufficient structures for public services, urban violence, lack of job opportunities.

Landscape of Northeast Brazil, one of the first works written in the social sciences in the Portuguese language referring to ecology. Freyre says:

In this case, one should understand the ecological criterion as a wide general criterion, not only scientific, but also philosophical and even aesthetic and poetic, of study and interpretation of a region; and not a rigid geometrical ecologism of an ecological or geographical sect, self-assured of being able to reduce problems of culture and human facts to facts of physics and natural history or to problems of geometry. (1985, p. xx)

The idea is to adopt a notion of social ecology understood in its wider sense--a sense which tries to extract, from the whole of interrelations and natural and cultural processes, the environmental reality which takes methods of economic exploitation into account, as well as problems like cleaning the environment and maintaining the beauty of the landscape.

THE ENVIRONMENTAL DRAMA IN THE HISTORICAL FORMATION OF BRAZIL

The country, discovered in 1500 (April 22) by the Portuguese Pedro Alvares Cabral, was initially called Land of the Holy Cross (Terra de Santa Cruz). But the original Portuguese placename did not last long. Soon it was discovered that the new land was rich in one tree whose timber had a strong red color and could be used as a dye and for cabinetwork. The color reminded one of a live coal (*brasa*, in Portuguese), hence the name *pau-brasil*, or brazilwood, in English. In the sixteenth century, the name "Brasil" prevailed, indicating a strong relationship between nature (the tree) and the way the Portuguese identified the new colony. This choice of denomination has a symbolic force (see Pádua 1987, p. 18), because brazilwood, more than a simple kind of plant, was a resource to be exploited within the framework of mercantilism then prevailing in Europe. In fact, the first Brazilian economic cycle is the brazilwood cycle. The resource, a gift of nature, was exploited to the greatest possible extent without any feeling for conservation. As José Augusto de Pádua says, "Brazil was an immense 'brazilwood.'" (1987, p. 18). The economic activity that developed along the first decades of the sixteenth century consisted of a rudimentary exploitation "which did not leave other noteworthy marks besides *the pitiless large scale destruction* of the native forests" (p. 19; italics mine). The result was the rapid end of the brazilwood cycle. The extraction of wood soon lost its economic interest (there was no replanting) and, what is worse, in a few decades the country was deprived of the best part of its coastal rain forests. The beginnings of Brazil's economic formation thus assumed the characteristics of "a project of predatory exploitation of nature" (p. 19), whose very foundation gave the country its name.

One century later, during another economic cycle—the sugar cane cycle—the first national historian, Frei Vicente do Salvador, commented, with respect to the predatory character of the civilization which was being installed in Brazil, that the problem was inherent to the colonizers, who, "in spite of any deep involvement they have with the land, and no matter how rich they are, they intend to take everything to Portugal . . . and this is not shown only by those who came from there, it is also shown by those who were born here, not as masters but as users, just to utilize [land] and leave it destroyed" (quoted by Pádua 1987, p. 18). The saga of the brazilwood was not the only native wood from Brazil to be taken to Portugal. According to Gilberto Freyre: "There is barely a noble building in Portugal which does not have a piece of virgin forest from Brazil, resisting with the hardness of iron the decadence which corrodes the old Portuguese civilization" (1985, p. 54).

The fact is that Portugal abusively took woods from Brazil—"fat wood and heavy wood which made the others look repulsive" (Freyre 1985, p. 53)—to build or repair convents, churches and palaces, and to build boats and ships. In a letter dated December 6, 1775, the Marquis of Pombal, who was a Portuguese minister in charge of the government at that time, demanded that only the best brazilwood should be sent to Portugal—not meager, "insignificant" or "bastard" logs.

The portentous exploitation of sugarcane in Brazil from the second half of the sixteenth century to the middle of the next is a noteworthy accomplishment of the Portuguese domination. Sugarcane production furthered the model of predatory economic activity in the colony by introducing black slave work which made the production of cane sugar possible under competitive conditions. Brazilian development, at that time, was a drama of a monoculture based on latifundia and slavery. Despising river waters, the rich monoculturalist in the Northeast of Brazil transformed them into "lavatories," as Freyre said (1985, p. 35), thus killing fish and extinguishing a source of food for poor people. This situation persists today in the Northeast's sugar zones. The colonial exploitive mentality was endorsed by the absence of administrative procedures that would have corrected or prohibited the ruinous monoculture practices. As a result, the white Portuguese settler sought to bend the environment, which he saw as an enemy to be destroyed. It is interesting to note that, much to the contrary, the black African settler-slave sought to adapt himself to the forest or to adapt the forest to his needs, whenever he escaped from slave work to form the so-called *quilombos* (villages of fugitive slaves) (Freyre 1985, p. 55). One of these, Palmares, lasted for a hundred years, from the seventeenth to the eighteenth century.

Brazil's historical record is a succession of cycles analogous to the two economic cycles described above. The European population that went to Brazil found a natural environment completely diverse from that of its own, and behaved in an alienated fashion, treating nature and the Amerindians brutally. It is interesting to mention that the so-called Philipian Regulations, decreed in Portugal at the end of in the sixteenth century, set punishment for the crime of cutting trees in the metropolis as banishment to Brazil (Castro 1977). In general, the colonial model adopted in Brazil divorced the white man from an attitude of respect or reverence for the environment that included a lack of respect for the Brazilian Indians. In the conquest of Brazil's hinterlands, unofficial expeditions of white settlers departing mostly from São Paulo in the sixteenth and seventeenth centuries, were aimed at capturing Indians to make them slaves. The so-called *bandeiras* (literally, "flags") hunted the indians unmercifully, and consequently caused the death of thousands of native people. It is easy to understand why Brazil's Indian population, which amounted to 8 million people at the time of its discovery, fell drastically during the first stages of colonization, reaching the current level of about 250 thousand people (Posey 1987, p. 12).

The perverse relationship between the colonial initiatives and, after independence, between the authorities from the new Brazilian Empire and the environment, is a datum that helps to understand the anti-ecologist culture that took shape in the country. This phenomenon was denounced by one of the founding fathers of Brazil, José Bonifácio, who saw in slavery the reason that, in his words, "the order of human vicissitudes" (quoted by Pádua 1987, p. 35) would be reversed outright in the country (i.e., luxury and corruption arising in Brazil before civilization and industry). Joaquim Nabuco, a great Brazilian abolitionist, politician and writer, also wrote, more than half a century later in 1883, how the destructive trajectory of the "work of slavery" produced through environmental damage, degradation of man himself. To Nabuco (1977, p. 155), "the environmental diagnosis of Brazil was very disturbing." Euclides da Cunha, author of the classical book *Rebellion in the Backlands (Os Sertões* in its original version), proposed similar conclusions at the end of the nineteenth century. To him, the various cycles of economic activity in Brazil were always an aggression to the land, to nature. In his view, the country was moving "to the future sacrificing the future, as if we were approaching deluge." (Cunha 1966, p. 44).

THE ENVIRONMENTAL TRAGEDY IN THE DEVELOPMENT OF THE AMAZON

Up to the mid-1960s, the Amazon region was a marginal area of the Brazilian economic landscape, in spite of the natural rubber boom that took place there from 1870-1914. Starting in 1966, with the creation of the Superintendency for the Development of the Amazon (SUDAM), and in 1970 with the National Integration Program, which included such projects as the Transamazon road, a set of initiatives aimed at transforming the observed panorama, took shape. The idea was "to integrate the vast Amazon region . . . in the national life through self-sustained development" (Ministério do Interior, SUDAM, 1967, p. 3). Such development effort coincided with the years of military rule that, during two decades of authoritarian government, imposed an ideology of accelerated growth on the country while ignoring its environmental costs. In 1972, at the United Nations Conference on the Environment, in Stockholm, the Brazilian delegation voiced apprehensions that the ecological costs of development were simply a veil to camouflage imperialist interests in blocking the progress of the Less Developed Countries (LDCs). Simultaneously, the Brazilian government was saying that industrial pollution was welcomed, since it would destroy a much less desired form of pollution—misery.

It is within that context that a policy of fiscal and credit incentives, with the goal of attracting private projects for the development of primary, manufacturing and mining activities in the Amazon, was formulated. Coincidentally, in 1966, enormous ore deposits were discovered in the Amazonian Carajás mountain range, in a virgin area inhabited only by Indian tribes. That discovery aroused plans for using the deposits on a massive scale. The plans began to be implemented a decade later through the so-called Great Carajás Program. It is interesting to see that already at the beginning of SUDAM's development plans the new human settlements established along the Transamazon road were not successful (La Penha 1988, p. 7).

The Amazon region has an area of 5.14 million square kilometers (about 1,270 million acres), corresponding to 60.4% of the Brazilian territory. (The Amazon's 1990 population of 8.9 million people, represents only 5.9% of Brazil's population.) 54.5 percent of the Amazon surface is dense forest; 20.4% is open forest; another 20.4% is made up of savanna-like vegetation; finally, 4.7% is marsh. Two-thirds of the world's forest reserves are found in the Brazilian Amazon. Such an environment is therefore a universe in itself, with peculiarities, a history, and an ethnology that defy understanding.

In 1987, information satellites monitoring natural resources recorded the most intense and widespread fires man has ever caused (Pinto 1990). Twenty million hectares (49 million acres), or around 2.5% of Brazil's area was burned (La Penha 1988, p. 2). This, obviously, aroused both national and international opinion, and prompted strong reactions from the ecological movement. The fires also underscored the ecological consequences of the Amazon development plans. The development of the Amazon is directed by the government in the form of of both public and private megaprojects that almost entirely neglect sound environmental concerns. In the Amazonian area where most projects are being undertaken, the process of land tenure structuring has been denounced as "simply unusual." It is based on a) concessions made by public authorities; b) simple de facto empowerment; and c) the immoderate practice of land conquest by people who manage to show spurious ownership rights (*grileiros*) (Lacerda de Melo 1989, p. 6).

On the other hand, the absence of an adequate technical basis for regional development, capable of transforming the unproductive forest (in a climax stage) into a productive forest, has contributed to the rapid predatory devastation of the Amazon ecosystems. This in turn causes soil desertification through compaction, destruction of the nutrient cycle, and loss of land productivity. When land can no longer be cultivated, it is abandoned, thus giving rise to the phenomenon of agricultural itinerancy. This is usually very destructive to forests, fauna and water resources, and soil (Lacerda de Melo 1989, p. 5; La Penha 1988, p. 20; Posey 1987, p. 12). Improper land use can be linked to lack of knowledge about Amazonian ecology and to miscalculations about the forest's productive potential.

Nevertheless, a millennial, meticulous and precise knowledge that fully respects life and adapts to tropical ecosystems exists within Brazilian Indian culture. The Amazonian development process which destabilized and reduced the indigenous populations, goes against the knowledge that was patiently accumulated over the ages. The Indians survived for centuries in a state of equilibrium in the Amazon that began long before the arrival of the Portuguese. "Their knowledge of ecosystems, of plant-man-animal relationships, and the manipulation of natural resources have evolved through uncounted generations, the product of trial and error from accumulated experience. It is incredibly small the knowledge our scientists have about the indigenous perception of ecology and utilization of natural resources." (Posey 1987, p. 13). Amerindian cultures offer a rich and untapped source of information on the natural resources of the Amazon. Transformation of this knowledge into modern technological knowhow can open new perspectives for an ecologically sound development of the region.

Many a researcher in the Amazon has learned the importance of indigenous knowledge about soil, microclimatic variations, botany, in a word, of ethnoecology. The knowledge is remarkable in the fields of ethnomedicine and ethnopharmacology. For instance, the Mebêngôkre classify over fifty types of diarrhea/dysentery, each of which is treated with specific medicines (Posey 1987, p. 24).

It is worrisome that almost all projects conceived for the development of the Amazonian economy have required very substantial modifications in local ways of life, to the point that it is said that "perhaps the greatest tragedy in Amazonia is the human one" (Posey 1987, p. 12). Big projects, like the Tucuruí hydroelectric plant, have forced the

Indians to abandon the land where they have always lived. In the case of the Tucuruí dam, the Kayapó Indians were removed (in 1984) from the territory that were legally theirs. The situation became even worse because it took seven years for the Eletronorte company—a state enterprise that undertook the project—to acknowledge the area as Indian land. In the meantime, the company worked without any constraints on their activities in the river basin. When the company finally recognized the Indians' rights, it paid them an indemnification. Simultaneously, the Kayapó group was transferred to a new reservation, the Mãe-Maria, which was already inhabited by another Indian group (Gavião). As if that were not enough, the new reservation was cut apart by a 150m wide and 19km long corridor for the deployment of a transmission line, without consulting or warning the Indians. When the Indians accidentally found out that the transmission line was being built, they attempted to change the project's blueprint, but all that they obtained was a new compensation payment. The result is that, at present, the two Indian groups are confined in the same reduced territory, surrounded by farms and suffering from the deforestation process being conducted in the region. They are also affected by the proximity of the Carajás railroad and a number of pig iron mills that utilize vegetable charcoal (Costa 1989).

The payment of indemnifications has been the trademark of the state's Indian policy. The policy is oriented toward giving precedence to the interests of capital over the rights of indigenous peoples. However, in the loss of territory by the Indians, there are political as well as symbolic aspects that must be considered. If the land is a supporting element of the Indians' sociocultural identity, then its value cannot be translated in accounting terms: "The concept of land as a merchandise belongs properly to our society. It does not have a correspondent valuation among [the Indians]" (Costa 1989, p. 75).

From the actual government point of view, Indian territories are to be reduced. In the delimitation, for instance, of the reservation of the Ianomãmi Indians, the area "given" to them (only 30% of their original territory) was fragmented into 19 discontinuous portions (Costa 1989, p. 76). In this process, the reasons why the Indians need a territory (for transit and wanderings) were ignored. The forest peoples are thus compressed into less and less space, while their land goes to ranchers, mining entrepeneurs and big companies.

Violence against so-called "primitive" peoples is added to violence against the natural environment. The Indians, without the tricks and arms of the civilized, cannot properly protect their ancestral holdings. The overall situation certainly is one of much violence. And violence has characterized relations with the Indians since Brazil's discovery. Brazil's situation is the same as the entire Latin American continent, where, in the first century of Iberian occupation, the native population declined from 80 million to 10 million people (according to Ansaldi 1989, p. 25; and Crosby 1986). This resulted in individuals like the Spanish Dominican Bartolomé de Las Casas (1474-1566) raising their voices in protest against the genocide that the white conqueror was practicing.

Another example of inadequate resource management in Amazonia is the hydroelectric dam of Tucuruí. At Tucuruí, a vast area had to be flooded for construction of the reservoir. Due to management problems in the organization charged with clearing the forest at the dam site, all the trees could not be cut before the area was filled. In fact, a substantial portion of the vegetation was standing when the waters covered the area. As a result, the organization went bankrupt, murders occurred in Rio de Janeiro, and the underwater power saw was invented. For a time, thousands of trees that remained in the bottom of the huge artificial lake (Lomba 1989) were cut. It is ironic that this happens when science announces that, among the ecological consequences of man's impact on the rain forest, the destruction of the genetic stock is certain to be the impact that outlasts all others, erasing millions of years of evolution. It is known that the actual biological inventory of the Amazon is far from being completely categorized, and that, at the same time, the Amazonian forest is an environmental monument to biodiversity (La Penha 1988, p. 30). To accept the permanent loss of biodiversity because of ill-conceived actions taken under the vague supposition that it is the price we have to pay for the costs of development, is something that the civilized conscience cannot tolerate.

However, we still have a long way to go to undo serious misconceptions concerning the Amazonian question. Among economists the notion of the desirability of development regardless of its ecological costs is still widespread, if the opinion of a distinguished member of the economics profession from Belém do Pará can serve as an example (Monteiro da Costa 1989, pp. 9-10):

"When intending without dissimulation to condemn the Amazon to remain unremittingly tied to backwardness, today's preservation international movement—or 'Amazonoia'—can block the region's industrial growth, against national and local interests, and relying on a basis characterized by emotionalism, faddism, personal promotion, international blackmail, minorities' gigoloism, and similar attitudes . . . environmental preservation is a means to secure human survival. It is not an end in itself; the Brazilian economy cannot dispense with the natural resource basis available in the Amazon; all historical evidence, here and elsewhere, reveals that the economic occupation/integration of a peripheral empty region, under the logic of industrial development, necessarily implies profound alterations in physico-environmental and socio-spatial conditions, as well as the annihilation of the old order."

It is disconcerting to accept reasoning like this in light of what is taking shape in the Amazon today. For example, the proposed pig iron production project, which falls within the framework of the Great Carajás Program, carries with it serious environmental implications. That project proposes to extract charcoal from the forest along the Carajás railroad using indiscriminate deforestation. The project would use the equivalent of 10.4 million tons of wood. That volume, measured in 420 kg dry logs, would form a 2,841 meter tall building (620 floors!) having a 100m x 100m base. In a normal-type forest, one would require 74,000 hectares per year (183,000 acres) to satisfy this demand (Fearnside 1988, p. 20). Alternatively, by adopting another plan to plant eucalyptus to produce charcoal, one would get a planted forest area "of unseen dimensions in the tropics . . . Plantations of such a magnitude [700,000 hectares] would become subject to substantial threats of diseases, insects, and soil degradation, as is the case with the Jari [Project]" (Fearnside, 1988, p. 19). This is a good illustration of the disaster that has been materializing under the aegis of government policies for Amazonian development goals; the Great Carajás Program is a pedagogic example (Fearnside 1988, p. 19).

CONCLUDING REMARKS: REASON FOR OPTIMISM?

What I attempted to do in this paper was not give an inventory of either the record of government policies aimed at development or the corresponding ecological concerns in Brazil. Instead, I wanted to illustrate how economic advancement has been pursued and how those plans look from an ecological perspective. The conclusion one can draw from the cases I have mentioned is that the rate and type of economic growth that has taken place at the expense of the environment is distressing. One could say that I chose only the worst development projects to strengthen my argument. However, the very extent of the aggression that has been committed in Brazil in the past, against both the environment and the native population, is enough to outweigh the successes that came from development policies. In any case, if one looks at Brazil's present situation, the ecological diagnosis is rather bleak (Viola 1987, pp. 82-83); it can be called a true eco-economic crisis. Predatory exploitation of nature which was considered "the price of backwardness" became "the price of progress" (Pádua 1987, p. 61). And in this conflict, the Brazilian *elites* —be they economic, political or intellectual—have generally tended to dismiss ecological concerns as exaggerations. The conclusion that the Brazilian version of capitalism can be called savage is not new. However, the so-called "price of progress"—the destruction of nature in Brazil—cannot be easily dispensed with, due to the grave social injustices found in the country. The elimination of these injustices remains a central topic in government rhetoric, superseding other priorities.

Nonetheless, Brazil has not been able, after four decades of development efforts, to suppress the tremendous degree of inequality that exists in the country. Thus, economic growth has taken place simultaneously with environmental degradation and unchanged poverty. Actually, as some people have already suggested (Lutzenberger, in Daly 1982, p. 415), Brazil has been constructing a First World for 10% of its population on the backs of the remaining 90%.

In spite of that situation, the prognosis should not be desperate. For one thing, population growth in Brazil has been slowing. At the present rate of increase, it will take over 70 years for the number of inhabitants to grow the same percentage that it did from 1940 to 1990. This decline is not much, but there are sound expectations that the rate of population increase will decline even more in the next decades. Another sign of hope is a new attitude that seems to be taking hold in official circles since 1988. In October of that year, then President José Sarney created a program which was called "Our Nature" (*"Nossa Natureza"* in Portuguese), aimed at the rational utilization of the Amazon. Also, President Fernando Collor, who was inaugurated on March 15, 1990, stressed in his inaugural speech his adherence to preservationist views, saying that he is part of a generation that considers the primal importance of the environment. A few days before, he had caused a great surprise by naming the well-known and controversial Brazilian ecologist José Lutzenberger to the post of Secretary of the Environment.

Lutzenberger has been an outspoken critic of the Amazonian policy of development. He has defended Indians' rights and opposes vigorously the concepts underlying the policy that leads to megaprojects. His perspective favors the idea that no system that depends on continuous growth can be ecologically viable (Lutzenberger 1984, pp. 7-9). President Collor knew all that when he made his choice of Lutzenberger. So it does not seem that the choice was intended simply to placate the international ecological movement—the so-called "Amazonoia"—where the nominee is a respected figure. In effect, it is fair to stress that President Collor's first foray out of Brasilia after his inauguration was to the Ianomãmi reservation in the northernmost state of Roraima. He went there with Lutzenberger, and seized the opportunity to proclaim that the Indians' land should be effectively protected. He also determined an ecological zoning of the Amazon, something that had been insisted on by scientists and ecologists (see Mendes 1989) as a prerequisite of Amazonian policy. Lutzenberger has also spoken against the use of mercury in gold mining in the region, pointing out its polluting character and potential of destruction of fish species. The importance of Collor's measures can be also reckoned from the fact that he suspended the scheme of incentives that promoted agricultural, industrial and mining activities in the Amazon in the past two decades. Of course, this is not enough to say that Brazil is approaching the adoption of policies pointing to the ideal of sustainable development. However, there seems to be getting momentum there a new attitude toward the environment—something that can be also observed in Brazil's new constitution which was written in 1988.

REFERENCES

Ansaldi, Waldo. 1989. La Nostalgia de la Beata por la Virginidad Perdida. a Propósito del Quinto Centenario de un (des)encuentro. *David y Goliath*, 18(54):24-28.

Castro, Cláudio de M. 1977. Ecologia: a Redescoberta da Pólvora. Unpublished. Programa ECIEL, Rio de Janeiro.

Costa, Vera Rita. 1989. Xingu: Hidrelétricas Coroam Quatro Séculos de Agressões. (Interview with anthropologists Leinad Santos and Lucia Andrade). *Ciência Hoje* 59: 74-75.

Crosby, A. 1986. *Ecological Imperialism: The Biological Expansion of Europe 900-1900.* Cambridge Cambridge University Press.

Cunha, Euclides da. 1966. *Obra Completa*, vol. 1. Rio de Janeiro: José Aguilar.

Daly, H. E. 1982. Interview with José Lutzenberger: Progress as if People Mattered, pp. 410-415. San Francisco: Friends of the Earth.

Fearnside, Ph. M. 1988. O Carvão de Carajás. *Ciência Hoje* 48: 17-21.

Freyre, Gilberto. 1985. *Nordeste: Aspectos da Influêcia da Cana sobre a Vida e a Paisagem do Nordeste do Brasil.* 5th ed. Rio de Janeiro; José Olympio, FUNDARPE, Recife. (1st ed., 1937.)

Jaguaribe, H. et. al. 1989. *Brasil: Reforma ou Caos*, Rio de Janeiro: Paz e Terra.

Lacerda de Melo, M. 1989. Notas sobre o Problema da Degradação Ambiental na Amazônia (comments for the Seminar on Tropicology). Joaquim Nabuco Foundation, Recife.

La Penha, Guilherme de, coordinator. 1988. Programa Nossa Natureza. Relatório Final do Grupo V-"Pesquisa". Presidência da República, Belém.

Lomba, Pedro Paulo. 1989. Dona Maria, a Louca, Na Floresta Amazônica. *Jornal do Brasil* Rio de Janeiro, p. 11.

Lutzenberger, José A. 1984. Preface. In Herman E. Daly, ed., *A Economia do Século XXI*, translated by Renato Souza. Porto Alegre: Mercado Aberto.

Mendes, Armando. 1989. A Nossa Amazônia dos Outros. Lecture delivered to the 6th Regional Meeting on Tropicology. Joaquim Nabuco Foundation-Goeldi Museum, Belém.

Ministério do Interior, SUDAM. 1967. 1o. Plano Qüinqüenal de Desenvolvimento, 1967-1971, SUDAM, Serviço de Documentação e Divulgação.

Monteiro da Costa and J. Marcelino. 1989. Perspectivas de crescimento industrial: o caso da Amazônia. Lecture delivered to the seminar on Economic Perspectives: Industrialization in the Brazilian Regions. Unpublished. Joaquim Nabuco Foundation, Recife.

Nabuco, Joaquim. 1977. *O Abolicionismo,* Petrópolis: Vozes. The 1st ed. is 1883).

Pádua, José Augusto. 1987. Natureza e Projeto Nacional: as Origens da Ecologia Política no Brasil. In José Augusto de Pádua, ed., *Ecologia e Politica no Brasil,* pp. 11-62. Rio de Janeiro, Espaço e Tempo-IUPERJ.

Pinto, Lúcio Flávio. 1990. A Cor do Verde. *Jornal do Commercio,* p. 4. (Caderno Cidades.) Recife.

Posey, Darrel A. 1987. *Alternatives to Destruction: Science of the Mebêngôkre.* Belém: Goeldi Museum.

Viola, Eduardo. 1987. O Movimento Ecológico no Brasil (1974-1986): Do Ambientalismo à Ecopolítica. In José A. Pádua, ed., *Ecologia e Política no Brasil.* pp. 63-110. Rio de Janeiro, Espaço e Tempo-IUPERJ.

32

TROPICAL MOIST FOREST MANAGEMENT: THE URGENCY OF TRANSITION TO SUSTAINABILITY

R. J. A. Goodland, E. O. A. Asibey, J. C. Post and M. B. Dyson*
The World Bank
The Environment Department
Washington, D.C. 20433 USA

ABSTRACT

An increasing number of citizens and foresters, the International Tropical Timber Organization (ITTO), and the vast majority of environmentalists realize that most current use of tropical moist forest is unsustainable. The environmental services of tropical forest and the rich biodiversity it harbors is being lost so fast that consumer boycotts and other trade constraints aim to reduce the rate of irreversible damage. On one hand, tropical moist deforestation benefits exceedingly few people, and only ephemerally. On the other hand, such deforestation permanently impoverishes, uproots or sickens millions of people, impairs local or global environmental services, and exacerbates global environmental risks. The World Resources Institute ranks commercial logging as one of the top causes leading to deforestation.

We present the case for an urgent transition to sustainability by first, improving forest management at least to "best practice"; second, by deflecting logging from primary to secondary forests; and third, a beginning a phased transition to plantations, especially rehabilitating degraded lands. The transition should be rapid in countries where forests are rapidly disappearing and where logging is a main cause (e.g., Ivory Coast, Nigeria, Ghana, Papua New Guinea). The transition is less urgent where tropical moist forests are extensive and stable.

INTRODUCTION

Today's irreversible and accelerating loss of biodiversity is our urgent chief reason for writing this paper. We are particularly concerned because the main global repository of

---

* The views presented here are those of the authors and should in no way be attributed to the World Bank.

biodiversity—tropical moist forests—is being consumed so unsustainably that boycotts and ecotage (environmental terrorism) (table 32.1) of use of tropical woods have been two of the more extreme public responses. Boycotts against unsustainable use dramatizes the vast difference of opinion within the world community concerning how such forests should be used. All possible uses of tropical moist forest may be ranked in order of environmental sustainability (table 32.2). We urge the most rapid possible transition to sustainability by concurrently improving forest management (tables 32.3 and 32.4) and creating and continuing to create more tree plantations.

Losses of tropical moist forest have many causes; these causes differ in type and importance in different regions. The underlying causes in all areas are perverse incentives such as greed and land tenure problems. Regarding more specific causes, although there is some disagreement as to precise ranking, many authors (e.g., Anderson 1989; Repetto 1990; Winterbottom 1990; Westoby 1987) agree that commercial logging and logging roads are among the most influential. In Amazonia, the chief causes are cattle ranching and road building inducing unplanned settlements; followed by slash-and-burn shifting agriculture. In parts of Africa, logging trails accelerate agricultural conversion and shifting cultivation. Myers (1989) contends that for every tree cut for timber in Zaire, 25 are cleared in building roads to get to it. Other forest loss is due to conversion for commercial agriculture for growing such products as rubber, oil palm, coffee, cacao, upland rice, or beans; flooding by hydroelectric reservoirs; and deforestation from commercial logging, particularly from settlement along logging roads.

In overpopulated parts of Central Africa and Southeast Asia, "the tropical timber industry is the most important cause of primary rainforest destruction." (Anderson 1989). Repetto (1990) ranks commercial logging as the top agent of devastation. The World Resources Institute ranks destructive commercial logging as the second direct cause of deforestation (Winterbottom 1990). FAO's Chief Forester (Westoby 1987) concluded: "Because exploitation has been uncontrolled, and management non-existent, marginal farmers, shifting cultivators and landless poor have followed in the wake of the loggers, completing the forest destruction."

FAO calculates that of tropical (i.e., dry- + wet-) forest wood extracted annually, 85% is used for fuel wood, 19% for local timber, and 5% as timber exports. We therefore clarify at the outset that this paper is not primarily on how to save the tropical forest, although that is certainly an aim (table 32.5). The paper also does not deal with the drier end of the tropical forest spectrum where practically all of the fuel wood comes from. This paper focuses on that relatively small part of the tropical moist forest from which hardwood is extracted. Although FAO calculates only 5% of tropical wood is exported, extraction of this 5% damages double or triple this amount of unextracted forest directly, and is responsible for even more secondary or induced damage from the logging roads, moreover encouraging settlement along them.

Globally, settlement along logging roads and peasant agriculture may be the main causes of tropical moist deforestation. Thus, imperfect though it undoubtedly is, commercial logging causes less direct damage in Amazonia, but more damage in the other two rain-forest areas of Africa and southeastern Asia. Hence the paradox of this paper:

The International Tropical Timber Organization (ITTO)[1] concludes that practically all tropical moist timber extraction is unsustainable. While this is unfortunately true, we assert that much could and should become more sustainable if "best practice" is followed (tables 32.3 and 32.4). We emphasize that controlled, selective logging is preferable to most exploitative alternatives. The paper discusses the conditions for sustainability and reviews successful systems.

We present the case for an urgent transition to sustainability of timber yield by means of three mutually supplementing actions. The first action needed is an immediate improvement in the management of tropical moist forests, at least to "best practice," for high quality hardwoods. The second is the deflection of commercial logging, particularly of commodity timber from primary forests to secondary forests. The third is a phased transition to tropical plantations (preferably on degraded lands) as a necessary, prudent and sustainable source of tropical timber. Today, much tropical timber is "commodity timber," which should and could be replaced by softwoods.

By commodity timber, we mean lower value construction wood, such as that used in poles and planks, rather than higher value, slower growing hardwoods that tend to be used widely for appearances (e.g., veneer) and durability. Temperate hardwoods and softwoods could alleviate pressures to cut tropical moist forests. We fully support FAO's independent review of the Tropical Forestry Action Plan[2] (Ullsten et al. 1990) in which they urge TFAP not to support the extension of logging that cannot be sustained (cf. Colchester and Lohmann 1990).

To the extent that successful forest management prevents more destructive exploitation, buys time for sustainability to be researched, and for tropical timber plantations to mature, or buys time for education of the benefits of intact forest by today's consumers, then controlled selective logging is much better than the alternatives. Although today's use of forest is unacceptable to many organizations and citizens (hence the boycotts), we feel it unnecessary to wait for total sustainability before forest management is improved.

---

[1] The International Tropical Timber Organization (ITTO) was created in 1985, after a decade of negotiations, by the signing by 41 (now 69) tropical timber consumer and producer countries of the International Tropical Timbers Agreement to regulate world tropical timber trade. ITTO was set up to administer the agreement. Three ITTO committees focus respectively on market intelligence, reforestation, and forest management and the development of the timber industry in developing countries. The main objective of ITTO is to support the tropical timber trade. ITTO is led by Dato Dr. B. C. Y. Freezailah, Executive Director, and is based in Yokohama. ITTO could play an enormously influential role in improving the sustainability of logging and of promoting the transition to plantations.

[2] The Tropical Forest Action Plan (TFAP) originated in FAO's Tropical Forests Action Programme, which in turn arose from FAO's International Year of the Forest (1985). The World Resources Institute together with FAO, UNDP and the World Bank launched the "Tropical Forests: A call for action" (3 vols) in 1985, followed by a major donors conference in Bellagio in 1987. FAO acted as secretariat fostering preparation of National Forestry Action Plans. About 10 NFAPs have been prepared so far, FAO had the first few years of operation reviewed by an independent panel (Ullsten et al. 1990), and the founding institutions are evaluating TFAP's achievements to date. TFAP and the various NFAPs could vastly help governments to promote sustainable logging practices, and to identify when, where and how large timber plantations should be.

TABLE 32.1 Tropical Timber Trade Restraints (Selected examples only)

1. West Germany, 1988: 200 city councils stopped using tropical timber.
2. West Germany, January 1989: The Minister for Building announced that the Government has stopped using tropical timber.
3. West Germany, 1989: The Timber Importers Federation introduced a code of conduct for timber importers.
4. Netherlands, February 1989: almost half local governments stopped using tropical timbers.
5. European Parliament,: July 1988: all member states to ban imports of Sarawak timber (later rejected by European Commission); quota system and compensation envisaged.
6. European Federation of Tropical Timber Trade Associations, February 1989: proposed a levy on tropical timber imports to the European Community.
7. United Kingdom, 1987: Friends of the Earth tropical timber campaign: cautious consumers encouraged to avoid tropical hardwoods.
8. United Kingdom, February 1990: Prince Charles calls for boycott of unsustainably produced tropical timbers.
9. United Kingdom, 1988: 200 retailers and timber users adopt policies of using only sustainably produced wood.
10. Japan, October 28, 1989: Osaka Royal Hotel: Former U.S. President Ronald Reagan raises possibility of boycott of Japanese products, mentioning tropical logging.
11. Hawaii, 1985: Ecoteurs firebombed a $250,000 wood-chipper which was chipping Kalapana-Kee-au Rainforest without a permit and in violation of a court order.
12. Hawaii, March 1990: Demonstration protesting the True Geothermal Energy Company's bulldozing 9000 acres of the 27,000 acre Wao Kele O Puna tract on Big Island's Kilauea volcano, the last remaining tropical rainforest in the United States.
13. ITTO, November 1989: Malaysia declined to adopt ITTO's proposal to label tropical logs as to the sustainability of their source.
14. Australia, April 1989: Federal Government considers banning importation of rainforest timbers.
15. India,1973: The Chipko women's tree hugging andolan (= movement), founded by Chandi Prasad Bhatt and Sunderlal Barhuguna; awarded UNEP's 1988 Global 500 Environmental Distinction award (Weber 1987).
16. Indonesia, 1990: Scott Paper Corporation Chairman Barry Kotek cancels clear-cutting of 850,000 ha of Irian Jaya forest for eucalyptus pulp because, he said, of NGO pressure.
17. Irian Jaya, February 1990: Indonesia's Environment Minister served two warning notices on PT Bintuni Utama (subsidiary of Marubeni Corp.) for illegally chipping 137,000 ha, mainly mangrove forest in Bintuni Bay, to supply 300,000 tons of chips annually.
18. Sarawak, February 1990: Mitsubishi subsidiary Daiya's 90,000 ha concession blocked by Iban and Penan peoples; many jailed.
19. United States, 1990: Senator Moynihan introduced S-822 legislation to ban the importation of teak from Myanmar (Burma).
20. Thailand, January 10, 1989: Prime Minister and Cabinet abolish all timber concessions and halts all logging nationwide.
21. ITTO, May 16-23, 1990: 8th meeting, Denpasar, Indonesia: agreement on the year 2000 target for all tropical timber in international trade to come from sustainably managed sources.

Wyatt-Smith (1987) recently concluded that "Modern timber exploitation methods and tropical moist forest conservation are no longer compatible and require separate consideration." Similarly, a study commissioned by and published for the ITTO found that tropical moist forests are managed sustainably only on one-fifth of one percent of their extent (Poore et al. 1990) in ITTO member countries. The study concludes that such forests are not managed sustainably worldwide (Anderson 1989; Webb 1989). Oldfield (1989) corroborates this by calculating that less than 0.2% of tropical forests are managed to produce a sustainable harvest of timber. The International Tropical Timber Organization (Poore et al. 1990) concludes that most, if not all, tropical forestry management is unsustainable, but that it could become sustainable under certain circumstances.

The circumstances that could lead to sustainability are the regeneration of the forest, and drastic revision of the economics of logging. If fine hardwood logging is done so that the environmental conditions of the forest are not significantly damaged, the yields will be lower and the extraction costs higher so that "commodity" logging would become less economic at current prices. This means that if sustainability is made a prerequisite for logging permits, tropical timber prices—currently so grossly undervalued—will have to rise substantially when full costs of production are internalized. In that case, plantations will become more attractive for wood production (of different species). The transition to sustainability that we advocate therefore means careful management of the forest for high quality hardwoods and a transition of logging from primary to secondary forests, with plantations preferably on degraded lands for the higher volume "commodity" timbers.

Action must be taken soon for the following reasons. In 1985, the World Bank estimated that of today's 33 tropical timber exporting countries, only 10 will have any timber left to export by the year 2000. Continuing "business as usual" would mean a massive loss of biodiversity and other irreversible, irreparable damage. It is already too late in many countries to reverse the loss of this major potentially renewable resource. In Thailand, Ivory Coast, Nigeria, Bangladesh, El Salvador, Peninsular Malaysia and Haiti for example, it is either too late or nearly so. Thailand, Indonesia, Philippines and Brazil have recently banned export of all raw logs, although it may be too late to restore renewability to seriously degraded former-forest areas. Ghana banned raw log export of primary species in 1979. These bans are partly to boost revenue by selling value-added, commercial commodities, and partly to conserve remaining forest.

SUSTAINABLE MEANS LOW YIELD

This section presents the case that any sustainable use of tropical moist forest perforce means low yield, hence low financial and commercial attraction (Janzen 1973). Sustainable timber yield means a larger area of forest would have to be logged to obtain the same amount of wood that can be extracted unsustainably from a smaller area. Sustainable use of tropical moist forest is here taken to mean that use of natural forest which indefinitely maintains the forest substantially unimpaired in the environmental services it provides, as well as in its biological quality. Clearly, any harvest must not exceed the regeneration rate of the resource nor impair the potential for similar harvest in the future.

TABLE 32.2 Environmental Sustainability Ranking of Use of Tropical Rain Forest

1. Intact Forest

1.1 Biological reserve; scientific repository; gene-pool, germ-plasm storehouse; phytochemical and ethnobotanical resources.

1.2 Environmental protection services; climatic buffer, watershed protection, protection of downstream activities.

1.3 Indigenous peoples and reservations based on natural, legal, and moral criteria.

1.4 Collecting, gathering, tapping, game- and fish-culling.

1.5 National park development; national and international tourism; recreation.

2. Utilization of Natural Forest

2.1 Dynamic sustained-yield management (Nigerian and Malaysian Shelterwood forestry).

2.2 Leaf protein, leaf chemicals, other chemicals.

2.3 Selective felling with careful removal.

2.4 Bole removal with slash, roots, stump, bark, and branches, left in situ, rather than whole-tree removal.

2.5 Enrichment planting, refining, liberation, reconstitution or directed regeneration.

2.6 Clear-cutting small tracts, leaving regeneration foci in strips or environs.

3. Tree Plantation

3.1 Mixed-species polyculture products (rubbers, oils, nuts, resins), over monoculture.

3.2 Mixed-species polyculture timber plus synergistic species and products; mixed-species timber; oligoculture timber.

3.3 Monoculture timber: veneer, dimension lumber, plywood, particle-board, timber, chips, fuel-wood, hogfuel.

4. Agri-Silviculture

4.1 Multiple-dimension forestry, "3-D" forestry of timber, products, synergists, browse, understory components, or graze.

4.2 Polycropping and intercropping, (rubber and synergists with understory and annuals).

4.3 Taungya: annuals and perennials planted simultaneously—becoming a tree plantation.

4.4 Treed pasture: wood and products plus synergists; browse and multispecies graze (legumes, forbs, grasses).

4.5 Subsistence rotational gardens, e.g., Mayan home garden, Kandy garden, chinampa, etc., of trees, perennials, and annuals with small livestock, fishponds, etc.

5. Agriculture

5.1 Long fallows, small areas, multivarieties of species, breed tolerance of pests and infertile soils, rotations.

5.2 Varzea management; naturally irrigated crops, Water Buffalo, capibara, turtles.

5.3 Perennial crops in preference to annuals; subsistence crops over export and cash crops.

5.4 Oligotrophic exports (hydrocarbons, carbohydrates) over eutrophic exports.

5.5 Multispecies pasture for mixed herbivores; small livestock and solitary stabled cattle.

5.6 Oligoculture pasture for monospecific herbivores (extensive ranching for cattle export): the worst option under prevailing low-management practices.

NOTE: This ranking runs in general from the top (preferred), gradually descending through all five overlapping categories, with the least desirable at the bottom. This is not an economic ranking. The actual mix of utilization adopted will be closely site-specific, given the enormous heterogeneity of the tropical rain forest. The most rational land use pattern is likely to be as diverse as the environment on which it depends (from Goodland 1980).

Sustainability has several components apart from wood production. These include: 1) the sustainability of forest dwelling peoples (such as Amerindians, Pygmies, and Penan), and other peoples depending on the forest; 2) the sustainability of extracted non-wood products (such as latex, oils, and nuts), 3) the sustainability of the environmental services provided by forest, 4) the sustainability of biodiversity.

The above important components are not dealt with here. This paper focuses rather on the sustainability of timber extraction. Demand management of wood (e.g., more efficient wood stoves, more efficient veneers, wafer board, particle board, and waste recycling) is mentioned only briefly in the recommendations section. Analyses are needed of how sustainability can also be achieved in these four components, but are not further discussed here. Most of these valuable forest uses can complement, rather than replace, wood production. It seems likely that if wood production can be sustainable, then many of these other values will also become sustainable. But it also is true that some selective logging, particularly where it involves heavy machinery, seriously degrades non-timber forest values. The idea of multiple use, for uses which are not mutually exclusive, merits more emphasis than it has received.

How can wood, the moist forest's major product, be harvested on a sustainable basis? First, we note that selective logging uses less area (at 4.5 million ha/yr) and is much less damaging than clear-cutting (at 7.5 million ha/yr) for cattle ranches, annual crops, etc. Therefore, selective logging damages smaller areas and its environmental effects are far less severe than those caused by clear-cutting. Highly selective extraction on long rotations with safeguards is the most sustainable commercial use so far devised for tropical forests. This may not be theoretically sustainable, but it is preferable to the alternatives. Unfortunately, it's use is not widespread. Considering that in present practices, yields of all wood products are so low after the first bonanza logging, can we accept the low continuous yield of hardwood which sustainable practices would produce from the start? Even selective logging without safeguards often leads to damage along logging trails and can lead to "commercial extinction," that is, the extinction of a species from much of its range, although the genetic materials (seeds) still occur, usually on the periphery of its range (e.g., *Virola surinamensis* and *Manilkara huberi* in the Amazon, and West African mahoganies). In addition, a substantial number of tropical timber species alive today are probably "the living dead"; that is, species for which living individuals can still be found, but where the population is not breeding (Janzen 1986).

DETECTION OF SUSTAINABILITY

There is little agreement on what sustainability of wood products means, and how to detect it (cf. Lugo 1988). Sustained yield tropical forestry can be achieved by first, longer than customary rotation (decreased perturbation frequency); second, highly selective extraction of a small number of trees per hectare, based on sound knowledge of growth and recruitment rates; and third, tight control over long periods, usually by the state. Theoretically, sustainability in any agricultural system, whether rice or teak, can be proven only after a minimum of three or more harvests. Poore et al. (1990) puts it well:

"It is not yet possible to demonstrate conclusively that any natural tropical forest any-where has been successfully managed for the sustainable production of timber. The reason for this is simple. The question cannot be answered with full rigor until a managed forest is in at least its third rotation." The first (bonanza) cut does not establish a base line against which any subsequent yield declines can be gauged. The third and subsequent rotations start to decline if this system is not sustainable.

In the case of trees, harvests are not precisely measured, so that minor yield declines in the third rotation are unlikely to be detected. This is further complicated because the trade accepts only a fraction of the available species today, but will have to accept more species at the time of the next harvest in a few decades. For example, in Ghana the number of primary species accepted doubled from 15 to 30 between 1979 and 1989. Growth rates of the forest remaining after the first few selective harvests may not differ greatly. Holes opened in the canopy by tree removal accelerate tree growth nearby, whereas damage to seedlings, saplings and soils by falling trees, skidders and vehicles retards growth of affected areas.

Detection of sustainability is difficult in the case of selective extraction, easier with slash-and-burn, and easiest with clear-cutting. First, selective logging of say 10% by volume leaves 90% unharvested, so that only careful measurement can detect changes. As much as half of the unharvested trees may have been damaged by the selective logging. Non-tree measures of sustainability also are difficult to analyze, such as the decrease in frequency of one of the dozen species of primates, or the decrease in fruit set of a species due to the decrease of a pollinator. When, as is usually the case, unplanned settlement follows logging trails, damage to the remaining intact forest is at least easy to detect, and at worst deplorable.

Second, in the case of slash-and-burn agriculture, where the forest is clear-cut and burnt, then declines in crop yield (i.e., lack of sustainability), the situation becomes apparent after three or so harvests. As much of the harvest is annual, this means sustainability or its lack can be detected after three years. Of course, this situation is very different from selective logging, particularly where slash-and-burn patches are clear-cut rather than planted underneath intact trees. Where the clear-cut patch is small and surrounded by intact forest, then organic matter, nutrients and propagules rain onto the plot, while conserving its humidity. Yields start to decline after three or four harvests, depending on site conditions, as the harvested crop is exported out of the ecosystem. Nutrient cycling is fundamentally important in sustainability.

Third, lack of sustainability is easiest to detect in the most damaging exploitation, namely clear-cutting (Fritz 1989). Regrowth—if the cut area was anything but a small clearing surrounded by intact forest—will not restore the original forest, except after many decades (some say centuries), again depending on site.

The nutrient cycle of the moist tropical forest is tightly closed and almost leak-free for two reasons. First, most (c. 90%) of the nutrients in the ecosystem are stored in the biomass, not in the soils as in most temperate ecosystems, where cold halts chemical reactions (especially leaching) for months of the year. Recent research suggests that more nutrients are in the litter and the next few centimeters below than was formerly thought.

Second, one of the few reliable generalizations is that almost all tropical soils under lowland moist forest are among the poorest in the world. In fact, many tropical forest soils are purer (more oligotrophic) than ground up window glass, and purer than the rainwater falling on the forest. Rain in many forests is a source of nutrients picked up from the atmosphere, from dust and from smoke even before it reaches the canopy. Rain becomes much more of a nutrient source when it picks up nutrients from the upper parts of the canopy and feeds them to the lower parts of the ecosystem. As soon as the protective forest is cut exposing the soils to sun and rain, the soils become even more infertile.

The bulk of ecosystem nutrients in the biomass are not uniformly distributed: less than 10% of the nutrients are in the wood; 90% are in the leaves etc. This clearly argues for the practice of leaving all the relatively eutrophic slash, leaves, branches and bark in the forest (i.e., removal only of the oligotrophic clean bole which is almost entirely C, H and O), which will help achieve sustainability from the nutrient budget's viewpoint. This increases the risk of fire—which, however, is usually minor compared with the major increased risks from logging, trails and settlement. It is said to have been important in the intensity of the Kalimantan fires. In any case, leaving slash in the forest is a necessary but far from sufficient condition for sustainability, because logging trails, their settlement potential and damage by skidders and non-target fallen trees are more decisive.

The number of harvests to detect yield declines should thus not be less than three, and probably more. To err on the conservative side, let us assume that declines could indeed be detected in the third rotation, and that the tree species of interest mature to marketable age in 30 years for the faster growing commercial species (e.g., *Cedrela*, *Cedrelinga*, *Schizolobium*, and possibly *Swietenia* mahogany), and 60 years for the slower growing species. Given that lack of sustainability (e.g., yield declines) can start to be detected only after three rotations or more, whatever sustained yield management of forest is tried, we will only be able to judge whether it is sustainable or not after at least 90 to 180 years. Until that time, we can say only that the system being tried is more likely to be sustainable than other alternatives, except conservation units (such as national parks) and "extractive reserves" (nuts, rubber, fruits), which also can be over-exploited.

As organized tropical forestry began only some 130 years ago (Poore et al. 1990), and tropical moist forests were not widely threatened until approximately 40 years ago, it is not surprising that sustainability became a goal only relatively recently. Meanwhile, as forest not under such selective management or other systems disappears to slash-and-burn or to cattle ranches, managed forest becomes more valuable and buys time for improvements to be introduced. Well-managed selective logging conserves most of the values of intact forest and preserves options open for the future; most alternatives do not.

THREE EXTERNALITIES

Population

Population stability is the first factor to be tackled in any approach to the sustainability of practically anything these days. This is because consumption per capita has an irreducible biological minimum (e.g., calories/day) and a higher culturally acceptable minimum. Therefore, the existence of more people means more total consumption— whether of wood or food or whatever else—and hence, in our present context, more demand on the regenerative capacities of the forest. Since populations in tropical countries are growing much faster than in temperate countries (annual increases of ca. 2.4% tropical versus 0.6% temperate) sustained yield from tropical forests will be increasingly difficult until populations stabilize. This can also be said of the sustainability of any resource use. But it is particularly important in forest management (selective logging) because, as we have seen, that is predicated on a much longer time frame than most alternatives.

Although sustainability cannot be expressed in absolute numbers, best practice and improved technology do not multiply yields convincingly. Human populations averaging 2.4% annual increase in tropical moist forest countries are projected to double in 29 years and balloon more than 8 times in 90 years, or 64 times in the 180 years before there is a chance that sustainability can be confirmed! Since the population link is not linear, this does not necessarily mean deforestation will multiply 14-fold, but demand for food will mushroom, and pressures to clear forest will intensify.

This link to population growth makes sustainability even less likely. Sustainability has two fundamentals: stabilizing populations to live within the carrying capacity of their resources, and stabilizing economic systems so that they do not increase strains on the finite environment (as do extraction of raw materials and assimilation of wastes). This is arguably the most crucial lesson the whole world has to learn—and fast.

The second population factor is that burgeoning populations are one of the main causes of deforestation. People impoverished from whatever cause are forced into short term behavior, such as cutting forest. Forest sustainability therefore means employment has to be created for the poor.

Unfortunately, neither forestry nor plantations create much employment, although both create more employment than ranches. This argues for the promotion of domestic forest industries, value added, knock-down (assembled by the buyer) furniture, and labor-intensive harvest and management, etc. Such tree plantation-based industries create more employment than any viable alternatives. Therefore, although forest management cannot be seen as the safety valve for all of the countries' restive poor, at least the income and employment can be sustainable and the people employed therein will not swell the local slums, as happens when other forest alternatives (such as cattle ranches) fail.

To the extent that long term forestry needs stability of population pressure on the forest over 90 to 180 years, or at least 2 to 4 human generations, long-term security of operation or control is essential, whether by the state, by community ownership, or by cooperatives. Secure control is a prerequisite to all long-term investments and particularly so for forestry use to become sustainable. As there is no population stability now—and, if there were, it would be uncertain to last for 90-180 years—the more area that comes under forest management and the less that goes into clearly unsustainable uses, the better will be the outcome for the country, for the resource-base, and for the people.

Tree plantations "absorb" more people (create more employment per investment unit) than natural forest management. On tree plantations, employee work in seed collection, germination, species trials, seedling care, propagation, planting, weeding, thinning, fire and pest control, and other labor-intensive activities. At the same time, plantations relieve pressures on remaining forests. A key element is to educate decision makers and people living outside the forest that long rotations are absolutely necessary to maintain fertility (de Camino 1987) and that land in long rotations is not available for the taking.

Sustainability of Institutions

From the above, we can see that a sustainable system must be supported by institutional stability—such as has rarely, if ever, been achieved for long periods in the past. When population pressures were negligible, institutional stability was not needed. Resources were sustainably managed by small populations until colonization wreaked such havoc. Now, institutional sustainability does not seem likely under prevailing circumstances of weakening governance, burgeoning populations and unprecedented national debt.

With timber as the source of much personal and political revenue, and where governments commonly hold power for—at most—five or ten years, there are fewer incentives for sustainable management than would otherwise be the case. So whether selective logging can be made technically possible is firmly related to political feasibility. The most plausible claims of approaching sustainability of timber exports (e.g., Ghana, Malaysia, Nigeria) were possible only because of rigorous martial control by colonial regimes.

Those approaches vanished soon after decolonization, and now the situation is worsening. The era of big government is weakening worldwide. Governance is slipping; instability is increasing. While the recent surge in nongovernmental organizations (NGOs) to fill the vacuum created by government abdication is encouraging, few have track records of more than a few decades; nor do they seem strong enough to ensure tight management over a forest for the required length of time. The recent (U.S.) Nature Conservancy approach of outright purchase of tracts to be conserved in perpetuity works well in North America where short-term pressures and values are not the same as in tropical countries.

Some governments recognize that short leases encourage rapid depletion, and are promoting longer leases, some of the order of 50 years, which are valuable to postpone more destructive uses. Strong state control, permanent usufruct or community tenure, or similarly long-term arrangements are fundamental for achieving sustainability. As such arrangements are rare at present, institution longevity is one of the most important prerequisites for sustainability, and deserves much more attention than it has received. Very long-term stability of society, incentives, taxes, value-added, reduction of logging and conversion waste, and log export constraints also are necessary preconditions.

Biological Productivity Rate

Regarding implications for public investment, it must be admitted that the net biological productivity of intact tropical moist forest is low and in principle is zero (Poore 1990, pers. comm.). What is produced is consumed or decomposes rapidly. This is one underlying reason why sustainable yield from moist forests is so difficult. Unless such forests accumulate humus or peat—and few do nowadays—all the carbon dioxide which is fixed is duly respired. Such a forest is in almost a steady state or dynamic equilibrium. The very few modern measurements of the chemical composition of the atmosphere above intact forest canopies (such as for a few months during one year in Amazonia) suggested that the forest was a minor sink for carbon dioxide around noon which is the peak of net carbon fixation. This measurement may not reflect the 24-hour-average (Fearnside 1990, pers. comm.). This means that photosynthesis exceeded respiration by a small margin at that time. Presumably, the same forest may be a minor CO_2 source in a different season. Timber exploitation opens the canopy to a modest extent, and removes boles, thus promoting rejuvenation and increasing the carbon sink value of managed forest over that of preserved forest.

Keeping the forest substantially intact while harvesting only the wood increment grown since the last cut provides such modest yields that they are not very attractive to most governments or investors. However, these small yields are essential for forest-dwelling ethnic minorities (e.g., Amerindians, Pygmies, Penan). Concessionaires profit by current unsustainable logging partly because sustainable logging is much more expensive, time consuming, and lower yield. According to ITTO (Poore et al. 1990), many concessionaires profit only because they do not comply with national regulations. This means that government interest could be strengthened by such schemes as renting land (E. Goldsmith 1980; J. Goldsmith 1989), maintaining carbon sink and coal displacement forests etc.

Even tropical tree plantations on prime sites do not grow particularly fast (3% overall annual volume increment may be reliable; 7% achievable in certain cases; rarely higher). So the rate of return on investment at today's high discount rates makes long rotation forestry and 20 to 30-year tree plantations economically less attractive than faster yielding alternatives, such as a two-year rice project.

Long-gestation projects also have greater scope for disaster, pests, fire, political instability, illegal felling and the risk that someone other than the investors' heirs will eventually benefit. Comparison of growth rates between biomass (wood) and rate of return on investment (money) suggests to some a lowering of discount rates for such public investment (see below). Lowering would not be arbitrary, but based on rates of return on alternative sustainable uses of capital.

Few truly sustainable renewable resource projects will yield 10% annually: even rice paddy agriculture depends today on diesel oil, biocides, and fertilizers. We are informed that lowering discount rates boosts the level of investment in general, expands the size of the economic subsystem relative to the fixed size of the surrounding ecosystem on which it depends, and increases throughput between the two. This means the already overstressed environmental source and sink functions would be further impaired by lowered discount rates. We therefore refrain from taking a position on the discount rate, except to note the need for clarification of the above issues.

EXAMPLES OF BEST PRACTICE

Forests were managed sustainably for thousands of years before the colonial era, by low population densities, by long fallows and taungya-type systems (Burmese for "fallow") (table 32.2; number 4 = Agrisilviculture), and by tight tribal community rules and customs. Since the start of the colonial era, sustainability has been constrained by social and other disorders outside the forest sector. Sustainability may have been approached for 30 years or more during colonial regimes in which the forestry sector was under strict, quasi-military control. These controlled systems yielded about 2.5 to 5 m^3/ha/yr (50 m^3/ha/20+yr rotation), and are outlined below.

Burma Teak Rotation: This is possibly the longest-lived of the recent approaches to sustainability. The British Navy used Burmese teak over 150 years ago, before annexing Burma as a colony. The seasonally moist forest is rich in teak to begin with, and by harvesting only the annual increment on long 30-year rotations, while leaving the growing stock intact, using non-damaging extraction methods (elephant and river transport only), yield declines were so small as to be scarcely detected. The annual increment was calculated as those trees whose growth-rate had just peaked. Research suggested this was when the tree had reached a girth limit of 30 inches (72.6 cm) diameter at breast height (dbh). Such trees were girdled and left standing to dry for three dry seasons. The fallow period of the 1939-1945 World War helped replenish the forest, which was perpetuated until the military government took over in 1962. Since then, the system appears to have been overexploited.

Malayan Uniform System: This started in the 1960s as a simplification of the earlier shelterwood system, which began in 1946. Tree harvesters planned to rotate every 70 years, but most of the flat and fertile lands were soon converted to oil-palm and rubber production. This meant that practice of the system was relegated to slopes where regeneration is less successful and more heterogeneous. The idea was to harvest all merchantable species at the start, and then by poisoning and other girdling, to promote even-aged regeneration of dipterocarps, especially *Shorea* spp. This illustrates the wider point that lands which could have been profitably managed for forestry in Peninsular Malaysia and some countries in East and West Africa have already been converted to agriculture, which is part of the reason the system failed. Now, practically all flat lowland moist forest has been converted, so that Malaysian timber exports (the largest in the world) now come almost entirely from the Malaysian states of Sabah and Sarawak. This has led to today's plight of the Penan, Iban and other forest-dwelling societies.

The areas which retain anything like the original forest are those on which little other than forestry can be successful. With regard to the number of harvests that are needed to detect unsustainability, there is little difference between farming trees or farming something else. Lack of sustainability becomes manifest after the third harvest or so: three years with crops of annuals, but later with perennials. If it is already accepted that farming something else is unsustainable, then the reasons why the nutrient balance in farming trees is expected to be more sustainable must be clarified beforehand (Iltis 1989): it may merely be that unsustainability takes longer to become noticeable in tree-based farming.

Nigerian Shelterwood System: This began in Nigeria in about 1920 and was extended to Ghana in 1946. The idea was to "liberate" the more valuable trees by opening the canopy around them five years or so before harvest, and meanwhile to kill less-desirable competitors by girdling and poisoning. This required reliable inventory and research, and enforced controls on yield to be extracted. What would have been the better areas for trials were already preempted by cacao and oil-palm plantations and the experiment was ruined by the 1964-1970 Nigerian civil war. The African Timber and Plywood Company in Nigeria had a mill run of 81 species for the sophisticated European market (Janzen 1990, pers.comm), and may have been truly sustainable.

Polycyclic Felling System (Improvement Thinning): As started in Ghana in the 1950s, this system consisted mainly of liberation thinning and poisoning, designed to be on a 30-year rotation (Asabere 1987). In spite of opposition on environmental grounds by professionals, the plan went ahead. A major factor in its economic success is market acceptance of the lesser-known timbers, and now the system is suffering a fate somewhat similar to the above three systems. Liberation thinning of surrounding weed species on 35,000 ha started in Sarawak in 1977 (Wadsworth 1990, pers. comm.) could shorten the cycle by 15 to 20 years.

Indian Systems: The valuable experience with the approximately 100-year rotation managed forests in India, said to be successful, is reported by Agarwal (1990). Therefore, this experience is not elaborated here. This work is particularly valuable for examples at the moist end of the spectrum.

Other Systems: The Carton of Colombia near Cali practice an interesting low-intensity harvest using aerial ropeways to minimize damage to the surrounding vegetation. This experiment is about 15 years old. In the 4-year old experiment in the Palcazu valley in Peru, narrow strips are cleared into which lateral remaining intact forest is expected to repopulate (Hartshorn 1989). Venezuela's 40,000 ha Ticoporo, Barinas, managed forest's 40-year concessions (established in 1955) reduces incentives for loggers to exceed sustainability. Brazil's Tapajos National Forest near Santarem has a few small plots well managed over the last decade or so. Suriname's CELOS system of strip planting appears to be sustainable, but the product is several times as expensive as wood that is logged from Suriname's abundant forest.

Enrichment planting appears likely to be less economic than plantations of the same species. ITTO's survey (Poore et al. 1990) describes successful examples such as that in Peninsular Malaysia (where there is so little commercial forest potential left that the industry has moved to Sabah and Sarawak), and Trinidad, where there has been little pressure on the forest largely because the economy was oil-fueled until recently, and plantation teak is now a main source of timber.

Forest Use with Wood as a Minor Product: Extractive reserves are forests from which a community extracts certain products, both for the market and for subsistence. The products are primarily rubber latex, fruits and nuts, but include a long list of other products extracted at a very low intensity (Asibey 1985; de Beer and McDermott 1989; Lamprecht 1989; Schwartzman 1989; Peters et al. 1989). Even the occasional tree may be ritually extracted for local consumption (for making canoes, huts, etc.), as can game animals (Asibey 1990; Asibey and Child 1990). Extractive reserves date from the end of the rubber era in South America and so are no more than 50 years old. These reserve systems are likely to be sustainable, provided both the community's population and the markets remain stable. They thus merit invigoration and support. But, however attractive and sustainable such systems can be, they are not panaceas. We do not examine these ideas in detail because the system does not produce wood, the main focus in this paper

The rattan trade from Asian forests is also likely to be sustainable, although rattan extraction exceeds regeneration by such wide margins that an increasing number of rattan species are commercially extinct (de Beer and McDermott 1989). The transition from rattan extraction to rattan plantation has started (in East Kalimantan). Widening the demand for non-hardwood products such as poles, charcoal and fuelwood is risky in that their sustainability is even more difficult to ascertain than that for hardwoods, and they are of lower value, hence less likely to be economic. In addition to more knowledge of extraction, we must ascertain how to increase a set of products without reducing other values below an agreed level and we must know how the value of the benefits (the mix of products) can be increased without reducing the objective below a certain specified level.

Agroforestry: Although not forest management, this is a system which also can be sustainable, assuming stable human populations and extraction rates (Arnold 1987; Lal 1988; Mendelsohn 1990). Agroforestry is highly appropriate and promising for the enormous areas of degraded or abandoned land that was formerly forested. Although some timber is produced from such systems, it is not the main commercial product; most agroforestry wood yield is for local consumption. As a private, small-scale alternative to large plantation estates, agroforestry has major benefits and lower costs. Agroforestry also is a low cost intermediate step in restoring low return or degraded agricultural lands back into tree crops, as in the case of privately owned woodlots (outgrower systems). Agroforestry also is compatible with outgrower systems for rubber, coffee and other tree crops, which should usefully form a buffer zone around forest reserves. Bush meat and other products can be sustainably harvested from such a system (Asibey 1990; Asibey and Child 1990).

Fuel Wood: Fuel wood collection is not a problem in moist forest regions, so is not discussed here. But, considering that 85% of tropical forest products go for fuel wood, peri-urban fuel wood lots must be part of agroforestry and need strenuous promotion by all users (such as hospitals, schools, barracks, prisons, steam users), as well as for sale to the private sector (such as bakeries, hotels, restaurants, laundries, and—above all—to households). This form of land use merits active encouragement (Teplitz-Semblitzky and Schramm 1989).

TROPICAL TIMBER TRADE

World exports of tropical timber peaked at US $8 billion in 1980, and are now around US $5 billion annually. About half this value derives from logs, 30% from sawnwood, 20% from plywood and 3% from veneer. Tropical timber accounts for about 11% of world forest timber trade (Erfurth 1988; Nectoux 1985, 1987, 1989). Only ten species account for well over half the trade in each of the three main areas: Africa, Tropical America and SE Asia. One tree family, the Dipterocarpaceae, accounts for about 70% of Southeast Asia's exports, particularly *Shorea* and *Parashorea*. Only three other genera provide more than 10% of regional exports, Triplochiton (Obeche) and Aucomea (Okoume) from Africa, and Virola from the Americas.

ITTO promotes the sustainable management of tropical forest (Poore et al. 1990). The nearest approach to this sustainable management is forest management on long rotations, under tight control. Apart from this, the two other acceptable sources for tropical timbers are timber plantations, and exploiting the wood salvage felled from forest in the rare instances where such conversion is thoroughly justified. We refrain from exemplifying these rare instances, but can imagine a case where the forest lost would be tiny and larger areas of the same forest are well conserved nearby. If the land is an exchange for an immensely powerful canyon-type hydroproject, and prevents construction of several nuclear power reactors, it might make such loss of forest justifiable.

Although tropical tree plantations are few and mainly young, they can supplement hardwoods extracted from the forest. Tropical trees grown in humid climates in poor soils and in pure stands will inevitably be attacked by pests and diseases (e.g., insects, fungi). Even Brazil's Jari project is now attacked by ants and phytophagous insects in spite of the dry season and the unpalatability of the planted *Gmelina* (Fearnside 1988). Plantations in areas with dry seasons long enough to control pests are likely to be more successful than plantations in perhumid areas, although the fire risks are greater.

Boycotts and other restraints (cf. table 32.1) emphasize the frustration of the consumers over deforestation, extinctions and deracination of vulnerable ethnic minorities. Consumers are requesting selectively for sustainably managed timber, thus creating a market demand and incentive to produce that product. Sustainably produced and so labelled supplies will prevent consumer rejection of all tropical hardwoods which could be difficult to reverse. Such consumer preference sends a signal, providing time for producers to adapt to this demand. A risk of boycotts is that if forests now controlled to some extent by timber corporations cease to be profitable, then uncontrolled deforestation may increase. Boycotts are likely to be more effective in promoting sustainability where logging and logging trails are the prime causes of deforestation, and where the timber is exported, than where other circumstances prevail. Wood boycotts cannot be effective where much hardwood is burnt or left to rot, as in much of Amazonia.

Until plantations can be vastly multiplied and expanded, only careful control and labor intensive selective extraction will reduce current levels of damage in forests. This will be difficult, but is preferable to today's alternative of increasing and irreversible damage. Only those areas or corporations which have already planted trees of acceptable types and areas, or who have posted a similar bond or escrow, should be permitted to continue transitional extraction (table 32.4). Meanwhile, management of secondary forest and extraction methods must be improved while plantations reach maturity.

Some fast-growing hardwood species are said to take less than 20 years to mature (Lamprecht 1989), but more species need to be investigated in this connection. While fast-growing species cannot substitute for slower growing hardwoods, they can satisfy some of the demand, thus buying time for hardwood plantations to mature. Fears of trade restrictions have almost doubled the price of Southeast Asian logs in the last year; for example, average world prices for Malaysian sawn timber jumped from US $276 per cubic meter in 1987, to $422 in 1989.

The tropical hardwood trade is unique in that it is the only major internationally traded terrestrial commodity extracted wild across the tropics, rather than cultivated. Plantation hardwoods today are not much more profitable than their forest-extracted counterparts, because logs are still allowed to be counted as basically free (i.e. no or low royalties), and less environmentally damaging) practices are not insisted on.

The time has long since vanished when we satisfied our desire for meat by hunting; it is now time to satisfy our need for tropical timbers from more sustainable sources. Human demand for such timbers long ago surpassed sustainable supply. The irresistible response has been to shorten rotations, to decrease selectivity, and to extract more than 10% of the standing timber at a time, thereby descending inevitably into clear unsustainability. Although trees can grow faster in plantations, tree maturity is measured in decades rather than months as for most crops. Therefore, during the transition highly selective, long rotation forestry is encouraged while plantations mature. Plantations should supplement removals from tropical moist forests (Grainger 1988). Global demand management for tropical timber clearly has to be an important element in the equations to approach sustainability (see below). Fast growing tropical trees such as eucalyptus, if necessary chemically treated to improve durability, are a useful interim measure.

ECONOMIC CONSIDERATIONS

This paper focuses mainly on environmental and other non-economic aspects, but we want to stress the two most influential cost factors in tropical forestry: time (and discount rate), and transport. These are important because ITTO (Poore et al. 1990) concludes that "The inability of tropical foresters to suggest ways of valuing the goods and services from the forest, which are meaningful to their colleagues in national treasuries and planning ministries, has been a major factor in the continuing loss of these forests" (Leslie 1987; Browder 1989).

Time and the Discount Rate

The long time factor involved in growing and marketing tropical timbers means the discount rate is more influential here than in most alternative investments, but there is no agreement that discount rates should be adjusted. Since resources are limited, some method of discounting futures will be needed. While it is true that high discount rates discourage investments with long term benefits and promote projects with high short-term benefits and low long-term costs, this does not mean the rate should be lowered. A 10% discount rate discourages investments in a 30-year forestry project in preference to a 3-year rice project. International real interest rates, which might be taken as indicating the opportunity cost of capital, have for decades been around 2-4%.

On the other hand, the discount rate is not generally the problem for encouraging sustainability, rather the failure comes when it fails to reflect full cost in the case of competing natural wood. If the latter were priced correctly, then presumably benefits would rise sufficiently to produce 10% returns on planting. In view of the extreme rapidity of irreversible loss of tropical forest (an estimated 1.8% of the total tropical forest area in 1989, according to Myers 1989), the transition to sustainability should be as short as possible. Business as usual until plantations planted today mature, in, say, 20 years' time, would mean that most of the remaining tropical moist forest will be lost.

A lower discount rate, as some advocate, would increase the scale of investments in general, both environmentally damaging ones, as well as beneficial ones. Herman Daly (1990, pers. comm.) feels we are relying on the discount rate as a policy tool for two dissimilar needs, namely for the scale of investments, as well as for allocation (the selection of priority or most profitable projects). The discount rate should be retained for allocative decisions, and another policy tool devised for the much needed scale decision. Natural resources, energy and severance taxes could be effective in maintaining scale at desired levels.

Subsidies or grants may be in order to foster social values not currently promoted in the market. The question becomes where best to allocate such subsidies—to conserve forest (the obvious choice to us), to reduce the discount rate during the transition to sustainability, or to promote plantations? Grants during the transition could be considered, and are greatly in the interest of industrial countries for other reasons (see "Carbon Sink Forest," below).

Transport Costs

Transport is the second major cost in forestry. In the case of the largest (temperate) rain forest in the United States, Congress spends $40 million every year from taxpayers mainly for the roads needed so that two corporations with 50-year concessions can export raw logs and pulp from Tongass, Alaska, to Japan. Now, 300-year old trees are being sold for less than the price of a hamburger (US $2). The attraction of plantations is that transport costs are substantially lower because the product is concentrated, and the distance to market is shorter. The Tongass example shows that perverse subsidies allow logging old growth forest to be more profitable than younger forest or plantations.

Highways and rural roads that pass through or near forests are critically important to the process of deforestation, because they facilitate the process of land clearing, unplanned settlement, hunting, fires, logging, mining or other major disturbances. Therefore, a most effective way of controlling deforestation is to control road building. Because most roads in tropical moist forests are to promote unsustainable human colonization, they increase both poverty and loss of biodiversity. Thorough assessment of all roads and their effects well before any decision to construct, is an essential and powerful mechanism for conserving forests.

Accelerating the transition to sustainability will be onerous for the nations concerned as well as for the few multinational timber corporations. The transition can be eased in several ways, such as Edward Goldsmith's rent scheme (1980), debt-for-nature swaps, and carbon sink forests. Goldsmith's rent scheme is the most elegant and practical. He proposes that countries wanting to conserve tropical forests should rent them from the sovereign owners who retain total control. The rent ceases as soon as the owner decides that alternative uses (deforestation) are more profitable or otherwise desirable than conservation. Clearly, the rent has to be set at a judicious level with a clause governing stability, and some renters may find the recurrent nature of the cost unattractive. A variant could be to rent with an option to buy. In this case, so many years of rent would accumulate until the forest would be purchased for the new owner to conserve as forest in perpetuity.

Debt-for-Nature Swaps

Such swaps multiply the effectiveness of modest conservation funds, depending on the degree to which the forest owner's debt is discounted on the secondary debt markets. Since the money needed for conservation is infinitesimal in comparison with the debt, such swaps need not be inflationary. Sovereignty issues are satisfied because the country retains total ownership. Since most remaining tropical moist forest is owned by deeply indebted countries, we feel there is much opportunity for debt-for-nature swaps. This major opportunity should be exploited without delay since it would benefit owning countries, and biodiversity, as well as the global commons (Oberndorfer 1990; Potvin 1990).

Carbon Sink Forests

Today's relative price for polluting the atmosphere with carbon dioxide is zero, which is clearly not optimal. Carbon sink forests are mechanisms to buy time for the internalization of the world's most pervasive negative externality, atmospheric accumulation of carbon dioxide, in large part creating the greenhouse hypothesis. Burning tropical forests contributes possibly as much as 30% of global CO_2 accumulation (Myers 1989). Growing trees sequester CO_2. Since trees grow faster in the tropics and land is cheaper there, tropical tree plantations offset CO_2 accumulation and reduce greenhouse risks.

Carbon sink forests were planted voluntarily in 1988 when a new 180 MW coal-fired utility in Connecticut agreed with the Government of Guatemala and others to plant the 15 million trees that were calculated as being necessary to offset the estimated amount of CO_2 emissions expected over the utilities 40-year life. This was followed in April 1990 by the Government of the Netherlands budgeting 1,875 million florins (ca. US $500 million) to plant 250,000 ha of tropical trees in Bolivia, Peru and Colombia. These carbon sink forests will offset the 6 million tons of CO_2 to be emitted by two new coal-fired electricity plants to be built between Amsterdam and Rotterdam. One reason adduced is that the cost of planting trees in the tropics is less than one-twelfth of that cost in the Netherlands.

It has been calculated that 465 million hectares of tree plantations would stabilize global CO_2 levels and mitigate greenhouse risks for three to five decades. All carbon sink forests will buy time; it is clearly not necessary to plant 100% of this total (Sedjo 1989, 1990). As the World Bank's chief economist, Vice President Stanley Fischer, put it in 1989, "the costs of inaction if the greenhouse hypothesis is proved correct, vastly exceed the costs of action if the hypothesis is proved false."

The most efficient, cost-effective means to reduce greenhouse risk in this context is to reduce deforestation. The problem is many plutocrats promote deforestation, whereas few are against tree planting. We predict that major CO_2 emitters will soon pay tropical countries to plant trees. Whether this is voluntary as noted above, or as part of some global CO_2 tax scheme is up to our leaders to indicate, and to all who elect them. If some of the carbon sink forests are converted into furniture, library books, houses, and veneer, then the carbon is sequestered longer.

The financial flows from carbon emitting nations to owners of tropical space available for plantations, will have to be set high enough so that those tropical countries with room find it profitable to accept the plantations promptly. Some researchers feel that even if greenhouse gas emissions cease overnight, the globe is set for climate change from the gases already released and on their way upwards. Although there is a useful demonstration effect in carbon sink forests, they are not as influential in ensuring sustainable use of the forest as the measures outlined in tables 32.3-32.5.

DEMAND MANAGEMENT

This paper focuses on the supply side and on improving the management of tropical forests. But there is a major need to conserve forests and biodiversity by managing demand (Smith 1989; USDA 1989). The world's biggest paper consumer, the United States, recycles less than one-third of its paper. Japan is the world's largest consumer of tropical timbers, importing about 30% of total world trade annually, followed by the United States and the EEC. The Nikkan Mokuzai Shimbun (Daily Timber News) emphasized in 1988, "the depletion of tropical timber resources in Southeast Asia has become a matter of reality today, so we have to look to Brazil for a new supply."

Tropical timber consumers could promote the transition to sustainability by controlling their demand. Some examples are: discouraging (ab)use of tropical hardwoods for throwaways (e.g., 25 billion pairs of waribashi chopsticks per year); creating substitutes for the $2 billion worth of tropical hardwood plywood concrete formers discarded after a couple of uses; decrease use (through punitive taxation) of tropical hardwoods for chemical feedstock (e.g., chipping tropical forest and mangroves for cellulose in Indonesia).

Practically all wood used by Japan is imported, although 68% of Japan is forested. Most of the annual 3.8 billion cubic feet imported is solid wood (more than 50% of the world's supply of tropical timbers), mainly logs which support Japan's 18,000 wood mills. ITTO and others are discussing labeling of woods as to the sustainability of origin, and levying a surcharge to create a forest conservation fund.

The even more important demand reduction need has been noted in previous sections: overconsumption by the affluent. The greedy rich destroy more than the hungry poor. While greed is difficult to prevent, at least consumers should pay the true value or cost of what they consume, which is far from the case at present with tropical timbers. To the extent that tropical timber exports are not sustainable—that they deplete natural capital— they should be taxed. This would promote efficiency, and decrease consumption, thereby safeguarding much biodiversity and the remaining richness of gene-pools.

Treating Forests as Non-Renewable

If a country elects to liquidate its forests, which this paper strongly advises against, then use of El Serafy's method (Ahmad et al. 1989) at least ensures the sustainability of income when the forest is mined out. El Serafy's paper concentrates on renewable resources, such as forests, which should be harvested sustainably, no faster than the rate at which the resource regenerates. If, however, the sovereign owner misguidedly mines the forest as if it were a non-renewable resource, then at least the receipts from decapitalizing the forest should be managed to provide income (sustainable by definition) rather than the rich owner becoming bankrupt soon after the last tree has been sold. El Serafy's elegant but parsimonious device is to divide the forest-mining receipts into separate income and investment streams. The investment stream should be invested in renewable assets, so that the income component can be consumed "in perpetuity." The return on the new (non-forest) renewable assets and the amount invested each year during the forest mining are set at the level such that when the forest is mined out, the renewable assets will be yielding an amount equal to the income component of the receipts.

The CITES Treaty

Significant biodiversity is lost by trade in rare species. Ratification and implementation of the international CITES treaty would help, especially if it were reinforced by ad hoc training of customs agents, raising fines and imposing other penalties. While this would be useful in reducing trade in rare species, we do not fully know what species to place on such restricted lists. The "endangered species list" is more meaningful for temperate and other biotas, than for the relatively unstudied tropical forest biotas. Even so, a list of all tropical endangered species commercially traded (e.g., pets, skins, ornamental plants) would not be unmanageably long. A step in the right direction was the 2-year (no parole) jail term given to Walter Sensen, notorious and persistent endangered species trafficker.

A few years ago, jail sentences for polluters were unheard of. At worst, the penalty for violating environmental laws was a minor fine: just another cost of doing business. Now, however, a single environmental violation can, in some countries, become a compounded felony. It ceases to be a normal cost of doing business when the cost is the executive's own neck. The protective wall between corporate actions and corporate executives is eroding. Jailing the heads and other responsible officers of a few major corporations for illegally and persistently destroying the environment, and seizing corporate assets, would surely at last "get the message across."

RECOMMENDATIONS

This paper has proposed a transition to sustainability of tropical forest use by improvements in current logging techniques, by deflecting logging to secondary forest, by extracting only high quality hardwoods from forest, and by promotion of tree plantations. Enough specifics are provided to show the general direction of what we recommend. However, we abjure from repeating or competing with the vast literature on either plantations or logging techniques (Wadsworth 1990; Wiersum 1984; Willan 1990). A necessary, but not sufficient, condition is demand management, which is also outlined.

Transition to Plantations

Plantations should be as heterogeneous as necessary to reduce pest and disease pressures, to mimic natural forest, and to hedge market bets. Since plantation productivity can be of the order of 10 times as great as extraction from forest, and as the plantation product is more uniform and hence more suitable for a sophisticated market, and moreover as transport and other management costs are lower in plantations than in forest, and fertilizer or at least limestone application costs are lower, plantations appear to be a more cost-effective source of supply. This contention will only hold if sustainable forest logging is insisted upon. In other words, forest logging is profitable partly because loggers are not forced to pay full costs or manage sustainably.

We thus feel that the costs of ensuring sustainability of forest logging would make timber plantations an attractive choice. Plantations would be partly site-specific and partly dependant on the externalities which are included in the cost stream. The more heterogeneous the species mix and the age classes, the more biodiversity will be conserved in the plantations. This tradeoff should be made explicit because the timber trade prefers monoculture stands of even-aged trees—despite the fact that costly plantation failures abound, due to inappropriate sites or species, and inability to manage them until maturity (Grainger 1988). Recent widespread indebtedness of governments, etc., makes financial incentives for plantations even less likely than they were formerly.

In choosing plantation sites, it should be borne in mind that two are apt to be most attractive for planting tropical hardwoods. The best is the site which lies well outside the moist forest that is currently being destroyed, where there is sufficient dry season to retard pests, but not sufficient to retard growth significantly, and distant enough from the forest for forest pests to have less chance than otherwise of finding the plantation. It should not, however, be expected that sites with all these characteristics are easy to find. Soils tend to be more fertile in the seasonal than in the wet tropics.

There is a role for exotics to be used in plantations, particularly where they are faster growing and less susceptible to pests than native species. Sites near markets or export corridors, and those nearer to sources of limestone and other low cost fertilizer should get preference. If the logging corporations want to recuperate their reputation and help conserve the forest, they will employ the settlers who are currently being forced into destructive short term behavior in the forest. By having settlers manage the plantations and intermediate buffer zones, they will relieve pressures on the forest. Since plantations can be sited on degraded or abandoned land, most original inhabitants may have already left.

The second best category of sites for tropical tree plantations are those vast and expanding tracts of degraded land such as abandoned pasture, formerly clear-cut forest, secondary forest, etc., which are not being used. Governments may want to offer incentives for the establishment of plantations in such areas. Because they are generally simpler in structure and species than primary forests, management of secondary forest is an important need. If tree plantations decrease population pressures inside forest, and act as buffer zones around intact forest, then the plantation becomes that much more useful and attractive. There is much support for the integrated management of plantations with conservation units (Ledec and Goodland 1988). The economic benefits stream can be augmented by including the benefits of carbon absorption (the carbon sink forests—see above), and all those environmental services which are provided by the plantation up to the time of harvest—such as climatic stabilization, weather buffering, dry season water augmentation, erosion control, and reduction of flash floods. Use of a range of plantation species would be prudent: the faster-growing ones could relieve pressure from demand until the slower-growing ones mature; other benefits of mixed plantations are indicated above.

Approach to Sustainability

Our recommendations to improve forest management are divided into specifics for countries (table 32.3) and for corporations (table 32.4). Additional expenditures could be financed from North to South by creative use of carbon sink forest incentives (Benchimol 1989). The international trade in tropical wood is worth more than US $5 billion annually, yet practically none is invested in ensuring sustainability of supply. Countries fulfilling the sort of criteria in table 32.3 would be more eligible for support by, for example, TFAP and ITTO. Corporations fulfilling such criteria would alone be eligible to enter the forest sector. A corporation could promote itself by converting an "on paper" park into a functioning conservation unit by an "adopt-a-park" approach. Corporations are capable of managing a conservation unit at the needed low level of inputs while managing long term forest fallows. In addition, corporations profiting from plantations of a tropical tree should conserve the ecosystem(s) supporting its wild relatives in perpetuity, wherever they are. For example, chocolate, coffee, oil palm and rubber corporations would thus conserve tropical forest tracts.

TABLE 32.3 Acceptable Practice for Countries
(selected examples only; from Goodland, Watson and Ledec 1985)

1. National goals of at least 10% of the total national territory, and a representative sample of all ecosystems to be in conservation status.
2. All (or at least 90%) of conservation units functional, rather than "paper parks"; a schedule for attaining 100%.
3. Comprehensive social, economic and environmental inventory and evaluation of all marketable and non-marketable forest values.
4. National forestry, agricultural, planning, and financial ministries actively investigating possibilities of, and implementing, effective plantations, especially by the private sector, testing fast growing species, and fostering carbon sink forests.
5. Governmental incentives favoring "best practice" logging from secondary forest.
6. Repeal of all incentives for cattle ranching in forest-induced pastures, for raw log exports, and for clear-cutting (Binswanger 1989).
7. Government ratification and enforcement of all relevant international environmental treaties, (CITES, Ramsar, Migratory Species, World Heritage, and Wildlife Preservation).
8. Government commitment to stabilize human population growth rate as soon as possible; urgent reduction of high growth (3%) countries down to current world average (1.8%); and to ensure longevity of existing institutions, to the greatest extent possible.
9. Government commitment to achieve national sustainability of natural resources (e.g., by severance taxes at point of extraction).
10. Abolition of all incentives to clear moist forest for unsustainable uses or for land speculation, or as a criterion to access credit or to prove ownership.
11. Promotion of incentives for intensification of agriculture, rather than forest conversion for extensification; remove tax on intact forest; tax extensive land use; secure land title.
12. Invite FAO (TFAP) and ITTO to help countries to complete forest inventory, monitoring and zonation systems (remote sensing); to assess the speed needed for the deflection of logging to secondary forest, and for the transition to plantations in each country; the size and nature of the plantations, and the logging improvements needed in the interim.
13. Raise stumpage fees or taxes to account for full opportunity cost and increase revenues to the forestry agency and to discourage wasteful logging practices (Openshaw and Feinstein 1989)
14. Forest dwellers areas to be adequate in size and protection, to buy time for them to choose their own future.

CONCLUSION

This paper has presented the case that sustainability of tropical hardwood production has not been achieved, would not be compellingly profitable financially were it to be enforced, and is unlikely to be achieved in the future unless major improvements (tables 32.3 to 32.5) are implemented promptly. Even so, sustainable management would be very profitable in environmental service benefits and in conserving options for the future. The main value of the forest does not lie in the timber. The timber must subsidize forest conservation for its environmental services and biodiversity values.

TABLE 32.4. Acceptable Practice for Timber Corporations
[Selected examples from Goodman, Watson and Ledec 1988]

1. Forceful promotion of thorough recycling, at all stages, of the industry concerned, and especially the sector involved; reduce packaging waste.
2. Proven track record of careful selective extractions, and of implementing "best practice."
3. Annual planting and management of an agreed area of tropical hardwood plantations, or defrayal of such costs when incurred for the corporation by state agencies.
4. Avowed corporate policy to shift from depletion to sustainability, from primary to secondary forest logging, and from extraction to plantations, within an agreed time.
5. Maximum value-added and domestic processing of forest industry.
6. Management of inviolate conservation cores protected by buffer zones of plantations and agroforestry.
7. Agreement to implement a schedule for the prompt restoration of areas previously degraded.
8. Effective policy to prioritize retraining of people displaced from extractive logging into careers with a future.
9. Promotion of efficiency in wood conversion and product use (e.g., thinner "wafer" boards and veneer, thinner or less wrapping paper, less bleach, employment of precision machines and use of "waste" such as sawdust).
10. Adoption of "best practice" logging methods: use of river floating, aerial cableways, blimps and elephants for log extraction and transport; less damaging road and trail construction than is commonly practiced, with improved maintenance; avoidance of slopes above agreed figures as to steepness, erodability and rainfall regime; use of more efficient sawmills and other processing technologies. [ITTO has adopted a Code of Best Practice]
11. In cooperation with government, prevention of unplanned activities along logging roads: by improving conditions outside the managed forest, by employing the locals in the managed forest or its value-added industries, and by convincing the locals that it is in their own self-interest to prevent incompatible settlement or other practices; maintenance of guard posts and patrols.
12. Redirect focus towards the huge areas of secondary forests to be put under sound management (where logging is less damaging and can be cheaper than plantations); promote underplanting where possible.

Logging leads to deforestation in some countries. Because deforestation-induced extinctions are irreversible, rapid, and increasingly create major environmental costs, we have proposed an urgent transition by taking three steps towards sustainability. First, improve the sustainability of logging (tables 32.3. and 32.4). Second, deflect logging to secondary forest. Third, accelerate the transition to plantations. The speed of this transition depends on the country; overdue in some, not immediately essential in others. Work is needed to calculate optimal transition speed in each country, and to ensure the plantations are the best that can be achieved. During the transition, until the plantations mature, influential improvements to reduce the damage of logging are readily available for both countries and for corporations, as suggested in tables 32.3 and 32.4. The tables suggest ways to at least approach sustainability while the transition to tree plantations is accelerated. We want to

TABLE 32.5 How to Save Tropical Forests [from Goodland 1990b]

REDUCE DEMAND	BOOST SUPPLY
1. *Essential Preconditions*	
Revamp orthodox economics; innovative Third World debt solution (Citicorp and Conservation International re: Beni, Bolivia); promote economic incentives for environmentally prudent behavior; assist vulnerable ethnic minorities (e.g., jungle dwellers). Reduce war and its environmental effects.	Ensure there are adequate conservation areas, national parks, World Heritage Sites, Biosphere Reserves; zoos and arboreta; gene, sperm and seed banks; museums, herbaria; rehabilitate degraded tracts; promote peace, environmental education, legislation; encourage legal profession; tout ethics (e.g., Taylor, Sagoff, Regan, Singer).
2. *Improve Development Projects*	
Multilateral development banks and bilateral development agencies (e.g., USAID, CIDA, SIDA, DANIDA, ODA), especially projects of forestry, rubber, oil palm, cocoa; alternatives to coca; promote scientific tourism; debt-for-nature swaps.	Plant rubber south of Amazon; promote cocoa pollinators; reactivate Asian "sleeping rubber"; encourage tropical agro-multinationals to conserve samples of the ecosystems they use (e.g., Hershey, Dole, Firestone, Goodyear, Cadbury, Coca Cola, R.J. Reynolds "Camel Adventure," Del Monte, Nabisco, Kentucky Chicken).
3. *Improve Land Settlement*	
Deflect transmigration to *alang* grassland and Polonoroeste to *cerrado* savanna; promote family planning; improve land tenure in existing settlements; demote highways and cars in forest; promote carts or bicycles.	Promote *cerrado*, grassland and savanna land settlements; upgrade slums; family planning; intensify existing agriculture; promote fluvial transport and airships; promote agrarian reform outside forest regions.
4. *Reduce Shifting Cultivation*	
Except for low population density, jungle dwelling ethnic minorities, alternatives to shifting cultivation should be encouraged.	Tropical Forest Action Plan; improve conventional land use; promote and improve fallows and subsistence economies; promote irrigation and fertilizers on agricultural land.
5. *Demote Tropical Ranching*	
Hamburger Connection; ethics, economics, employment, sustainability (extinctions); raise health issues, (e.g., cholesterol); Burger King, McDonalds.	Promote cattle on natural rangeland; promote non-cattle meat (capybara, agouti); Raise fish (rivers, ponds); encourage trends: red-to-white meat and poultry-to-fish, or some vegetarianism; cows to pigs and goats.
6. *Improve Tropical Timber Trade*	
Tax tropical undressed log exports; consumer pressure against tropical lumber imports (e.g., Japan, USA, EEC); promote sustainable forestry; reduce export incentives.	Tree plantations near markets; buffer zones around conservation areas; agroforestry; recycle paper; promote value-added to tropical wood exports.

(See also: Caufield 1985; Goodland 1990a, c; Gradwohl and Greenberg 1988; Hecht and Cockburn 1989; Lal 1986; McDermott 1988; McNeely et al. 1990; Office of Technology Assessment 1984; Poore and Sayer 1987; Secrett 1985; WWF 1990.)

emphasize that suddenly halting tropical logging worldwide, abandonment of such interests by forestry departments and timber corporations, would, in our opinion, probably lead to the virtual elimination of forestry as a land use option, and to an explosion of deforestation even worse than we are currently experiencing.

Incentives should discourage log exports, but promote value added wood products exports. Critics point out that, although well intentioned, indiscriminate tropical timber boycotts add yet another constraint to the already difficult and low yield timber sector. That is why effective consumer preferences are designed to create a demand for sustainable timber, rather than an indiscriminate boycott.

Also effective in conserving forest would be a tropical (ex-forest) ranch beef boycott (the worst and least sustainable use of tropical forest). Reinforcing consumer preference against unsustainably produced tropical timber could be effective. The tropical timber trade needs improvement, but sudden cessation of all sales is less likely to promote improvement than continual pressure in the direction of sustainability.

Solid political/sociological evidence is needed to clarify this area, as well as on the details on the efficacy of the "pressures." Improving the sustainability of logging buys time for plantations to mature. The hope is that plantations will create enough employment to reduce pressures on forest which will be relieved from logging. In addition, value added and the suggestions outlined in tables 32.3 and 32.4 will create more permanent jobs. At the same time, to be realistic, forests will continue to dwindle. As forests become scarcer, they will be valued more—both for conservation and as a commodity.

Sustained yield of wood from moist tropical forests, although modest, is a useful start for governments and timber corporations to capitalize reforestation and timber plantations which yield higher returns, create more jobs, and are more easily sustainable. The major benefits of forest management therefore need to be more effectively communicated. Tropical tree plantations are a necessary prudent source of tropical timbers, and are part of the solution to loss of tropical forests. Plantations currently provide less than 5% of total tropical hardwood production, and unless the transition is accelerated, this is likely to continue until at least 2020, by which time the problem will be even more intractable.

Apart from selective logging under the conditions suggested and promoting plantations, the sustainability of tropical moist forests can be promoted directly by establishing state reserves, exploiting fallow forest (of which there is probably some ten times as much as of accessible intact pristine forest), and using the timber from forest when it is cleared for other purposes (although much clearing cannot be justified). The problem is not only the sustainability of harvest, but also the social and economic whole context in which it has to be considered. In the context of the whole, even selective logging which is not fully sustainable can be better justified than the other relentlessly likely economic alternatives. The paradox is that of the best being enemy of the good.

Acknowledgments

An earlier version of this article appears in *Environmental Conservation.* In addition to the assistance provided by our World Bank colleagues, we want to thank Philip

Fearnside, Jose Flores-Rodas, Alan Grainger, Dan Janzen, George Ledec, Larry Lohmann, Duncan Poore, Jeff Sayer, Lee Talbot, Koy Thomson, Anthony Whitten, and Frank Wadsworth for their most generous comments on the drafts.

REFERENCES

Agarwal, A., ed. 1991. *The Economics of the Sustainable Use of Forest Resources*. New Delhi: Center for Science and Environment.

Ahmad, Y.J., S. El Serafy, and E. Lutz, eds. 1989. *Environmental Accounting for Sustainable Development*. Washington, D.C.: The World Bank and United Nations Environment Programme.

Anderson, P. 1989. The Myth of Sustainable Logging: The Case For a Ban on Tropical Timber Imports. *The Ecologist* 19(5):166-168.

Arnold, J. E. M. 1987. Economic Considerations in Agroforestry. In H. A. Steppler and P. K. R. Nair, eds., *Agroforestry*, pp. 173-190. Nairobi: ICRAF.

Asabere, P. 1987. Attempts at Sustained Yield Management in the Tropical High Forests of Ghana. In F. Mergen, ed., *Natural Management of Tropical Moist Forests*, pp. 47-70. New Haven, Conn.: Yale School of Forestry.

Asibey, E. O. A. and G. Child. 1990a. Wildlife Management for Rural Development in Sub-Saharan Africa. *Unasylva* 42(161):3-10.

Asibey, E. O. A. 1991. In press. *Bushmeat Production in West Africa*. Washington, D.C.: National Wildlife Federation.

Asibey, E. O. A. 1985. *Forestry, Rural Development and Food Security in Africa*. Document WFD/85/03. Accra, FAO Regional Office.

Benchimol, S. 1989. *Amazonia: Planetarizao e Moratoria Ecologica*. Sao Paulo: CERED.

Binswanger, H. 1989. Brazilian Policies that Encourage Deforestation in the Brazilian Amazon. Environment Department Working Paper no. 16. Washington, D.C.: World Bank.

Browder, J. O. 1989. Social and Economic Constraints on Market-Oriented Extraction of Tropical Rainforest Resources. Unpublished. Blacksburg, Va: Virginia Polytechnic Institute.

Caufield, C. 1985. *In the Rainforest: Report from a Strange, Beautiful Imperilled World*. London: Heinemann.

Colchester, M. and Lohmann, L. 1990. *The Tropical Forestry Action Plan: What Progress?* Penang: World Rainforest Movement; and Sturminster Norton, Dorset: The Ecologist.

de Beer, J. H. and M. J. McDermott. 1989. *The Economic Value of Non-Timber Forest Products in South-East Asia*. Amsterdam: Netherlands Committee of IUCN.

De Camino Velozo, R. 1987. *Incentives for Community Development in Conservation Programmes*. FAO Conservation Guide No.12. Rome.

Erfurth, T. 1988. Tropical Countries as Suppliers of Timber to International and Domestic Markets. In R. Johnson and W. R. Smith, eds., *Forest Products Trade*, 198-209. Seattle: University of Washington Press.

Fearnside, P. 1988. Jari at age 19: Lessons for Brazil's Silvicultural Plans at Carajas. *InterCiencia* 13(1):12-24;13(2):95.

Fritz, E. C. 1989. *Clearcutting: a Crime Against Nature*. Washington, D.C.: The Wilderness Society.

Goldsmith, E. 1980. World Ecological Areas Programme (WEAP): A Proposal to Save the World's Tropical Rain Forests. *The Ecologist* 10(1-2):1-5.

Goldsmith, J. 1989. Goldsmith Plan to Save Rainforest. *The (Manchester) Guardian*, August 31, p. 1.

Goodland, R. J. A. 1980. Environmental Ranking of Amazonian Development Projects in Brazil. *Environmental Conservation* 7(1):9-26.

Goodland, R. J. A., ed. 1990a. *The Race to Save the Tropics*. Washington, D.C.: Island Press.

Goodland, R. J. A. 1990b. Environmental Sustainability in Economic Development. *Land Degradation and Rehabilitation* 1:311-322.

Goodland, R. J. A. 1991. Tropical Moist Deforestation: Ethics, Solutions and Religions. Environment Working Paper no. 43. Washington, D.C.: World Bank.

Goodland, R. J. A., C. B. Watson, and G. Ledec, 1985. *Environmental Management in Tropical Agriculture.* Boulder Colo: Westview Press.

Gradwohl, J. and R. Greenberg 1988. *Saving the Tropical Forests.* London: Earthscan.

Grainger, A. 1988. Future Supplies of High-Grade Tropical Hardwoods from Intensive Plantations. *Journal of World Forest Resource Management* 3(1):15- 29.

Hartshorn, G. 1989. Sustained Yield Management of Natural Forests: [Peru] Palcazu Production Forest. In J. O. Browder, ed., *Fragile Lands of Latin America,* pp. 130-138. Boulder, Colo: Westview Press.

Hecht, S. B. and A. Cockburn. 1989. *The Fate of the Forest: Developers, Destroyers and Defenders of the Amazon.* London: Verso.

Iltis, H. 1989. *Tropical Deforestation and the Fallacy of Agricultural Hope.* [United Nations Environment Programme: Tokyo conference on sustainable development]. Madison,Wisc: Botany Department

Janzen, D. H. 1973. Tropical Agroecosystems. *Science* 182:1212-1219.

Janzen, D. H. 1986. The Future of Tropical Ecology. *Annual Review Ecology Systematics* 17:304-324.

Lal, R. 1986. Conversion of Tropical Rainforests: Agroeconomic Potential and Ecological Consequences. *Advances in Agronomy* 39:173-264.

Lal, R. 1988. *Agroforestry as a Possible Sustainable Farming System in the Humid Tropics.* Agriculture Working Paper. Washington, D.C.: World Bank.

Lamprecht, H. 1989. *Silviculture in the Tropics: Tropical Forest Ecosystems and their Tree Species and Methods for their Long Term Utilization.* Eschborn: Deutsche Gesellschaft für Technische Zusammenarbeit.

Ledec, G. and R. J. A. Goodland. 1988. *Wildlands: their Protection and Management in Economic Development.* Washington, D.C.: World Bank.

Leslie, A. J. 1987. A Second Look at the Economics of Natural Management Systems in Tropical Mixed Forests. *Unasylva* 155(39):46-58.

Lugo, A. E. 1988. Tropical Forest Management with Emphasis on Wood Management. In A.E. Lugo, J. R. Clark, R. D.Childs and J. M. Savage, eds., *Ecological Development in the Humid Tropics: Guidelines for Planners,* pp. 1690190. Morrilton, Ark.: Winrock International.

McDermott, M. J., ed. 1988. *The Future of the Tropical Rainforest.* Oxford: Oxford Forestry Institute.

McNeely, J., K. Miller, W. Reid, R. Mittermeier, and T. Werner. 1990. *Conserving the World's Biodiversity.* Washington,D.C.: World Bank.

Mendelsohn, R. 1990. The Promise and Pitfalls of Extractive Reserves. Unpublished. New Haven, Conn.: Yale University Forestry School.

Myers, N. 1989.. *Deforestation Rates in Tropical Forests and Their Climatic Implications.* London: Friends of the Earth.

Nectoux, F. 1985. *Timber: an Investigation of the UK Tropical Timber Industry.* London: Friends of the Earth.

Nectoux, F. and N. Dudley. 1987, *A Hard Wood Story.* London: Friends of the Earth (and Earth Resources Research).

Nectoux, F. and Y. Kuroda. 1989. *Timber from the South Seas: An Analysis of Japan's Tmber Trade and Its Environmental Impact.* Gland, Switzerland: World Wide Fund for Nature.

Oberndorfer, D. 1990. Schutz der Tropischen Regenwalder (Furchwalder) durch Okonomische Kompensation. *Freiburg, Ber. Naturf. Ges.* 80:225-261.

Office of Technology Assessment. 1984. *Technologies to Sustain Tropical Forests* Washington, D.C.: Government Printing Office.

Oldfield, S. 1989. The Tropical Chainsaw Massacre. *New Scientist* 23 (Sept.):54-57.

Openshaw, K. and C. Feinstein. 1989. *Fuelwood Stumpage: Financing Renewable Energy for the World's Other Half.* Industry and Energy Working Paper 270. Washington, D.C.: World Bank.

Peters, C., A. Gentry and R. Mendelsohn. 1989. Valuation of an Amazonian Rainforest. *Nature* 339:655-656.

Poore, M. E. D., P. Burgess, J. Palmer, S. Rietbergen and T. Synnott. 1990. *No Timber Without Trees: Sustainability in the Tropical Forest.* London: Earthscan.

Poore, M. E. D. and J. Sayer. 1987. *The Management of Tropical Moist Forest Lands: Ecological Guidelines.* Gland, Switzerland: IUCN.

Potvin, J. 1990. *Debt-for-Conservation: A New Assessment Framework and Some Practical Options.* LAC Environment Division. Washington, D.C.: World Bank.

Repetto, R. 1990. Deforestation in the Tropics. *Scientific American* 262(4):36-45.

Schwartzman, S. 1989. Extractive Reserves: The Rubber Tappers Strategy for Sustainable Use of the Amazon Rainforest. In O. Browder, ed., *Fragile Lands of Latin America*, pp. 150-165. Boulder, Co.: Westview.

Secrett, C. 1985. *Rainforest: Protecting the Planet's Richest Resource.* London: Friends of the Earth and Russell Press.

Sedjo, R. 1989. Forests to Offset the Greenhouse. *Journal of Forestry* (July):12-14.

Sedjo, R. 1990. Is Tropical Forest Management Economic? *Journal of Business Administration* (in press).

Smith, W. R. 1989. The Use of Pacific Northwest Wood Products in Japan. Seattle: University of Washington, Graduate School of Business. *Pacific Northwest Executive*, (October), 5(4):23-29.

Teplitz-Sembitsky, W. and G. Schramm. 1989. *Woodfuel Supply and Environmental Management.* Industry and Energy Department Working Paper no. 19. Washington, D.C.: World Bank.

Ullsten, O., S. M. Nor and M. Yudelman, 1990. *Tropical Forestry Action Plan: Report of the Independent Review.* Rome: Food and Agriculture Organization.

United States Department of Agriculture. 1989. Tropical Hardwood Imports. *Foreign Agriculture Service: U.S. Trade Data.* Washington, D.C.: Government Printing Office.

Wadsworth, F., ed. 1991. *Silviculture and Management in Tropical Rain Forests.* San Juan, Puerto Rico, Institute of Tropical Forestry, Golden Jubilee Congress.In press.

Webb, L. J. 1989. Statement to Rebut Sustained Yield Arguments by Forestry in Northern Queensland, February 12. (Griffith University, Nathan, Queensland 4111, Australia and personal communication 1990).

Weber, T. 1987 *Hugging the Trees: The Story of the Chipko Movement.* Delhi: Penguin.

Westoby, J. 1987 *The Purpose of Forests: Follies of Development.* Oxford: Blackwell.

Wiersum, K. F., ed. 1984. *Strategies and Designs for Afforestation, Reforestation and Tree Planting.* Wageningen, Dutch Center for Agricultural Publishing and Documentation (PUDOC).

Willan, R. G., ed. 1991. *Management Systems in the Tropical Moist Rorests of Africa.* Rome: FAO Forestry paper series (In press).

Winterbottom, R. 1990.. *Taking Stock: The Tropical Forest Action Plan After Five Years.* Washington, D.C.: World Resources Institute.

World Wildlife Fund. 1990. *Tropical Forest Conservation.* Gland, Switzerland: World Wildlife Fund International, Position Paper 3.

Wyatt-Smith, J. 1987. Problems and Prospects for Natural Management of Tropical Moist Forests.(5-22) in F. Mergen and J. R. Vincent, eds., *Natural Management of Tropical Moist Forests.* pp. 5-22. New Haven, Conn: Yale School of Forestry.

Index